Ecosystem Collapse and Recovery

There is a growing concern that many important ecosystems, such as coral reefs and tropical rain forests, might be at risk of sudden collapse as a result of human disturbance. At the same time, efforts to support the recovery of degraded ecosystems are increasing, through approaches such as ecological restoration and rewilding. Given the dependence of human livelihoods on the multiple benefits provided by ecosystems, there is an urgent need to understand the situations under which ecosystem collapse can occur and how ecosystem recovery can best be supported. To help develop this understanding, this volume provides the first scientific account of the ecological mechanisms associated with the collapse of ecosystems and their subsequent recovery. After providing an overview of relevant theory, the text evaluates these ideas in the light of available empirical evidence, by profiling case studies drawn from both contemporary and prehistoric ecosystems. Implications for conservation policy and practice are then examined.

ADRIAN C. NEWTON is Professor of Conservation Ecology in the Centre for Ecology, Environment and Sustainability at Bournemouth University, United Kingdom. His research examines human impacts on the environment, with a particular focus on biodiversity loss and its consequences. His most recent projects have analysed the collapse and recovery of a range of different ecosystem types, including forests, grassland and heathland, and their implications for economic development.

ECOLOGY, BIODIVERSITY AND CONSERVATION

General Editor
Michael Usher, *University of Stirling*

Editorial Board
Jane Carruthers, *University of South Africa, Pretoria*
Joachim Claudet, *Centre National de la Recherche Scientifique (CNRS), Paris*
Tasman Crowe, *University College Dublin*
Andy Dobson, *Princeton University, New Jersey*
Valerie Eviner, *University of California, Davis*
Julia Fa, *Manchester Metropolitan University*
Janet Franklin, *University of California, Riverside*
Rob Fuller, *British Trust for Ornithology*
Chris Margules, *James Cook University, North Queensland*
Dave Richardson, *University of Stellenbosch, South Africa*
Peter Thomas, *Keele University*
Des Thompson, *NatureScot*
Lawrence Walker, *University of Nevada, Las Vegas*

The world's biological diversity faces unprecedented threats. The urgent challenge facing the concerned biologist is to understand ecological processes well enough to maintain their functioning in the face of the pressures resulting from human population growth. Those concerned with the conservation of biodiversity and with restoration also need to be acquainted with the political, social, historical, economic and legal frameworks within which ecological and conservation practice must be developed. The new Ecology, Biodiversity, and Conservation series will present balanced, comprehensive, up-to-date, and critical reviews of selected topics within the sciences of ecology and conservation biology, both botanical and zoological, and both 'pure' and 'applied'. It is aimed at advanced final-year undergraduates, graduate students, researchers, and university teachers, as well as ecologists and conservationists in industry, government and the voluntary sectors. The series encompasses a wide range of approaches and scales (spatial, temporal, and taxonomic), including quantitative, theoretical, population, community, ecosystem, landscape, historical, experimental, behavioural and evolutionary studies. The emphasis is on science related to the real world of plants and animals rather than on purely theoretical abstractions and mathematical models. Books in this series will, wherever possible, consider issues from a broad perspective. Some books will challenge existing paradigms and present new ecological concepts, empirical or theoretical models, and testable hypotheses. Other books will explore new approaches and present syntheses on topics of ecological importance.

Ecology and Control of Introduced Plants
Judith H. Myers and Dawn Bazely

Invertebrate Conservation and Agricultural Ecosystems
T. R. New

Risks and Decisions for Conservation and Environmental Management
Mark Burgman

Ecology of Populations
Esa Ranta, Per Lundberg, and Veijo Kaitala

Nonequilibrium Ecology
Klaus Rohde

The Ecology of Phytoplankton
C. S. Reynolds

Systematic Conservation Planning
Chris Margules and Sahotra Sarkar

Large-Scale Landscape Experiments: Lessons from Tumut
David B. Lindenmayer

Assessing the Conservation Value of Freshwaters: An International Perspective
Philip J. Boon and Catherine M. Pringle

Insect Species Conservation
T. R. New

Bird Conservation and Agriculture
Jeremy D. Wilson, Andrew D. Evans, and Philip V. Grice

Cave Biology: Life in Darkness
Aldemaro Romero

Biodiversity in Environmental Assessment: Enhancing Ecosystem Services for Human Well-Being
Roel Slootweg, Asha Rajvanshi, Vinod B. Mathur, and Arend Kolhoff

Mapping Species Distributions: Spatial Inference and Prediction
Janet Franklin

Decline and Recovery of the Island Fox: A Case Study for Population Recovery
Timothy J. Coonan, Catherin A. Schwemm, and David K. Garcelon

Ecosystem Functioning
Kurt Jax

Spatio-Temporal Heterogeneity: Concepts and Analyses
Pierre R. L. Dutilleul

Parasites in Ecological Communities: From Interactions to Ecosystems
Melanie J. Hatcher and Alison M. Dunn

Zoo Conservation Biology
John E. Fa, Stephan M. Funk, and Donnamarie O'Connell

Marine Protected Areas: A Multidisciplinary Approach
Joachim Claudet

Biodiversity in Dead Wood
Jogeir N. Stokland, Juha Siitonen, and Bengt Gunnar Jonsson

Landslide Ecology
Lawrence R. Walker and Aaron B. Shiels

Nature's Wealth: The Economics of Ecosystem Services and Poverty
Pieter J. H. van Beukering, Elissaios Papyrakis, Jetske Bouma, and Roy Brouwer

Birds and Climate Change: Impacts and Conservation Responses
James W. Pearce-Higgins and Rhys E. Green

Marine Ecosystems: Human Impacts on Biodiversity, Functioning and Services
Tasman P. Crowe and Christopher L. J. Frid

Wood Ant Ecology and Conservation
Jenni A. Stockan and Elva J. H. Robinson

Detecting and Responding to Alien Plant Incursions
John R. Wilson, F. Dane Panetta and Cory Lindgren

Conserving Africa's Mega-Diversity in the Anthropocene: The Hluhluwe-iMfolozi Park Story
Joris P. G. M. Cromsigt, Sally Archibald and Norman Owen-Smith

National Park Science: A Century of Research in South Africa
Jane Carruthers

Plant Conservation Science and Practice: The Role of Botanic Gardens
Stephen Blackmore and Sara Oldfield

Habitat Suitability and Distribution Models: With Applications in R
Antoine Guisan, Wilfried Thuiller and Niklaus E. Zimmermann

Ecology and Conservation of Forest Birds
Grzegorz Mikusiński, Jean-Michel Roberge and Robert J. Fuller

Species Conservation: Lessons from Islands
Jamieson A. Copsey, Simon A. Black, Jim J. Groombridge and Carl G. Jones

Soil Fauna Assemblages: Global to Local Scales
Uffe N. Nielsen

Curious About Nature
Tim Burt and Des Thompson

Comparative Plant Succession Among Terrestrial Biomes of the World
Karel Prach and Lawrence R. Walker

Ecological-Economic Modelling for Biodiversity Conservation
Martin Drechsler

Freshwater Biodiversity: Status, Threats and Conservation
David Dudgeon

Joint Species Distribution Modelling: With Applications in R
Otso Ovaskainen and Nerea Abrego

Ecosystem Collapse and Recovery

ADRIAN C. NEWTON
Bournemouth University

CAMBRIDGE
UNIVERSITY PRESS

University Printing House, Cambridge CB2 8BS, United Kingdom

One Liberty Plaza, 20th Floor, New York, NY 10006, USA

477 Williamstown Road, Port Melbourne, VIC 3207, Australia

314–321, 3rd Floor, Plot 3, Splendor Forum, Jasola District Centre, New Delhi – 110025, India

79 Anson Road, #06-04/06, Singapore 079906

Cambridge University Press is part of the University of Cambridge.

It furthers the University's mission by disseminating knowledge in the pursuit of education, learning, and research at the highest international levels of excellence.

www.cambridge.org
Information on this title: www.cambridge.org/9781108472739
DOI: 10.1017/9781108561105

© Adrian C. Newton 2021

This publication is in copyright. Subject to statutory exception and to the provisions of relevant collective licensing agreements, no reproduction of any part may take place without the written permission of Cambridge University Press.

First published 2021

Printed in the United Kingdom by TJ Books Limited, Padstow Cornwall

A catalogue record for this publication is available from the British Library.

ISBN 978-1-108-47273-9 Hardback
ISBN 978-1-108-46020-0 Paperback

Cambridge University Press has no responsibility for the persistence or accuracy of URLs for external or third-party internet websites referred to in this publication and does not guarantee that any content on such websites is, or will remain, accurate or appropriate.

People are suffering. People are dying. Entire ecosystems are collapsing. We are in the beginning of a mass extinction. And all you can talk about is money and fairytales of eternal economic growth. How dare you!
Greta Thunberg

Contents

Preface	*page* xi
1 Introduction	1
1.1 The IUCN Red List of Ecosystems	2
1.2 What Is an Ecosystem?	9
1.3 What Is Ecosystem Collapse?	18
1.4 What Is Ecosystem Recovery?	21
1.5 The Role of Metaphor	23
1.6 Ecosystem Collapse: A Useful Focus?	26
1.7 Structure of the Book	29
2 Ecological Theory	33
2.1 What Is an Ecological Theory?	35
2.2 Disturbance Theory	40
2.3 Succession	47
2.4 State-and-Transition Models	52
2.5 Dynamical Systems Theory	56
2.6 Planetary Boundaries	71
2.7 Critical Loads	75
2.8 Food Webs, Ecological Networks and Extinction Cascades	78
2.9 Resilience and Recovery	84
2.10 Conclusions and Propositions	92
3 Case Studies from Prehistory	95
3.1 The 'Big Five' Mass Extinctions	96
3.2 Extinction of Australian Megafauna	120
3.3 Megafauna Extinctions in the Late Quaternary	127
3.4 Holocene Examples of Ecosystem Collapse	138
3.5 Evaluation of Propositions	154

4 Contemporary Case Studies — 161
4.1 Coral Reefs — 161
4.2 Marine Fisheries — 177
4.3 Freshwater Ecosystems — 199
4.4 Forests — 222
4.5 Other Ecosystems — 264
4.6 Evaluation of Propositions — 276

5 Synthesis — 286
5.1 Understanding Ecosystem Collapse and Recovery — 287
5.2 Living with Ecosystem Collapse and Recovery — 323

6 Conclusions — 377
6.1 Answers — 380
6.2 Summaries — 387
6.3 Provocations — 395
6.4 Coda — 396

References — 399
Index — 472

Preface

In early 2016, while I was developing plans for this book, a major ecological catastrophe was under way. Australia's Great Barrier Reef, the world's largest living structure, experienced the worst coral bleaching event in its history, with more than 90% of surveyed reefs affected (Hughes et al., 2017a). The researcher who led these field surveys, Professor Terry Hughes of James Cook University in Queensland, summarised the results as follows: 'I showed the results of aerial surveys of bleaching on the Great Barrier Reef to my students. And then we wept' (*The Washington Post*, 20 April 2016).

It is very unusual for a scientific researcher to admit to an emotional reaction such as this. Scientists are supposed to be cool-headed, rational folk, who demonstrate calm objectivity and precision rather than emotional outbursts. In reality, of course, scientists are people too and are subject to the same feelings and emotions as everyone else. What this example illustrates is that some ecologists – perhaps most – are motivated not only by the intellectual pleasures of their chosen subject but also by a love of the natural world. When the ecosystems they study and admire are seriously damaged, perhaps beyond repair, they will suffer grief and a deep sense of loss. I feel the same way about ecosystems with which I am familiar, such as the ancient beech woodlands in southern England, which are currently suffering dieback as a result of climate change.

I mention this because scientific texts typically give scant regard to the emotional import of their subject matter. I have written this book because of growing concern about the condition of ecosystems throughout the world. Everywhere ecosystems are being degraded by multiple environmental pressures, most of which originate from human activity. Human impacts on the biosphere are now so completely pervasive that the functioning of the Earth system itself has been altered profoundly (Crutzen, 2006). Examples such as the bleaching of the Great Barrier

Reef show us that we cannot take ecosystems for granted: there may come a point at which any ecosystem can no longer cope with the pressures to which it is being subjected. The ecosystem might then collapse.

This text therefore follows scientific convention, by focusing on ecosystem collapse as an ecological phenomenon. I have sought to provide an overview of our current understanding of this phenomenon, and the mechanisms involved, from a purely intellectual rather than an emotional standpoint. Even when considering the implications of ecosystem collapse to human society, I have here focused on socio-economic impacts rather than the potential implications of the loss of aesthetic, cultural or spiritual values associated with nature. This is not to decry such values or undermine their importance: the cultural identity of many human societies is profoundly linked to the ecosystems on which they depend. Even those of us who live in relatively developed or urban societies depend, to a surprising degree, on contact with nature for our health and well-being (Maller *et al.*, 2006). Ecosystem collapse is therefore always a cause for grief.

Or is it? It is a persistent theme in ecological theory that profound ecosystem change can sometimes provide opportunities for innovation and novelty. One needs only to think of our own group of mammals, whose evolutionary diversification is partly attributable to a meteorite strike that eliminated their dinosaur competitors, to recognise the truth in this suggestion. Theories suggest that collapse is part of the natural dynamic of some ecosystems, and further, many ecosystems have demonstrated a remarkable ability to recover after major disturbance events. Perhaps ecosystem recovery is the *yang* to the *yin* of collapse; maybe the two phenomena are complementary and interconnected. At the very least, our growing understanding of how ecosystems can recover following perturbation offers a measure of hope. A focus on such sources of hope coincides with a growing movement encouraging greater optimism in conservation (Balmford and Knowlton, 2017).

I therefore hope that the reader will forgive my focus here on the science of ecosystem collapse and recovery, while neglecting what these phenomena might mean to us as people, either as individuals or collectively. As a scientist, I believe that if we are going to solve a problem, we first need to understand it. As an ecologist, I cannot think of a more important problem to understand than how ecosystems collapse and how they might recover afterwards. When I started thinking about this book, against the background of what was happening to the Great Barrier Reef,

I was struck by many questions, such as: What causes such ecosystems to collapse? Are some ecosystems more at risk of collapse than others? If one ecosystem collapses, might others follow? What happens after an ecosystem has collapsed – can it recover? Do all ecosystems recover in the same way? My main reason for writing this book was to try and answer some of these questions, to help improve my own understanding of the problem. I don't pretend to have come up with all the answers. But I very much hope that this text will encourage others to do so.

A big thank you to everyone who so kindly answered my impertinent questions or who generously sent me literature or other information, particularly Peter Petraitis, Peter White, Sally Keith and Peter Bellingham. Needless to say, the errors are all mine. I also thank the wonderful medical practitioners who helped me through a period of ill health that I experienced while writing this book, especially Bonnie Southgate and Rob Patterson. I'm also deeply grateful to my wife, Lynn, and son, Arthur, for all their love and support. Special thanks to Lynn for checking the manuscript prior to submission and for preparing the index.

1 · *Introduction*

Ecosystem collapse has become the focus of international attention in recent years, reflecting growing concern about the impact of human activities on the biosphere. Global environmental assessments such as the Global Biodiversity Outlook (Secretariat of the Convention on Biological Diversity, 2014) and the Global Environmental Outlook (UNEP, 2012) refer repeatedly to the collapse of specific ecosystems, focusing primarily on coral reefs and marine fisheries. Global trends in biodiversity loss documented by WWF's Living Planet Report have similarly been interpreted as increasing the risk of ecosystem collapse (WWF, 2006, 2016). It has been suggested that the entire global ecosystem might soon transition irreversibly to another state as a result of human influence (Barnosky *et al.*, 2014), although this has been contested (Brook *et al.*, 2013). Human civilisation itself may be at risk of collapse at the global scale owing to environmental degradation (Ehrlich and Ehrlich, 2013), although it has to be admitted that the researchers making this contention have a somewhat uneven record in predicting the future (Sabin, 2013). At the same time, as a response to widespread environmental degradation, international policy commitments are increasingly focusing on supporting ecological recovery, as illustrated by the target of restoring at least 15% of degraded ecosystems adopted by parties to the Convention on Biological Diversity (Aichi Target 15; CBD, 2012).

What is meant, though, by terms such as 'ecosystem collapse' and 'recovery'? How do these terms relate to associated concepts such as thresholds, tipping points, stability and resilience? Are these operational scientific concepts that can be applied in a practical context to support environmental management and conservation, or are they simply metaphors? Unfortunately, ecologists seem to delight in defining concepts in multiple alternative ways, which can generate confusion and hinder the development of scientific understanding (Peters, 1991). One need only think of the 163 definitions of 70 different stability concepts discovered by Grimm and Wissel (1997) in the ecological literature to appreciate the magnitude of the problem (and there have been many further definitions proposed in the two decades since that paper was published). I therefore consider here how ecosystem collapse and recovery are defined in the literature, along with associated concepts. I then examine the key features of both collapse and recovery, including their potential causes. Finally I describe how the rest of this book is structured. First, though, I provide an overview of how ecosystem collapse is being addressed by a new initiative attempting to develop an IUCN Red List of Ecosystems. The debate generated by this initiative provides a valuable entry point for considering what ecosystem collapse entails.

1.1 The IUCN Red List of Ecosystems

The IUCN (International Union for Conservation of Nature) Red List of Threatened Species™ (www.redlist.org) is widely recognised to be the most authoritative global assessment of the extinction risk of species (Mace *et al.*, 2008; Rodrigues *et al.*, 2006). Although principally designed to evaluate the extinction risk of individual species throughout their geographic ranges, the IUCN criteria are also used to develop regional, national and local lists of threatened species. Red List assessments have been widely used to inform the development of conservation strategies and plans and to develop indicators of biodiversity loss, including at the global scale (Butchart *et al.*, 2010; Mace *et al.*, 2008). During Red List assessments, species are assigned to one of a series of categories by applying five quantitative criteria (IUCN, 2001). The criteria are principally based on a declining or small population size, or the declining geographic range of a species. Taxa that meet the appropriate threshold for at least one of the five criteria may be categorised either as Extinct (EX) or as Critically Endangered (CR), Endangered (EN) or Vulnerable (VU); those failing to meet the thresholds may be categorised as Near Threatened (NT), Least Concern (LC), or Data Deficient (DD) (IUCN, 2001).

Recently this assessment approach has been extended to ecosystems. The IUCN Red List of Ecosystems protocol has been developed to provide a tool for assessing the conservation status of ecosystems, which can be applied at a variety of scales (Bland *et al.*, 2017a). The approach closely parallels that developed for Red List assessments of species, with five rule-based criteria (A–E) used to assign ecosystems to a risk category. These categories relate to the risk of ecosystem collapse (Figure 1.1), mirroring how categories in Red List assessments of species relate to the

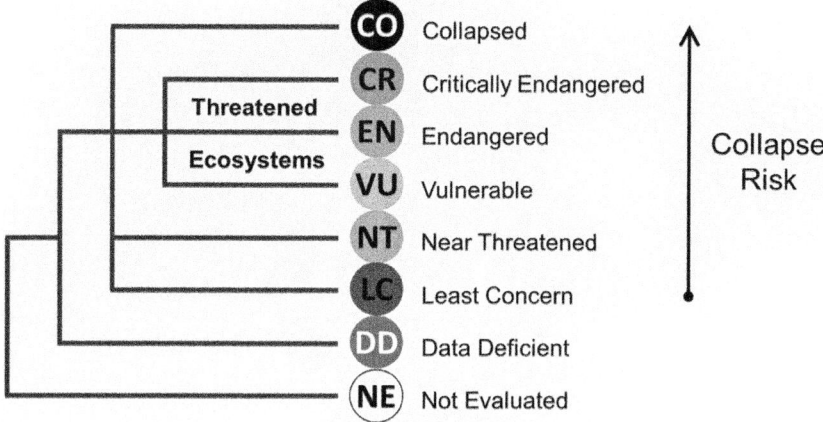

Figure 1.1 Structure of the IUCN Red List of Ecosystems categories (from Bland et al., 2017a). Note how 'Collapse' of an ecosystem is considered as a category, equivalent to 'Extinct' in the Red List assessment of species.

risk of extinction. Two of the criteria assess spatial symptoms of ecosystem collapse, namely declining distribution (A) and restricted distribution (B), whereas two criteria assess functional symptoms of ecosystem collapse, namely environmental degradation (C) and disruption of biotic processes and interactions (D) (Figure 1.2). The final category (E) is based on producing quantitative estimates of the risk of collapse using an appropriate modeling approach (Bland et al., 2017a).

While describing the Red List protocol, Bland et al. (2017a) make a number of valuable points about ecosystem collapse and its relationship to recovery:

- Ecosystem collapse is not directly analogous to species extinction, as, after undergoing collapse, ecosystems do not typically disappear but transition into some other type of ecosystem. This ecosystem may retain some of the species characteristic of the original ecosystem, but the abundance, interaction and ecological functions of these species may change after collapse.
- Many characteristic features may disappear from an ecosystem undergoing collapse, long before the last characteristic species disappears from the ecosystem.
- Ecosystem collapse is therefore a transformation of identity, a loss of defining features and/or replacement by a different ecosystem.
- An ecosystem is considered as collapsed when, after it loses its defining biotic or abiotic features, it can no longer sustain its characteristic native

4 · Introduction

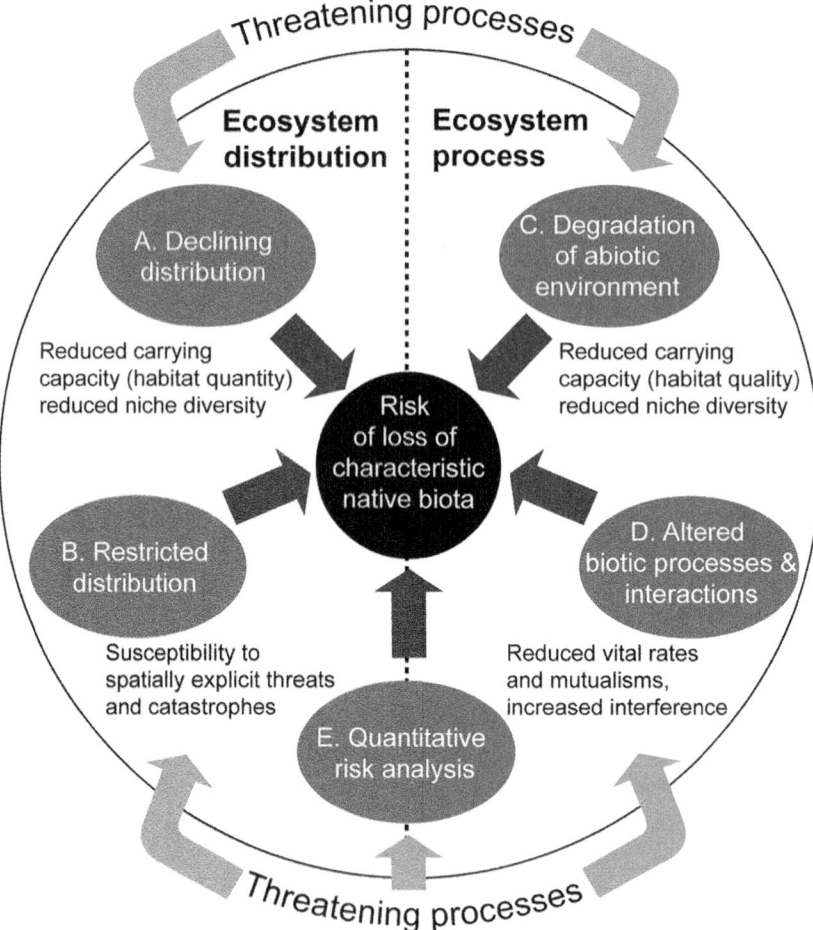

Figure 1.2 The mechanisms of ecosystem collapse and symptoms of collapse risk (from Keith *et al.*, 2013). Note that the five ellipses labelled A–E represent the five criteria used in the IUCN Red List of Ecosystems protocol, based on different thresholds of decline (Bland *et al.*, 2017a; Keith *et al.*, 2013).

species, and it has moved outside its natural range of spatial and temporal variability in composition, structure and/or function.
- Ecosystem collapse may result from a variety of different threatening processes (Table 1.1) and occurs through a variety of different pathways. Such pathways include trophic cascades, loss of foundation or keystone species, environmental degradation and climatic forcing. The

Table 1.1 *Selected definitions of some key terms*

Term	Definition	Reference
Degradation	The incremental and progressive impairment of an ecosystem on account of continuing stress events or punctuated minor disturbances that occur with such frequency that natural recovery does not have time to occur.	Clewell and Aronson (2013)
	Direct damage to an ecosystem's biotic and/or abiotic biological condition.	Salafsky et al. (2008)
	A state or process in which ecosystem resources or attributes are reduced relative to some reference state or goals owing to human disturbance.	Ghazoul and Chazdon (2017)
Disturbance	Any relatively discrete event in time (or space) that disrupts ecosystem, community or population structure and changes resources, substrate availability or the physical environment.	White and Pickett (1985)
	Anything that causes disruption to a system.	Resilience Alliance (2010)
Driver or pressure	The ultimate factors – usually social, economic, political, institutional or cultural – that enable or otherwise add to the occurrence or persistence of proximate direct threats. There is typically a chain of drivers behind any given direct threat.	Bland et al. (2017a)
Ecosystem	The whole system (in the sense of physics), including not only the organism-complex but also the whole complex of physical factors forming what we call the environment of the biome – the habitat factors in the widest sense.	Tansley (1935)
	A biotic community or assemblage and its associated physical environment in a specific place. The main components of the ecosystem are its abiotic and biotic features and the interactions between them.	Pickett and Cadenasso (2002)
	Complexes of organisms and their associated physical environment within	Bland et al. (2017a)

(cont.)

6 · Introduction

Table 1.1 (*cont.*)

Term	Definition	Reference
	a specified area. They have four essential elements: a biotic complex, an abiotic environment, the interactions within and between them and a physical space in which these operate.	
	All the organisms and the abiotic pools with which they interact. Ecosystem processes are the transfers of energy and materials from one pool to another.	Chapin *et al.* (2002)
	A unit comprising a community (or communities) of organisms and their physical and chemical environment, at any scale, desirably specified, in which there are continuous fluxes of matter and energy in an interactive open system.	Willis (1997)
Ecosystem collapse	A transformation of identity, a loss of defining features and a replacement by a different ecosystem type.	Bland *et al.* (2017a)
	An ecosystem is collapsed when all occurrences lose defining biotic or abiotic features no longer sustain the characteristic native biota and have moved outside their natural range of spatial and temporal variability in composition, structure and/or function.	
	A theoretical threshold, beyond which an ecosystem no longer sustains most of its characteristic native biota or no longer sustains the abundance of biota that have a key role in ecosystem organisation (e.g. trophic or structural dominants, unique functional groups, ecosystem engineers, etc.). Collapse has occurred when all occurrences of an ecosystem have moved outside the natural range of spatial and temporal variability in composition, structure and function. Some or many of the pre-collapse elements of the system may remain within a collapsed ecosystem, but their relative	Keith *et al.* (2013)

Table 1.1 (*cont.*)

Term	Definition	Reference
	abundances may differ and they may be organised and interact in different ways with a new set of operating rules.	
	An abrupt and undesirable change in ecosystem state.	Lindenmayer *et al.* (2016)
	Major changes in ecosystem conditions are widespread and are either irreversible or very time- and energy-consuming to reverse.	Lindenmayer and Sato (2018)
Ecosystem recovery	The process by which an ecosystem returns to a previous condition after being in a degraded or disrupted condition.	Based on Elliott *et al.* (2007)
	An ecosystem has recovered when it contains sufficient biotic and abiotic resources to continue its development without further assistance or subsidy. It will sustain itself structurally and functionally.	Society for Ecological Restoration (2004)
	A pathway of ecosystem redevelopment towards a less compromised state, or the attainment of a fully functioning system comparable to 'target' reference sites.	Simenstad *et al.* (2006)
Threatening process, or threat	A tractable agent, mechanism or process that causes either a continuing decline in distribution, continuing environmental degradation or continuing disruption of biotic interactions or a future decline in those factors that is likely to occur in the near future (i.e. within 20 years).	Keith *et al.* (2013)
	Direct threats are the proximate activities or processes that have impacted, are impacting, or may impact the status of the ecosystem being assessed. Threats can be past (historical), ongoing and/or likely to occur in the future. Natural phenomena are also regarded as direct threats in some situations.	Bland *et al.* (2017a)

process of collapse may be gradual or sudden, linear or non-linear and deterministic or stochastic.
- The process of ecosystem collapse may be influenced by abiotic or external factors (e.g. weather patterns or human disturbance), internal biotic processes (e.g. competition or predation), historical legacies (e.g. climatic history, extinction debts or history of exploitation) and spatial context (e.g. the location, size and connectivity of ecosystem remnants).
- Symptoms of collapse may differ between ecosystems depending on their particular ecological characteristics, the nature of threatening processes and the pathways of collapse.
- Ecosystem collapse can be evidenced by time series data for relevant variables, or it could be inferred by comparing occurrences of the ecosystem where defining features have been lost with other areas where such losses have not occurred. Major changes in functionally similar ecosystems can also provide insights into the process of collapse in ecosystems of interest.
- Ecosystem collapse may be reversible. After sufficient time has passed, or characteristic species and/or ecological function have been reintroduced, an ecosystem could potentially recover. However, in many cases recovery will not be possible.

According to the IUCN Red List protocol, collapse therefore refers to a process of transformation of an ecosystem, leading to its conversion to a different ecosystem type (Table 1.1). This raises the question of how much change in the ecosystem is required to have occurred before collapse is considered to have taken place. This is defined in the thresholds associated with the Red List criteria, which are used to define a category of risk. For example, thresholds for the application of criteria A and B are typically defined as 100% loss of spatial distribution of the ecosystem type, for an ecosystem to be classified as collapsed (Bland et al., 2017a).

The Red List of Ecosystems represents the first time that ecosystem collapse has formally been incorporated into an environmental assessment process, and it is being used as a basis for classifying the conservation status of ecosystems. Development of the process has been supported by a growing body of scientific literature, aiming to ensure that the protocol is objective, transparent and rigorous. The idea of adapting the Red List approach to systematically assessing the conservation status of ecosystems has been under discussion for some years. For example, Rodríguez et al.

(2007) provided some initial suggestions regarding how a Red List assessment of ecosystems might be conducted, which were further developed by Rodríguez et al. (2011), although neither publication explicitly considered the phenomenon of ecosystem collapse. In relation to the Red List, collapse is first mentioned by Rodríguez et al. (2012), who refer to it as analogous to the extinction of a species. Use of ecosystem collapse in this context is further elaborated by Keith et al. (2013), in their review of the science underpinning the Red List initiative.

Keith et al. (2013) highlight a number of issues and challenges in developing a risk assessment of ecosystems. These include the definition of units of assessment, which they refer to as ecosystem types; developing an operational definition of ecosystem collapse as the endpoint of environmental degradation; identifying the relationships between threatening processes, mechanisms and symptoms of ecosystem collapse; and identifying thresholds of collapse, on which the assessment criteria might be based (Figure 1.2). The suggestions made by these authors were challenged by Boitani et al. (2015), who pointed out that there is no consistent approach to ecosystem classification that might be used for assessing their conservation status and that there is limited scientific support for the criteria and thresholds that were proposed. They also question the use of ecosystem collapse in this context. This debate usefully highlights a number of issues that are relevant to the theme of this book and are consequently considered a little later in detail.

1.2 What Is an Ecosystem?

I once spent an entertaining international flight discussing ecology with a fellow researcher I'd met at a meeting that we'd both been attending. She admitted that she had never really liked the term 'ecosystem', nor fully understood what an ecosystem actually was. This was all the more remarkable, given that the event we had just attended was part of the Millennium Ecosystem Assessment (MEA). With hindsight, publication of the MEA (2005a) turned out to be a landmark event, which firmly established the concept of ecosystem services – the benefits provided by ecosystems to people – in the international policy arena. It also stimulated renewed interest in ecosystems as a focus of research. At the preparatory workshops, a large international community of researchers and policy-makers convened to design the assessment, which provided the first global appraisal of the condition and trends in the world's ecosystems and the services that they provide. For me, as a participant in some of the meetings, the sight of this community discussing how best to conduct the assessment was enormously entertaining and instructive. I vividly remember calls for the MEA to produce a comprehensive map of global ecosystems being roundly rejected despite this being seen as a logical first step by a number of participants. I ended up sympathising with my colleague on the plane, as I shared her confusion. (Interestingly, her semantic uncertainty didn't prevent her from going on to play a leading role in the development of ecosystem service science.)

It may seem surprising that a concept so fundamental to the science of ecology might still be under discussion, but my colleague was not alone in wondering what an ecosystem actually is. For example, in the 1990s, the *Bulletin of the Ecological Society of America* published a series of commentary articles debating the definition of an ecosystem (Blew, 1996; Fauth, 1997; Marín, 1997; Rowe, 1997; Rowe and Barnes, 1994). These authors agreed that there is a lack of consensus on how the 'troublesome' term 'ecosystem' should be defined and that this confusion is hindering both the development of ecological theory and its practical application. Semantic uncertainty and conceptual fuzziness are nothing new in ecology, and although they do indeed impede progress (Peters, 1991), they also illustrate how the world can appear differently depending on how you look at it. Rowe and Barnes (1994), for example, highlighted the different perspectives from those approaching ecology from the point of view of organisms ('bio-ecologists') *versus* those coming from an Earth science background ('geo-ecologists'). For bio-ecologists, with their organism-centred view, ecosystems are biotic communities of plants and animals, plus the features of the abiotic environment that are used by organisms as resources. From this perspective, an ecosystem is flexible in time and space, depending on the location of the organisms of interest. The challenges of rigidly applying the ecosystem concept when the focus is an individual, highly mobile species are beautifully illustrated by a quote from Drury in 1969: 'I have struggled unsuccessfully with the problem of defining ecosystems into which a seagull can be fitted' (Rowe and Barnes, 1994).

This relates to the issue of whether ecosystems can be meaningfully mapped, the same problem that I encountered at the MEA workshop. For geo-ecologists, an ecosystem is a defined area of the earth's surface, defined by abiotic factors such as landforms, topography and climate (Rowe and Barnes, 1994). For bio-ecologists, whose organisms of interest may not respect the boundaries between such areas as they disperse and migrate, the limits of an ecosystem will necessarily be much less well defined. This is highlighted by Fitzsimmons (1996), who criticised ecologists for 'uncritical acceptance of ecosystems as tangible objects that should be managed, protected and considered a part of biological diversity', describing them as being 'geographic fabrications devised to facilitate an analysis ... rather than being entities in their own right'. As the author notes, this reflects the lack of a standardised classification system for ecosystems, which can be thought of as being of any size or shape, ranging from the size of a drop of water to the whole planet (Fitzsimmons, 1996). While recognising their value as a research tool, Fitzsimmons (1996) asserts that ecosystem concepts do not offer a coherent way of dividing landscapes, either for developing environmental policy or for informing land management. This is because an infinite number of different ecosystems can potentially be identified for any part of the earth's surface.

So are ecosystems a fundamental unit of study for ecologists, as some authors have suggested (e.g. Golley, 1993), or are they just fuzzy human constructs? This question is addressed by Post et al. (2007), who highlight the overriding importance of understanding boundaries for any conceptualisation of an ecosystem. Specifically, the boundaries of an ecosystem in time and space need to relate to the process being studied and the question being asked. In some ecosystems, such as streams and lakes, there is a close coincidence between the physical boundaries of the system and the ecological processes being investigated. It may be no coincidence, therefore, that much of the early theoretical progress in ecology arose from studies of freshwater ecosystems (Golley, 1993). However in many situations, ecological processes may operate at different scales than those implied by physical boundaries. For example, salmonid fish that spend part of their lives in marine environments can transfer marine-derived nutrients to rivers in such amounts that the trophic productivity of the rivers is increased, supporting the development of juvenile salmon (Buxton et al., 2015). In this case, the scale of the ecosystem is unclear because there is a lack of a relationship between processes such as resource supply and nutrient cycling, and the physical boundaries of the system (Post et al., 2007). Such problems are often encountered when attempts are made to relate the structure of food webs to ecosystem processes.

As noted by Post et al. (2007), these difficulties of defining and applying ecosystem concepts are most apparent at the interface between different approaches to ecology. Many authors have noted that there are two main subdisciplines in ecology, one focusing on populations and communities and the other on ecosystems (Loreau, 2010a; O'Neill et al., 1986). Population and community ecology focus on the dynamics of populations and their interactions with other populations, either of the same or of other species. Key ecological processes include survival, reproduction, mutualism, competition and evolution. In contrast, ecosystem ecology examines the functioning of the overall ecosystem, including both biological organisms and their abiotic environment, with key processes of interest including the flow of materials and energy through the system. Although the difference between these two traditions might seem trivial to an outsider, it represents a profound division in the development of ecological science. Ecosystem ecology dominated the field in the 1960s, but subsequently it fell into disfavour. By the 1980s, it had become largely marginalised, with ecological textbooks of that era either playing down ecosystem concepts and theory, or ignoring them altogether (de LaPlante, 2005).

The reasons for the decline in ecosystem ecology relate to a perceived incompatibility with evolutionary theory (de LaPlante, 2005), which, being closely aligned with population ecology, came to dominate ecological science in the 1970s. In addition, there was a backlash against the idea that ecosystems are stable, equilibrial and homeostatic systems, behaving in ways that are analogous to machines, which characterised much of the early literature on the topic (O'Neill, 2001). Conservation scientists, in particular, criticised the 'black box' approaches that were widely used by ecosystem scientists to measure the flows of material or energy through an ecosystem, which imply that energy flow or nutrient cycling are more important than the identity of the organisms that perform these functions (Post et al., 2007). Furthermore, the suggestion that ecosystems are closed, self-regulating systems that are characterised by stable states, leading to a nonscientific idea of the 'balance of nature', was roundly rejected as being a myth (Soulé and Lease, 1995) and for failing to provide a suitable basis for conservation (Pickett et al., 1992).

An ecosystem, then, is not an empirical observation about nature, but a specific way of looking at nature, which focuses on some of its properties but de-emphasises others (O'Neill, 2001). Concepts of ecosystems have evolved over time (Willis, 1997); they are now increasingly seen as disequilibrial, open, hierarchical, spatially patterned and scaled (O'Neill, 2001). There have been substantive attempts to refine our understanding of ecosystem properties based on such developments (Patten et al., 1997, 2011) and in this way, some of the criticisms levelled at earlier ecosystem science have been addressed. Most recent investigations recognise that the biotic and abiotic components of an ecosystem are intimately linked and that the behaviour of organisms and species identity matter (Jones and Lawton, 1995; O'Connor and Crowe, 2005). However, some of the original conceptions of ecosystems still pervade current literature on ecosystem science, as will be explored in Chapter 2.

Post *et al.* (2007) note that many investigations avoid scaling problems by focusing on areas that are well bounded; in other words, where there is a strong coincidence between the interactions among organisms and the fluxes and flows of energy or material. It is in situations where boundaries are less clear that problems of scale become more difficult to resolve. Key issues include the relationship between the boundary and the process being studied; the origin of the boundary; its permeability to the process or species being studied and its grain, extent and dimensionality (Post *et al.*, 2007). In their very useful summary of ecosystem boundaries, Post *et al.* (2007) note the following:

- As ecosystem processes are scale-dependent, the choice of boundaries is of profound importance to the definition of an ecosystem and to the questions that can be asked about it by researchers. An over-reliance upon physical or structural boundaries can obscure those boundaries that are related to important ecological processes and interactions.
- Landscape ecologists often recognise structural boundaries in a landscape, which are often based on visible or measurable discontinuities at physical boundaries (e.g. aquatic–terrestrial boundaries) or changes in the biotic composition of two adjacent communities (e.g. shifts in dominant plant species; Table 1.2). Although they are used to support much environmental planning and management, such 'tangible' boundaries do not always coincide with the ecological processes of interest, highlighting the need to use them with care.
- Functional definitions relate to the ability of boundaries to mediate interactions and exchanges among units of study (e.g. communities, populations, habitat patches, watersheds) and are central to defining ecosystems. Such ecosystem boundaries may be set by discontinuities or steep gradients in the flux and flow of material and energy.
- From a population or community ecology perspective, ecosystem boundaries might be set by discontinuities or steep gradients in rates of immigration and emigration, or in species interactions. Such boundaries may be governed by geomorphic forms or processes, or could be independent of the physical landscape.

Table 1.2 *Attributes of ecosystem boundaries (adapted from Post* et al.*, 2007)*

Types of definition	
Structural	Based on physical boundaries (e.g. watersheds, aquatic-terrestrial boundaries such as between lakes or rivers and surrounding land).
Functional	Based on changes in the rates of interactions and exchanges among the units of study.
Relationship to process of interest	
Structural	Structural boundaries can be process-independent.
Functional	By definition, functional boundaries depend upon the process of interest.
Origin	
Structural	
Geomorphic	Topographical boundaries, including watersheds, aquatic-terrestrial boundaries, continental shelf *versus* deep sea and so forth.
Physiochemical	Thermoclines and chemoclines, e.g. freshwater-saltwater boundaries in estuaries.
Dimensional	Surface-related boundaries, including surface *versus* soil, benthic *versus* pelagic, ground *versus* canopy.
Biological	Physical boundaries among habitats, e.g. the boundary between old fields and forest.
Functional	
Material and energy flow	Ecosystem boundaries defined by steep gradients in the flow of material and energy, including resource sheds, nutrient spiralling and discontinuities in nutrient or energy exchange. Often moderated by structural boundaries that limit exchange between ecosystems.
Species interactions	Community boundaries defined by the location of weak(er) species interactions. At times mediated by structural boundaries that limit interactions between species.
Movement of organisms	Population boundaries set by limits to immigration or emigration and gene flow. Often mediated by structural boundaries that limit migration and gene flow.
Actions	
Transmission	The boundary is semipermeable and allows only a fraction of organisms or material to pass, or reduces the strength of species interaction.
Transformation	The boundary changes the state of material or species interactions, e.g. N transformation at the soil-stream interface.

Table 1.2 (*cont.*)

Types of definition	
Absorption or reflection	The boundary is impermeable and either stops or redirects interactions among species or the flow of organisms and material.
Neutral	The boundary does not affect the flow of material or species interactions. Can only apply to structural boundaries.

- Delineating ecosystem boundaries is relatively straightforward where strong associations occur among resource flow, community composition and physical boundaries (e.g. in lakes and islands), because of the convergence of functional and structural characteristics. Such ecosystems can be considered as well bounded. In such situations, interactions among organisms will typically be stronger and cycling of material and energy will typically be tighter within than across the boundaries of these ecosystems. The boundaries may also be characterised by physical or geomorphic features that hinder exchange of organisms or resources across the boundary.
- In contrast to well-bounded ecosystems, a variety of relatively open systems exist, including most terrestrial habitats, estuaries and streams, for which boundaries are often unclear. In such cases, boundaries may not be fully delineated by well-defined physical features, and there may be little congruence among the physical boundaries, community composition and ecosystem processes (Figure 1.3). In this situation, the functional and structural attributes of ecosystem boundaries are segregated, and different processes and mechanisms may operate at different spatial scales. The question of interest may determine the appropriate scale of analysis, which should be stated explicitly. As a consequence, different definitions of ecosystem boundaries may be required for addressing different research questions. Furthermore, for questions that bridge population/community and ecosystem approaches to ecology, such as many questions in food web ecology, the appropriate measure of ecosystem boundaries may not depend on the question being asked but rather on the answer to that question.

The problems identified by Fitzsimmons (1996) and Post *et al.* (2007) relating to the difficulty of mapping ecosystems have essentially been overlooked in the drive towards ecosystem-based land management,

16 · Introduction

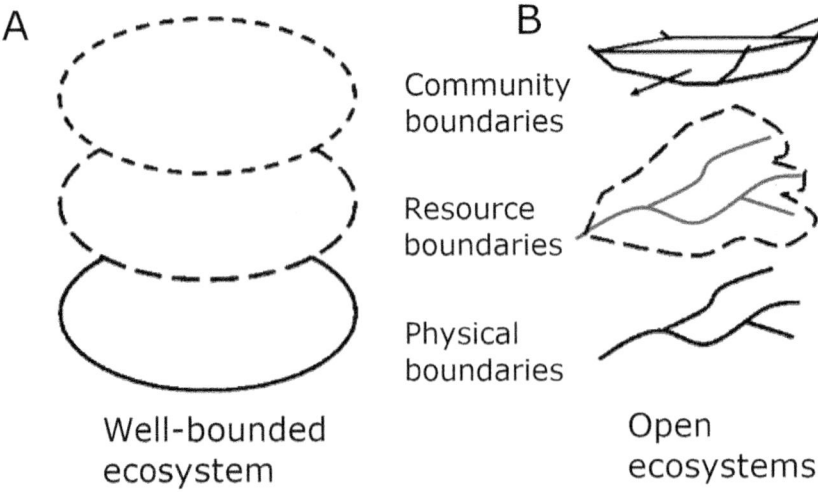

Figure 1.3 Spatial correspondence among physical boundaries, community boundaries and the boundaries of ecosystems. In well-bounded ecosystems (A), a strong spatial correspondence occurs among all three sets of boundaries, for example at the aquatic–terrestrial boundaries of many lakes. In open ecosystems (B), little correspondence exists among boundaries. The example illustrated here is the catchment of a stream. Reprinted from Post *et al.* (2007), with permission from Elsevier.

which has occurred in the wake of the MEA. Efforts to map the spatial distribution of ecosystem services, and the ecosystem types that underpin them, are now widespread, reflecting their increasing incorporation in environmental policy (Crossman *et al.*, 2013). The rapid growth of ecosystem service science has encouraged something of a renaissance in ecosystem ecology, but challenges remain in applying the ecosystem concept in practice. In fact, contrasting views regarding the nature of ecosystems are rooted in different philosophical standpoints, such as ecological organisms *versus* individualism, which date back to the very origins of ecology as a science (Kirchhoff *et al.*, 2010). These are considered further in Chapter 2.

These issues are also considered by Keith *et al.* (2013) in their review of the scientific foundations of the IUCN Red List of Ecosystems. Here, ecosystems are defined as complexes of organisms and their associated physical environment occurring within an area. Interestingly, other terms that are widely used in conservation ecology, such as 'ecological communities', 'habitats', 'biotopes' and 'vegetation types', are considered as synonyms of 'ecosystem types' (Keith *et al.*, 2013). Although this is a very pragmatic suggestion, it overlooks the fundamental differences between conceptions of ecosystems and ecological communities outlined earlier. As noted by Boitani *et al.* (2015) in their critique of Keith *et al.* (2013), application of the ecosystem concept involves the study of ecological processes and function, and understanding the flows and interactions among the components of the system, which goes well beyond measures of species richness and composition that are typically used to characterise ecological communities. As a result, there is no single way in which ecosystems can be consistently defined to support conservation management, because their definition will vary at different scales and in different ecological contexts and according to the specific features under consideration (Boitani *et al.*, 2015). This is why no globally applicable standardised list of ecosystem types is currently available.

In their response to these criticisms, Keith *et al.* (2015) draw a distinction between the definition of an ecosystem and classifications of ecosystem types for application of the Red List criteria. While recognising that the development of suitable typologies of ecosystem types is challenging, they highlight two main sources of evidence suggesting these challenges can be overcome. First, a number of risk assessments have already been conducted at national and regional scales in a number of locations, usually based on systematic typologies. These examples indicate that ecosystem types can be reliably interpreted so long as their defining features are adequately described. Second, Keith *et al.* (2015) suggest that the Red List of Ecosystems can follow similar procedures to the Red List of Threatened Species (RLTS) when addressing uncertainties in classification of assessment units. This overlooks the fact that the taxonomy of species is subject to rigorous scientific processes using internationally agreed protocols, which is not the case for ecosystem types. Boitani *et al.* (2015) also highlight that the proposed criteria for assessment described by Keith *et al.* (2013) are designed to apply only at very specific scales, namely at the landscape scale (hectares to square kilometres) in relation to space, and at the scale of decades in relation to time. This would mean that the assessment is not applicable to all ecosystems at any scale or complexity, a criticism not explicitly addressed by Keith *et al.* (2015).

So what should we conclude from this debate? Is the concept of an ecosystem analogous to a naked Danish Emperor, as Fitzsimmons (1996) suggested? Or, following Smith *et al.* (2012), is it true to say that ecology is a systems science, almost by definition? After all, ecologists have identified the ecosystem as their most important concept (Willis, 1997).

As noted previously, different ecosystem concepts and definitions reflect deep historical divisions among different groups of ecological researchers. These reflect contrasting approaches to investigating the subject, which are ultimately grounded in different philosophical world views (Kirchhoff et al., 2010). Efforts to integrate different ecological perspectives, such as community ecology and ecosystem ecology, continue to attract the attention of leading researchers (Loreau, 2010a; Matthews et al., 2011). Yet no consensus has emerged on how best to achieve this integration, and it remains elusive. Perhaps the diversity of approach is even a strength of ecology as a discipline, which should be further encouraged (Peterson et al., 1997).

In this book, unless otherwise stated, the definition of ecosystem that is employed follows that given by Pickett and Cadenasso (2002), namely a biotic community or assemblage and its associated physical environment in a specific place (see Table 1.1). This definition is closely based on that of Tansley (1935). This originates from an organism-centred viewpoint and recognises that ecosystem boundaries are flexible rather than geographically determined. However, it should be remembered that when other work is described or cited in the text, this may be based on a variety of other concepts or definitions of an ecosystem.

1.3 What Is Ecosystem Collapse?

Ecosystem collapse is one of those terms that appears to have been in long-term use without ever having been formally defined. Palaeobiologists have been using the phrase for at least 30 years without saying precisely what they mean by it (McGhee, 1988; Twitchett et al., 2001). Interest in the phenomenon was greatly stimulated by Jared Diamond's best-selling and highly entertaining book entitled *Collapse*, although the focus was human civilisations rather than ecosystems (Diamond, 2005). Here, collapse was defined as a 'drastic decrease in human population size and/or political/economic/social complexity, over a considerable area, for an extended time'. According to this definition, Diamond (2005) recognised that the identification of situations that qualify as a collapse is essentially arbitrary; collapse can be seen as an extreme form of milder types of decline.

In their consideration of the dynamics of human-environment systems, Young and Leemans (2006) similarly define a collapse as 'a failure or sudden end to something, or a sudden reduction or decrease

1.3 What Is Ecosystem Collapse? · 19

in value'. Further, these authors define collapse as any situation where the rate of change to a system:

- has negative effects on human welfare, which, in the short or long term, are socially intolerable;
- will result in a fundamental downsizing, a loss of coherence and/or significant restructuring of the constellation of arrangements that characterise the system and
- cannot be stopped or controlled via an incremental change in behaviour, resource allocation or institutional values.

Similarly, in their examination of social-ecological systems, Cumming and Peterson (2017) propose four criteria to define collapse:

1. The identity of the social-ecological system must be lost. Key actors, system components and interactions must disappear.
2. Loss of identity should happen fast relative to regeneration times and turnover rates of identity-defining components of the system.
3. Collapse involves substantial losses of social-ecological capital. Definition of the word 'substantial' depends on the criteria used to define system identity.
4. The consequences of collapse must be lasting.

Obviously, given their focus on human society, such definitions are strongly anthropocentric in focus. However, at least some of these features might be considered relevant to the collapse of ecosystems. These definitions imply a change in a change in the structure and/or function of the system, as well as its size, and that individual aspects or properties of the system may be unable to prevent collapse from occurring. Further to this, Lindenmayer *et al.* (2016) suggest that for an ecosystem to be considered collapsed, at least three key conditions must be met: collapse must be (1) irreversible or time- and energy-consuming to reverse, (2) widespread and (3) undesirable in terms of impaired provision of ecosystem services or associated with major losses of biodiversity. These criteria again imply an anthropocentric focus, but inclusion of terms such as 'undesirable' implies the need for value judgements, which are difficult to apply consistently (Boitani *et al.*, 2015).

The definition presented by Keith *et al.* (2013), which was further refined by Bland *et al.* (2017a), apparently represents the first attempt to formally define the concept of ecosystem collapse (Table 1.1). In doing so, they highlight a number of challenges in defining a suitable endpoint for environmental degradation. Chief among these is the issue of uncertainty. As there is uncertainty in the boundaries of ecosystem types, there will necessarily be uncertainty in the endpoints of degradation, which may be more accurately portrayed as bounded ranges rather than discrete points. Further, if environmental degradation causes an ecosystem to transform into a different ecosystem type, this process and the associated transition points will also be inherently uncertain (Keith et al., 2013).

Boitani et al. (2015) highlight a number of problems with the definition of collapse presented by Keith et al. (2013). They note an important difference between species and ecosystems when conducting risk assessments: while extinction of a species has a clear meaning, in that no individuals of the taxon remain, the endpoints for an ecosystem can be far more ambiguous. Ecosystems can change rapidly in time and space, but major changes in ecosystem properties can occur without major changes to the biotic component. Conversely, the composition and structure of biological communities can undergo profound change without greatly affecting ecosystem properties (Chapin et al., 2002). As a result, there are multiple possible endpoints for an ecosystem that is being degraded, and there may be no consensus on what is desirable or undesirable (Boitani et al., 2015). Further, Boitani et al. (2015) highlight the difficulty of quantifying the natural range of variability in an ecosystem, while also noting that the concept of collapse is both vague and ecosystem-specific. As a consequence, comparison between ecosystems will be very difficult, undermining the concept of a standardised approach to ecosystem assessment. Rather, these authors suggest that it will be necessary to define collapse separately for each type of ecosystem, using a variety of attributes and threshold values (Boitani et al., 2015).

In response to these criticisms, Keith et al. (2015) referred to recent theories on ecosystem resilience and the concept of multiple stable states (Folke et al., 2004). In this context, a key issue is how to determine whether transitions between ecosystem states are part of the natural variability within an ecosystem type, or represent a process of collapse and replacement by a different ecosystem type. Keith et al. (2015) note the uncertainty associated with this determination, but suggest the answer will depend on whether transitions to different states involve loss of the defining features (characteristic biota and processes) that explicitly describe the ecosystem type. However, neither Keith et al. (2015) nor Boitani et al. (2015) consider the fact that the theory of multiple stable states in ecosystems has itself been challenged (Petraitis, 2013). This issue will therefore be considered in greater depth in Chapter 2.

This text will follow the definition of ecosystem collapse presented by Bland *et al.* (2017a), unless stated otherwise (Table 1.1). It is used here as a working definition, noting the shortcomings highlighted by Boitani *et al.* (2015). Whether this definition stands up to further scrutiny will be revisited in later chapters of this book.

1.4 What Is Ecosystem Recovery?

Given the importance of ecological recovery in conservation, it is surprising that its definition has received so little explicit consideration from researchers. In their review of species recovery, Westwood *et al.* (2014) note how the concept of recovery is poorly defined both in the academic literature and in legislation, which may have contributed to the relatively low success rates of species recovery plans. These authors highlight the need to differentiate between the process or recovery and the end-state, when recovery has been achieved, and suggest that different stages of recovery might usefully be identified based on analogies with medical practice (Westwood *et al.*, 2014). More simply, Redford *et al.* (2011) describe the recovery of species as 'moving in the opposite direction from extinction'. At the scale of ecosystems, recovery might therefore be described analogously as moving in the opposite direction from collapse.

Definitions of ecosystem recovery typically focus on returning to the conditions that prevailed prior to a disturbance event or perturbation (Table 1.1). As a result of external natural or anthropogenic disturbances, ecosystems can either remain fundamentally unchanged or be damaged or degraded in some way. If the source of this degradation is removed, ecosystems may subsequently: (i) fully recover to their condition prior to the disturbance, (ii) partly recover to a reduced or altered condition or (iii) remain in a damaged state indefinitely (Lotze *et al.*, 2011).

Our understanding of the process of ecosystem recovery is largely attributable to the science and practice of ecological restoration, which have developed rapidly in recent decades. Ecological restoration is commonly defined as a management approach that is designed to support the recovery of ecosystems that have been degraded, damaged or destroyed (Society of Ecological Restoration, 2004). In this context, ecosystem recovery implies returning to a historically defined target, such as the situation that prevailed before the degradation took place. Ecological restoration can therefore be understood as a process of guiding ecosystem recovery towards a historically based reference ecosystem (Balaguer *et al.*, 2014). In practice, particularly in areas with a long history of human modification, the identification of such historical references can be very challenging. For this reason, extant ecosystems that are relatively undisturbed are often used as references or restoration targets, in relation to their composition, structure, function and/or dynamics (Bullock *et al.*, 2011). Clearly, the choice of reference and the ecosystem variables of interest will have a significant bearing on the extent to which recovery is judged to have taken place.

Bullock *et al.* (2011) highlighted a number of generalisations about ecosystem recovery: the rate of recovery is generally lower in less productive ecosystems, in ecosystems that are more degraded and in situations where environmental factors constrain recovery. In their review of recovery rate in degraded ecosystems, Jones and Schmitz (2009) found that the recovery rate may be independent of perturbation magnitude but instead may be idiosyncratic to the ecosystem type. These authors also found that recovery in aquatic systems was generally more rapid than in terrestrial systems, independently of disturbance magnitude. This was attributed to the lower longevity of the longest-living species and more rapid cycling between nutrient pools in aquatic, rather than terrestrial, systems (Jones and Schmitz, 2009). In addition, these authors noted that recovery rate depends critically on the type of variable measured.

Ecosystem recovery during ecological restoration has been conceptualised as a non-linear trajectory, which describes the change in indicators of restoration progress over time (Clewell and Aronson, 2013). Earlier research tended to suggest that recovery trajectories typically display a smooth increase over time, until the pre-disturbance or reference state is reached (e.g. Bradshaw, 1984). More recent research has indicated that recovery trajectories are often complex and do not necessarily follow simple, monotonic recovery pathways that return the system to a pre-disturbance state (Bullock *et al.*, 2011; Suding, 2011). Furthermore, some forms of degradation, such as eutrophication and overfishing, may not be fully reversible (Duarte *et al.*, 2015). As a result, recovery trajectories can be very difficult or impossible to predict and may differ markedly from the trajectories of degradation (Duarte *et al.*, 2015; Newton and Cantarello, 2015; Matthews *et al.*, 2009).

In addition to the variation observed in recovery trajectories, there is a growing awareness that in the current era of rapid environmental change, many ecosystems are being transformed into new configurations that do not have a historical analogue (Hobbs *et al.*, 2009; Keith *et al.*, 2009). As a consequence, it may not be possible to specify a single endpoint of ecosystem recovery. Rather, multiple possible endpoints might need to be explicitly recognised in restoration planning and management (Suding, 2011; Suding and Gross, 2006), which may include the creation of entirely novel ecosystems (Hobbs *et al.*, 2009). This has led to suggestions that conservation and ecological restoration practice will need to shift away from a focus on existing or historical assemblages of species and instead embrace the new – which has been described rather ominously as a 'new ecological world order' (Hobbs *et al.*, 2006, 2009).

The concept of novel ecosystems, and what it might mean for conservation practice, has been the focus of a lively – even acrimonious – debate (Miller and Bestelmeyer, 2016). For some authors, novel ecosystems represent current and future reality, which environmental managers need to address, perhaps becoming more efficient in the process (Perring *et al.*, 2013, 2015). Conversely, Murcia *et al.* (2014) suggest the term 'novel ecosystem' should not be used, as it is an ill-defined theoretical construct, which could potentially undermine the practice of ecological restoration. Balaguer *et al.* (2014) argue strongly that the idea of a historically based reference remains a cornerstone concept in ecological restoration despite the widespread ecological changes that are currently taking place. Similarly, Lotze *et al.* (2011) suggest that despite the problems of shifting baselines and the complexity of human-induced changes, increasing understanding of the natural dynamics of ecosystems and their past histories can help to identify meaningful reference points that can assist in assessing ecosystem recovery. I will return to this debate in Chapter 2, as it has major implications for what we mean by ecosystem recovery and how we might measure it.

1.5 The Role of Metaphor

In their analysis of ecosystem concepts, Pickett and Cadenasso (2002) examine the role of metaphor, which may be defined as a figure of speech by which we understand one thing in terms of another (Larson, 2014). Considerations of an ecosystem as a machine or as a superorganism essentially represent different structural metaphors; similarly, whether ecosystems are viewed as either resilient or vulnerable to human activity refers to different metaphors of ecosystem behaviour (Pickett and Cadenasso, 2002). Metaphors and analogies are seen as a useful part of the scientific method, helping to stimulate new thinking. For example, initial ideas about successional changes in vegetation were informed by insights into the development of individual organisms (Pickett and

Cadenasso, 2002). For Peters (1991), the use of metaphors is simply part of the development of scientific ideas and is not therefore something that should be particularly despised. However, metaphors can be misleading or false; in this context, Peters (1991) refers to the analysis of ecological mechanisms based on the consideration of ecosystems as machines. For Pickett *et al.* (2007), metaphors are useful for fostering communication between scientific disciplines or between scientists and the general public. Yet they can also foster misunderstanding, as illustrated by the widespread belief in the existence of a 'balance of nature' despite scientific evidence to the contrary (Pickett *et al.*, 2007).

The main problem with the use of metaphor in science is that the choice of metaphor is typically laden with values, which are often not made explicit. Larson (2011) examines this issue in detail, with reference to examples such as DNA barcoding and invasional meltdown. DNA barcoding is a method for identifying species based on analysis of short sequences of DNA (Hebert *et al.*, 2003). Although use of the technique is now widespread, the metaphor of DNA barcoding is grounded in consumerism, and by communicating these materialistic values, it risks contributing to the objectification of nature and the reduction of living organisms to commodities that can be purchased and traded (Larson, 2011). The term 'invasional meltdown' was developed to describe the process by which non-indigenous species might interact to promote the invasion of each other, leading to the replacement of native communities, perhaps beyond the point of no return (Simberloff and Von Holle, 1999). Use of this metaphor has been criticised for evoking concern and fear, by employing imagery associated with nuclear accidents. In this way, it implies that drastic change is either inevitable or very difficult to prevent (Larson, 2011). This metaphor was also criticised for employing military or martial imagery and implying a bias against non-native species; for limiting understanding; and for hindering the development of appropriate management responses to invasive species (Larson, 2005). As noted by Simberloff (2006), similar concerns could be levied about many – if not all – metaphors in science. The metaphor of keystone species, for example, was initially widely applied in ecology, but over time its meaning became so extended and obscure that its application to conservation management was considered dangerous and inappropriate (Simberloff, 2006).

The concepts that are the focus of this book – ecosystem collapse and recovery – can therefore be viewed as metaphors. In noting this, it is also important to recognise that they implicitly incorporate values. For example, the use of the term 'ecosystem collapse' to describe the endpoint of environmental degradation (Keith et al., 2013, 2015) implies a state of profound ecological change or dysfunction, which might be difficult to reverse. It could be argued that the term is emotionally loaded and is likely to engender concern or even fear among people who encounter it. It is widely recognised that alarmist language can help generate media coverage. In this way, the use of the term 'ecosystem collapse' might help raise awareness about ecosystem degradation and stimulate further research and action as 'invasional meltdown' did for the issue of invasive species (Larson, 2011; Simberloff, 2006). However, research into perceptions of climate change indicate that while fear can attract people's attention, it is generally ineffective in motivating genuine personal engagement (Chapman et al., 2017; O'Neill and Nicholson-Cole, 2009). Use of non-threatening imagery may be more effective in achieving behavioural change and motivating action, particularly if people can be given something positive that they can do (Larson, 2011). More positive metaphors around ecological restoration — such as ecosystem recovery — might therefore be preferable.

If ecosystem collapse and recovery are metaphors, what is their truth — or their 'truthiness'? As suggested by Proctor and Larson (2005), metaphors are perhaps best interrogated in terms of the understandings they provide, rather than in terms of whether or not they are true. In other words, are concepts of ecosystem collapse and recovery useful for understanding the impacts of human activity on the environment? This is a question that underpins the text throughout this book and will be considered in depth in Chapter 6.

Does the use of the terms 'ecosystem collapse' and 'recovery' imply a particular world view and its associated values? The answer is surely yes. Development of the IUCN Red List of Ecosystems (Keith et al., 2013, 2015) is based on the assumption that ecosystems are of intrinsic value and that environmental degradation is something negative that can lead to undesirable consequences for nature, the ultimate expression of which is ecosystem collapse. The term 'ecosystem recovery' similarly implies that human disturbance can cause negative impacts on ecosystems, from which recovery might be deemed a desirable outcome. In making these values explicit, is the objective rigour of science being put at risk? Or would ecological science be more effective in terms of conservation outcomes if it became more explicitly and consciously value-laden, as some have suggested (Larson, 2014)?

The problems associated with the choice of metaphor and its associated values are illustrated by the concept of ecosystem services, which currently dominates research on human–environment relationships. Although beginning as a metaphor, ecosystem services have increasingly become a central approach in environmental policy and management despite its tendency to obscure the complexities of human–environment interactions (Norgaard, 2010). The concept of ecosystem services focuses on direct use of ecosystem benefits by people and the economic quantification of these values. It has been suggested that this approach can detract from effective environmental management by obscuring other understandings of human–environment relationships; rather than dogmatic application of this single concept, a variety of different metaphors should be explored (Raymond et al., 2013). These issues will be explored further in Chapter 5.

1.6 Ecosystem Collapse: A Useful Focus?

Most ecologists would agree that supporting ecosystem recovery through ecological restoration is a worthwhile endeavour, even if they disagree about what the precise endpoint or management objective should be. Ecosystem collapse is less well established, either as a scientific concept or as something that conservation management should be explicitly designed to prevent. So how useful is ecosystem collapse as a concept?

It could be argued that the main purpose of conservation is to prevent species from becoming extinct, as extinction is forever, and we have a moral duty to prevent this from happening. This tallies with the fact that much conservation action has targeted individual species, particularly charismatic vertebrates such as birds and mammals. Yet the need to also conserve ecosystems is well established. For example, Franklin (1993) notes that there are simply too many species to conserve by a species-by-species approach and that larger-scale methods that focus on ecosystems are the only way to conserve the vast majority of species. This is particularly true for relatively cryptic species, such as many invertebrates, fungi and bacteria, which play an important role in ecosystem function. Similarly, Noss (1996) highlights the value of focusing conservation not just on entities, such as species, but also on ecological processes, such as interactions between species and nutrient cycling. Persistence of any species requires that these processes be maintained, and therefore a conservation focus on ecosystems might actually be more effective and efficient than a focus on species.

As explored in Section 1.2, the ecosystem concept has a number of weaknesses. It can be argued that ecosystems are artificial constructs or

1.6 Ecosystem Collapse: A Useful Focus?

metaphors rather than real entities; they are characterised by fuzzy or open boundaries, and they are often defined arbitrarily. Yet Noss (1996) suggests that these weaknesses may actually be strengths in a conservation context, by providing a flexible way of addressing the wide diversity of conservation problems that exist, at a range of different scales. Although Noss (1996) recognises the problems of mapping ecosystems, and that any ecosystem classification or map is an abstraction of reality, he notes that such maps are very useful in a practical management context.

If ecosystems provide a legitimate focus for conservation efforts, then logically the collapse or loss of an ecosystem should be the focus of conservation concern. The global biodiversity crisis that we are currently experiencing represents more than the decline or loss of species and a loss of ecological complexity, interactions between species and ecological functions and processes. The collapse of a coral reef resulting from a bleaching event represents more than the loss of some individual fish or coral species, but the breakdown of a complex, interacting system that has developed over the millennia. This book is therefore based on the contention that something is happening out there in the real world, which is more than simply a loss of species or a global extinction crisis. It is loss of assemblages of species and the interactions between them and a loss of ecological processes and the functioning of ecological systems. Such losses could potentially have major implications for human society, and for this reason if no other, understanding how and why ecosystems collapse surely merits urgent consideration.

Given this, how should we view the recent debate about ecosystem collapse in the literature (Boitani *et al.*, 2015; Keith *et al.*, 2013, 2015)? As noted by Boitani *et al.* (2015), the concept of ecosystems incorporates more than species richness and composition and includes ecological processes and functions, flows of energy and material and interactions between the components of the system. Keith *et al.* (2013) suggested that ecosystems can be described by identifying and measuring key processes and interactions, but Boitani *et al.* (2015) are surely right to point out that in practice, it will be very difficult to do this in a consistent way in different parts of the world. This is because the parameters that define ecosystems vary depending on scale, ecological context, functional relationships and species assemblages. As a practical solution to this problem, Keith *et al.* (2015) propose to focus the IUCN Red List of Ecosystems on an assessment of ecosystem types, which can be defined using proxies such as vegetation composition. Such proposals are clearly pragmatic and are similar to the solutions proposed by Boitani *et al.* (2015), but these authors are correct to point out that an assessment of vegetation types (or other proxies) is not the same as an assessment of ecosystems – and therefore should not be described as such.

Fortunately, the focus in this book is not on the Red List of Ecosystems but on ecosystem collapse. As someone sympathetic to the goals of the IUCN Red List, I wish them well in their efforts. Development of the Red List of Ecosystems has helped put ecosystem collapse on the scientific map; it is increasingly becoming the focus of research interest (Bland *et al.*, 2018a; Lindenmayer *et al.*, 2016; MacDougall *et al.*, 2013; Sato and Lindenmayer, 2017; Valiente-Banuet and Verdú, 2013). The Red List of Ecosystems is surely a worthwhile endeavour, but as recognised on both sides of the debate, the operational challenges are substantial. Here, though, the focus is on the collapse and recovery of ecosystems, rather than purely proxies, such as vegetation communities. A really interesting question is how collapse and recovery of the biological characteristics of an ecosystem, such as species richness and composition, relate to trends in other ecosystem properties, such as function and service provision. Given the long controversy about the nature of the relationship between species richness and ecosystem function (Hooper *et al.*, 2005; Tilman *et al.*, 2014), these linkages are likely to be complex; they will be examined further in Chapter 5.

A further issue featured in the debate relates to multiple states of ecosystems. As described by Keith *et al.* (2015), when undertaking an assessment of collapse risk, a key task is to differentiate between transitions between ecosystem states that might occur as part of the natural variability within an ecosystem type or as a process of collapse and replacement by a different or novel ecosystem. According to the analytical framework proposed by these authors, determination of this will depend on whether transitions to particular states involve loss of the defining features (characteristic biota and ecological processes) that explicitly describe the ecosystem type. In practice, this could be very challenging. The process of vegetation succession, for example, typically involves the successive development of plant communities that can differ markedly in terms of species composition and structure – such as the transition from lichen heath to conifer forest. Should this be considered as a form of ecosystem collapse, or not?

Boitani *et al.* (2015) conclude that given the difficulties with defining ecosystem collapse, it will often be necessary to define collapse separately for each ecosystem considered, using a variety of attributes and threshold values. For conducting a global assessment of collapse risk, as proposed in the Red List of Ecosystems, this does indeed present a substantial difficulty. For the purposes of this text, however, this implies a richness in the concept that deserves attention. Does collapse represent something quite different, in different types of ecosystem? If so, does this imply that it is difficult to develop generalisations about ecosystem collapse? And if that is the case, where does it leave ecological theory? These fascinating questions will be addressed in the subsequent chapters, but for now, the implication of this line of thinking is that the best way to consider ecosystem collapse is on a case-by-case basis. For this reason, this book will focus on profiling empirical data and case studies drawn from a wide variety of different ecosystems. The central assertion made by Boitani *et al.* (2015), that it is not in fact practicable to assess collapse risk in a consistent way across different types of ecosystem, will be reexamined in the light of this empirical evidence, in Chapters 5 and 6.

Yet this book is not solely about ecosystem collapse; it also considers ecosystem recovery. This resonates well with the spirit of conservation optimism, which is currently fashionable. It is surely true that doom and gloom won't save the world (Knowlton, 2017). If you, like me, feel in need of some positivity regarding what conservation efforts can achieve, I recommend that you read Andrew Balmford's excellent book *Wild Hope* (Balmford, 2012), which provides a detailed profile of some recent conservation successes. (Hopefully, unlike me, you will not feel more depressed after having read it, as you come to realise just how difficult it is to design and implement a successful conservation project.) Anybody who works on ecological restoration knows that many ecosystems possess a quite startling capacity for recovery, if provided with a suitable opportunity to do so. Personally, I find this to be the most potent source of wild hope and conservation optimism. Researchers may argue about what the objectives of ecological restoration should be and how one should go about it, but nevertheless, this is an approach that demonstrably works and can deliver enormous benefits to both wildlife and people (Nellemann and Corcoran, 2010).

1.7 Structure of the Book

Although ecosystem collapse is arguably an issue of outstanding societal and scientific importance, there is great uncertainty about what exactly it comprises, how and when it might occur and what the consequences of it might be. Similarly, there is a lack of understanding of how ecosystem collapse relates to the process of subsequent recovery. The aim of this

Table 1.3 *Some scientific questions about ecosystem collapse and recovery*

- What is ecosystem collapse, and how should it be defined and assessed?
- How, why and when does ecosystem collapse occur?
- If one ecosystem collapses, might others follow?
- What are the different mechanisms that can cause ecosystem collapse, and are these the same in different types of ecosystems?
- Are some ecosystems more at risk of collapse than others? If so, why is this?
- Is it possible to provide early warning of imminent collapse, and, if so, how?
- What are the implications of ecosystem collapse for biodiversity, ecosystem function and the provision of ecosystem benefits to people?
- How does collapse relate to ecosystem recovery? Can recovery occur after collapse, and if so, how?
- Are the trajectories and mechanisms of recovery related to those of collapse?
- What are the mechanisms of ecosystem recovery, and are these the same in different types of ecosystem?
- How can ecosystem collapse be prevented?
- How can ecological recovery be supported by appropriate environmental management?

---

book is therefore to attempt to answer some key questions about both ecosystem collapse and recovery (Table 1.3) by providing an overview and evaluation of available evidence. At the same time, I highlight some current gaps in knowledge and areas of uncertainty, while also identifying additional questions that might usefully guide future research. In so doing, I apologise in advance for any inappropriate personal asides, annoying sidetracks, unwarranted speculations, needless flowery language and attempts at humour. Any such efforts represent an attempt to avoid being consistently boring (following Sand-Jensen, 2007), which is surely a laudable goal, even if I have failed to achieve it.

A key objective here is to identify the mechanisms underlying both ecosystem collapse and recovery. This includes an analysis of what can be learned from the pattern of collapse and recovery (such as rate and trajectory) and the ecological processes that influence them. The approach is grounded in an appraisal of empirical evidence, based on analysis of real-world case studies. Examples of both collapse and recovery are drawn from a wide range of different ecosystem types, from a variety of geographical locations. The selection of case studies is deliberately broad and includes examples from terrestrial, marine and freshwater environments. This requires integration of evidence from different sub-disciplines of ecological science, which are often quite separated in the

1.7 Structure of the Book · 31

literature. Such synthesis is required to determine whether the phenomena of ecosystem collapse and recovery are equivalent in different types of ecosystem and whether they are underpinned by similar mechanisms in different ecological contexts. An attempt to identify generalisations through an appraisal of empirical evidence also provides opportunities to evaluate and test available theory.

Consequently, Chapter 2 provides an overview of relevant ecological theory, including areas of dynamical systems theory and succession theory, and associated concepts such as planetary boundaries, tipping points, ecological thresholds, resilience and multiple stable states. This chapter also provides an overview of relevant empirical theories, including theories of ecological disturbance and the processes by which ecosystems are degraded. By presenting this theoretical context, this chapter provides a framework for evaluating the empirical evidence presented in the following chapters, which will be used to help identify those theoretical concepts and ideas that offer most potential for understanding the phenomena of ecosystem collapse and recovery.

Chapters 3 and 4 present real-world case studies of both collapse and recovery, focusing on those examples that are relatively well supported by empirical data. Chapter 3 focuses on examples drawn from deep time, evidenced by analysis of the fossil record, whereas Chapter 4 focuses on contemporary examples. Some of the best documented examples of ecosystem collapse are those evident in the fossil record, such as the 'Big Five' mass extinctions that define geological periods and the mass extinction that occurred at the end of the Pleistocene (Brannen, 2017). These mass extinctions provide a powerful example of how entire communities of species can be wiped out, but it is also informative to consider what happened afterwards. Did these ecosystems recover, and, if so, how? Palaeoecological evidence can therefore potentially provide insights into ecosystem recovery as well as collapse, which will be explored in Chapter 3. Similarly, Chapter 4 will examine both collapse and recovery by profiling relevant case studies, but here the focus will be on recent or current examples in a wide range of different ecosystem types (including coral reefs, marine fisheries, freshwater rivers and lakes, forests, savanna etc.).

Chapter 5 provides an overall synthesis, by integrating the empirical evidence presented in the preceding two chapters and evaluating its relationship to the theory presented in Chapter 2. The aim here is to identify some generalisations regarding the mechanisms of ecosystem collapse and recovery and the different ecological processes involved, to provide an overview of current understanding. This chapter will also explore how this understanding might help inform the identification of appropriate management and policy responses. Key questions include: How can ecological collapse be prevented? And how can ecological recovery be supported by appropriate environmental management? Both avoiding ecosystem collapse and supporting recovery represent significant challenges to conservation management, whether the objectives are the conservation of species, the maintenance of ecosystem function and/or the provision of ecosystem services to people. Relevant issues here include the identification of appropriate early warning indicators of collapse and how to evaluate the effectiveness of ecological restoration actions. Chapter 6 provides a brief summary of the key findings of the preceding chapters, together with some suggestions for future research priorities.

2 · *Ecological Theory*

Like many other scientists of my generation, I was taught that science is fundamentally about developing and testing relevant theory. For scientific research to be rigorous, there should be an underlying theoretical framework, which is used to generate hypotheses that can potentially be falsified by collecting data. Over time, I have come to appreciate that not everybody shares this view and that there are alternative ways of making a significant contribution to scientific knowledge. There are, indeed, many different ways of doing science. Providing an accurate description of some ecological phenomenon, such as ecosystem collapse, might arguably be considered as a valid scientific endeavour in its own right, without needing any underlying theory. In the current era of 'big data', it has been suggested that entirely new ways of doing science are developing that are not dependent on theory at all, involving induction from large numbers of observations (Mahootian and Eastman, 2009). Such approaches are transforming how ecological science is conducted (Hampton *et al.*, 2013) and are potentially changing the role of theory in the process.

If we consider the role of theory in science, we enter the realm of philosophy. The idea that science should be conducted by developing and testing hypotheses, namely, by following the hypothetico-deductive method, is strongly associated with the work of Karl Popper (Popper, 1959). His suggestions have since attracted a great deal of criticism; for example, Nagel (1979) describes Popper's work as being 'close to being a caricature of scientific procedure', whereas Hansson (2006) argues that falsificationism depends on an incorrect view of the nature of scientific inquiry and is not a tenable research methodology. Is there then such a thing as a single 'scientific method', or a method of reasoning that applies to all areas of science? Interestingly, philosophers disagree about the answer to this question (Sober, 2015). Even more interestingly, philosophers also disagree about whether or not philosophers should even attempt to identify the methods that scientists *ought* to use (Sober,

2015). Despite these debates, which are perhaps best left to philosophers, hypothetico-deductive approaches have long been popular in ecology (Mentis, 1988) and remain so today.

However, ecologists have a rather fraught relationship with theory. There is no shortage of theoretical ideas in the literature, but most ecological publications do not provide any reference to the theoretical context of their work (Marquet et al., 2014; Scheiner, 2013). Use of theory also appears to be declining, perhaps because researchers increasingly underappreciate the role that theory can play, or because of the increasing use of big data (Marquet et al., 2014). There is a sense that ecology is not developing rapidly enough as a science and is failing to develop generally applicable theory that can be used to address environmental issues (Belovsky et al., 2004). A fear has been expressed that as a result, ecologists will all be washed out to sea on an immense tide of unrelated information (Watt, 1971). Similar concerns have been levied at conservation science, where available theory has been criticised as being too weak to be usefully predictive, or for having turned into dogma, stifling further theoretical development (With, 1997). Does this make ecology a weak science? Or are ecologists just suffering from 'physics envy' (Egler, 1986)? Perhaps ecologists should console themselves with the fact that ecology is, after all, far more difficult than physics – or, at least, rocket science (Hilborn and Ludwig, 1993).

I strongly encourage any early career researchers to reflect on the role of theory in ecology and ideally to have a go at writing a paper about it. I know from personal experience that this can provide valuable life lessons for any young scientist. My interest in this area began while attending a particularly dull conference, which featured an entirely descriptive series of talks. As I was shaking my head in despair at the lack of any theoretical context, a sympathetic person on the row behind me whispered in my ear, 'Go and read this book'. The book he introduced me to was *A Critique for Ecology* by R. H. Peters (1991). I immediately left the lecture hall, bought the book and spent much of the rest of the conference reading it. It is rare to encounter a researcher who has been so scathingly critical of the scientific discipline in which they work. As someone who had also come to suspect that not all was well in the science of ecology, I found Peters's critique to be deeply refreshing.

Inspired by Peters, I then had the bright idea of writing a paper about the lack of theory in conservation biology, which I duly drafted and submitted. Subsequently, I received from a journal editor the most interesting rejection letter I have ever received, essentially stating that theory is a prickly subject and one not worth exploring because of the opprobrium that this would likely attract, particularly from more established scientists. This provided a fascinating insight into how conservation science works. However, this did not prevent the same journal publishing a paper on precisely the same subject very shortly thereafter, by another author. This provided a second valuable lesson. I am pleased to say that the author of this paper did a far better job than I would have done (With, 1997). I then made the mistake of presenting my unpublished ideas in a talk at another institution, whereupon a member of the audience was stimulated to write a paper on the same subject, which he thoughtfully sent me shortly after publication. Yet another valuable lesson on how science works!

Perhaps the reader will therefore understand why I approached this chapter with a degree of trepidation. Theory is a subject that appears to elicit strong opinions among many researchers, who might disagree about not only whether a particular idea has scientific merit but also whether it qualifies as science at all. Remarkably, there is also a lack of consensus about what an ecological theory actually is. Even though the beliefs I inherited from my teachers are now clearly outdated, I still think that theory is an important part of science – including conservation science. At the very least, it provides a way of organising scientific ideas and a useful framework for evaluating empirical information, helping to turn data into knowledge. Theory also provides a means for making explicit the assumptions that underlie research and helping us interpret what we find (Scheiner, 2013). This chapter therefore provides an overview of ecological theory that is relevant to understanding ecosystem collapse and recovery, after first briefly considering what a theory comprises.

2.1 What Is an Ecological Theory?

Given that theory is such a fundamental component of scientific endeavour, it is surprising that ecologists disagree about what it entails (Marquet et al., 2014; Table 2.1). For Peters (1991), a theory is any construct that makes potentially falsifiable predictions. This definition is bracing in its simplicity and implies that any regression equation used to make a projection would qualify as a theory. Further, Peters (1991) differentiates predictive constructs from what he beautifully describes as a 'prescientific soup' of logical devices, memory aids, inspirational prods, incentives to thought, political opinions, personal ideals, half-formed notions and odd

Table 2.1 *Selected definitions of ecological theory*

Definition	Source
A hierarchical framework that contains clearly formulated postulates, based on a minimal set of assumptions, from which a set of predictions logically follows.	Marquet et al. (2014)
A logical construction comprising propositions, some of which contain established information (axioms) while others define questions (postulates). The working part of a theory provides the information and logical basis for making generalisations.	Ford (2000)
Any construct that makes potentially falsifiable predictions.	Peters (1991)
A system of conceptual constructs that organises and explains the observable phenomena in a stated domain of interest.	Pickett et al. (2007)
A set of assumptions, generalisations of facts and rules for model building. A test can include work that confirms those assumptions, generalisations of facts or rules that are currently not well supported, as well as efforts to disconfirm them directly.	Scheiner (2013)
Hierarchical frameworks that connect broad general principles to highly specific models.	Scheiner and Willig (2011)

beliefs. With such a diverse list of ingredients, this soup sounds somewhat indigestible, but his aim here is to differentiate the many activities, concepts and constructs that contribute to the process of doing science, from those constructs that are actually predictive. Peters's suggestions are both pragmatic and logical and were motivated by a laudable desire for ecology to be useful to solving real-world practical problems. Yet his ideas were highly controversial, not least because according to his definition of theory, a substantial part of ecology would not qualify as science at all, because it does not provide any predictive models.

The book on ecological understanding produced by Pickett et al. (2007) can perhaps be considered as a response to Peters's ideas, and here the concept of theory is quite different. There is no mention of falsifiable predictions as a benchmark for what constitutes a theory; rather, the role of theory is focused on organisation and explanation (Table 2.1). Pickett et al. (2007) take a much more inclusive approach to what constitutes a theory, identifying 11 different components (Box 2.1). The essential difference between these authors boils down to the role of causal explanations; in contradiction of Peters (1991), Pickett et al. (2007) state that prediction is only valuable to science if it helps increase understanding.

> **Box 2.1** *Components of ecological theory, according to Pickett et al. (2007)*
>
> *Domain.* The scope in space, time and phenomena addressed by a theory; specification of the universe of discourse for a theory.
>
> *Assumptions.* Conditions or structures needed to build the theory.
>
> *Concepts.* Labelled regularities in phenomena.
>
> *Definitions.* Conventions and prescriptions necessary for the theory to work with clarity.
>
> *Facts.* Confirmable records of phenomena.
>
> *Confirmed generalisations.* Condensations and abstractions from a body of facts that have been tested or systematically observed.
>
> *Laws.* Conditional statements of relationship or causation, statements of identity or statements of process that hold within a universe of discourse.
>
> *Models.* Conceptual constructs that represent or simplify the structure and interactions in the material world.
>
> *Translation modes.* Procedures and concepts needed to move from the abstractions of a theory to the specifics of application or test or vice versa.
>
> *Hypotheses.* Testable statements derived from or representing various components of theory.
>
> *Framework.* Nested causal or logical structure of a theory.

Maurer (2000) makes the same point more bluntly: empirical relationships based on linear statistical models provide only approximate or temporary solutions to environmental problems, and they do not contribute to an understanding of the relative importance of different problems, nor how they might best be addressed. Yet for Peters (1991), it is precisely the search for 'explanation', 'cause', 'mechanism' and 'understanding' that has led ecology astray as a science, making it 'a discipline of insoluble questions' rather than a source of useful information about the environment.

Pickett *et al.*'s list of theory components helpfully makes a distinction between models and theory, which are often confused in the literature (Marquet *et al.*, 2014). According to Pickett *et al.* (2002), models are useful to help translate ecological concepts or observations into usable tools. Such models may be verbal, diagrammatic, graphical, physical or quantitative. Models may be simplified, partial statements of theories and may be used to explore the influence of different assumptions on which the theory is based; however, they do not constitute theories themselves (Marquet et al., 2014). Put another way, models are where the theoretical rubber meets the empirical road (Scheiner and Willig, 2011).

Another rather vexed question in the literature is whether ecological theory needs to be expressed mathematically. Ecologists often think of theory as equivalent to mathematical models; many textbooks devoted to theoretical ecology simply present a compilation of such models (Kolasa, 2011). According to Marquet et al. (2014), it is generally preferable to represent theories mathematically, because the logic is more transparent, they are less ambiguous and they are more amenable to rigorous empirical testing. However, these authors note that purely verbal theories can be highly successful, as illustrated by the theory of evolution by natural selection, as developed by Darwin and Wallace. Suggestions that the only good theory is a mathematical one are essentially another manifestation of physics envy (Pickett et al., 2007).

A further issue is the relationship between theory and empirical data. For Kolasa (2011), the boundaries between theory and empirical research are fuzzy, as empirical research is often permeated by theoretical constructs, and any theory has an empirical content. Yet some of the theories employed by ecologists are purely mathematical in origin, particularly those appropriated from other scientific disciplines, such as complexity science or physics. At the same time, there has recently been a rapid growth in the synthesis and analysis of large empirical data sets in the ecological literature, employing techniques such as meta-analysis (Lortie and Bonte, 2016; Vetter et al., 2013). What is the contribution of such empirical studies to theory? Kolasa (2011) suggests that the role of empirical generalisations is different from that of theoretical frameworks or models; their particular value is that they may stimulate questions about the underlying causes of the patterns observed, which may subsequently give rise to theoretical explanations. The neutral theory of species diversity (Hubbell, 2001), a personal favourite of mine, provides an example of a theory that was developed in response to an empirical generalisation.

Can analyses of empirical data be viewed as a theory in their own right? For some researchers, the answer is yes. Rigler (1982) described 'empirical theories' as tools that describe the environment and that can be used to predict patterns. These he differentiated from 'explanatory theories', which similarly enable predictions to be made, but also provide insights into why a system behaves the way that it does. Over time, development of empirical theory might lead to the development of explanatory theory, as has arguably occurred with neutral theory. Pickett et al. (2007) describe Rigler's proposals as 'popular but erroneous', because of the lack of conceptual constructs, linked by a framework. Essentially, these criticisms are similar to those these authors levy at Peters (1991), whose ideas are closely related to those of Rigler (1982).

What then is an ecological theory? I have to confess some sympathy with the ideas of Rigler (1982) and Peters (1991), particularly as they are oriented towards using empirical data to solve practical problems in the real world, including those relating to conservation. Surely the ability of a theory to generate testable predictions is a useful measure of its scientific value, even if it is lacking an underpinning conceptual framework despite this being valued more highly by Pickett et al. (2007). Yet as ecological scientists, despite what Peters (1991) suggests, we want to be able to do more than make successful predictions; we want to understand what is going on in the world. As suggested by Pace (2001), prediction should therefore be coupled with mechanistic understanding, perhaps as an iterative process. Statistical analyses, meta-analyses, experiments and mechanistic models can all usefully contribute to this process. Although I accept that theories formulated mathematically may be more rigorous and precise, purely verbal theories can also be valuable, as Darwin showed. A pluralist and inclusive approach to what constitutes ecological theory is therefore merited, and that is what is adopted here.

As with many other areas of ecology, ecosystem collapse and recovery are arguably theory-rich. Even though no theory has been explicitly developed to address these phenomena directly, many theoretical ideas and concepts drawn from other ecological domains could potentially contribute to their understanding. In the brief overview that follows, I evaluate each of these ideas by applying two simple criteria:

- Does the theory make testable predictions?
- Does it help us to understand the underlying mechanisms?

Before we dive into the world of ecological theory, let us pause briefly for a moment of Zen. The synthesis that follows will involve integration and acceptance of multiple terminologies, diverse concepts and a range of different perspectives. By offering a way of being that incorporates togetherness, Zen can potentially be useful for tasks such as this (Lortie and Bonte, 2016). With appropriate mindfulness, you may be able to embrace this diversity, and perhaps even see the entire universe in your breakfast cereal (Allendorf, 1997). Such an approach might help overcome the lack of trust, understanding and interaction that currently exists between ecological empiricists and theorists (Haller, 2014), which is reflected in some of the literature considered in the following sections.

2.2 Disturbance Theory

The impacts of disturbance on ecological communities have long been the focus of research interest. This is illustrated by the lengthy debate regarding the relationship between diversity and stability, which spans nearly a century, during which time it has attracted substantial attention from theoreticians (Ives, 2007). Increasing recognition of the role of natural disturbance in shaping the composition and structure of ecological communities has contributed to a 'paradigm shift' in ecology (Pickett and White, 1985) and the development of what has been referred to as 'disturbance theory'. Over time, this has developed from an initial focus on the impacts of disturbance on the species richness and composition of communities, to consider impacts on attributes of ecosystems such as trophic structure and function (Pickett *et al.*, 2007). Similarly, there has been a shift from a primary focus on natural forms of disturbance to a greater consideration of anthropogenic sources (Newton and Echeverría, 2014). Although there has been progress in developing mathematically formalised theory relating to disturbance impacts, for example on forests (e.g. Clark, 1989, 1991), development of disturbance theory has largely been conceptual and is still at an early stage (Pickett *et al.*, 2007).

Most disturbance theory focuses on how ecosystems respond to disturbances. In relation to this, a wide variety of different ideas have been proposed (Pulsford *et al.*, 2016). Some of these are considered in the sections that follow, including those addressing succession and resilience. Here, I focus explicitly on characterisation of the disturbance regime itself and how these characteristics may link with ecosystem properties to determine impacts.

Disturbances vary in their frequency, extent, intensity, duration and predictability; the type of disturbance is also important for determining its effects on an ecosystem (Villnäs *et al.*, 2013). A disturbance regime therefore consists of many different disturbance events, each of which

> Box 2.2 *Key features of disturbance in ecological systems (adapted from Grimm et al., 2017 and Peters et al., 2011)*
>
> 1. Disturbance is a structural disruption, caused by factors or threatening processes that disrupt system process and may thereby lead to structural disruptions.
> 2. Explicit definition of a system, in terms of its boundaries, composition and interactions, is key to understanding disturbance impacts, including effects over time and space.
> 3. A disturbance event may be characterised by its intensity or magnitude, spatial extent, duration and timing, independently of the impacts of that event.
> 4. Disturbance regimes describe the spatial and temporal patterns of different disturbances across landscapes or seascapes, including the type, frequency, timing, spatial extent, intensity and interactions among different disturbance types.
> 5. A disturbance event can interact with other events and with ecosystem properties to feed forward to generate additional disturbance.

has its own characteristics. These events interact with the properties of an ecosystem to determine how it is affected by the disturbance and how it recovers afterwards (Sousa, 1984). An ecosystem may be affected by a number of different types of disturbance simultaneously, which may potentially interact with each other. Also, a particular disturbance event may have much greater impact if it follows another disturbance from which the ecosystem has not yet recovered (Resilience Alliance, 2010). For example, the impacts of drought on forest ecosystem may be much greater if the trees have already been weakened by a pest or disease outbreak. Furthermore, the characteristics of a disturbance regime can vary dynamically over both time and space, and these dynamics can have a major influence on the composition, structure and functioning of an ecosystem (Box 2.2).

Disturbances can usefully be classified as either 'pulse' disturbances, which are relatively short-term events that occur at a discrete event in time, or as 'press' disturbances, which are more continuous, cumulative or gradual (Bender *et al.*, 1984; Resilience Alliance, 2010). Although this distinction is still widely employed (Cantarello *et al.*, 2017; Grimm *et al.*, 2017), it can sometimes be difficult to disentangle these different

42 · **Ecological Theory**

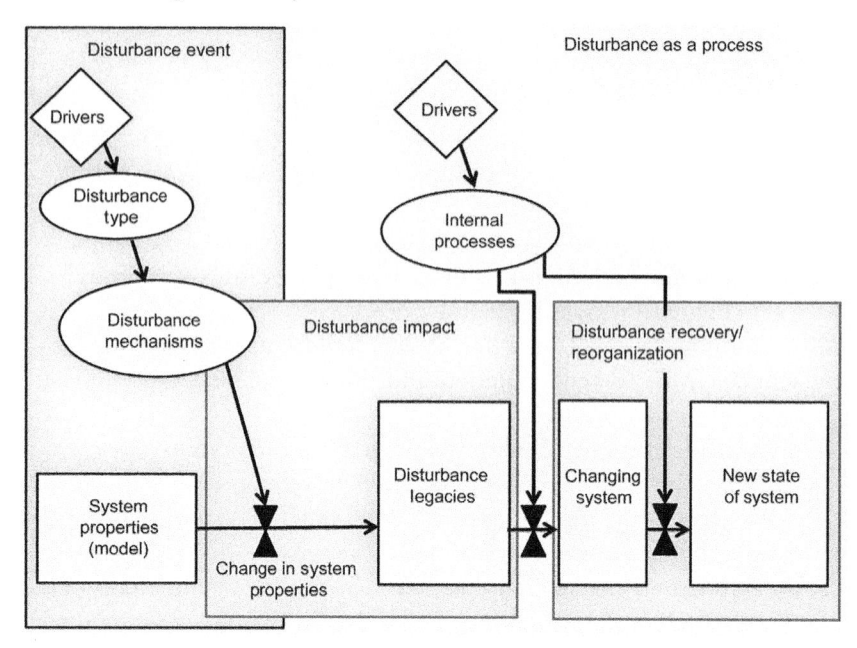

Figure 2.1 A conceptual framework to support the comparative analysis of disturbance impacts on ecosystems, from Grimm *et al.* (2017) and adapted from Peters *et al.* (2011). Examples of drivers include stocking rate of animals for grazing disturbances, and temperate and precipitation for drought events. Ecosystem properties include biomass, soil moisture and soil temperature, which can interact with drivers to determine disturbance outcomes. Potential mechanisms include defoliation, consumption, mortality and growth and so on.

disturbance types in field situations. For example, a 'press' disturbance such as climate change might result in an increased frequency of 'pulse' disturbances, such as fires or flooding events. A further important distinction is whether disturbances result from some factor or driver that is external to the system, or from interactions or processes that occur within it (Grimm *et al.*, 2017; Schröder, 2009). Differentiating between these two possibilities depends critically on how the boundaries of the system are defined (Box 2.2).

Peters *et al.* (2011) presented a conceptual framework to support the quantitative comparison of disturbance impacts across different types of ecosystem, to assist in the development of theory in this area. This framework was based on the disaggregation of three measurable components (Figure 2.1):

2.2 Disturbance Theory · 43

(1) environmental drivers and their associated characteristics, which interact with:
(2) initial properties and spatial structure of a given ecological system, to determine:
(3) physical and biological mechanisms, which result in a change in system properties.

According to this framework, these components result in abiotic and biotic legacies, which can interact with subsequent drivers and successional processes to influence the response of the ecosystem (Peters *et al.*, 2011). At any time, a driver of sufficient magnitude can create a new disturbance event, which can potentially push the system beyond its historical pattern of variability (Peters *et al.*, 2011). A further element of this framework is the need to specify the amount of disruption of an ecosystem property relative to some form of baseline or reference state, which could be obtained through analysis of time-series data or comparison with an undisturbed system. Following this framework, drivers and mechanisms interacting with initial system properties could result in the removal or addition of structural components, which could disrupt the functioning of the system so that it differs from the reference state (Peters *et al.*, 2011). This offers one potential way of conceptualising the process of ecosystem collapse and for exploring empirical data.

Smith *et al.* (2009) note that a key characteristic of anthropogenic disturbances is that they tend to be chronic (or press) perturbations rather than the relatively discrete 'pulse' disturbances that generally characterise natural disturbance regimes. Such perturbations can alter the availability of resources in ecosystems. To provide insights into the mechanisms underlying ecosystem responses to changes in resource availability, these authors propose the hierarchical-response framework (HRF), which integrates three major levels of response – individual, community reordering and immigration or loss of species. Based on the HRF, Smith *et al.* (2009) present the following hypotheses:

- The *dominance hypothesis* is based on the observation that dominant species have a major influence on species interactions and the distribution of resources. Resource change can favour new dominant species, the traits of which could influence the overall response of the ecosystem and its new structure and function. For example, ecosystems dominated by long-lived species with slow demographic turnover, such as trees, will be relatively slow to respond to resource alterations.

- The *biogeochemical hypothesis* predicts that sizes of key carbon and nutrient pools and turnover rates will determine the rate and magnitude of ecosystem responses to resource alterations. For example, ecosystems with limited nutrient availability would be expected to respond rapidly to alterations in nutrient inputs.
- The *biodiversity hypothesis* states that the number and traits of species will constrain the extent to which species reordering and immigration will occur, and the nature and rate of the resultant response.
- The *trophic hypothesis* predicts resource alterations may cause asynchrony in species interactions (e.g. pollinators, predators, pathogens), which will create complex and transitory dynamics in ecosystems.

In the conservation literature, disturbances that might lead to loss of biodiversity are commonly referred to as threats or threatening processes. There have been a number of attempts to classify or group different threats, most of which are anthropogenic in origin. For example, Salafsky et al. (2002) described a preliminary taxonomy of direct threats to biodiversity, which differentiated between those that can result in elimination of an ecosystem (e.g. conversion to agricultural land, harvesting ecosystem elements) *versus* those that can result in degradation of an ecosystem (e.g. pollution or introduction of exotic species; Box 2.3). A standardised classification of threats was proposed by Salafsky et al. (2008), which has since been further developed and implemented by IUCN.

This approach was criticised by Balmford et al. (2009), who highlighted the value of differentiating the threat mechanism from its source, rather than conflating them into a single measure. For these authors, the *threat mechanism* is the process that has a direct negative effect on the state of an ecosystem or other conservation target (e.g. habitat destruction, invasive species, exploitation or pollution), whereas the *source* is the cause of that mechanism. For example, the poor condition of a marine ecosystem may be attributable to the introduction of an exotic species, which could be considered as the threat mechanism. The source of this threat could be international shipping, which could be intensifying because of an underlying driver, namely, growth in international trade (Balmford et al., 2009). The term 'threatening process,' which is widespread in the literature (e.g. Westoby and Burgman, 2006), is synonymous with threat mechanism. Characteristics of threats such as severity or intensity, frequency and probability can be used to describe a threat regime, in a similar way to other types of disturbance regime, and threats may also usefully be mapped (Battisti et al., 2016).

Box 2.3 *A preliminary taxonomy of direct threats to biodiversity (from Salafsky et al., 2002)*

Examples of specific direct threats[2]

General threats[1]	Forest	Savanna, grasslands, deserts	Freshwater	Marine
Ecosystem elimination[3]				
Conversion to agricultural land	Swidden plots, farms, plantations, ranches	Farms, ranches	Farmland reclamation	Aquaculture
Economic development	Roads, dams, urban areas, settlements	Roads, dams, urban areas, settlements	Dredging, diking, filling, urban areas	Dredging, diking, filling, urban areas
Harvesting ecosystem elements	Clear-cut logging, chip and pulp mills	Severe overgrazing	Extensive water diversion	Intensive coral mining, bottom trawling, drift netting
Mineral extraction	Mining, oil drilling	Mining, oil drilling	Mining, oil drilling	Mining, oil drilling, deep sea mining
Climate change	Severe fires, drought, hurricanes	Severe fires, drought	Drought, salinisation	Temperature fluctuation, sea-level fluctuation
Ecosystem degradation[3]				
Partial conversion	Selective logging	Grazing	Water diversion	Coral mining
Pollution	Acid rain, toxic chemicals, litter, radioactive fallout	Radioactive fallout	Acid rain, sewage, toxic chemicals, flotsam	Sewage, sediment, toxic substances, oil spills, radioactive fallout

(*cont.*)

(cont.)

General threats[1]	Examples of specific direct threats[2]			
	Forest	Savanna, grasslands, deserts	Freshwater	Marine
Human presence	Tourism, off-road vehicles, war and military activity	Tourism, off-road vehicles, war and military activity	Commercial and recreational boats, war and military activity	Dive tourism, cyanide and bomb fishing, war and military activity
Ecosystem disruption	Fragmentation, fire, fire suppression, predator removal	Fragmentation, fire, fire suppression, predator removal	Change of salinity patterns	Coral bleaching, predator suppression
Exotic species	Introduction or escape of plants and animals	Introduction or escape of plants and animals	Ballast water, introduction or escape of plants and animals	Ballast water, introduction or escape of plants and animals
Species decline and elimination				
Overexploitation of species	Hunting, gathering	Hunting, gathering	Fishing, hunting	Fishing, hunting
Physical disturbance	Disruption of nesting, disruption of migration	Disruption of migration	Disruption of migration, power-plant intakes	Disruption of migration, disruption of reproduction
Pathogens	Disease and pollution effects	Disease and pollution effects	Disease and pollution effects	Disease and pollution effects

Notes: [1]Direct threats in a generic sense. [2]Examples of the threat in different types of biomes. [3]Rows are not completely mutually exclusive: for example, there is obviously a grey area between ecosystem elimination and degradation.

In their consideration of the impacts of threats on ecosystems, Keith *et al.* (2013) refer to a number of different theories, including:

- *species-area relationships*, which imply that a decline in the area of an ecosystem will reduce its ability to sustain its characteristic biota;
- Caughley's (1994) *small population paradigm*, which refers to the impacts of demographic and environmental stochasticity, and processes such as inbreeding and genetic drift, occurring in small populations;
- *biodiversity and ecosystem function relationships*, indicating that biodiversity loss can reduce the capacity of ecosystems to capture resources, produce biomass, decompose organic matter and recycle carbon, water and nutrients (Cardinale *et al.*, 2012);
- *simulation models*, which allow the probability of ecosystem collapse to be estimated and the role of different mechanisms to be explored.

Boitani *et al.* (2015) challenged the application of these theories to develop quantitative criteria for assessing the risk of ecosystem collapse, as proposed by Keith *et al.* (2013). For example, Boitani *et al.* (2015) suggested that there is little scientific ground for the application of species-area relationships and small population processes to assessing the status of ecosystems. Further, they suggested that the relationship between ecosystem area and ecosystem functions has empirical support at small scales (such as individual landscapes) but not at larger scales (cf. Huggett, 2005). Other areas of theory mentioned in this context include resilience theory (Folke *et al.*, 2004) and state-and-transition models, which are considered in Section 2.4.

To summarise, theories relating to ecological disturbance do not comprise a single consolidated body of theory, but rather represent a set of interconnected ideas. Conceptual frameworks such as those described by Peters *et al.* (2011) and Grimm *et al.* (2017) provide a potential way of organising and exploring empirical data, which could usefully be integrated with classifications of different types of disturbance or threat to identify generalisations about underlying mechanisms. For making testable predictions, approaches are required that enable the potential impacts of disturbance on ecosystems to be forecast, an aspect that is considered further in the following sections.

2.3 Succession

Succession is one of the oldest and best established ecological theories, which has been intensively tested in both terrestrial and aquatic

> Box 2.4 *Suggested propositions of succession theory (adapted from Pickett et al., 2013)*
>
> 1. Succession is driven by the interactions of organisms with each other and the physical environment.
> 2. Successional patterns in communities result from the interactions of individuals.
> 3. Multiple trophic levels participate in the driving interactions.
> 4. Succession results from processes of disturbance, differential availability of species to a site and differential performance of species within a site.
> 5. Successional causes can operate on any timescale.
> 6. The possible outcomes of interaction between individuals are tolerance (no effective interaction), inhibition or facilitation.
> 7. The species composition of a site tends to equilibrate with the environment of that site.
> 8. The specific form of a successional trajectory is contingent on starting conditions, and the stochasticity of invasion and controls on species interactions.
> 9. Succession produces temporal gradients of the physical environment, biotic communities and the interaction of the two.

ecosystems (Pulsford *et al.*, 2016). It can be defined as the change in structure or composition of a group of organisms of different species at a site through time (Pickett *et al.*, 2013). The process of succession typically involves directional change in species composition, which may also be associated with changes in the physiognomy or structure of the communities present, as occurs for example with a shift from vegetation dominance by herbaceous plants to trees. A set of propositions of succession theory were identified by Pickett *et al.* (2013) (Box 2.4), which highlight the importance of interactions between individual organisms in determining how ecological communities respond to disturbance. The understanding of successional processes has also been strengthened by development of *hierarchy theory* (Ahl and Allen, 1996), which considers interactions between individual organisms as nested within a hierarchy of patterns and interactions occurring at community and ecosystem scales. Influences can also occur from the top-down within the hierarchy, for example from that of the ecosystem or landscape scale to populations and individuals during succession (Pickett *et al.*, 2013).

2.3 Succession · 49

Specific mechanism | **General mechanism** | **Outcome**

- Disturbance / Resource gradients → Differential site availability
- Immigration / Survival → Differential species availability
- Resource availability / Physiology / Life history characteristics / Competition / Allelopathy / Consumption / Mutualism / Plasticity and acclimation → Differential species performance

→ Change in community composition or structure

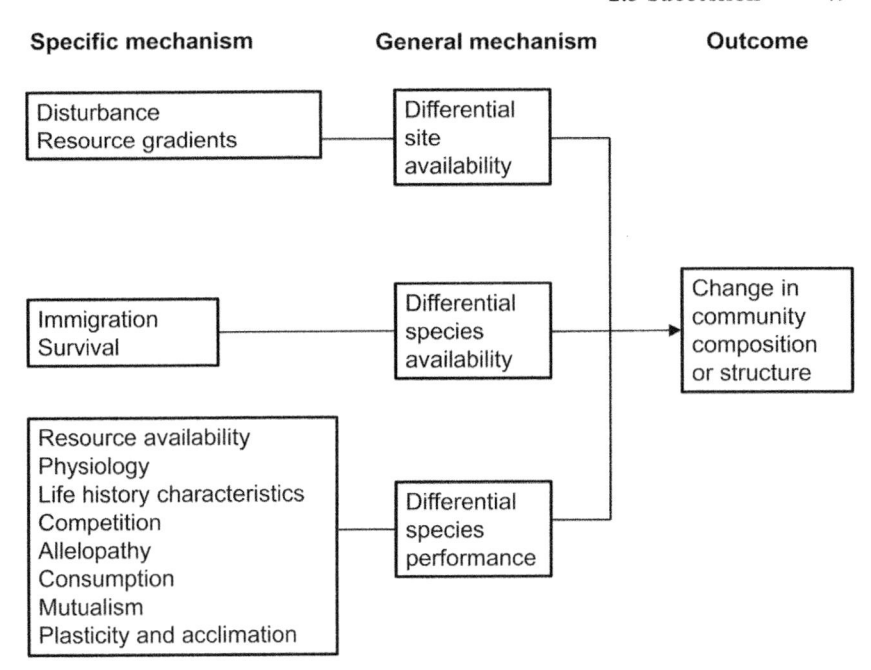

Figure 2.2 Mechanisms of ecological succession. Adapted from Pickett *et al.* (2013).

Successional processes can be attributed to the response of individual species to disturbance, which will depend on the type and characteristics of the disturbance event and the attributes or life-history traits of the species present (Figure 2.2). For example, species differ in their traits relating to dispersal ability, establishment, growth, survival, reproduction and mortality on a particular site (Prach and Pyšek, 1999), which will influence their ability to colonise and persist in an area following disturbance. Analysis and classification of life-history traits has proved to be a successful approach for predicting responses of communities to disturbance, particularly in the case of terrestrial vegetation (Pulsford *et al.*, 2016). Such approaches therefore offer a way of understanding how ecosystems recover.

Succession was originally conceived as a progressive directional change in species composition, leading towards the creation of a stable climax community. Current conceptions recognise that succession is inherently variable and may follow many different pathways on a particular site (Box 2.4), owing to variation in factors including the characteristics of the disturbance event, the existence of resource gradients, the order in which species colonise and the landscape context (Pickett *et al.*, 2013). While the mechanism's underlying succession may cause the composition or structure of a community to equilibriate over time, often the disturbance regime prevents equilibrium from being established, leading to the widespread acceptance of a 'non-equilibrium' view of ecology (Mori, 2011).

Although succession is usually considered to be progressive, associated with an increase in biomass over time, it can also be *retrogressive*. In such cases, biomass or other ecosystem attributes subsequently decline after an initial increase (Pulsford *et al.*, 2016). The mechanisms of retrogressive succession are much less well understood. The best example comes from palaeoecological investigations of vegetation development during the postglacial period, which indicate an initial increase in ecosystem productivity and biomass followed by a phase of decline, occurring over a timescale of millennia (Birks and Birks, 2004). For example, in a study of six vegetation chronosequences in five different countries using space-for-time substitution, Wardle *et al.* (2004) showed a decline in tree basal area occurring within 1,000 to 10,000 years after the onset of primary succession. Ratios of nitrogen to phosphorus (N:P) and carbon to phosphorus (C:P) increased in all six chronosequences, together with an increase in the N:P ratio of the leaf litter. This suggests that loss of available P over time, caused by the leaching of soils, may eventually limit tree growth and lead to biomass declines.

Can retrogressive succession lead to ecosystem collapse? The examples described by Wardle *et al.* (2004) presented a gradual process of change occurring in forest ecosystems over long timescales. However, in one of the six cases, Franz Josef in New Zealand, the final phase no longer supported trees. Other studies of retrogression have also highlighted marked declines in species richness, with losses of up to 78% of species (Richardson *et al.*, 2004). This suggests that retrogressive succession could indeed be a mechanism contributing to ecosystem collapse in some cases, even if the process is gradual.

It has long been known that vegetation succession is not necessarily linear, but can also be circular, when pairs of species tend to mutually replace each other. This was first documented in heathland and bog ecosystems by the classic work of Watt (1947, 1955) and has subsequently been observed in a range of other ecosystem types, including shrublands and forests (Glenn-Lewin and van der Maarel, 1992). More recently, Vera (2000) has developed a theory of the 'cyclical turnover of vegetation', focusing on the role of large herbivores in determining the

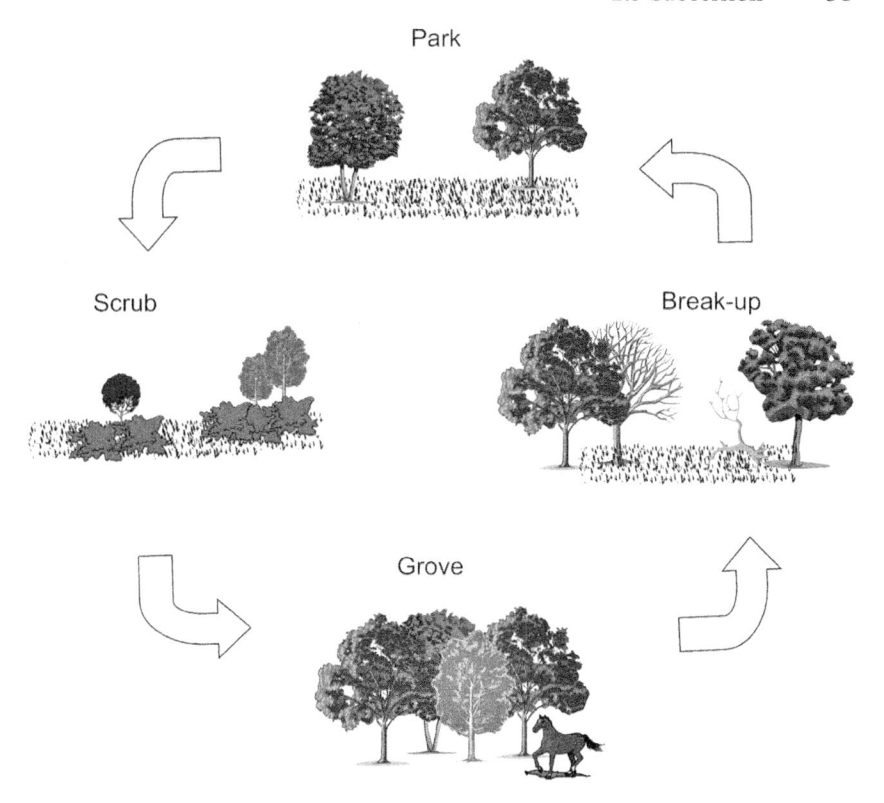

Figure 2.3 Illustration of Vera's theory of the cyclical turnover of vegetation. Adapted from Kirby (2004) with permission of Oxford University Press.

structure and composition of forest vegetation. According to this theory, intense browsing pressure exerted by populations of large herbivores can maintain extensive areas largely free of tree cover, in areas that would otherwise develop dense forest as a result of succession. Under such conditions, Vera (2000) proposed that vegetation dynamics are cyclical, as tree species primarily regenerate outside woodland through protection of seedlings from herbivory by spiny shrubs. Groves of trees would therefore become established within shrub vegetation, providing an example of tree regeneration by facilitation. Such groves would mature over time and eventually collapse, owing to the prevention of tree regeneration under a forest canopy by herbivory. Groves of trees would therefore be replaced by grassland, which would subsequently be colonised by shrubs, re-initiating the vegetation cycle (Figure 2.3). The result would be a dynamic park-like mosaic of woodland and grassland, a pattern that would be created and maintained by populations of large wild herbivores (Vera, 2000).

Vera's theory has had a major influence on conservation policy and management. In particular, it is a significant contributor to the concept of 're-wilding', which can involve the (re)introduction of populations of large herbivores that are allowed to roam freely to provide 'naturalistic grazing' (Pereira and Navarro, 2015). Examples of large-scale naturalistic grazing initiatives include the Oostvaardersplassen (Vera, 2009) in the Netherlands and the Knepp Estate and Ennerdale in the United Kingdom (Taylor, 2009). These have parallels in the concept of 'Pleistocene rewilding' currently being explored in both North and South America (Galetti, 2004; Rubenstein et al., 2006). Vera's hypothesis has stimulated a great deal of debate among both researchers and conservation practitioners, much of which has focused on examining whether the early postglacial vegetation of northern Europe was densely forested, as traditionally believed, or more open in character. Palaeoecological evidence generally provides little support for Vera's theory during the Holocene, although the situation was apparently different in previous interglacials, when a richer fauna was present (Sandom et al., 2014). Analysis of contemporary vegetation dynamics under conditions of high herbivore pressure similarly provides mixed support for Vera's theory. Whereas facilitation of tree establishment by spiny shrubs can readily be observed in the field, the role of large herbivores in converting woodland to grassland is far more difficult to demonstrate (Newton et al., 2013a,b). Nevertheless, field evidence suggests that high herbivore pressure can contribute to the dieback of forest stands in some situations, for example in combination with other pressures such as climate change and disease (Cantarello et al., 2017; Evans et al., 2017; Martin et al., 2015, 2017). Vera's theory therefore provides an example of a successional theory that incorporates a potential mechanism of ecosystem collapse.

2.4 State-and-Transition Models

In their discussion of the scientific rationale for the IUCN Red List of Ecosystems, Keith et al. (2013, 2015) relate their concept of ecosystem collapse to transitions among alternative states of an ecosystem. In their critique of the Red List of Ecosystems, Boitani et al. (2015) similarly indicate that most ecosystems exist in alternative stable states. Further, Keith et al. (2015) note the relevance of state-and-transition frameworks for understanding such transitions (Hobbs and Suding, 2009). In this context, Keith *et al.* (2015) highlight the need to differentiate between transitions that occur as part of the natural variation of an ecosystem type, from those that represent a process of collapse and replacement by a different ecosystem type. In other words, some transitions between ecosystem states can be viewed as ecosystem collapse, whereas others cannot.

State-and-transition models were first developed by rangeland ecologists in the late 1980s, reflecting a growing recognition of the limitations of traditional successional theory, with its focus on a single trajectory towards a 'climax' community. Such models are based on the identification of discrete vegetation states and the processes that cause transitions between them. For example, Friedel (1991) noted that in rangelands, trajectories of degradation and succession do not necessarily coincide; in an overgrazed system, the reduction or removal of grazing pressure does not necessarily restore the rangeland to what it was previously. Furthermore, certain combinations of weather, grazing, fire and local variability may push a system into a new state that is not easily reversed, such as an eroded soil surface that can develop following heavy grazing coupled with torrential rain (Friedel, 1991).

In an influential paper, Westoby *et al.* (1989) described the main features of state-and-transition models, in which transitions between discrete states of the vegetation are triggered by natural events (such as weather or fire) or by management actions (such as a change in the stocking rate of grazing animals). Such transitions can occur very quickly (for example as the result of a fire) or over a prolonged period. Westoby *et al.* (1989) note that vegetation states are essentially abstractions that encompass a degree of variation both in space and time and cite perennial saltbush shrubland or grassland in temperate Australia as an example. Here there are persistent shrub-dominated and grass-dominated states, with a transition from shrub- to grass-dominated driven by grazing. However, removal of grazing does not result in a transition from a grass-dominated state back to domination by shrubs. A further example is semi-arid woodland in eastern Australia, which can occur with either a grass or shrub understorey; the grass understorey is favoured by fire and provides fuel, whereas the shrub understorey suppresses fuel and is sensitive to repeated fires (Westoby *et al.*, 1989).

Although state-and-transition models are typically conceptual or qualitative rather than quantitative, they have had a significant impact on rangeland management and are widely used among both researchers and practitioners, partly because of their flexibility and ready applicability

(Suding and Hobbs, 2009). Originally developed for rangelands, they have subsequently been used to investigate a number of other types of terrestrial ecosystem, such as forests (Czembor and Vesk, 2009). Analysis of state-and-transition models from the perspective of network structure has highlighted a number of different types, including a *linear structure* associated with classical ecological succession (state A→B→C); a *cyclical structure* (A→B→C→A); *radiation*, where a single state can transition to or from several other states; and *maximum connectivity*, where any state can transition to any other (Phillips, 2011). This highlights a potential method of developing generalisations about the types of transitions that can occur.

In their review of state-and-transition models, Bestelmeyer *et al.* (2017) highlighted a number of problems associated with their implementation. First, it can be difficult to define the characteristics and boundaries of the 'site' under consideration, for example in terms of its climate, soil and topographic position. Here there is a risk of concluding that a set of plant communities are alternative states of a specific site when in fact they occur on different sites; clearly, this depends critically on how the site is defined. Second, Bestelmeyer *et al.* (2017) highlight the importance of distinguishing between dynamics that are transient from those that represent persistent transitions between alternative states. According to these authors, transient dynamics represent temporary changes in vegetation that can be reversed over a timescale of a few years to several decades, whereas state transitions involve more persistent changes that require several decades to recover to their former state, or they never recover (Figure 2.4). This distinction between transient and persistent transitions is essentially arbitrary, and the authors note that it may be difficult to distinguish them in practice. Yet this distinction is highly pertinent to deciding whether or not an ecosystem has collapsed.

Following the definitions of Keith *et al.* (2015), a persistent transition might be considered as ecosystem collapse, whereas a transient change would not be. Interestingly, this discussion also highlights the linkage between ecosystem collapse and recovery. According to Bestelmeyer *et al.* (2017), the criteria used to distinguish between transient dynamics and state transitions depend on the length of time needed for recovery and their relationship to management time frame. These authors note that if recovery does not occur within an acceptable management time frame, it is often categorised as a state transition. This raises the question of what an 'acceptable' time frame might be, particularly for ecosystems

2.4 State-and-Transition Models · 55

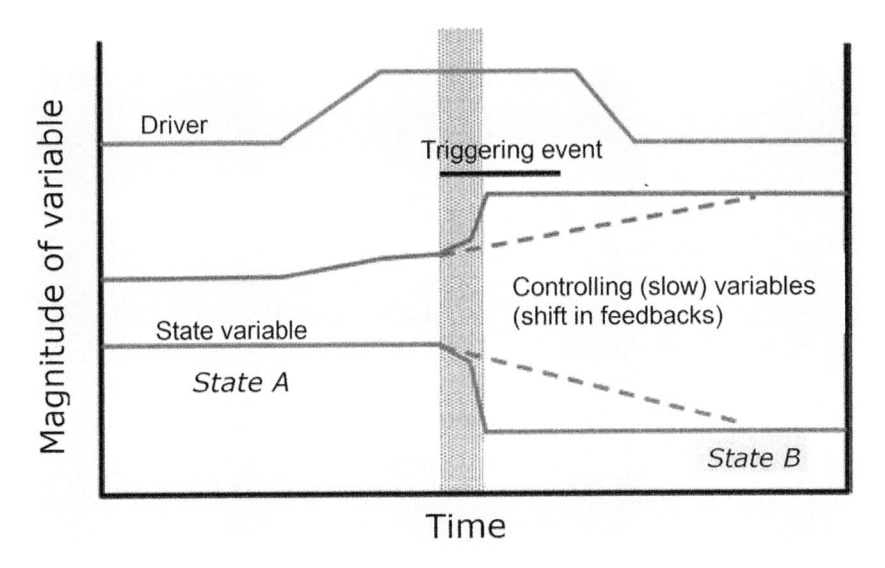

Figure 2.4 A schematic illustrating the pattern and interaction of variables over time involved in state transitions. Reprinted from Bestelmeyer *et al.* (2017) with permission from Springer Nature.

such as forests that can take many decades to recover from even transient changes.

Do state-and-transition models make testable predictions? For Pulsford *et al.* (2016), such models can provide broad predictive capabilities to assess and estimate potential future changes, but their predictive powers are low and their ability to deal with uncertainty is limited. Similarly, Rodriguez Iglesias and Kothmann (1997) suggest that they are very limited as theories because they do not generate predictive rules of broad relevance, owing to their lack of inclusion of ecological processes or underlying mechanisms. For Bestelmeyer *et al.* (2017), their value lies in providing a flexible method of organising information about ecosystem change in a way that can support environmental management, rather than acting as predictive theories.

Despite their limitations, state-and-transition models have contributed to a shift in thinking about the nature of change in ecological systems. As a result, it is now widely acknowledged that vegetation change: (1) does not necessarily follow a single successional trajectory but may produce multiple different communities; (2) is not necessarily reversible; and (3) can be discontinuous and sudden (Bestelmeyer *et al.*, 2017).

2.5 Dynamical Systems Theory

The theories most often associated with ecosystem collapse are those arising from dynamical systems theory, which encompasses a range of different mathematical and conceptual ideas. I apologise wholeheartedly for failing to describe these ideas here at the level of detail that they certainly deserve; this is only a brief overview of what is a substantial body of scientific work. I also apologise for not representing these ideas mathematically, which is where their real power resides; my aim here is to provide a simple and accessible introduction to what is a deep and complex subject (Box 2.5). Those interested in exploring this topic in greater depth are referred to the outstanding texts that have been produced by mathematically minded ecologists working in this area, such as Scheffer (2009), Levin (1999) and Petraitis (2013).

I have grouped together here several related areas of theory that have all impacted on current thinking about ecosystem collapse. These theories originate from mathematics but have subsequently been applied to ecological problems. A fascinating history of how these different theories developed, and the relationships between them, is provided by Aubin and Dahan Dalmedico (2002), complete with topological monsters. These authors highlight the development of *bifurcation theory*, which describes a situation when the state of a non-linear system bifurcates or branches at some critical value of the parameters on which the state of the system depends. Mention is also made of *catastrophe theory*, which entertainingly became the focus of a bitter controversy in the late 1970s, when its proponents were accused of using arcane mathematical techniques that nobody else understood (which seems a rather unfair criticism, in retrospect). As a result of this debate, the researcher principally responsible for developing catastrophe theory, René Thom, subsequently described it as a 'shipwreck'. Yet it was only a partial wreck, as some of the topological approaches developed lived on to inform subsequent research (Aubin and Dahan Dalmedico, 2002). Catastrophe theory can be considered as a branch of bifurcation theory, which examines situations when small changes in certain parameters of a non-linear system can cause equilibria to appear or disappear, or to change from attracting to repelling or vice versa. This leads to large and sudden changes of the behaviour of the system.

Other relevant theories include *chaos theory* (Box 2.5), which is a branch of mathematics focusing on the behaviour of dynamical systems that are highly sensitive to initial conditions. This became very popular in

Box 2.5 *A simple guide to dynamical systems, chaos and complexity (adapted from Rickles et al., 2007)*

- A *system* is an object studied in some field and might be linear or nonlinear, simple or complicated, complex or chaotic. *Complex systems* are systems built up from very large numbers of mutually interacting subunits whose repeated interactions result in rich, collective behaviour that feeds back into the behaviour of the individual parts. *Chaotic systems* can have very few interacting subunits, but they interact in such a way as to produce very intricate dynamics.
- Systems possess properties that are represented by *variables*. The values taken by a system's variables at an instant of time describe the system's *state*. A state is often represented by a point in a geometrical space (phase space), with axes corresponding to the variables.
- A *dynamical system* is a system whose state (and variables) evolves over time, according to some rule. How a system evolves over time depends both on this rule and on its initial conditions – that is, the system's state at an initial time.
- A *non-linear system* is one for which inputs are not proportional to outputs: a small change in some variable may result in a large change in the system.
- Following an intervention in a system (changing the value of some variable), it can take time for it to settle down into its "normal" behaviour. The phase space points corresponding to this 'normal' state comprise the system's *attractor*. The phase space can have a number of attractors. The set of points that are 'pulled' towards a particular attractor are known as the *basin of attraction*. Complex systems have phase spaces that evolve over time and a range of possible attractors.
- Complex and chaotic systems are both examples of non-linear dynamical systems. Chaos is the generation of complicated, aperiodic, seemingly random behaviour from the iteration of a simple rule. Complexity is the generation of rich, collective dynamical behaviour from simple interactions between large numbers of subunits. Chaotic systems are not necessarily complex, and complex systems are not necessarily chaotic.
- Some systems have a property known as *criticality*. A system is critical if its state changes dramatically given some small input.

the 1970s and 1980s, but 'chaos science' has since largely been supplanted as a research endeavour by other approaches such as complexity science (Aubin and Dahan Dalmedico, 2002). Chaos theory, catastrophe theory and other early theories of dynamical systems all address deterministic systems, where a set of equations is used to determine how a system moves through its state space over time (Anderson, 1999). In contrast, complexity science is based on a different way of modeling system behaviour, which focuses on the interactions between individuals that are connected together in *complex adaptive systems* (CAS) (Anderson, 1999). This represents an extension of traditional systems theory. According to this approach, system behaviour is a property that emerges from interactions between individuals at a lower level of aggregation. The structures of CAS are complex, as they involve dynamic networks of interactions, and these systems are adaptive, in the sense that individual and collective behaviours mutate and self-organise in response to changing conditions or events.

According to Levin (1998), the study of CAS examines how complicated structures and patterns of interaction can arise through simple rules that guide change. Essential elements of CAS are (Levin, 1998):

- sustained diversity and individuality of components;
- localised interactions among those components; and
- an autonomous process (such as natural selection) that selects from among those components, based on the results of local interactions, a subset for replication or enhancement.

Complexity theory is now highly interdisciplinary and has been applied to a wide variety of different fields, including management and organisation, finance, economics, medicine, social science and computing. Consequently, complexity theory is not a single consolidated theory but encompasses a range of ideas drawn from different disciplines.

In ecology, systems theory has been applied to address a wide variety of different problems. Much early research focused on representing the natural world as a set of stocks and flows of energy and materials, regulated by a variety of different feedback processes, which can be explored using a number of different mathematical techniques. This approach is perhaps best encapsulated by the enormously influential work of Howard Odum, who encouraged his readers to think holistically, by viewing an ecosystem through a macroscope as well as a microscope (Mitsch and Day, 2004; Odum, 1971). Early work on processes organising community dynamics and patterns, including predator–prey

2.5 Dynamical Systems Theory · 59

interactions, also employed systems concepts (Hartvigsen *et al.*, 1998). A key landmark was the publication of two reviews by Robert May (1976, 1977), which showed how a variety of different elements of dynamical systems theory, including bifurcation theory and chaos, could be applied in an ecological context. The indication that simple non-linear differential equations can produce a wide range of different behaviour, from stable endpoints to apparent chaos, had a major impact on ecological thinking, the legacy of which is still being felt. May (1977) also showed how ecosystems could exist in alternative states and could undergo rapid and dramatic changes between them, citing examples including outbreaks of spruce budworm, crashes of commercial fisheries and human host-parasite systems. This publication therefore represents an important milestone in the literature on ecosystem collapse, by highlighting how major ecological shifts could result from minor perturbations – as if 'the hinge of history turned on the length of Cleopatra's nose' (May, 1977; Petraitis, 2013).

Since the 1970s, researchers have been able to progressively increase model complexity and perform simulations at larger spatial and temporal scales and at higher resolution. The ability to explicitly represent space in models was an important step forwards (Hartvigsen *et al.*, 1998). Furthermore, the development of CAS theory enabled systems theory to be extended to incorporate elements such as diversity and heterogeneity, which tended to be simplified in traditional approaches. CAS also enabled the role of adaptation in system dynamics to be explored (Hartvigsen *et al.*, 1998). In the past 20 years, CAS approaches have developed rapidly from both theoretical and applied perspectives and have made a major contribution to understanding ecological dynamics (Hagstrom and Levin, 2017). Together with probability theory and dynamical systems theory, CAS theory has led to a greater emphasis on non-linear dynamics and stochasticity and has led to the development of a substantial literature on tipping points, resilience, critical transitions, alternative stable states and associated concepts (Hagstrom and Levin, 2017).

How do these theories relate to ecosystem collapse and recovery? One of the most relevant contributions to this body of literature is the review by Scheffer *et al.* (2001) on catastrophic shifts in ecosystems. Building on the theoretical contribution of May (1977), this publication suggests that field evidence indicates the existence of alternative stability domains in different ecosystems and seeks to explain them. First, Scheffer *et al.* (2001) note that external pressures (or conditions) on ecosystems such as climate

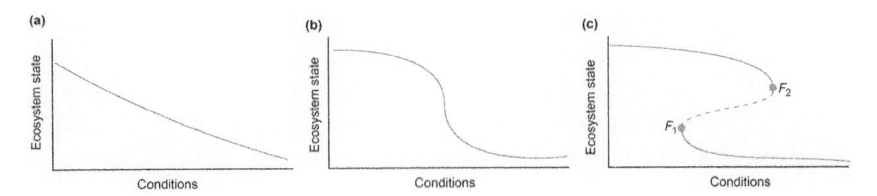

Figure 2.5 Different ways in which an ecosystem can respond to changes in conditions, according to the theory of alternative stable states. Reprinted from Scheffer and Carpenter (2003), with permission from Elsevier.

change, nutrient inputs, habitat fragmentation or harvesting of species often change gradually with time. In response, some ecosystems may change in a smooth, continuous way (Figure 2.5a), whereas others may be relatively unresponsive, until conditions reach a particular critical value (Figure 2.5b). These two situations are contrasted with third possibility, when the ecosystem response curve is 'folded backwards' (Figure 2.5c). This describes a situation where the ecosystem can exist in two alternative stable states, which are separated by an unstable equilibrium between the 'basins of attraction' of the two states (Scheffer *et al.*, 2001). If external conditions change sufficiently, the state of the ecosystem may cross a threshold value (referred to as a 'saddle-node' or 'fold' bifurcation, Figure 2.5c), which results in a catastrophic transition to another state. A second key feature of this model is that for the state of the ecosystem to return to its initial value, it is not sufficient simply to restore the environmental conditions that prevailed before the collapse. Rather, it may be necessary to return to a different transition point, from which the ecosystem can recover to its original state. This pattern of response is referred to as *hysteresis* (Scheffer *et al.*, 2001). The authors then describe a number of real-world examples in relation to this model, including shallow lakes, coral reefs, woodlands/grasslands and deserts.

In this paper, Scheffer *et al.* (2001) do not provide a detailed account of the underlying theory. However, these ideas arise from dynamical systems theory, specifically bifurcation theory. The ideas were elaborated further in a series of publications, by these and other authors. For example, Scheffer and Carpenter (2003) note that the term 'stable states' incorrectly implies a lack of dynamics, and therefore the term 'regimes' may be preferable when referring to real-world ecosystems. A sudden transition in an ecosystem might therefore appropriately be referred to as a 'regime shift'. These authors note that according to theory, a regime shift can be triggered either by internal processes acting within the system or by an external perturbation or by some combination of the two. A system is considered to possess alternative attractors if it can be in more than one regime or 'stable state' for the same value of an external condition, such as nutrient load or climate (Scheffer and Carpenter, 2003). A change among attractors also implies a shift in the dominant feedbacks controlling the dynamics of the system, providing an insight into the mechanisms underlying shifts between alternative states. In other words, once a critical threshold has been exceeded, positive feedback can propel the system towards a contrasting state (Scheffer *et al.*, 2009). However, positive feedback loops by themselves are not sufficient to guarantee the existence of multiple stable states (Angeli *et al.*, 2004), and they are not an essential feature of regime shifts.

A detailed account of critical (including catastrophic) transitions is provided by Scheffer (2009), together with a description of underlying theory. Here the author points out that many models of dynamical systems converge to a cycle (such as a limit cycle) or to a more complex dynamic regime rather than to a single stable state. An example of such complex dynamics is chaos, which tends to a state of continual change in which the same pattern is never repeated exactly (Scheffer, 2009). In a chaotic system, small variations in the initial state can increase exponentially with time; as a result, the system is inherently unpredictable, and the current state can never be determined precisely. Weather provides a powerful example of such a system. Further, Scheffer (2009) points out that in most models of dynamical systems, conditions essentially remain constant. This is clearly a simplification of the situation in nature, which is characterised by continual variation. For this reason, 'stable states' are best appreciated as a feature of models rather than a phenomenon encountered in the real world, which is why the term 'dynamic regime' is preferred in such situations. This highlights the important distinction between models of dynamical systems and their application to real-world situations. As Scheffer (2009) nicely puts it, 'we can never map the large complex systems of nature and society on the mirror world of math'; in other words, the phenomena observed in mathematical theory do not necessarily correspond directly to what is observed in nature. For example, regime shifts in ecosystems do not necessarily indicate the existence of alternative basins of attraction, or shifts between alternative stable states, but could result from a large sudden change in environmental conditions. Scheffer (2009) limits the term 'critical transitions' to those regime shifts that are driven by positive feedback, which drives a system to a different state once a threshold is passed (see also Box 2.6).

> Box 2.6 *Criticality and disturbance*
>
> Although the focus here is on critical transitions between alternative ecosystem states, there are other aspects of criticality that are relevant to ecosystem collapse. Pascual and Guichard (2005) provide a valuable overview of criticality and disturbance of ecosystems, from a spatial perspective. In addition to classical phase transitions between states, these authors also highlight two other types of criticality: self-organised (SOC) and 'robust'. Disturbance phenomena such as forest fires and disease epidemics may demonstrate SOC, in which there is no sudden shift associated with environmental perturbations crossing a critical point. Instead, the system may take itself to a critical state through its own internal dynamics, for example through the growth rates of trees and the frequency of fires. This is analogous to the most familiar example of SOC, where sand grains are added to a sand pile, which eventually collapses when it reaches a critical angle. 'Robust' criticality has been identified in models of disturbed mussel bed dynamics, where the spatial pattern of the mussels can change abruptly even though the cover or abundance of the species does not. These other forms of criticality therefore provide an additional possible mechanism of ecosystem collapse, which, as documented by Pascual and Guichard (2005), will depend on the temporal and spatial scales of both disturbance and recovery.

The application of these ideas in ecology has not been without controversy. For example, in their review of examples in the literature, Connell and Sousa (1983) concluded that these provided no evidence of the existence of multiple stable states in unexploited natural populations or communities (although they did accept evidence in managed livestock grazing systems). In many cases in the literature, species assemblages occurring on a particular site were compared before and after some change in the physical environment, such an increase in nutrient loading resulting from a pollution event. Connell and Sousa (1983) suggest that these examples do not qualify as examples of multiple stable states, because the environmental conditions have actually changed, and therefore the states are not directly comparable. In response, Scheffer (2009) suggested that this represents a failure to consider the abiotic environment as an intrinsic part of ecosystems. Other flaws highlighted by Connell and Sousa (1983) included the artificial maintenance of one of the states by human intervention, or a failure to conduct observations over a sufficient timescale to determine whether the communities were stable or not.

More recently, Petraitis (2013) has provided a detailed examination of both empirical evidence and the underlying theory relating to the occurrence of multiple stable states in natural ecosystems. Here the author highlights some interesting differences between how theoretical and experimental ecologists view the world. First, for theorists, perturbations to a model means moving the state variables (such as the productivity of an ecosystem) off the equilibrium point and observing what happens, without further intervention. This equates to what experimentalists would term the 'pulse' type of disturbance, discussed earlier (Bender *et al.*, 1984). Many experiments have been conducted examining this type of disturbance, such as those involving experimental burning or cutting of vegetation, although these perturbations tend not to be introduced to communities that are at or near to equilibrium (Petraitis, 2013). Experimentalists also examine 'press' perturbations, for example if herbivores are excluded from vegetation by fences. Secondly, experimental ecologists usually consider any disturbance event originating from outside the system of interest as being *extrinsic*. For example, a fire, flood or pest outbreak is often considered in this way. Yet in the world of modeling, disturbance may often be incorporated in a model as parameters (e.g. grazing or nutrient load) and may therefore be considered as *intrinsic* to the system. In other words, perturbations in dynamical systems are metaphors for disturbances in ecological communities (Petraitis, 2013). It is precisely this kind of difference in perspective that can hinder communication between experimentalists and theoreticians and produce the kind of heated debate one sometimes encounters in the literature.

Petraitis (2013) summarises the underlying theory in this way: multiple stable states mean that there are two or more stable points, or basins of attraction, for a single set of parameter values in a system. He goes on to highlight the following implications that can be inferred from available theory (Petraitis, 2013; Petraitis and Dudgeon, 2004):

- According to the theory of alternative stable states envisioned by May (1977) and Scheffer *et al.* (2001), small changes in parameter values cause a catastrophic shift in state variables. Such shifts occur in dynamical systems that have a bifurcation fold and show hysteresis. However, other types of dynamical system may also demonstrate catastrophic shifts. Furthermore, not all systems with multiple stable states will necessarily demonstrate hysteresis.

- Hysteresis can arise in any system in which at least one process shows a threshold over a range of parameter values. A system with hysteresis may not show recovery, and changes in such systems may lag behind changes in environmental conditions. It is also possible that a system may have the potential for hysteresis but not demonstrate it in nature, because of the limited range of environmental variation encountered. In this type of system, it could appear as if a change in ecosystem state is irreversible.
- When considering ecological communities, bifurcation folds require at least one population process, such as birth rate, death rate or recruitment rate, to display a threshold response with a change in an environmental parameter.
- The presence of an environmental threshold does not guarantee that a system has alternative states. Conversely, the existence of multiple stable states does not necessarily imply a threshold response. In ecology, thresholds are perceived as dramatic changes in some aspect of an ecosystem and have been widely documented (Groffman et al., 2006; Horan et al., 2011; Huggett, 2005). Although the existence of thresholds has often been interpreted as evidence for the existence of multiple stable states, this is not necessarily true, as demonstrated by Petraitis and Hoffman (2010) in a grazing model.
- It is possible for alternative states to be persistent rather than stable. For example, consider the endemic insect fauna in a relict forest that is no longer able to replace itself, because of recruitment failure of the tree species present. The existence of the insect community depends upon habitat provided by the tree species. The insect fauna itself may be self-sustaining, but long-term persistence is not possible because the trees are not self-sustaining.
- The presence of environmental variation means that identical perturbations may not give the same response every time or at every place. Identical disturbances might cause an ecosystem to return to the original state or to diverge to an alternative state.
- The basin of attraction may vary in its shape and location in state space with changes in parameters, but different equilibrium states under different parameter conditions cannot be considered as alternative stable states. Therefore, demonstrations that an ecosystem can show abrupt shifts either in equilibrium conditions or in ecological processes (such as recruitment) with relatively small changes in environmental parameters are not sufficient tests of the existence of alternative stable states.

2.5 Dynamical Systems Theory

The best kind of evidence for the existence of multiple basins of attraction is that provided by field experiments. Such experiments would need to demonstrate whether a particular system has more than one basin of attraction. Although relatively few experiments have been performed with this specific objective, many of those that have been conducted employ a set of criteria proposed by Peterson (1984), namely:

(i) the experiment should be performed in a single environment (i.e. at the same site);
(ii) the site should be shown to have the potential to be occupied by two or more different communities;
(iii) the communities should be self-replicating.

In addition, Petraitis (2013) suggests that the experiment should only use pulse perturbations that mimic a natural event. As noted previously, in practice many experimenters also examine press perturbations, for example by holding constant the densities of one or more species. This approach has the result of changing a state variable (such as species density) into a constant parameter. Many of the models that have been used as examples for multiple stable states (e.g. May, 1977; Scheffer *et al.*, 2001) have used this approach, enabling the interactions between two species (such as a grass and a herbivore) to be examined using a single-species model. In such situations, the herbivore is effectively uncoupled from the dynamics of the grass, as it is considered as a parameter and not as a state variable (Petraitis, 2013). An ecologist simulating a press perturbation by means of such a model would vary the herbivore density parameter and observe the impact on the system. However, this is very different from what a mathematician means by perturbation of a system, which refers only to changing the value of a state variable. It is for this reason that Petraitis (2013) suggests that experimental tests of multiple stable states should employ pulse perturbations, as it is these that are incorporated in the underlying mathematical theory.

In practice, it is very difficult to meet all the requirements listed by Peterson (1984) or those suggested by Connell and Sousa (1983) (see also Sousa and Connell (1985) and Sutherland (1990) for more on this debate). For example, in field situations, comparison is often made between occurrence of a particular type of ecosystem in different places or at different times. Yet it is very difficult to prove that the environment is in fact the same at different times or in different places. As a consequence, there is limited experimental support for the existence of multiple stable states of ecosystems, particularly in freshwater and marine environments (Capon *et al.*, 2015; Knowlton, 2004; Petraitis, 2013). For example, in their review of available evidence, Schröder *et al.* (2005) found that alternative stable states were found in only 38% of systems tested experimentally, indicating that this is only one type of ecosystem behaviour among many possibilities. Such evidence will be considered further in subsequent chapters.

Petraitis (2013) also attempts to rescue catastrophe theory from obscurity, noting that most ecologists have tended to ignore it. This may reflect the controversy alluded to earlier, resulting from the overblown claims made by its proponents and its 'metaphysical' application in the social sciences, such as analysing self-pity in humans (Deakin, 1990; Loehle, 1989). Perhaps most ecologists have found the mathematics too difficult to understand. Or maybe they have agreed with the contention of May (1977), that ecological applications of catastrophe analysis usually represent post hoc window-dressing. Either way, Petraitis (2013) and Petraitis and Dudgeon (2015) suggest that catastrophe theory can be of potential value by placing models of multiple stable states in a broader theoretical context. However, there are still very few examples of applying this theory in the ecological literature. Loehle (1989) provides a review of these, noting that the existence of catastrophic behaviour in a system does not necessarily mean that catastrophe theory is applicable. Neither does catastrophic behaviour necessarily imply rapid or sudden change. The theory depends on the existence of multiple stable states and hysteresis and so is not applicable to phenomena such as sudden declines of individual species (e.g. the passenger pigeon), unless there is evidence for hysteresis. Ecological situations where the theory has been applied include predator-prey models, outbreaks of insect pests and management of fish populations (Loehle, 1989). In a more recent review, Petraitis and Dudgeon (2015) note nine documented examples in aquatic systems.

No discussion of dynamical systems theory would be complete without some consideration of ball and cup diagrams. These are widely used as a way of illustrating multiple stable states, their use in ecology originating from the work of Lotka (1956) and Lewontin (1969). In this conceptual model, different states or stability domains are represented by the cups or valleys in the diagram (Figure 2.6). The position of the ball, representing the system state, can potentially move from one valley to another but comes to rest within an individual valley (or 'basin of attraction') until the system is perturbed in some way. The width of the valley affects the size of the perturbation required to move the ball out of one valley and into another (Beisner *et al.*, 2003; Gunderson, 2000). In

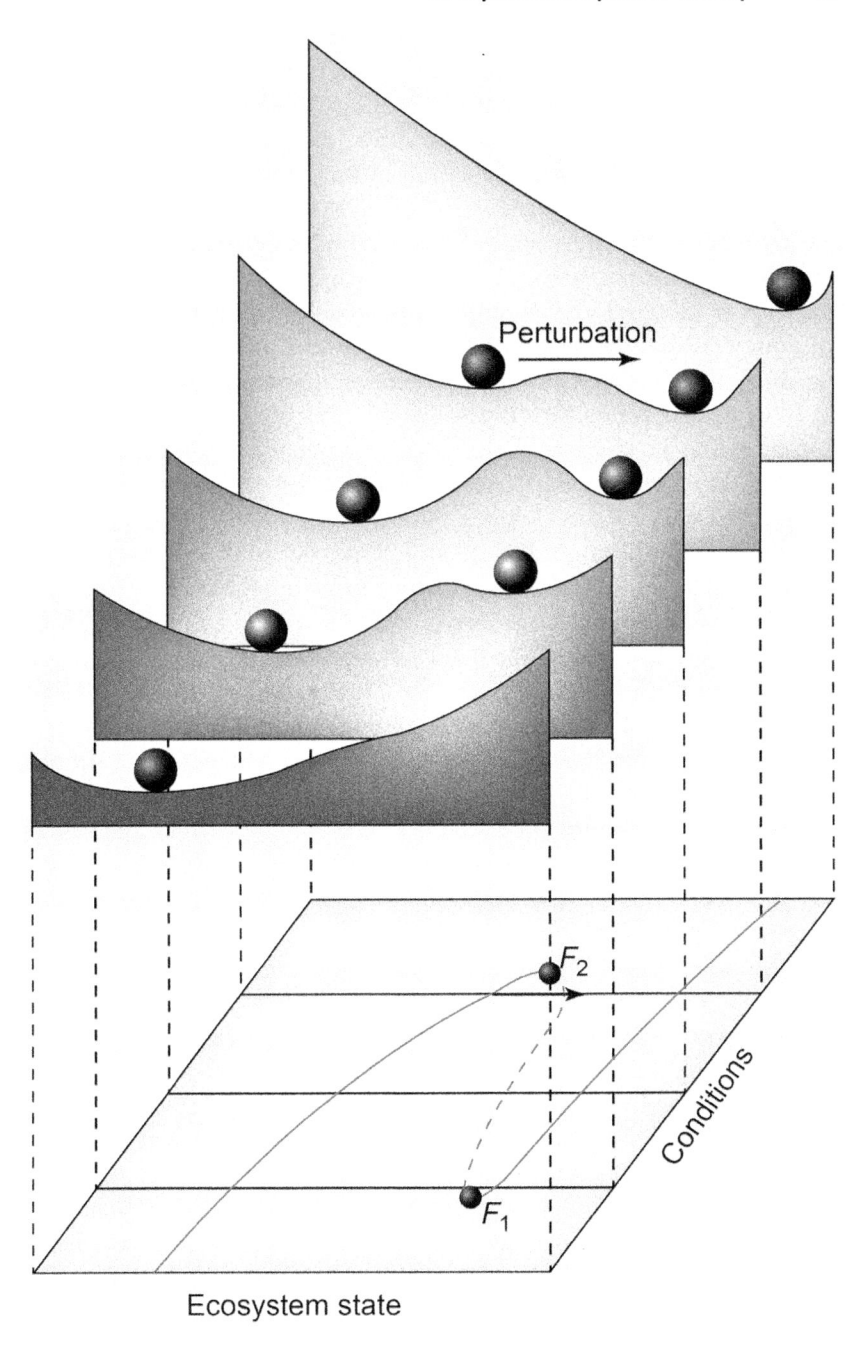

Figure 2.6 Illustration of how ecosystem state may vary with conditions, according to the theory of alternative stable states. Reprinted from Scheffer and Carpenter (2003), with permission from Elsevier.

addition, the landscape of valleys can change over time; repeated perturbations might reduce the steepness of the slopes, enabling a smaller disturbance to move the ball between valleys (Beisner et al., 2003). As a result, the system would shift more readily between states.

There has been a great proliferation of ball and cup diagrams in the recent literature. Unfortunately these have sometimes added to the confusion about the underlying theory, rather than helping to clarify it. Often it is not clear what the axes represent, as they are not labelled (e.g. Beisner et al., 2003). Even when the nature of the axes is indicated, this is not done consistently; in some publications the x axis is labelled as time (e.g. Briske et al., 2006), in others as external conditions (Scheffer et al., 2001) or as ecosystem state (Standish et al., 2014) or as the degree of ecosystem change (Ghazoul et al., 2015). What the ball represents also varies. In one of my own personal favourites of the genre, the ball is described as 'biodiversity', which mysteriously mutates into 'changed biodiversity' after having rolled down a slope, helpfully changing colour in the process (Secretariat of the Convention on Biological Diversity, 2010). Personally I have also been somewhat mystified by those diagrams featuring a series of valleys (or states) arranged in a series along a slope (e.g. Bellwood et al., 2004; Standish et al., 2014) – what is causing the ball to roll down the hill, or even up it? Is it being pulled or pushed? Or both?

As Scheffer (2009) notes, these diagrams are a metaphor and should not be taken too literally. In fact, the diagrams are rather misleading; to more accurately represent the mathematical theory, one needs to imagine the balls rolling through a very viscous fluid, such as treacle, so that there is no momentum. Attraction basins should actually be structures in many dimensions, rather than one, and sometimes they might be circular channels rather than isolated depressions. At the same time, the landscape of the hills and valleys should be viewed not as static but as vibrating or wobbling structures (Scheffer, 2009). Given the mental challenge of trying to imagine balls rolling around a multidimensional wobbling landscape covered in treacle, it is perhaps understandable that simpler two-dimensional diagrams have proved to be more popular. Petraitis (2013) also highlights how ball and cup diagrams have been widely misinterpreted. For example, the hills or ridges between basins of attraction are often assumed to show an unstable equilibrium point between two stable equilibria, but this is not necessarily the case; further, the units on the x axis depend on how the slice is taken through the phase diagram. The representation of a single ridge between valleys also obscures the fact that there may be many alternative pathways from

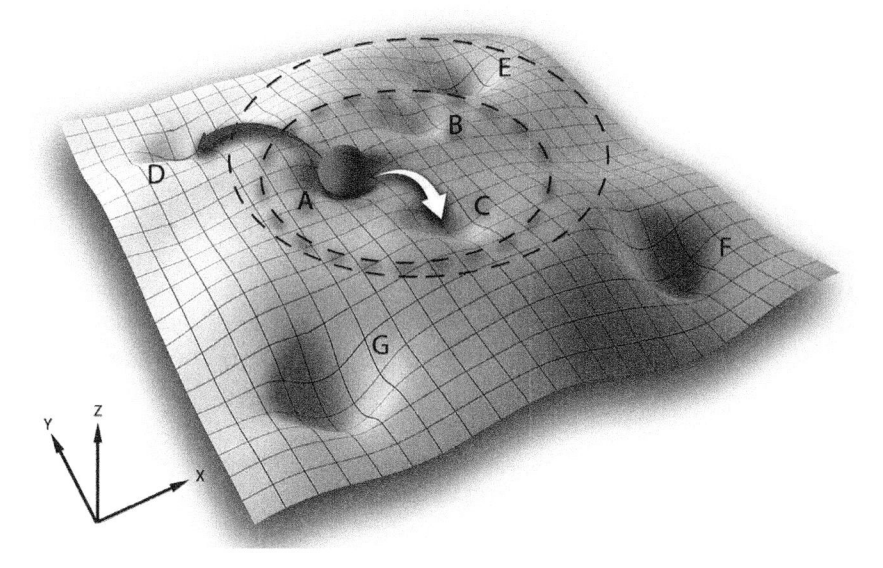

Figure 2.7 Illustration of a model of ecosystem dynamics based on the theory of alternative stable states. States A–G are defined by two state variables represented on the X and Y axes. The vertical axis (Z) represents potential for change. The two broken lines represent alternative interpretations of ecosystem collapse. For the inner line, transitions between states A, B and C (e.g. white arrow) represent natural variability without loss of key defining features, while transitions across broken lines (e.g. grey arrow) to states D, E, F and G represent collapse and replacement by novel ecosystems. The outer broken line represents an alternative interpretation of ecosystem collapse in which state E is included within the natural variation of the ecosystem type. Reprinted from Keith *et al.* (2015).

one basin of attraction to another. For these reasons, Petraitis (2013) concludes that the representation of stability as a ball on the surface of a landscape is little more than a cartoon and may communicate a misleading impression of the behaviour of the system.

Nonetheless, Keith *et al.* (2015) provide an illustration (or cartoon) of how ecosystem collapse may be conceptualised in relation to the occurrence of multiple stable states, using a three-dimensional ball and cup diagram (Figure 2.7). This contrasts transitions between ecosystem states that form part of the natural variability of the system without loss of key defining features, from a situation where a transition has occurred to a novel ecosystem that lacks these features. The latter situation is considered to represent ecosystem collapse.

In summary, dynamical systems theory has undoubtedly had a major impact on ecological thinking and has provided many valuable insights.

Key among these is the recognition that there can be a substantial change in the state or condition of an ecosystem in response to a small change in environmental conditions. This implies that ecosystems could perhaps collapse suddenly, without much warning, even if environmental conditions change only gradually. Further, systems models highlight the importance of positive feedback mechanisms in causing an ecosystem to shift from one state or regime to another. This provides an important insight into how ecosystems might collapse. Specifically, according to theory, there are two different ways in which a system can move to an alternative state: (i) a change in external conditions (which in models are represented by parameters), or (ii) a change in the state of the system itself (which in models is represented by state variables) (Van Nes et al., 2016). In either case, positive feedback processes then drive the transition to another state. Dynamical systems theory has therefore made a significant contribution to our understanding of ecosystem dynamics, even if testing predictions from the theory is often difficult (Petraitis, 2013; Scheffer, 2009).

Nevertheless, there remain challenges in applying the theory to the real world. As with any other approach to modelling, systems modeling approaches represent a simplification of real-world systems. For example, describing alternative ecosystem states as a binary choice, such as between corals or macroalgae, greatly simplifies the situation in actual ecosystems, which can vary markedly in the identity of dominant groups (Petraitis and Dudgeon, 2015). Furthermore, the theory that has been developed generally focuses on closed systems, in which future trajectories depend only on the current state of the system (Ashwin et al., 2012). Yet real-world systems are never closed, because they have inputs (such as environmental pressures) that can change system trajectories. As noted earlier, the way in which modelers generally address this problem is to view external perturbations (or press disturbances) by varying parameters over time; other approaches include incorporation of stochasticity or use of control theory to design inputs that control a system's outputs in a desired way (Ashwin et al., 2012). In the real world, ecosystems are typically subjected to multiple types of disturbance simultaneously, and these can interact, such as when the risk of fire increases following drought. Such interactions are generally not incorporated in dynamical systems models. In addition, treating perturbations as parameters fails to capture the complexity seen in nature, where feedbacks can potentially occur between perturbations and state variables. For example, grazing of vegetation may increase the dominance of plants that can tolerate grazing, such as grasses (McNaughton, 1984); similarly, increased frequency of fire may lead to dominance of plant species that are fire-promoting (Brooks *et al.*, 2004). Such feedbacks are not always captured by dynamical systems models, yet potentially they provide an additional mechanism of ecosystem collapse.

From a practical management or policy perspective, it is important to know the mechanisms responsible for ecosystem collapse so that appropriate response measures can be developed. Unfortunately, the available theory does not really help to identify what the feedback mechanisms might be. Further, the theory is silent on the timescales involved and, therefore, whether an ecosystem transition is likely to be transient or persistent. In addition, it should be remembered that evidence of a transition between ecosystem states does not automatically indicate that dynamical systems theory is applicable. Ecosystem collapse could potentially occur because of a rapid change in environmental conditions, because of interactions between environmental drivers or because some critical threshold of ecosystem viability has been passed, without any internal feedback processes. This raises the question of whether theories describing multiple stable states actually relate to the alternative states of ecosystems that we observe in nature. While it is widely assumed in the literature that these represent one and the same thing, this is not necessarily the case. Multiple stable-state concepts are based on the assumption that the same underlying processes and mechanisms apply across all states, whereas this might not be true in reality. For example, when an ecosystem is severely degraded, it might actually become an entirely different system altogether, as when a forest is flooded and becomes a lake. These issues will be explored further in Chapter 5.

2.6 Planetary Boundaries

Rockström *et al.* (2009a,b) presented a new framework for examining human impacts on the biosphere, referred to as 'planetary boundaries'. This is based on the idea that elements of the Earth system may react in a non-linear, abrupt way to anthropogenic pressures and, therefore, that there may be critical threshold values of some key variables. If these thresholds are crossed, then parts of the Earth system could shift into a new state, which may have profound implications for human society. In their original paper, Rockström *et al.* (2009a) identified nine Earth-system processes and associated boundaries, namely climate change, stratospheric ozone, land-use change, freshwater use, biological diversity, ocean acidification, nitrogen and phosphorus inputs to the biosphere and oceans, aerosol loading and chemical pollution. Three of these variables, namely climate change, rate of biodiversity loss and interference with the nitrogen cycle, were judged to have already crossed their boundaries. These suggestions were subsequently updated with the addition of new variables such as the intriguingly named 'novel entities' and replacement of biological diversity with 'biosphere integrity' (Steffen *et al.*, 2015). As an attempt to raise concerns about human impacts on the biosphere, and even to build political momentum to address the issue, the planetary boundaries concept has been highly successful.

Unsurprisingly, given that it is essentially based on expert opinion, it has also generated much scientific debate. For example, Lewis (2012) highlighted two potential flaws in the concept. First, the proposed variables include some that are absolute limits rather than thresholds, citing the availability of rock phosphate as an example. Second is an issue of scale; some of the thresholds (such as climate change) are global, whereas others (such as nitrogen run-off) are local or regional and are not integrated at a global scale. Similarly, Brook *et al.* (2013) suggested that the key drivers of terrestrial ecosystem change, namely climate change, land-use change, habitat fragmentation and biodiversity loss, are unlikely to cause shifts in the state of the biosphere at the global scale. Such shifts could only be produced if there were spatial homogeneity in these drivers and ecosystem responses, and interconnectivity at the continental scale, which they argue is not the case. Terrestrial ecosystems are therefore likely to respond heterogeneously to these drivers rather than demonstrating abrupt shifts at the global scale. These authors also highlight the problem of assuming that the underlying mechanisms and processes are necessarily transferable across local, regional and global scales (Brook *et al.*, 2013).

More recently, Montoya *et al.* (2018) provided what can only be described as a scathing critique of the planetary boundaries concept, describing it as 'vague', 'deeply flawed', 'ill-founded' and 'implausible' science that is not supported by any evidence. Criticisms of the framework made by these authors include a lack of clear definitions, a failure to specify units, inclusion of variables such as 'biosphere integrity' that are impossible to measure meaningfully and the arbitrary nature of the proposed boundaries. It is actually quite unusual to see an idea dismissed so aggressively in the scientific literature; one can only conclude that these authors felt strongly about the issue! Unsurprisingly, these criticisms were rejected by the developers of the planetary boundary concept, who

2.6 Planetary Boundaries · 73

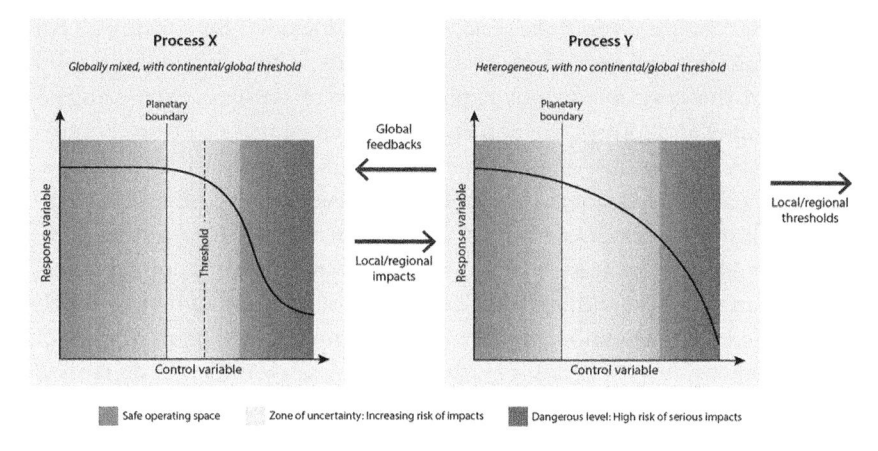

Figure 2.8 The conceptual framework of the planetary boundary approach, showing the safe operating space and the area of high risk. Reprinted from Steffen *et al.* (2015), with permission.

described the article as 'vitriolic', 'confrontational' and 'opinionated' (Rockström *et al.*, 2018). These are also strong words. In fact, having checked, I can say this is the first time that an author of a paper in this journal has accused another of being 'vitriolic'. Debates among ecologists don't often get as heated as this (although optimal foraging theory provides another example; according to Perry and Pianka (1997), this was once described in print as a 'complete waste of time'!).

According to Rockström *et al.* (2018), at least some of this disagreement is attributable to a misunderstanding of what the planetary boundary concept entails, specifically in relation to the occurrence of tipping points and thresholds at the global scale. Steffen *et al.* (2015) note that as originally defined, a planetary boundary is *not* equivalent to a global threshold or tipping point. Such thresholds may exist in some Earth-system processes, but not necessarily in all; and in cases where such a threshold does exist, the proposed planetary boundary is not placed at the position of the threshold but slightly 'upstream' of it (Figure 2.8). This reflects the idea underpinning the framework that the 'safe operating space' refers to a situation before any thresholds have been reached so that society can potentially react in time to avert a process of abrupt change (Steffen *et al.*, 2015).

So what will happen if planetary boundaries are crossed? Montoya *et al.* (2018) interpret this as leading to planetary collapse. This suggests that the planetary boundary concept is relevant to an understanding of ecosystem collapse, even if the scale under consideration is explicitly global rather than that of individual ecosystems. However, the authors who developed the concept do not explicitly refer to collapse. According to Rockström *et al.* (2009a), the consequence of crossing a planetary boundary could be 'irreversible and, in some cases, abrupt environmental change, leading to a state less conducive to human development'. Explicit reference is made to elements of the Earth system shifting into a new state, which could have 'deleterious or potentially even disastrous consequences for humans'. This would arguably qualify as collapse, as defined by Keith *et al.* (2015). The references cited by Rockström *et al.* (2009a) in support of such statements (e.g. Scheffer *et al.*, 2001) are those grounded in dynamical systems theory, so here we see that the theoretical ideas underlying the planetary boundaries concept are partly those associated with transitions between alternative stable states, such as bifurcation theory. Similarly, as described by Steffen *et al.* (2015), the further a planetary boundary is transgressed, 'the higher the risk of regime shifts, destabilized system processes, or erosion of resilience'. Again, this implies a shift to a different state and potentially a reduction in the capacity for ecological recovery.

One of the valuable aspects of the planetary boundary concept is its consideration of underlying mechanisms, while acknowledging that these remain poorly understood and are consequently associated with a high degree of uncertainty. Particular emphasis is given to positive feedback mechanisms that might result in non-linear trends or threshold responses (Figure 2.8). A number of these mechanisms have been identified for the global climate system (Lenton *et al.*, 2008). Other feedbacks referred to by Scheffer *et al.* (2015) include the suggestion that land-cover change beyond a certain value can activate positive feedbacks that lead to land-cover change across a much larger area; reduction of evapotranspiration resulting from the conversion of tropical rain forest to cropland or grazing land; and changing fire regimes in boreal forest. These authors also make an interesting suggestion in relation to the scaling issue highlighted by Brook *et al.* (2013), namely that processes demonstrating threshold behaviour at local and regional scales can potentially generate feedbacks that affect processes at larger scales. For example, a regional decline in natural carbon sinks could affect the climate system at the global scale and increase the likelihood of large-scale thresholds occurring, such as loss of the Greenland ice sheet (Scheffer *et al.*, 2015). A key question therefore is whether a sufficient number of local-scale regime shifts could cause a threshold response at a larger scale and thereby initiate a system-wide collapse (Hughes *et al.*, 2013).

In their consideration of how a planetary boundary for biodiversity might best be conceived, Mace *et al.* (2014) suggest that positive feedbacks may exist between biodiversity and other types of boundary. Although biodiversity has no known threshold close to its current state, these authors note that it does have a positive feedback on itself, owing to interdependencies between species. For example, loss of a species at one trophic level may result in further losses of species at other trophic levels, in what is referred to as a *trophic cascade* (Estes *et al.*, 2011) (see Section 2.8). Mace *et al.* (2014) suggest that biodiversity loss is likely to increase the likelihood that other boundaries are crossed, mostly because of the associated loss of adaptive capacity. However, evidence for this is not presented, and any such relationship is likely to differ markedly between boundaries. It is difficult to envisage much impact of biodiversity loss on boundaries relating to stratospheric ozone depletion, atmospheric aerosol loading or introduction of novel entities, for example. Conversely, boundaries such as land-system change, climate change, ocean acidification, biogeochemical flows and freshwater use are likely to have large impacts on biodiversity. Mace *et al.* (2014) state that on the basis of current knowledge, exceeding any of these boundaries would not trigger a biodiversity feedback strong enough to precipitate a threshold being crossed or lead to an accelerating cascade of biodiversity loss. Barnosky *et al.* (2012) suggested that we may indeed be approaching a state shift in the earth's biosphere, partly because of biodiversity loss, but this suggestion has been roundly rebuffed (Brook *et al.*, 2013).

2.7 Critical Loads

The critical loads approach provides a method for evaluating the potential impacts of pollutants on ecosystems. A critical load can be defined as 'the quantitative exposure to one or more pollutants below which significant harmful effects on sensitive elements of the environment do not occur, according to present knowledge' (Nilsson and Grennfelt, 1988). The approach has principally been applied to deposition of elements such as nitrogen and sulphur, which when deposited to excess can cause ecosystem acidification, eutrophication of waters, changes in community composition, increased susceptibility to pests and diseases and a decline in ecosystem condition (Burns *et al.*, 2008).

Critical loads are perhaps best understood as a policy and management tool, having been used extensively in Europe since the 1980s, initially as a response to the phenomenon of acid rain (Payne *et al.*, 2013). The approach is increasingly being applied in other parts of the world, including North America (Pardo *et al.*, 2011) and parts of Asia (Hettelingh *et al.*, 1995). Critical loads are typically employed by policy-makers to help set emissions standards and to inform emissions control programs, whereas environmental managers use them to evaluate the potential impact of pollution sources and to set benchmarks for ecosystem protection (Burns *et al.*, 2008; Pardo *et al.*, 2011). They are also used for mapping pollution impacts, controlling individual pollution sources and informing international negotiations on transboundary air pollution, such as the 1988 Sofia Protocol on the control of emissions of nitrogen oxides (Burns *et al.*, 2008; Payne *et al.*, 2013).

According to Pardo *et al.* (2011), there are three main approaches to estimating critical loads: *empirical*, *steady-state mass balance*, and *dynamic modeling*. Empirically determined critical loads are based on field observations of ecosystem responses to pollutant deposition, such as those provided by experiments or long-term monitoring data, or on the basis of expert judgement (Payne *et al.*, 2013). Steady-state mass balance modeling focuses on estimating the net loss or accumulation of pollutant inputs and outputs over the long term, based on the assumption that the ecosystem is at steady state with respect to inputs (Pardo *et al.*, 2011; Skeffington, 1999). Dynamic models also employ a mass balance approach but may incorporate temporal dynamics and typically require extensive data sets for parameterisation and testing (e.g. Belyazid *et al.*, 2006). Empirical approaches are generally preferred as they are based on measurable ecosystem responses to pollutant inputs, although this method can overestimate the critical load (i.e. set it too high) if the system has not reached a steady state. In such situations, a similar response could occur at a lower deposition rate over a longer period, and to avoid this problem, the steady-state mass balance approach is sometimes preferred (Pardo *et al.*, 2011).

The concept of critical loads is based on the idea that it is possible to define a level of pollution that does not harm the natural environment (Payne *et al.*, 2013). Alert readers will detect some similarity with the concept of planetary boundaries discussed earlier, although clearly the latter differs in its explicit focus on the global scale. In fact, critical loads are cited by Rockström *et al.* (2009a) as one of the approaches on which the planetary boundaries concept is based. One of the aspects shared by the two approaches is the idea of ecological thresholds; it is once the threshold associated with a critical load is crossed that damage to an ecosystem ensues. An important difference between the two approaches is that critical loads are grounded very firmly in empirical evidence; there is no reference to dynamical systems theory or the possibility of transitions between different ecosystem states. In this sense, therefore, critical loads represent a form of empirical theory rather than an idea that is mathematical in origin. On the other hand, the existence of ecological thresholds is an assumption on which the critical loads concept is based, rather than an empirically determined phenomenon (Skeffington, 1999).

Is it in fact possible to define a level of pollution that does not harm an ecosystem? This question was addressed by Payne *et al.* (2013), who conducted the first systematic attempt to verify a critical load with field survey data, by examining European grasslands along a gradient of nitrogen deposition. Results showed that approximately 60% of the species change points that were detected occurred at or below the range of critical load currently defined in policy, suggesting that the underlying principle of 'no harm' in pollution policy should perhaps be modified to a decision regarding how much harm might be acceptable. A number of other authors have highlighted potential limitations in the critical load approach, such as the uncertainty regarding judgements regarding what constitutes a 'significant harmful effect' (Rockström *et al.*, 2009a). Often, direct evidence is lacking regarding the full effects of pollutants on biota, and there is also uncertainty regarding the rates of change of ecosystems with changing pollutant loads, especially in relation to their recovery ability (Bull, 1995). Pardo *et al.* (2011) highlight other sources of uncertainty, including time lags and the possibility of interactions with other stressors. The impacts of pollutants on ecosystems can vary with climatic, hydrologic and soil conditions; for example, increased precipitation increases the critical nitrogen load for lichens (Pardo *et al.*, 2011). This highlights the fact that an understanding of the factors that affect the critical load is crucial for a complete mechanistic understanding of ecosystem responses. For Skeffington (1999), critical loads should not be viewed as unambiguous thresholds above which damage occurs, but as highly uncertain estimates of the relative risk of deleterious environmental effects on a given ecosystem.

Owing to such uncertainties, identification of critical loads is not always supported by a thorough understanding of underlying mechanisms, although a substantive research effort has documented a wide variety of pollutant impacts on ecosystems (Bobbink *et al.*, 2010; Greaver *et al.*, 2012). These impacts include the phenomenon of nitrogen saturation, which has been identified as a factor leading to tree mortality and forest stand dieback, an example of ecosystem collapse (Aber *et al.*,

1989). The critical load concept also offers some predictive ability, in that adverse impacts on ecosystems are anticipated once the threshold has been exceeded, even if the details of these impacts are not fully specified. The concept also communicates some important ecological insights, which are relevant to understanding the process of ecosystem degradation, and how this might affect the capacity for recovery. Key among these is the fact that the impact of pollutants on ecosystems are more strongly determined by cumulative annual loading than by short-term atmospheric concentrations (Lovett, 2013). The approach also highlights the need to determine the dose–response relationships between deposition loads of pollutants and ecosystem responses, which can be achieved through field surveys such as those conducted by Payne *et al.* (2013). Critical loads also have value as a framework for organising, simplifying and applying information on the ecological impacts of air pollution and communicating these to both land managers and the general public (Lovett, 2013).

Other concepts have also been developed to capture the effects of pollutants and other stressors on ecosystems. Some of these have also been incorporated into policy. Examples include *ecosystem integrity* and the associated (or identical) concept of *ecosystem health*. These refer to the ecological processes and structures that underpin the capacity of an ecosystem to self-organise and to provide ecosystem services to people (Kandziora *et al.*, 2013), a theme that will be examined further in Chapter 5. These two concepts have been the focus of some theoretical development (e.g. see Jørgensen *et al.*, 2007). However, I leave the final word on them to Wicklum and Davies (1995), who described them as 'ecologically inappropriate', as they are 'not inherent properties of an ecosystem and are not supported by either empirical evidence or ecological theory'. Ouch.

2.8 Food Webs, Ecological Networks and Extinction Cascades

Food webs represent the trophic linkages between producers and consumers and between consumers and their predators (McDonald-Madden *et al.*, 2016). Analysis of the structure and function of food webs has a long history in ecological science and is widely used to explore the relationships between species in an ecosystem. Increasingly, research into food webs has begun to examine how the loss of a species may affect the

2.8 Food Webs, Ecological Networks and Extinction Cascades

entire community. A number of studies have investigated how the effects of removing a species from a location may cascade through a food web, in relation to different network structures. This research has been extended to examine the characteristics of species in relation to food web dynamics, to identify those species that could cause extinction cascades if lost from the system, potentially leading to ecosystem collapse (McDonald-Madden et al., 2016).

Research into food web ecology has shown that complex food web structures can generate counter-intuitive outcomes; for example, suppression of a predator may actually cause a reduction in prey numbers rather than an increase. Further, the magnitude of flows of material and energy through particular groups of species does not necessarily relate closely to food web stability; trophic interactions that are associated with small material flows can have a major impact on the stability of a food web (Hastings et al., 2016). The relationship between food web connectance and stability has been widely studied, but the results have been contradictory: whereas some studies have found a positive relationship, others have found the converse, or no relationship at all (van Altena et al., 2016).

While food webs are based on trophic interactions, many organisms are also part of other types of ecological network, characterised by different interactions. Examples include networks based on mutualistic interactions, such as pollinator or seed disperser networks, and those composed of mycorrhizal fungi that characterise many forest ecosystems. Analysis of mutualistic networks has shown that high connectedness can promote network stability and increase species richness, although again exceptions have been found (Lurgi et al., 2016). Many species are simultaneously part of a number of different networks characterised by different types of interaction, leading to consideration of a 'network of networks' (Lurgi et al., 2016). Recent research in this area has shown that the relative abundance of different interaction types, such as the proportion of trophic *versus* mutualistic interactions, may affect the stability of the community and its relationship to network structure (Mougi and Kondoh, 2012).

On the basis of the research that has been undertaken, can any general rules be identified regarding the collapse or recovery of food webs and other types of ecological network? This question was addressed by Bascompte and Stouffer (2009), while noting that current knowledge is very fragmented and partial, and therefore any answer will necessarily be tentative. In particular, it is unclear how the assembly of ecological networks relates to the process of collapse and whether such collapse is reversible (the authors concluding that such reversibility is unlikely, but this is untested). Despite this uncertainty, these authors make the following suggestions (Bascompte and Stouffer, 2009):

- Although detailed information is lacking on how ecological networks develop over time, available evidence suggests that newly appearing species tend to interact with already well-connected species.
- Network assembly is often modular, in that species within a module interact more frequently with each other than with species from other modules. Assembly often proceeds by linkage of these trophic or mutualistic modules to core species. Some species may be network hubs, being typified by a large number of connections within their modules and by connecting different modules together.
- Because of their structure, ecological networks tend to be relatively robust to the extinction of specialist species, but are more vulnerable to the loss of more generalist species, or species close to the bottom of food webs.
- The disassembly of ecological networks tends to be characterised by thresholds, leading to network collapse. In other words, the consequences of species loss become amplified and self-reinforcing as more and more species are extirpated.

Other results reported by researchers that may have wide applicability include:

- highly connected species, top predators and/or trophically unique species can occupy 'keystone' roles or positions within food webs. Some plant species, parasites and species at intermediate trophic levels can also be keystone species. Loss of such species can cause many secondary extinctions. However, the traits that determine whether or not a species acts as a keystone are context-dependent, and they change during the process of food web collapse (Jonsson *et al.*, 2015).
- Keystone species have been found in all major habitats. Evidence suggests that carnivores are more likely to act as keystones in aquatic ecosystems, whereas herbivores are more likely in terrestrial environments (although this could be an artefact given the current rarity of terrestrial carnivores). Terrestrial carnivores and top predators are exploited by human hunters at median rates 4–10 times higher than other species, whereas in marine environments, exploitation rates by humans are 11–15 times greater than the median rate for other species.

2.8 Food Webs, Ecological Networks and Extinction Cascades · 81

For these reasons, humans can be considered as a hyperkeystone species (Worm and Paine, 2016).

- Food web robustness to species loss is enhanced by the presence of multiple independent pathways from primary producers to top predators. Furthermore, trophic links can be classified as either functional or redundant, based on their contribution to robustness. Empirical evidence suggests that hub species are not necessarily the most important for network robustness, as many of their links may be redundant. Network robustness can be reduced by species extirpation even when such losses do not result in any secondary extinctions. This provides a potential mechanism for tipping points in the collapse of ecosystems (Allesina et al., 2009).
- Communities affected by species loss in the past are likely to be more robust against further loss, compared to communities that have not been subjected to previous extirpations. This may be attributable to the loss of fragile interactions present in the intact community. In contrast, network structures promoting robustness, such as asymmetric links connecting specialist species to generalist species, are more likely to be preserved during the disassembly process (Jonsson et al., 2015; Kaneryd et al., 2012).
- Ecological networks may be more robust to perturbations that they have experienced historically than to novel ones, suggesting that new environmental challenges might collapse otherwise robust natural ecosystems (Strona and Lafferty, 2016).
- Certain types of ecosystems may be especially sensitive to disruption of biotic processes and interactions, namely ecosystems with strong top-down trophic regulation, systems with many mutualistic or facilitation interactions, systems that are strongly dependent on mobile links and systems where disturbance regimes impose top-down regulation and positive feedbacks operate between the biota and the disturbance (Keith et al., 2013).

One of the most interesting food web phenomena, reflected by its high popularity among researchers, is the *trophic cascade*. This refers to a situation where a change in the abundance of a top predator propagates or cascades down food chains, ultimately having an impact on vegetation (Ripple et al., 2016a). The concept perhaps dates back to Charles Darwin, who observed that in villages where domestic cats were more abundant, the populations of mice were reduced. This had an impact on the populations of bees, which were eaten by the mice. Consequently

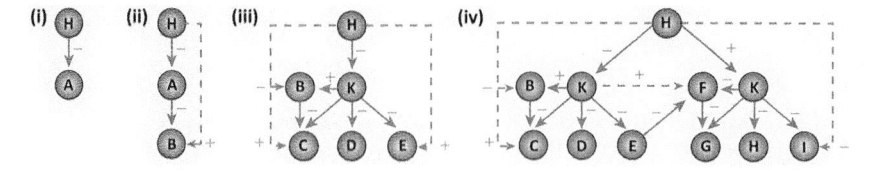

Figure 2.9 Effects of humans on ecological networks according to food web theory. Arrows denote direct (solid lines) or indirect (dashed lines) effects, which can be positive (+) or negative (−). Shown are (i) direct effects of humans H on a target species A, (ii) direct and indirect effects on a target A and non-target species B linked through a trophic cascade, (iii) direct effect on a keystone species K with indirect effects on several non-target species B–E and (iv) direct effects of humans on two keystone species operating in different interaction webs, with resulting indirect effects within and across food webs. Reprinted from Worm and Paine (2016), with permission from Elsevier.

the amount of clover pollination by bees was higher in locations with more cats (Ripple *et al.*, 2016a). Since then, trophic cascades have been observed throughout the world, in terrestrial, freshwater and marine systems (Estes *et al.*, 2011). A classic example is that documented by John Terborgh and colleagues in the islands of Lago Guri in Venezuela. In islands where vertebrate predators were absent, densities of rodents, howler monkeys, iguanas and leaf-cutter ants were 10–100 times higher than on the mainland, demonstrating how powerfully predators normally limit the population sizes of these species (Terborgh *et al.*, 2001). In some examples of trophic cascades, the influence of top predator is to suppress herbivory and to increase the abundance and production of plants. Examples include the sea otter/kelp forest system in the North Pacific Ocean and the wolf/ungulate/forest system in temperate and boreal North America (Estes *et al.*, 2011). Conversely in other ecosystems, such as the largemouth bass/planktivore/zooplankton/phytoplankton system in US Midwestern lakes, presence of the top predators reduces the abundance and production of plants (Estes *et al.*, 2011).

Are trophic cascades more likely in some types of ecosystem than others? Initially it was believed that trophic cascades were more likely in aquatic than terrestrial food webs, as the latter were believed to be more extensively interconnected, but current evidence does not support this distinction (Ripple *et al.*, 2016a). A meta-analysis of 114 studies suggested that the strongest cascades occurred in association with invertebrate herbivores and endothermic vertebrate predators (Borer *et al.*, 2005). In another review, Shurin *et al.* (2002) found that the effects of predators differed markedly among systems but were strongest in lentic

2.8 Food Webs, Ecological Networks and Extinction Cascades

and marine benthos and weakest in marine plankton and terrestrial food webs. The effects of predators on herbivores were generally found to be larger and more variable than on plants. Other factors that have been identified as contributing to strong trophic cascades include high system productivity, distinct metabolic requirements of organisms within a system and high nutritional quality of primary producers (Casey et al., 2017).

The widespread evidence of trophic cascades suggests that loss of top predators could lead to major changes in ecosystem composition, structure and function, and therefore provides a potential mechanism for ecosystem collapse (Bland et al., 2018a). A trophic cascade that causes persistent changes in an ecosystem can be considered as a regime shift (Pershing et al., 2015), which, as noted earlier, provides one definition of ecosystem collapse. According to Estes et al. (2011), the transitions in ecosystems associated with trophic cascades are often abrupt, are sometimes difficult to reverse and commonly lead to marked changes in patterns of energy and material flow.

Analysis of ecological networks has also provided insights into another type of cascade – an extinction cascade. Increased mortality rate and decreased abundance of a given species can lead to extirpation of another species, through their effects on ecological interactions. Evidence suggests that there may be critical abundance thresholds of species, which, if crossed, will lead to cascading extirpation of one or more other species (Sellman et al., 2016). Modeling studies involving alteration of mortality rates in ecological communities suggest that the first species to be lost is often not the one whose mortality rate is increased but another species in the network. Such secondary extirpations are particularly likely if small increases in mortality rate or small decreases in abundance occur in large-bodied species at the top of food chains (Säterberg et al., 2013). When a species essentially becomes too rare to fulfil its ecological role in an ecosystem, it is considered to be functionally or ecologically extinct. Evidence suggests that a species can become functionally extinct well before it becomes numerically extinct (Sellman et al., 2016).

The generalisations noted previously relating to secondary extinctions provide insights into the occurrence of extinction cascades. For example, the degree of connectance between species, trophic position, the structural properties of the community (e.g. network structure, omnivory, and interaction strength distributions) and properties of the species (e.g. their functional response, their ability to switch to new resources and density dependence) play a role in determining the likelihood of secondary extinctions and therefore extinction cascades (Petchey *et al.*, 2008). In the absence of demographic and environmental variation, extinction cascades and community collapses should be less likely to occur in species-rich multi-trophic communities than in species-poor ones. Conversely in a highly variable environment, the risk of cascading extinctions is likely to be higher in species-rich food webs than in species-poor ones. This is because the mean population density of primary producers decreases with increasing species richness, as a result of the increased intensity of inter-specific competition (Kaneryd *et al.*, 2012).

2.9 Resilience and Recovery

Although much of this chapter has focused on ecological theories relating to ecosystem collapse, many of these theoretical ideas are also relevant to understanding how ecosystems recover from anthropogenic disturbance. Concepts such as succession and the assembly of ecological networks, for example, are of fundamental importance to the process of ecological recovery. This section builds on this foundation by considering some additional theories that have been explicitly developed to provide insights into this recovery process.

A key concept in this context is resilience. This essentially refers to the ability of an ecosystem to tolerate a disturbance event and to recover afterwards. In recent years, resilience has increasingly been adopted as a component of environmental policy, supported by a growing research effort (Newton, 2016). Unfortunately, resilience is another of those ecological concepts that is shrouded in semantic uncertainty. A large number of different definitions have been proposed (Brand and Jax, 2007; Grimm and Wissel, 1997), and, as a result, the literature on the topic is rather confused. A useful distinction can however be drawn between *resistance* and *resilience*. The former relates to the ability of a system to remain essentially unchanged despite the presence of disturbance (Grimm and Wissel, 1997), whereas the term 'resilience' can usefully be restricted to the capacity to recover after disturbance (Nimmo *et al.*, 2015). Much previous literature fails to disaggregate these two components, yet there are powerful reasons to do so, as the mechanisms underlying them can be different (Nimmo *et al.*, 2015). For example, resistance of forest ecosystems to wind disturbance was found to be related to initial biotic conditions, whereas recovery was influenced by light availability (Bruelheide and Luginbűhl, 2009). Similarly, in intertidal macroalgal communities, resistance to heat-stress was found to be influenced by the cover of dominant species, whereas resilience was affected by species composition and the degree of heating (Allison, 2004).

Two different types of resilience are often differentiated in the literature. *'Engineering resilience'* refers simply to the time required for an ecosystem to return to an equilibrium point following a disturbance event (Pimm, 1984), whereas *'ecological resilience'* can be defined as the amount of disturbance that a system can absorb before transitioning to another stable state (Brand and Jax, 2007). The concept of engineering resilience is based on the assumption that during recovery, an ecosystem will return to a structure and function that prevailed prior to the disturbance event. However, as noted in Sections 2.3 and 2.4, it is now recognised that ecological recovery may pursue a different trajectory, leading to an ecosystem state different from the one that prevailed originally. The concept of ecological resilience was developed to reflect this possibility. A variety of different definitions of ecological resilience have been proposed, focusing variously on the amount of disturbance an ecosystem can absorb before it changes state, the ability to return to an original state after disturbance, the degree to which the system is capable of self-organisation and the capacity for reorganisation and adaptation (Brand and Jax, 2007).

Much of the recent research on resilience has focused on the ecological resilience concept and has drawn primarily from the dynamical systems theory that was considered in Section 2.5 (Gunderson, 2000). As noted by Grimm and Calabrese (2011), rather than addressing the process of recovery, this research focuses on the ability of ecosystems to 'absorb' the effect of disturbances. Potentially, the mechanisms that buffer the effect of disturbances can be lost, representing a decline in ecological resilience and leading to an abrupt regime shift or transition to an alternative state. But what are these buffering mechanisms? Standish et al. (2014) focus on the role of functional traits of species, which can provide information about the ecological strategies that are representative of particular ecosystem states. For example, traits relating to nitrogen fixation, seed dispersal and growth rate may influence the ability of terrestrial plant communities to respond to disturbance. Oliver et al. (2015) explore this idea further, focusing specifically on resilience of ecosystem functions and noting that this is likely to be determined by multiple factors acting at various scales of biological organisation, namely, species, communities and landscapes (Table 2.2). These scales are

Table 2.2 *Mechanisms underpinning the resistance and recovery of ecosystem functions to environmental perturbation*

Species (intraspecific)	Community (interspecific)	Landscape (ecosystem context)
Sensitivity to environmental change (RES)	Correlation between response and effect traits (RES)	Local environmental heterogeneity (RES)
Intrinsic rate of population increase (RES/REC)	Functional redundancy (RES/REC)	Landscape-level functional connectivity (RES/REC)
Adaptive phenotypic plasticity (RES/REC)	Network interaction structure (RES)	Potential for alternative stable states (RES/REC)
Genetic variability (RES/REC)	–	Area of natural habitat cover at the landscape scale (RES/REC)
Allee effects (RES/REC)	–	–

The abbreviations RES, REC and RES/REC indicate the importance of each mechanism for resistance, recovery or both, respectively.
Source: Reprinted from Oliver *et al.* (2015), with permission from Elsevier

interconnected so that changes at a particular scale can affect other scales in the same system; for example, responses of individual species to disturbance may affect other species, thereby affecting community structure and composition as well as the distribution of functional traits. Such changes may be mediated by the spatial context, such as landscape-scale heterogeneity or habitat connectivity, to influence the resilience of ecosystem function (Oliver et al., 2015). Standish et al. (2014) similarly note the importance of habitat connectivity and the temporal and spatial components of disturbance for determining resilience.

Ecological resilience has not been without controversy. There have been accusations of systematic bias within this field of research, resulting from the selection of ecosystems for study that exhibit responses consistent with theory (Boettiger and Hastings, 2012; Spears *et al.*, 2017). Grimm and Calabrese (2011) note that ecological resilience is not very tractable analytically, owing to the difficulty of formalising these ideas mathematically, and suggest that there is a lack of consideration of underlying mechanisms in the literature. The problems described in

2.9 Resilience and Recovery · 87

Section 2.5 relating to the application of dynamical systems theory, such as testing the existence of alternative stable states (Petraitis, 2013), also apply to ecological resilience concepts. Kirchoff *et al.* (2010) go further and suggest that the 'resilience approach' is based on an organismic view of ecosystems that has a largely cultural origin and is not universally accepted (see Chapter 1). Such concerns arise from the application of resilience approaches to socioecological systems, of which humans are a part (e.g. Liu *et al.*, 2007), a development that has proved particularly controversial, especially among social scientists (e.g. Olsson *et al.*, 2015; Tanner *et al.*, 2015). For these reasons the adoption of resilience as a policy goal has also been questioned (Newton, 2016). Standish *et al.* (2014) make the important point that resilience is not always a good thing, as is generally assumed by policymakers; ecosystem states degraded by human activities can have greater resilience to disturbance than those that are less degraded.

Many theoretical ideas about ecosystem recovery have also been developed by researchers investigating *ecological restoration*, which is the science and practice of assisting the recovery of ecosystems. Key among these ideas is the process of ecological succession, considered in Section 2.3, which has been examined in relation to numerous observations of recovery trajectories made by restoration researchers (Choi, 2004; Walker and del Moral, 2003). Results of such studies indicate that recovery trajectories are often difficult to predict and may differ substantially from the kind of 'orderly process' envisaged in early versions of successional theory (Choi, 2004; Suding *et al.*, 2004). Traditional ecological theory viewed ecosystems as complex, self-regulating systems, where following a perturbation, processes such as succession would bring the system back to a near-equilibrium state (O'Neill, 1999).

However, more recently it has been recognised that there may be multiple equilibria, with different recovery trajectories leading to alternative ecosystem states (Sections 2.4 and 2.5) (Hobbs and Norton, 1996; Suding *et al.*, 2004). Alternatively, according to non-equilibrium theories, recovery trajectories may be divergent, cyclic or arrested and may never arrive at an equilibrial state owing to the frequency and intensity of disturbance (Suding and Gross, 2006). On this basis, Suding and Hobbs (2009) proposed three kinds of model describing ecosystem recovery: (i) *continuum models*, where a change in the environmental controlling variable is proportional to the system response; (ii) *stochastic models*, which describe highly variable non-equilibrium relationships between system response variables and environmental drivers and (iii) *threshold* or *regime*

88 · Ecological Theory

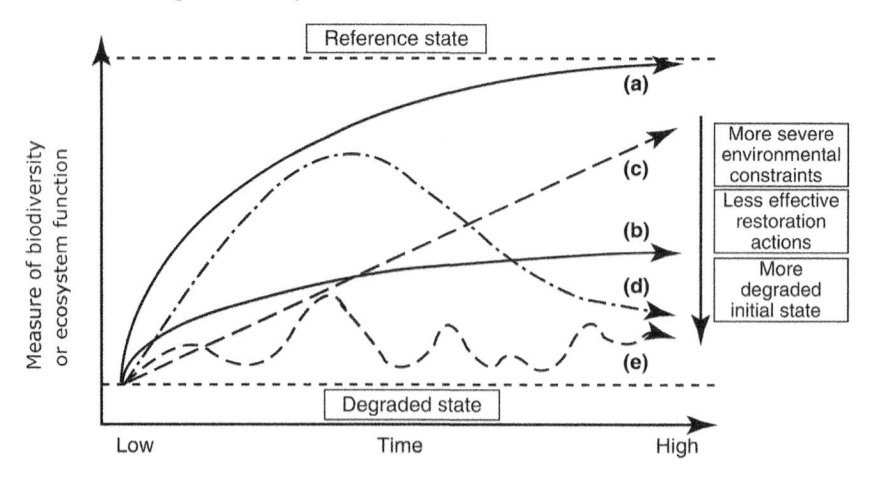

Figure 2.10 Trajectories of ecosystem recovery. Analysis of empirical restoration trajectories of individual measures indicates they can be asymptotic (a, b), linear (c), unimodal (d) or stochastic (e) over time. In addition to the form of the trajectory, these examples differ in whether or not they recover to a reference state, or the situation that prevailed prior to degradation occurring. Adapted from Bullock *et al.* (2011) with permission from Elsevier.

shift models, which describe abrupt non-linear changes in an ecosystem with small changes in environmental conditions.

According to theory, what form should a recovery trajectory take? This is a surprisingly difficult question to answer, as there is no single theory that enables all aspects of ecosystem recovery to be predicted. I discovered this when drafting a review paper on the topic with colleagues; our initial discussions among ourselves revealed a surprising diversity of opinion regarding what form the trajectory should theoretically take. Our solution to this problem was to consult empirical evidence in the literature, where we quickly found a wide range of different examples (Bullock *et al.*, 2011). Some trajectories were asymptotic; others were linear, unimodal or highly stochastic (Figure 2.10). We also found that different measures of ecosystem structure, function and composition can display a range of different trajectories within a single restoration. Clearly, the rate of recovery is likely to be lower in less productive ecosystems if the initial state is more degraded, if certain environmental factors constrain recovery or if less effective restoration methods are employed (Bullock *et al.*, 2011). Such factors could also potentially affect the form of the recovery trajectory. We examined this issue further by conducting meta-analyses of restoration projects, which, interestingly,

Table 2.3 *Mechanisms of ecosystem recovery, organised according to scale*

Speed/scale	Level	Structure	Process
Fast/	small Mortality, growth,	Individual reproduction	Physiology, behaviour, size
	Population	Density, structure (age, size, genetic)	Evolution, extinction
	Community	Diversity, composition, functional groups	Coexistence, competition, mutualism, predation
	Ecosystem	Nutrient pools/production	Resistance, resilience, nutrient flux/retention
	Landscape	Exogenous disturbance, propagule pressure	Connectedness, colonisation
Slow/large	Region	Temperature, precipitation	Pollution inputs, climate change

(Attributes of change)

Source: Adapted from Suding and Gross (2006), by permission of Island Press, Washington, DC; © 2006 Island Press

tended to show asymptotic responses of both biodiversity and ecosystem function measures, in both tropical (Martin et al., 2013) and temperate (Spake et al., 2015) forests.

Understanding the mechanisms that influence how ecosystems recover is challenging because of the need to relate dynamics occurring at a range of different scales, in terms of space, time and ecological organisation. One way of addressing this complexity is to consider these dynamics along a gradient from small to large scales (Table 2.3). This can provide insights into how ecosystem recovery might be supported by different management interventions conducted at different scales, which ideally would break any feedbacks that maintain the ecosystem in a degraded state (Suding and Gross, 2006).

Another theoretical idea that has greatly interested the ecological restoration community is that of *assembly rules* (Temperton et al., 2004), which can be thought of as the principles by which ecological communities are constructed. This term was originally introduced by Diamond (1975), based on his analysis of community composition of fruit-eating birds in New Guinea. Based on his observations, he identified interspecific interactions, particularly competition, as key factors in determining the structure of these bird communities. The concept was subsequently broadened to include any ecological process that might determine local community composition (Keddy, 1992). Subsequently, Hubbell (2001) drew a distinction between *niche-assembly* and *dispersal-assembly* rules, suggesting that spatial distribution patterns of trees in tropical forests are mostly attributable to the latter. This suggestion has proved highly controversial; the debate regarding the relative role of factors such as competition, dispersal and chance in shaping the composition of ecological communities continues to this day.

Belyea and Lancaster (1999) identified two classes of mechanism, namely resource dynamics and spatial organisation, which underpin assembly rules operating at different scales of community organisation (Table 2.4). Further, Keddy (1992) suggested that the factors determining community assembly act on functional traits rather than species, an idea that has been supported by subsequent research (e.g. Mouillot *et al.*, 2007). Potential mechanisms for community assembly rules include *niche filtering*, which is based on the idea that coexisting species are more similar to one another than expected by chance, because environmental conditions act as a filter that allow only a narrow spectrum of traits to persist (Mouchet *et al.*, 2010). Conversely, according to the competitive exclusion and limiting similarity principles, coexistence of species that are functionally different is more likely (Mouchet *et al.*, 2010). Alternatively, the neutral theory developed by Hubbell (2001) suggests that variation in functional traits between species has no influence on their ability to coexist, as individuals and species are essentially equivalent. Potentially, these different mechanisms can co-occur in the same ecosystem or can operate sequentially along environmental gradients (Mouchet *et al.*, 2010). Another important factor is scale; for example, environmental filtering is assumed to be stronger at the regional scale, whereas species interactions (such as competition) are likely to operate at the local scale (Mouchet *et al.*, 2010).

What is the empirical evidence regarding assembly rules? Götzenberger *et al.* (2012) reviewed available evidence for plant communities and concluded from their meta-analyses that non-random co-occurrence of plant species is not a widespread phenomenon. The authors of this study expressed disappointment that 'a long history of assembly rules study has failed to reach more definitive conclusions', noting that this finding may simply reflect the lack of robust evidence and methodological limitations of some of the studies that have been conducted. However, this tentative evidence suggests that Hubbell (2001) may have been right to question the role of interspecific competition and niche differentiation in structuring plant communities.

Table 2.4 *Assembly rules and underlying mechanisms*

Rule and mechanism	Resulting patterns
Resource use within guilds tends to increase owing to interspecific competition	Traits associated with resource competition are overdispersed; niche overlap is minimised
Resource use across guilds or functional groups tends to increase owing to interspecific competition	Species tend to be drawn from a different functional group until each group is represented and then the pattern repeats
	The proportion of species from different guilds remains constant
Spatial or temporal variance/covariance tends to increase within guilds through mechanisms operating at establishment	Intraspecific aggregation is greater than interspecific aggregation within guilds
	Traits associated with resource competition are underdispersed
Per capita predation risk tends to decrease owing to apparent competition among prey for predator refuges	The surviving prey species is that which supports the largest population of generalist predator; traits associated with competition for predator refuges are overdispersed among prey
	Less competitive prey species can invade if the population of a superior competitor has been depleted by a specialist predator
Total predation pressure tends to increase and per capita predation risk tends to decrease through exploitation competition among predators for prey and apparent competition among prey	Predator–prey ratios remain constant
Spatial aggregation of prey (or host) populations tends to increase owing to localised predation or parasitism	Prey (or host) populations are spatially aggregated, and predator–prey ratios are highest around the margins of these patches
A greater proportion of available energy and nutrients is retained in the biomass, owing to selection of species that cycle energy and nutrients most effectively	Generalist species tend to be replaced by specialist species
	Small, short-lived species tend to be replaced by larger, longer-lived species
Non-random spatial patterns of communities tend to develop owing to positive feedback involving local alteration of environmental constraints	Communities are spatially organised in a non-random pattern

Source: Reprinted from Belyea and Lancaster (1999), with permission

2.10 Conclusions and Propositions

As ecosystem collapse has only recently become the focus of significant research attention, it is perhaps not surprising that there is no consolidated body of theory that has been explicitly developed to address it. The same could also be said of ecosystem recovery. Yet this chapter has provided ample evidence that ecology is indeed theory-rich – there are many ideas and concepts derived from different ecological subdisciplines that provide insights into these phenomena. Of these, dynamical systems theory has been particularly influential and is often given particular emphasis in this context (e.g. Keith *et al.*, 2013, 2015). However, this theory has its limitations: it is difficult to test experimentally, and it fails to incorporate the complex interactions between multiple pressures that are currently affecting many ecosystems. These limitations can potentially be addressed, at least partly, by considering dynamical systems theory together with the wide variety of other ideas that can contribute to an understanding of ecosystem collapse and recovery.

One can therefore imagine that a theory of ecosystem collapse might be developed by integrating some of these different components. How might this best be achieved? Following Ford (2000), one potentially useful approach would be to identify a set of propositions (or axioms) based on the results of previous research. Such propositions could help summarise existing knowledge, while also providing a basis for the development of further questions or hypotheses, thereby guiding future research. As a contribution to this endeavour, and in the spirit of Pickett *et al.* (2013), I provide some tentative propositions in Box 2.7. These are not intended to be definitive but are provided to stimulate further thinking and critical evaluation, while also providing a framework for subsequent chapters. When considering them, it is helpful to bear in mind the salutary words of Ginzburg and Jensen (2004), in their examination of how ecological theories should be judged: 'Natural selection of ideas makes science different from astrology. The strength of this selection depends on how important a particular field is'. Sadly, theoretical ecology is rather unimportant to wider society, and as a result, ecological theories evolve and proliferate without a strong selection pressure. Consequently, many ecological theories flourish without ever being adequately tested. The reader is therefore encouraged to apply stringent selection pressure to the ideas presented here, because ecosystem collapse is surely a topic that merits deep societal concern.

2.10 Conclusions and Propositions · 93

> Box 2.7 *Some tentative propositions based on existing theories of ecosystem collapse and recovery*
>
> 1. Ecosystem collapse can be caused either by extrinsic or intrinsic factors, or by a combination of the two. Extrinsic factors refer to the external disturbance regime affecting an ecosystem and intrinsic factors to interactions of organisms between each other and with the physical environment.
> 2. Any ecosystem can potentially collapse, if subjected to disturbance of an appropriate type and occurring at sufficient frequency, extent, intensity or duration. Novel disturbances are more likely to cause collapse.
> 3. Disturbance regimes characterised by interactions between multiple types of disturbance are more likely to cause ecosystem collapse. A particular disturbance event is more likely to result in collapse if it follows another event from which the ecosystem has not yet recovered.
> 4. Anthropogenic pressures are more likely to cause ecosystem collapse than natural forms of disturbance, because they are often continuous ('press') disturbances rather than one-off 'pulse' events. Impacts of press disturbances can be cumulative.
> 5. Some ecosystems can exist in more than one state or regime, such as different successional stages. Transitions between these states can form part of the natural dynamics of an ecosystem. When such transitions are persistent rather than transient, they can become of concern to human society.
> 6. A persistent ecosystem transition, or collapse, can arise when ecological recovery is impeded. This can occur if there are stabilising feedback processes that maintain an ecosystem in a degraded state or when the processes of ecological recovery fail. Understanding these reasons for lack of recovery is key to understanding collapse.
> 7. Collapse can be caused by feedbacks in the internal ecological processes of an ecosystem, which can drive a system to a different state. As a result of these feedbacks, major ecological shifts can result from minor perturbations. Such shifts can occur when external conditions reach a critical value.
> 8. Collapse may be more likely if disturbance events cause the loss of generalist species, those that are highly connected to other species, top predators and/or trophically unique species. Loss of such species can cause many secondary extinctions.

9. The disassembly of ecological networks tends to be characterised by thresholds. In such situations, the consequences of species loss become amplified and self-reinforcing as more and more species are extirpated, leading to network collapse.
10. Certain types of ecosystems may be especially sensitive to disruption of biotic processes and interactions, namely those with strong top-down trophic regulation or with many mutualistic or facilitation interactions, those that are strongly dependent on mobile links and where positive feedbacks operate between the biota and the disturbance.
11. Extinction cascades and community collapses should be less likely to occur in species-rich multi-trophic communities than in species-poor ones, unless the environment is highly variable.
12. The ability of ecosystem to recover is critically dependent on intrinsic factors, namely interactions of organisms between each other and with the physical environment. The rate or extent of recovery can be limited by intrinsic and/or by extrinsic factors, such as the disturbance regime and the extent of degradation.

3 · *Case Studies from Prehistory*

Near to my home on the south coast of England, there is a wonderful museum dedicated to a unique group of fossils (www.theetchescollection.org/). They are the personal collection of Steve Etches, which features more than 2,000 specimens found during a lifetime spent on fossil hunting; his first discovery of an echinoid made at the age of five is proudly displayed among the exhibits. The museum is unusual in focusing on a relatively narrow window of time, spanning about 8 million years (Myr) in the late Jurassic period. This is the Kimmeridgian stage, named after the Dorset village where the museum is located. Here a sequence of fossil-rich mudstones and oil shales provide a remarkable insight into a tropical marine ecosystem that occurred here around 150 million years ago (Ma). What makes this museum particularly special is the way the fossils have been presented: as if they were a community of modern-day animals, busily interacting with each other. Evidence of these interactions include a pliosaur limb bone bitten in half by a marine crocodile, a plesiosaur scarred by pliosaur teeth, an ichthyosaur with fragments of reptile bone in its jaw and fish bones in its stomach, bite marks on ammonites made by large squid and the ink sacs of squid preserved inside the remains of fish stomachs. From such meticulous observation of these wonderfully preserved fossils, it has even been possible to create a putative Jurassic food web, with reptilian megapredators consuming ichthyosaurs and squid (as well as each other) and with ammonites, coelacanths and other fish species located further down the food chain.

Although relatives of some of these species survive today, such as the coelacanths, this is a community that has essentially become extinct. So what happened to these fascinating fauna? Within a few million years, the earth experienced a series of major disturbance events, including at least three large bolide impacts, one of which may have been bigger than the end-Cretaceous Chicxulub impact; major climatic change, including both global warming and an aridity episode; a substantial drop in global sea level; oceanic anoxia and stagnation; and some of the largest volcanic episodes in the history of the planet. These included the production of large-scale flood basalts and eruption of the Shatsky Rise supervolcano – one of the largest volcanoes in the solar system (Tennant *et al.*, 2017). Unsurprisingly, these perturbations were associated with major changes in Earth's biota, with many taxa becoming extinct. Plesiosaurs seem to have suffered particularly badly, yet other groups, such as the ichthyosaurs, survived relatively unscathed. These changes mark the boundary with the Cretaceous Period at around 145 Ma, and were accompanied by the loss of a number of diverse lineages. Although losses of species were significant, these were not sufficient for this event to qualify as one of the 'Big Five' mass extinctions that have been identified in the fossil record. The early Cretaceous was also associated with the origins of many major extant groups, such as birds and sharks (Tennant *et al.*, 2017).

Major extinction events provide a powerful example of what happens when large numbers of species are simultaneously lost. Despite its many limitations, including temporal, geographic and taxonomic biases, the fossil record therefore provides some of the best evidence of ecosystem collapse. As it also provides evidence of what happened after the collapse, it can afford insights into the process of subsequent recovery. In the case of much of the Kimmeridgian fauna, we know that extinction was forever; for such examples, it might indeed be justifiable to consider the process of ecosystem collapse as analogous to the extinction of a species (Boitani *et al.*, 2015; Keith *et al.*, 2013). In this particular case, recovery of diversity and ecosystem function involved evolutionary processes, leading to the development of new forms and lineages over timescales of millions of years (Tennant *et al.*, 2017). This chapter therefore examines the fossil record, to explore what can be learned from palaeontological evidence regarding both ecosystem collapse and recovery. To evaluate the theoretical ideas presented in Chapter 2, this evidence will then be considered in relation to the propositions presented at the end of that chapter.

3.1 The 'Big Five' Mass Extinctions

Collapse

Although the events that mark the end of the Jurassic period were large-scale and dramatic and were associated with a major loss of life, they do not qualify as a true mass extinction event. In palaeontological circles,

3.1 The 'Big Five' Mass Extinctions · 97

mass extinctions are defined as substantial biodiversity losses that are taxonomically broad, global in extent and relatively rapid (Jablonski, 1994). While a large number of extinction events have been documented in the fossil record, only a small number of these have been documented as global-scale phenomena. The Big Five mass extinction events are those associated with the end of the Ordovician, Devonian, Permian, Triassic and Cretaceous periods, each of which is associated with an estimated loss of more than 75% of species over a period of 2 Myr or less (Table 3.1; Barnosky et al., 2011).

Each mass extinction event has been attributed to a distinct set of causes (Table 3.1), but when the Big Five events are considered together, a surprisingly consistent picture emerges. Individual factors that have repeatedly been identified as a cause of a significant loss of species include large-scale volcanism, climate change (often associated with changes in atmospheric CO_2 concentration), changes in the oxygen concentration of the oceans, and bolide impacts. Some of these factors can be attributed to tectonic changes in the earth and, in some cases, to the evolution of life itself. For example, the Ordovician event was characterised by large-scale glaciation associated with a decline in atmospheric CO_2 concentrations, which has been attributed to the uplift and weathering of the Appalachian Mountains (Barnosky et al., 2011). As the glaciation ended, widespread anoxia developed in deeper ocean waters, perhaps attributable to the influx of glacial meltwater (Brannen, 2017).

Glaciation and ocean anoxia have also been identified as causal factors in the Devonian event, again associated with an increase in rock weathering and a decline in atmospheric CO_2 concentration. However, in this case, the diversification of land plants and the development of deep root systems appear to have been an underlying driving factor, perhaps accounting for the increase in weathering (Brannen, 2017). The Late Devonian event is sometimes referred to as a mass depletion, rather than a mass extinction, as it appears to have been driven by low speciation rates (Bambach et al., 2004), and loss of taxa was relatively gradual (Fan et al., 2020). Global warming associated with an increase in atmospheric CO_2 concentrations attributable to large-scale volcanism were key factors in both the Permian and Triassic extinction events. In the Permian event, ocean anoxia has also again been implicated (Penn et al., 2018), which may have been caused by shifts in ocean currents linked to global warming and was perhaps accompanied by an increase in hydrogen sulphide concentrations (Kump et al., 2005) (Figure 3.1). At the end of the Triassic, the increase in atmospheric

Table 3.1 The 'Big Five' mass extinction events

Event	Time (approximate end date)	Duration (estimated range)	Estimated loss of genera (%)	Estimated loss of species (%)	Proposed causes
Ordovician	443 Ma	3.3 Myr to 1.9 Myr	57	86	Glaciation, marine transgressions, changes in atmospheric and ocean chemistry, including reduction in atmospheric CO_2.
Devonian	359 Ma	29 Myr to 2 Myr	35	75	Global cooling and global warming, possibly tied to the diversification of land plants, reduction in atmospheric CO_2, widespread deepwater anoxia, marine transgressions. Possible role of bolide impact.
Permian	251 Ma	2.8 Myr to 160 kyr	56	96	Volcanism, global warming, widespread deepwater anoxia, elevated CO_2 concentrations, ocean acidification. Possible role of bolide impact.
Triassic	200 Ma	8.3 Myr to 600 kyr	47	80	Volcanism, elevated CO_2 concentrations, global warming.
Cretaceous	65 Ma	2.5 Myr to less than a year	40	76	Bolide impact, global cooling, volcanism, widespread deepwater anoxia and eutrophication, CO_2 concentration elevated before extinction and declined during extinction.

Source: Reprinted from Barnosky et al. (2011), with permission from Springer Nature.

3.1 The 'Big Five' Mass Extinctions · 99

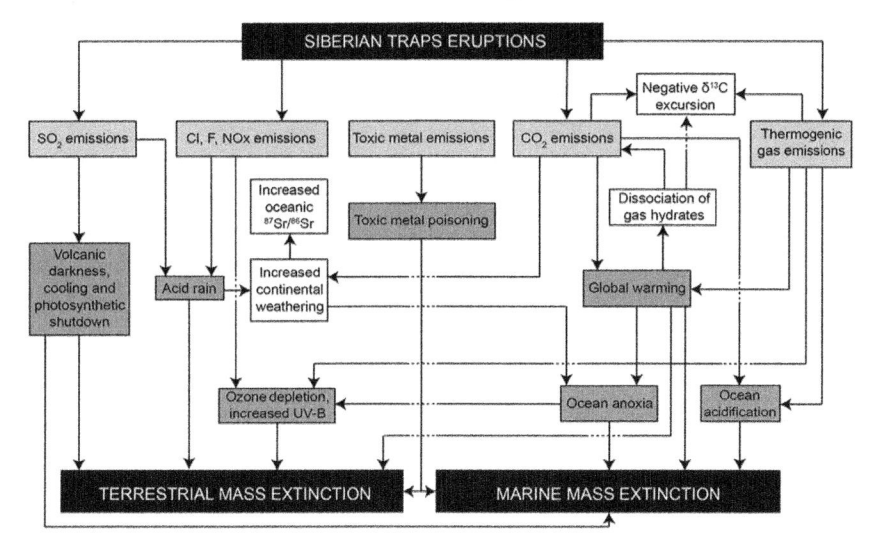

Figure 3.1 Proposed relationship between factors causing the end-Permian extinction. Reprinted from Bond and Grasby (2017).

CO_2 led to an increase in ocean acidity, which caused an almost complete loss of the world's coral reefs (Brannen, 2017).

Owing to outstanding research by Walter and Luis Alvarez (Alvarez *et al.*, 1980), the extinction event at the end of the Cretaceous Period is now widely attributed to the impact of the largest asteroid known to have hit any planet in the solar system within the last 500 Myr (Schulte *et al.*, 2010), and the ensuing global cataclysm. The impact crater has been located at Chicxulub in southern Mexico and has recently been analysed through a drilling project (King *et al.*, 2017). The idea that the dinosaurs were wiped out by a collision with an object bigger than Mount Everest travelling at 64,000 km h^{-1}, which released as much energy as 100,000,000 Mt of TNT (4.2×10^{23} J), has unsurprisingly captured the imagination of both the general public and researchers (just try and picture it for a moment!). The impact could have caused catastrophic environmental effects, such as thermal damage to the biosphere, extended darkness, global cooling and acid rain, thereby providing a range of different extinction mechanisms (Schulte *et al.*, 2010). However, the role of the asteroid in the extinction event has been the focus of intense debate, especially as it was accompanied by a major volcanic episode, namely the eruption of the Deccan flood basalts in India. Recent research suggests that the Chicxulub impact could have generated enough seismic energy to trigger volcanic eruptions worldwide, including those of the Deccan Traps (Richards *et al.*, 2015), implicating both bolide strike and volcanism in the extinction event (Schoene *et al.*, 2019). The understanding of the ecological impact has been wonderfully illuminated by the recent discovery of some remarkable fossils, which appear to be animals killed by the Chicxulub impact itself. The fossil fish even include ejecta spherules in their gills (DePalma *et al.*, 2019). Such evidence has led to suggestions that bolide impacts might also have contributed to some other mass extinctions, such as the Devonian and Permian events, but this is controversial (Barnosky *et al.*, 2011; Racki, 2012).

Can we identify any generalisations relating to the causes of mass extinction events? Although most literature has focused on the individual circumstances of each event, some authors have detected some common patterns. For example, Arens and West (2008) tested the hypothesis that extensive volcanism combined with bolide impact could offer a general mechanism of mass extinction. Analysis of fossil extinction percentages suggested that periods of elevated extinction were associated with the co-occurrence of bolide impact and continental flood basalt volcanism. Neither factor was found to be statistically significant in isolation. These authors also noted the different ecological features of these extinction mechanisms: bolide impacts, like marine anoxic incursions, can be considered as pulse disturbances that are sudden and catastrophic. In contrast, volcanism, climate change and sea-level change can be considered as press disturbances, which alter community composition (Arens and West, 2008). For these authors, it is the coincidence of both press and pulse disturbances, not just bolide impact and volcanism, which are responsible for the largest mass extinction events. The two types of disturbance differ in their impacts; whereas pulse disturbances may cause significant mortality in the immediate aftermath of an event, press disturbances can result in loss of species over much longer time scales. Furthermore, press disturbances can contribute to species extinction by reducing population sizes and reproduction, and destroying habitat or contracting geographic ranges, rather than increasing mortality directly (Arens and West, 2008).

Based on a review of more recent data, Bond and Grasby (2017) provide a contrasting perspective, highlighting the absence of a close temporal linkage between bolide impacts and extinctions. In fact, Chicxulub at the end-Cretaceous is the only example of an impact that has been shown to occur at the same time as a major extinction event.

These authors also ask the important question of how bolide impacts and volcanism actually affect ecosystems. In other words, what are the actual mechanisms of a mass extinction? There is limited understanding of how these factors act as environmental disturbances, and it is unclear why they have sometimes caused a profound collapse in much of the biosphere (such as at the end of the Permian), whereas at other times they didn't. Bond and Grasby (2017) further note the difficulty of understanding ecosystem-wide responses to multiple environmental changes even in modern situations (e.g. Queirós et al., 2015). The challenge for ancient ecosystems consisting of species that are no longer extant is substantially greater. Nonetheless, consideration of the physiological impact of these disturbances can provide valuable insights, supported by experimental research on extant species. Bond and Grasby (2017) examine such information for the three main mechanisms – the 'big three' killers – associated with mass extinctions:

- *The effects of increased CO_2 and reduced pH on organisms.* Numerous deleterious organismal responses to these factors have been reported, including hypercapnic stress, biomineralisation crises and interference with chemical receptors and sensory systems. While elevated atmospheric CO_2 concentrations can directly result in reduced ocean pH, these are two separate forms of disturbance. Hypercapnia refers to poisoning by CO_2, which decreases the capacity of tissues to oxygenate and can result in decreases in growth rate, reproduction and survival. Organisms with a carbonate skeleton may experience reduced biomineralisation, which can negatively affect development and growth. Ocean acidification affects fish behaviour by impairing olfactory functions and by altering neurotransmitter function. Reduced pH can also disrupt chemical signalling in invertebrates, reducing reproductive capacity and increasing the risk of predation.
- *Thermal stress.* Rising temperatures increase aerobic metabolism in animals, with metabolic rate increasing exponentially with temperature. An increase in metabolism increases the requirement for oxygen, which may exceed the ability of the organism to supply it. Active organisms have a higher aerobic scope than less active (e.g. sessile) organisms, which may account for the greater survival of more active marine invertebrates in many extinction events (Clapham, 2016). It should be noted that global cooling has also been implicated in some mass extinction events, such as the end Ordovician. Although either large-scale warming or cooling is likely to be deleterious to most

ecosystems, evidence is lacking to determine whether thermal stress has been a significant cause of mortality in mass extinction events or other factors that can accompany temperature change were responsible.
- *Deoxygenation.* Increasing global temperatures are often accompanied by the expansion of areas of anoxic conditions in water bodies, and marine anoxia has been repeatedly identified as a key factor in many extinction events. Depletion of oxygen can cause death by asphyxia, even though many animal groups have evolved specialist adaptations that enable them to tolerate oxygen-poor conditions. Even in the extinction event with the most complete record of anoxia, namely the end-Permian, the timing, magnitude and extent of anoxia varied greatly between locations. Complete anoxia of the oceans is very unlikely to have occurred, and therefore anoxia must have acted in synergy with other factors to cause mass extinction.

Bond and Grasby (2017) also highlight the need to understand the combined effects of multiple disturbances on complex ecosystems, particularly how different stressors might interact. Such interactions can form self-reinforcing mechanisms that hasten extinction. Analysis of contemporary extinction dynamics has highlighted a large number of such interactions, for example between the effects of habitat destruction, fragmentation, fire, climate change, and the spread of invasive species and diseases (Brook *et al.*, 2008). The extent to which such interactions contributed to previous mass extinction events is unclear, although factors such as the spread of invasive species have been implicated in some events (McGhee *et al.*, 2013; Stigall, 2012).

Based on evidence both from studies of fossils and contemporary species, Brook *et al.* (2008) propose the following generalisations about extinction dynamics:

- Extinction occurs most often when novel threats emerge that are outside of the evolutionary experience of species or occur at a rate that outpaces adaptation.
- Population size is important; small populations are more likely to become extinct as a result of chance effects (referred to as the small population paradigm).
- Long-term persistence is influenced by geographical range and dispersal ability. Widespread species experience fewer range-wide catastrophes and so have lower extinction probabilities. This has been examined in the fossil record through analysis of benthic marine invertebrates; survivorship was found to be associated with geographic range

throughout the Phanerozoic. However, the relationship was found to be less strong during mass extinction events, which was attributed to widespread environmental disturbance (Payne and Finnegan, 2007).
- The severity of threatening processes together with life-history traits can strongly affect extinction risk. For example, traits such as ecological specialisation and low population density act synergistically to increase extinction risk. Similarly, large-bodied, long-generation and low-fecundity species are particularly vulnerable to extinction caused by environmental change.
- Extinction does not necessarily happen quickly but can take from decades to millennia to conclude.

Roopnarine (2006) points out that many of the species that became extinct during mass extinction events may have been lost because of changing interactions with other species, rather than the direct effects of abiotic factors. Similarly, Brook *et al.* (2008) note that threatening processes may not directly result in species extinction by themselves but can initiate a suite of secondary processes and synergistic feedbacks that can eventually cause extinction. For example, if organisms at the base of a trophic pyramid became extinct, then this might eventually cause the collapse of an entire ecosystem, through an extinction cascade (see Section 2.8). Such cascades involve the secondary extinction of a species, which can occur as a result of perturbations to other species to which it is linked trophically. Extinction cascades may be either 'bottom-up' or 'top-down', reflecting whether perturbations are focused on the base or the apex of the trophic pyramid (Quince *et al.*, 2005). The disruption of primary production has repeatedly been invoked as a cause of mass extinction and has been linked with cascades – or even avalanches – of secondary extinctions at higher trophic levels (Vermeij, 2004). Evidence of such disruption includes anomalous shifts in carbon stable isotope ratios at the end of the Permian and the end of the Cretaceous, loss of abundant photosynthetic species at the end of the Cretaceous and spikes in fungal abundance at the end of the Cretaceous and possibly also the Permian (Roopnarine, 2006)

Whether such disruption of primary production leads to a cascade of extinctions through a trophic network depends on a range of factors, including taxonomic richness, functional or guild diversity, the pattern and relative strengths of trophic links between species and comparative species' richness among guilds of similar trophic function but different composition (Roopnarine, 2006). Unfortunately, the precise mechanisms and pathways of secondary extinctions are difficult to determine in fossil communities, but there is evidence that ancient ecosystems changed dramatically over time. Examples include: (i) the evolution of major new ecological roles (such as heterotrophy), (ii) variation of species diversity both overall and within guilds, (iii) the increase in the energetics of biotic interactions and (iv) the evolution of species with increased metabolic rates and complexity (Roopnarine, 2006). This suggests that extinction cascades driven by bottom-up processes such as decline in primary production could have made a significant contribution to previous mass extinction events.

Analysis of this phenomenon using modeling approaches suggested that the increasing number of connections among guilds that developed during the Phanerozoic, reflecting increases in functional diversity, may have delayed the onset of extinction cascades (Roopnarine, 2006). Furthermore, model results consistently indicated the preferential survival of competitively inferior taxa in such extinction events. This has empirical support from the survival of opportunistic generalist species, which have been observed to increase in terms of representation and geographic range after mass extinction events (Roopnarine, 2006). Such species have been referred to as 'disaster' taxa; the increase in stromatolites after the end-Permian extinction provides an example (Schubert and Bottjer, 1992). Ecologically, such species might be described as early colonisers or as 'weedy' species.

From the data obtained from the fossil record, it is unclear how different disturbances associated with mass extinctions differentially affected ecosystem components, such as different trophic levels. Further, it is unclear whether the factors implicated in mass extinction events pose the greatest threat to those organisms at the base of food chains (e.g. zooplankton and phytoplankton) or to those at a higher trophic levels (e.g. molluscs, crustaceans, fish) that are more active and have higher metabolic rates. In the end-Cretaceous event, it is known that in marine ecosystems, all levels of the trophic pyramid were affected, from the plesiosaurs, mosasaurs and fish at the top of food chains, all the way down to plankton (Bond and Grasby, 2017). Yet it is not clear whether the same was true for all other mass extinction events or whether any extinction cascades were consistently top-down or bottom-up. Another issue that is difficult to determine in fossil communities is the role of behaviour. Organisms at higher trophic levels typically demonstrate more complex behaviour, which is key to understanding the impacts of environmental change on communities and ecosystems

(Nagelkerken and Munday, 2016). Animals first respond to environmental change predominantly by changing their behaviour, which, in turn, affects species interactions and ecological processes. Differential capacity for behavioural change may therefore have contributed to the contrasting survival of different animal groups through mass extinction events.

As palaeontological investigations of mass extinctions have tended to focus on evolutionary and taxonomic issues, relatively little is known about their ecological impacts. Researchers have begun to apply some innovative approaches to address this problem. Most information is available for the global carbon cycle, which can be examined using techniques such as carbon isotope analysis. Given the role of large-scale volcanism in many mass extinction events, it is no surprise that substantial impacts on the carbon cycle have been detected. For example, at the end of the Triassic, the partial pressure of CO_2 may have increased by as much as fourfold, requiring the release of 8,000 Gt of volcanic CO_2 and a further 5,000 Gt of methane. (By comparison, since the Industrial Revolution, humans have so far released about 1.5 Gt of CO_2.) Carbon isotope analysis of this event indicates episodes where carbon was released rapidly from Earth's mantle, which generated marine anoxia and led to an increase in organic carbon burial (van de Schootbrugge *et al.*, 2008). Similarly, a negative shift in carbon isotope ratios at the end of the Permian was found to be associated with a turnover in the vegetation and proliferation of fungal spores indicating increased decomposition and perhaps ecosystem collapse (Cui *et al.*, 2017).

Other researchers have attempted to assess the relative impact of different mass extinction events in terms of their ecological impacts. For example, Droser *et al.* (2000) compared the end Devonian and Ordovician events by considering different scales of change, ranging from community-level changes to structural changes within an ecosystem and then to the complete appearance or disappearance of an ecosystem. Although these two extinctions were similar in terms of their taxonomic impact, the ecological importance of species lost differed between the two events. Whereas the Ordovician event resulted in only slight permanent ecological change, the Devonian extinction resulted in the complete restructuring of marine ecosystems. For example, tabulate corals were unable to recover after the Devonian extinction because of the loss of reef-building stromatoporoids, on which they depended for structural habitat. This shows that in terms of retaining ecological

Table 3.2 *Comparison of mass extinction events in terms of ecological and taxonomic severity, where the latter is measured by familial diversity loss*

Extinction event (ranked by ecological severity)	Familial diversity loss (%)		
Event	Event	Marine	Terrestrial
1. End-Permian	1. End-Permian	−47.5	−61.5
2. End-Cretaceous	2. Late Devonian	−27.8	−43.6
3. End-Triassic	3. End-Ordovician	−24.3	n.a.
4. Late Devonian	4. End-Triassic	−23.4	−21.7
5. End-Ordovician	5. End-Cretaceous	−14.7	−6.3

Note: n.a.: not applicable.
Source: Reprinted from McGhee *et al.* (2004), with permission from Elsevier

structure after a mass extinction, some taxa are more important than others (Droser *et al.*, 2000). This approach was extended by McGhee *et al.* (2004), who compared the ecological impacts of all the 'Big Five' extinctions. Results highlighted the unusual characteristics of the end-Cretaceous event, which was the least severe in terms of taxonomic diversity loss but the second most severe ecologically (Table 3.2). In contrast, the end-Ordovician failed to eliminate any functionally key taxa or evolutionary traits, so its ecological impact was minimal despite the major loss of diversity. From this analysis, McGhee *et al.* (2004) concluded that the selective removal of specific biological traits or higher taxa may result in much larger ecological changes than if traits or taxa are eliminated at random. Changes in functional diversity across mass extinction events have also been identified for some marine invertebrates (Christie *et al.*, 2013). In addition, attempts have been made to examine shifts in the structure of palaeocommunities over time, which have again highlighted the relatively large ecological impact of the end-Permian event (Muscente *et al.*, 2018).

At first glance, it would seem that the mass extinctions in the fossil record provide the best available examples of ecosystem collapse. Certainly, palaeontologists tend to see them that way (Field *et al.*, 2018; McGhee, 1988; Twitchett *et al.*, 2001). Massive bolide impacts, large-scale volcanism, global warming and marine anoxia clearly wreaked havoc with the world's ecosystems at a number of times in the past, and profoundly affected ecosystem structure and function as well as species composition. It is not easy to drive more than three-quarters of the world's species to extinction, so the ecological magnitude of these mass extinction events is beyond question. However, some caution is required in interpreting palaeontological evidence and relating it to the collapse of contemporary ecosystems. The biases in the fossil record are well established. For example, there is a close relationship between measures of species richness in the fossil record and the surface area of exposed rock, suggesting a strong sampling effect; furthermore, sampling and preservation have not been uniform over time (Smith, 2001). Most fossil evidence refers to taxa that produce a heavily mineralised skeleton or shell, and to marine benthic fauna in particular. It is not for nothing that palaeontologists write papers with striking but misleading titles such as 'When bivalves took over the world' (Fraiser and Bottjer, 2007), which, for me, evokes memories of the 1950s science-fiction B movies. Soft-bodied terrestrial forms, including diverse groups such as fungi and many invertebrates, are very poorly represented as fossils. Furthermore, stratigraphic processes, such as hiatuses in sediment deposition and changes in water depth, have a major influence on the fossil record (Holland and Patzkowsky, 2015). Many mass extinctions are associated with major reductions in the area of continental shelf, owing to marine transgressions. As a result of such issues, it has been suggested that extinction peaks are enhanced or even largely generated by sampling artefacts (Smith, 2001). In other words, mass extinction events might be as imaginary as the world being taken over by bivalves.

A further issue is spatial and temporal scale. Mass extinctions, by definition, are global in scale and represent an impact on the entire biosphere rather than an individual ecosystem. Such events involve massive species loss in both terrestrial and marine ecosystems, which requires some form of teleconnection between these environments, or shared extinction processes. Few abiotic factors have this global reach, which is why climate change is usually invoked as a principal mechanism in mass extinctions (Bond and Grasby, 2017). Yet, at the scale of individual ecosystems, other factors could potentially have played a leading role. It is even conceivable that biosphere-scale impacts might have involved a cascade of collapse, with the loss of one ecosystem resulting in the decline of another, in a manner analogous to the cascades of species extinctions. In terms of timescale, mass extinction events were typically characterised by durations of tens of thousands or even millions of years, although the end-Cretaceous event was apparently much more rapid (Table 3.1). Such time scales indicate that while factors such as volcanism and climate change can cause profound changes to ecosystems, their full effects may require geological timescales to become fully apparent. Caution is therefore required when applying these mechanisms to contemporary ecosystems at timescales relevant to human lifespans. This point is emphasised by Vermeij (2004), who suggests that most mass extinctions may have been caused by bottom-up interruptions to food chains, involving a decline in photosynthesis. This contrasts with contemporary loss of species, which is often driven by top-down causes such as the gain or loss of consumer species, together with habitat loss and fragmentation (Vermeij, 2004).

Do mass extinction events provide any insights into the collapse of individual ecosystems, at relatively limited spatial and temporal scales? This can best be explored by examining what happened to terrestrial vegetation during these events. Mass extinctions of terrestrial faunas parallel those of marine biota, with similar timing and magnitude of species losses. However, taxonomic losses among plants above the rank of genus are relatively rare at the global scale during these events (Table 3.3; McElwain and Punyasena, 2007). This fascinating observation highlights an ability of a plant of higher taxa to persist in the face of extreme environmental change, even if population sizes were very restricted. This resilience has been attributed to a number of different plant characteristics, including the ability to regrow from roots or stumps, even if the above-ground plant is destroyed; production of seeds or spores that can persist in the soil; and high dispersal ability (Knoll, 1984).

Despite the limited loss of higher taxa, evidence is accumulating that some plant communities underwent major changes in terms of structure and composition, leading to ecosystem collapse at local to regional scales (Table 3.3). It should be noted that there were also peaks in plant extinction that were apparently not accompanied by faunal mass extinction, such as in the early Oligocene (Wing, 2004). Vegetation changes that have been identified in association with mass extinction events include the following (McElwain and Punyasena, 2007):

- *End-Permian*: Evidence from Greenland indicates a relatively sudden dieback of the dominant gymnosperm vegetation coincident with the mass extinction event. In parts of Gondwana (now modern Antarctica), the dominant broadleaved deciduous glossopterid forests were replaced by needle-leaved conifers (*Volziopsis*) and subsequently by pteridosperms. Similarly, Australian locations indicate extinction of dominant glossopterid vegetation and its replacement by shrubby pteridosperms, needle-leaved conifers and lycopsids. The widespread high

Table 3.3 *Comparison of percentage extinctions between plants and animals in mass extinction events*

Extinction event	Magnitude of global faunal extinction (family)		Magnitude of floral extinction	
	Terrestrial (%)	Marine (%)	Global (family)	Regional (family and/or genera) (%)
End-Ordovician	n.a.	22–24	n.a.	n.a.
Late Devonian	44	21–28	Negligible	30–40
End-Permian	62	47–50	Negligible	19
End-Triassic	22	20–23	Negligible	17
End-Cretaceous	6	14–15	Negligible	18–30

Note: n.a.: not applicable.
Source: Reprinted from McElwain and Punyasena (2007), with permission from Elsevier.

abundance of spores (*Reduviasporonites*) is thought to indicate a peak in saprotrophic or pathogenic fungal activity associated with the mortality of trees, supporting a scenario of extensive forest loss at the extinction boundary. While there have been suggestions that these spores may be of algal origin, recent chemical and morphological analysis has confirmed their fungal nature (Sephton *et al.*, 2009; Visscher *et al.*, 2011).

- *End-Triassic*: Plant extinctions were localised events and not global in scale (Lucas and Tanner, 2015). Remarkably, only one plant family (Peltaspermaceae) became globally extinct at this time. However, evidence suggests a major disturbance of plant communities. Up to 90% of species became locally to regionally extinct across North America, Greenland and Europe. Significant changes in the dominance and diversity structure of Triassic forests have been recorded in Greenland and in parts of North America, where high-diversity pollen assemblages were replaced by lower diversity assemblages dominated by *Classopollis*, a pollen type associated with hot and arid conditions. In Greenland, high-diversity communities dominated by conifers and bennettites were replaced by lower diversity forests dominated by taxa that had previously been relatively rare.
- *End-Cretaceous*: The magnitude and extent of change in fossil floras across North America is consistent with sudden ecosystem collapse. In this region, 18–30% of plant genera and families and up to 57% of plant

species disappeared at this time. This was associated with loss of ecological dominants within floras, loss of specialised insect taxa, rapid loss of plant species richness and major compositional changes in plant communities, including a shift from angiosperm-rich to gymnosperm-dominated communities. Evidence that this occurred at the same time in both northern and southern hemispheres indicates global ecological upheaval and rapid ecosystem failure (Vajda et al., 2001). Extinction risk may have been higher among insect-pollinated than among wind-pollinated species and higher among evergreen than deciduous species. A peak in abundance of fungal spores suggests that large-scale regional and perhaps even global dieback of terrestrial vegetation occurred. Despite the initial massive decline in species diversity, few flowering plant families were lost, which is again deeply surprising.

Overall, a common pattern was the temporary or permanent loss of ecological dominants from terrestrial plant communities at local to regional spatial scales, which took place over thousands to millions of years (McElwain and Punyasena, 2007). Relatively abundant, large plants tended to be most vulnerable to decline (Wing, 2004). Typically, those species that flourished immediately after the mass extinction events were present at low abundance prior to the event, suggesting that they were probably poor competitors. While plant families and orders did not undergo 'mass extinction' (Table 3.3; Cascales-Miñana and Cleal, 2011), major ecological changes did occur across the events. These are likely to have significantly impacted ecosystem function at local to regional scales and may have contributed to faunal losses by altering primary productivity and biogeochemical cycles (McElwain and Punyasena, 2007). Evidence of changes in ecosystem function include dramatic changes in river drainage patterns at the end-Permian in Antarctica, Australia and South Africa, and increased particulate and nutrient flux from terrestrial to marine ecosystems at the end-Permian and end-Devonian events (McElwain and Punyasena, 2007), as well as the end-Cretaceous (Mizukami et al., 2013). In this way, vegetation changes could have contributed to faunal extinction in marine ecosystems, for example by increasing the incidence of marine anoxia.

Recovery
While mass extinctions provide dramatic examples of ecosystem collapse, they also provide useful insights into the process of subsequent recovery.

Table 3.4 *Selectivity of taxa in mass extinction events*

Event	'Losers'	'Winners'
End-Ordovician	Strophomenid and rhynchonellid brachiopods, nautiloids, trilobites, crinoids, conodonts, graptolites	Siliceous sponges, tabulate corals
Late Devonian	Stromatoporoids, tabulate corals, trilobites, cricoconarids, eurypterids, brachiopods, ammonoids, agnathans, placoderms	Chondrichthyans, actinopterygians (ray-finned fishes)
End-Permian	Brachiopods, crinoids, ammonoids, trilobites, tabulate and rugose corals, basal tetrapods	Bivalves, gastropods, malacostracans, echinoids, scleractinian corals, archosaurs
End-Triassic	Calcareous sponges, scleractinian corals, brachiopods, nautiloids, ammonites	Siliceous sponges, dinosaurs
End-Cretaceous	Non-avian dinosaurs, ammonites, calcareous plankton, mosasaurs, pterosaurs, rudist bivalves	Birds, mammals, spiny-rayed fishes

Notes: 'Losers' are those groups that suffered total extinction or heavy losses, whereas 'winners' displayed evolutionary radiation or an increase in abundance or dominance following the extinction event. Note that not all survivors are winners (Jablonski, 2005).
Source: Reprinted from Hull (2015), with permission from Elsevier.

This issue is examined in detail in an outstanding review by Hull (2015), on which the following account is based.

A first key point is that mass extinctions are usually selective, in that certain organisms, such as taxonomic groups or clades, are affected more than others (Table 3.4). As a result, diverse groups can be entirely lost during an extinction event, as occurred for example with the non-avian dinosaurs at the end of the Cretaceous. Selectivity can arise from the factors influencing extinction risk, including geographic range, body size, ecological characteristics and physiological traits. For example, during the end-Ordovician and end-Cretaceous events, geographic location was an important factor in determining which taxa became extinct. In the end-

Permian and end-Cretaceous events, large-bodied species of terrestrial faunas were preferentially lost, along with marine predators; carnivores and herbivores were also more affected than detritivores during the end-Cretaceous. This selectivity influenced the ecological characteristics of surviving communities. It also had major structural impacts on ecosystems, as illustrated by the loss of coral reefs at the end of the Triassic. Such impacts persisted for hundreds of thousands to millions of years and led to permanent changes in both ecosystem structure and function. However, mass extinctions also provided new opportunities for groups of taxa that were previously limited in abundance or distribution (Table 3.4); many extinction events were followed by periods of evolutionary radiation and innovation.

A second key point is that following mass extinction events, novel ecosystems were often created, which afterwards persisted for periods of up to millions of years. Such novel ecosystems differed functionally than those that preceded the event and those that developed subsequently. This could have arisen as a result of selectivity during the extinction event, which resulted in the survival of species with particular morphological, physiological or ecological characteristics. An example is the loss of large-bodied taxa associated with many mass extinction events, which has been referred to as the 'Lilliput effect' (Twitchett, 2007). As body size is an important functional trait, widespread dwarfing will have changed ecosystem processes such as patterns of energy flow. Extinction selectivity also resulted in the preferential loss of particular morphologies, trophic levels or functional types, which would have reduced the number of viable life-history strategies following an extinction. Subsequently, entirely new ecological interactions and strategies might evolve. An example is provided by reptiles following the end-Permian extinction, which evolved from sprawled into upright predators on land and also colonised the oceans in the Early Triassic (Hull, 2015). Other evidence of major ecosystem change following the end-Permian includes (i) the replacement of coral and metazoan reef systems by microbial carbonate mounds for up to 6 Myr; (ii) loss of key marine functional types including macroalgae, metazoan suspension feeders, mobile predators and deposit feeders for at least 1 Myr; and (iii) substantial decline in the burrowing of benthic sediments for $\sim$4 Myr (Hull, 2015). These examples show that the effects of collapse can persist for extraordinarily long periods of time.

Following mass extinction events, the rate of recovery was low, complex ecosystems typically requiring many thousands or even millions of years to become re-established (Figure 3.2). This reflects the fact that

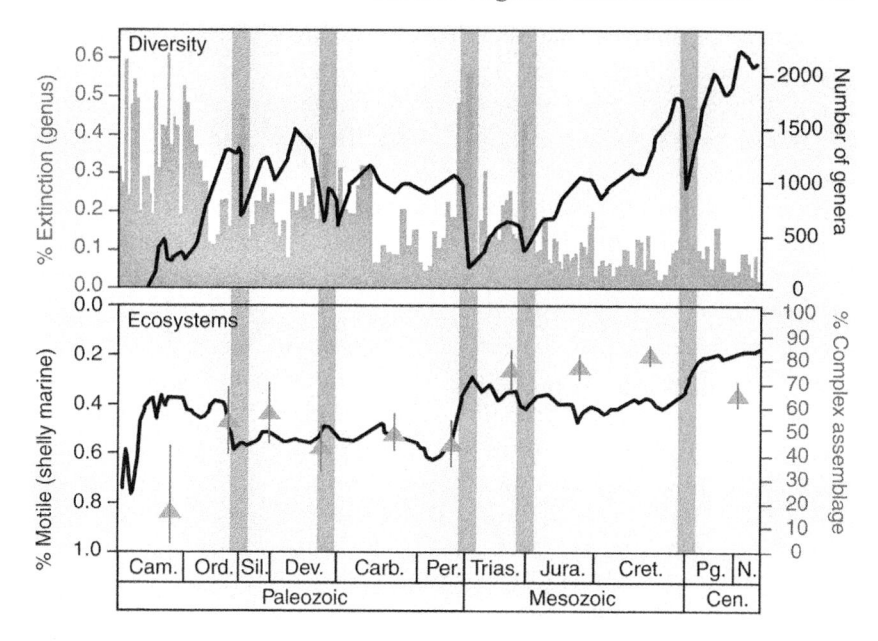

Figure 3.2 Mass extinctions are characterised by increased extinction rates relative to background values (upper panel) and long-lasting ecosystem change (lower panel). During the Phanerozoic, background rates of extinction in shelly marine invertebrates have generally declined (grey bars, upper panel), with the Big Five mass extinctions (vertical bars) associated with major declines in diversity (black line). Step changes in community structure (as illustrated by a single example, black line, lower panel) and complexity (grey triangles, lower panel) coincide with the largest of the Big Five mass extinctions. Reprinted from Hull (2015), with permission from Elsevier.

recovery was at least partly dependent on evolutionary diversification and the development of new forms. Rates of diversification differed across taxa, environments, habitats and locations. For example, following the end-Permian, recovery of burrowing in deep-sea sediments (bioturbation) varied between localities by timescales of up to millions of years and depended on the amelioration of anoxia or high-temperature conditions (Twitchett et al., 2004). Similarly, recovery of the diversity of benthic fauna took millions of years longer than that of pelagic groups such as ammonites, conodonts and marine tetrapods. In terrestrial ecosystems, some ecological strategies recovered very rapidly, whereas others, such as small-bodied insectivory and large-bodied herbivory, required more than 15 Myr to re-establish (Benton et al., 2004), a pattern that affected food-web structures (Roopnarine et al., 2007).

In the phase immediately following mass extinction events, ecosystems tended to be highly dynamic (Figure 3.2). In some cases, this has been attributed to a continuing influence of the factors responsible for causing the extinction event. For example, after the end-Permian event, the impacts associated with the volcanic eruption of the Siberian traps, including volcanic outgassing, soil erosion, ecosystem productivity and marine anoxia, continued for millions of years after the extinction. This indicates that the unusual ecosystem function and behaviour observed following the Permian event was partly caused by volcanism, as well as altered ecosystem structure and composition. In contrast, the end-Cretaceous extinction was caused by a relatively short-lived perturbation. Ecosystem impacts are illustrated by the two main calcifying groups of organisms in the open ocean, coccolithophores (marine algae) and planktonic foraminifera (heterotrophic protists). More than 90% of species in these groups became extinct during the event. For the next million years or so, both groups displayed a succession of short-lived, low-diversity communities, each of which were dominated by a single taxon. The reasons for these continual shifts in the relative abundance of taxa are unclear, as many successive communities appear to be functionally identical (Hull, 2015). This pattern of compositional change is reminiscent of ecological succession (see Section 2.3). However, these 'successional' changes in the fossil record occur over much longer timescales than those attributable to successional mechanisms in contemporary ecosystems. Hull (2015) suggests that perhaps a new type of succession needs to be identified for such situations, which she refers to as 'Earth system succession' (Box 3.1).

Earth system succession describes the pattern of change in the biosphere or the geosphere following either a disturbance event or an evolutionary innovation (Figure 3.3). The process differs from ecological succession both in spatial and temporal scales and also in terms of underlying mechanisms; the evolution of new species and life history strategies can be a major contributor. The concept, like traditional views of ecological succession, is equilibrial, in that processes such as evolution will eventually return the Earth system towards a dynamic equilibrium. The possibility of Earth system succession has implications for ecosystems, as it suggests that there may be limits on the rate of re-establishment of geologically stable ecosystems after global biotic disturbances (Hull, 2015). These limits would be determined by the size of biogeochemical reservoirs and the exchange rates between them and the extent to which they were perturbed. Earth system succession may also provide a general

Box 3.1 *Earth system succession (adapted from Hull, 2015)*

The structure and function of ecosystems affects key Earth system processes, such as rates of weathering, soil formation, organic carbon sequestration, nutrient availability and recycling, and the availability of key substrates such as soils and reefs. These processes transfer different elements between Earth system reservoirs, for example from the ocean to the atmosphere or from soils to rivers. At a global scale, the time required for elements to move between reservoirs can range from 100 years to 10 Myr. Fluxes in Earth system reservoirs can in turn affect ecosystems, by altering nutrient availability and environmental conditions. These effects will depend on the relative size of the flux, element amounts and their distribution between different reservoirs.

'Earth system succession' refers to sequential changes in biogeochemical reservoirs and fluxes that may occur in the Earth system after perturbation. Mass extinctions may cause lasting changes to ecosystem function or affect biogeochemical reservoirs directly. The return to a geological stable equilibrium can take long periods of time. For instance, Earth system sequestration of CO_2 after the Paleocene-Eocene Thermal Maximum ($\sim$55 Ma) took between 120,000 and 220,000 years, during which time temperatures were higher, ocean carbonate saturation state declined and the structure and function of ecosystems changed. Some biogeochemical perturbations or ecosystem changes appear to have had a global effect on environmental conditions. For example, after the end-Cretaceous extinction, the reduction of pelagic carbonate producers (e.g. coccolithophores) led to a very high carbonate saturation state throughout the ocean. Similarly, the end-Triassic extinction was followed by a silica deposition boom, driven by increased weathering of new basalts and siliceous sponges outcompeting carbonate corals following an ocean acidification event.

Long-term changes in ecosystem composition and function may also, in turn, lead to changes in global environmental conditions, providing a potential feedback mechanism that has been invoked as a cause of mass extinctions. Earth system succession might also contribute to understanding the effects of environmental change or key biological innovations, such as the evolution of oxygenic photosynthesis, pelagic calcifiers or grasses. In each case, as the novel organisms increased in abundance, biogeochemical cycles were fundamentally changed.

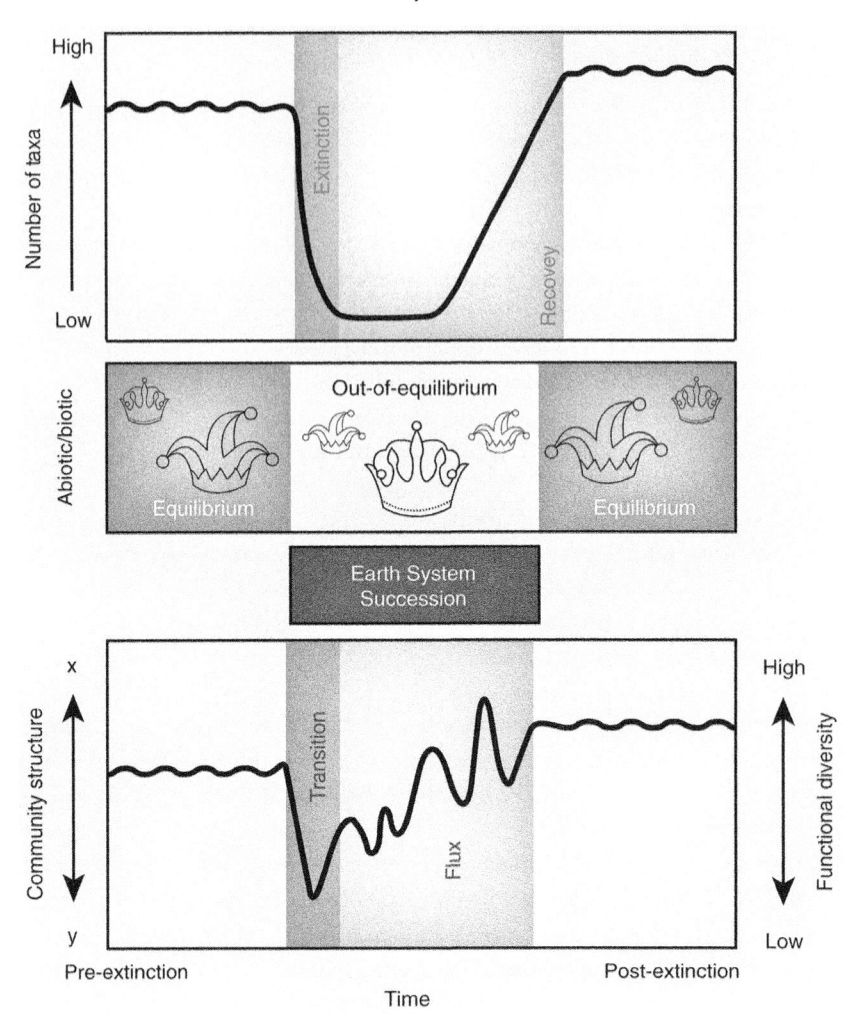

Figure 3.3 The concept of Earth system succession, as described by Hull (2015). Diversification of organisms following mass extinctions coincides with widespread changes in global ecosystems. Environmental disturbance and ecological change during the extinction event (top and bottom) perturbs the dynamic equilibrium of the Earth system (middle). As species evolve and ecosystems develop, Earth system feedbacks affect the direction, nature and timing of ecosystem turnover (bottom). Reprinted from Hull (2015), with permission from Elsevier.

explanation for why some ecological and evolutionary changes take so long to happen (Hull, 2015).

The similarities between ecological succession and recovery following mass extinctions are further explored by Solé *et al.* (2002). These authors again note the common presence of opportunistic groups of species just after the extinction events, followed by their gradual replacement by non-opportunistic species. Exploration of an evolutionary trophic model predicted progressively slower recovery from primary producers through herbivores to predators, a finding supported by available fossil evidence (Solé et al., 2002). These authors also explored the relationship between the magnitude of extinction and the rate of recovery. Although such a relationship has been proposed (e.g. Erwin, 2001), there is little evidence for it in the fossil record. Model outputs suggested that recovery time increases linearly with an increase in the percentage of primary producers that undergo extinction, until a critical threshold is reached. Above this value, recovery times increase exponentially. Solé et al. (2002) also note the likely importance of trophic dynamics in post-extinction recovery, although this was not explored in their model. Mass extinction events were typically followed by a period of restructuring of ecosystems, involving an increase in productivity by primary producers and consumers over time. This subsequently generated ecological niches at higher trophic levels, which led to an increase in the complexity of ecosystem structure (Chen and Benton, 2012).

Wei et al. (2015) highlight the considerable variation in the rate of recovery following the 'Big Five' mass extinctions. Defining recovery on the basis of re-attainment of species richness equal to or exceeding pre-event values and re-development of stable, well-integrated trophic systems, the duration of the recovery interval was estimated at ~4 Myr for the end-Ordovician crisis, ~10 Myr for the Late Devonian, ~5 Myr for the end-Permian, ~2.5 Myr for the end-Triassic, and ~3 Myr for the end-Cretaceous event (Wei et al., 2015). These recovery periods are generally longer than the extinction events that preceded them (Jablonski, 2005). Variation in recovery time has been attributed to the disturbed or harsh environmental conditions that followed each extinction event (Wei et al., 2015). For example, recovery after the Late Devonian event was halted or reversed by two glaciation episodes during which environmental conditions deteriorated (Chen et al., 2005). Similarly, after the end-Permian event, tropical sea-surface temperatures remained persistently high for at least 2 Myr. In both cases, environmental conditions also fluctuated markedly during these intervals of adverse environmental conditions (Wei et al., 2015). These observations support suggestions that press disturbances may have impeded ecosystem recovery after mass extinction events (Arens and West, 2008).

Two recent studies provide further insights into the process of recovery following the end-Cretaceous event. Henehan *et al.* (2019) used boron isotopes in foraminifera to examine changes in oceanic biogeochemical cycling across the event. Results indicated a rapid decline in the pH of ocean waters following the Chicxulub impact, supporting impact-induced ocean acidification as a mechanism for collapse of marine ecosystems. These findings also suggested a ~50% reduction in global marine primary productivity as a result of the event, which recovered within a few tens of thousands of years. Surface ocean pH recovered within 40 kyr of the end-Cretaceous boundary, returning to pre-event values after an additional ~80 kyr. However, impacts on the carbon cycle in the deep ocean lasted much longer: up to a million years. These patterns were attributed in part to the loss of plankton functional groups such as calcifying species.

Lyson *et al.* (2019) describe a remarkable set of fossils from Colorado, including reptiles and mammals, which provides an insight into the process of recovery in a continental location for the first million years after the end-Cretaceous extinction. Within about 100,000 years, the taxonomic richness of mammals doubled, and the maximum body mass of mammals reached values close to those prior to the event. Around 300,000 years after the event, a threefold increase in maximum mammalian body mass was observed, associated with an increase in the species richness of plants. Further increases were observed around 700,000 years. Overall, these results showed that mammal diversification and body-mass shifts were associated with intervals of relatively warm climate and associated changes in vegetation.

So what happened to plants? At the end of the Cretaceous, there was a peak in fungal spores, apparently associated with regional or even global dieback of vegetation, which was quickly followed by a 'fern spike' (McElwain and Punyasena, 2007). This peak in fern spores suggests a high abundance of weedy generalist and herbaceous species, including ferns, which were recolonising a de-vegetated landscape. Genetic evidence suggests that macrolichens also prospered (Huang *et al.*, 2019). Fossil floras continued to be relatively species-poor for more than 1 Myr thereafter, suggesting a long recovery process (McElwain and Punyasena, 2007). In Colorado, Lyson *et al.* (2019) reported low-diversity forests dominated by ferns and palms after the event, which lasted for up to 100,000 years. The subsequent 200,000 years was associated with a period of ecosystem recovery, during which the diversity of larger plants increased steadily. This process continued for another 400,000 years, when new plant taxa such as legumes appear in the record.

The recovery period after the end-Permian extinction was even longer. This event was also characterised by large-scale dieback of woody vegetation, which, in the more humid climatic zones of Pangaea, resulted in the widespread disappearance of peat forests (Looy et al., 1999). In semi-arid equatorial areas, the dominant conifer forests suffered a massive dieback and were replaced by lycopsid vegetation for a period of 4–5 Myr. This shift seems to have been associated with extensive soil erosion, creating oligotrophic conditions favouring groups such as lycopsids and bryophytes. This vegetation probably assisted a gradual process of soil recovery, which eventually supported re-establishment of conifer forests from isolated refugia through a process of succession and migration (although the rate was much lower than contemporary ecological succession). Overall, the recovery of both semi-arid conifer forests and humid peat forest seems to have occurred at a similar rate to that of marine ecosystems, emphasising the fact that recovery after the end-Permian was exceptionally slow in a number of different ecosystem types (Looy et al., 1999). The reasons for this are unclear; it may be that the length of the recovery interval was related to the severity of the extinction event (Wing, 2004), although modeling research suggests this is unlikely (Kirchner and Weil, 2000). Alternatively, it may reflect continuing environmental stress; both the end-Triassic and end-Permian events were followed by prolonged ecosystem instability (McElwain and Punyasena, 2007).

One cannot leave the topic of mass extinctions without briefly considering the current biodiversity crisis. In this era of the Anthropocene (Crutzen, 2006), there is no doubt that human activities are having an immense impact on the world's biota (Dirzo and Raven, 2003). But does this qualify as a sixth mass extinction? The answer to this question has proved controversial. The message from assessments of rapidly declining groups of taxa such as amphibians is that a mass extinction may be imminent, if it is not already occurring (Wake and Vredenburg, 2008). Ceballos *et al.* (2015) are even less equivocal, stating that a sixth mass extinction is already under way, based on their estimate that the mean rate of vertebrate species loss over the past century is up to 100 times higher than the background extinction rate. This interpretation was further emphasised by subsequent analysis of population decline and range contraction in terrestrial vertebrates (Ceballos *et al.*, 2017). For some authors (e.g. Cafaro, 2015), the sixth extinction is self-evident. Others (e.g. Briggs, 2014) have suggested that current extinction rates are not exceptionally high, and there is no evidence for a sixth mass extinction event. The issue was examined in detail by Barnosky *et al.* (2011), who compared current extinction rates with those observed in the fossil record. These authors concluded that while current rates of species loss are dramatic, they do not qualify as a mass extinction as used in the palaeontological sense when referring to the 'Big Five'. In recent centuries, only a few per cent of described species have gone extinct (although there may have been much higher losses of species that have never been described). However, future loss of those species that are currently endangered could result in a mass extinction event, which might resemble previous mass extinctions in the occurrence of multiple ecological stressors, including rapid climate change and highly elevated atmospheric CO_2 concentrations (Barnosky *et al.*, 2011).

I will give the final word on the sixth mass extinction to Brannen (2017), who recounts a discussion on the issue with Smithsonian palaeontologist Douglas Erwin. For Erwin, arguments for the sixth mass extinction are based on junk science, because of a lack of understanding among contemporary ecologists about how bad the mass extinctions in the fossil record actually were. Erwin illustrates the incompleteness of the fossil record by referring to the passenger pigeon, which was once one of the most abundant species on Earth until it was driven to extinction by human action in the nineteenth century. Yet loss of this species is barely detectable in the fossil record. Most of the palaeontological evidence for mass extinctions comes from abundant, widespread and skeletonised marine invertebrates, such as those Earth-conquering bivalves. Virtually none of these have become extinct in the modern era; it is actually very difficult to drive such hardy and ubiquitous organisms to extinction. But it happened. For Erwin, it was the devastating chains of secondary extinctions, leading to the collapse of entire food webs and ecosystems, which characterised and drove mass extinction events that killed almost everything on the planet. If this is how mass extinctions unfold, then the current biodiversity crisis might lead to something similar in future, perhaps by crossing a threshold or tipping point (Brannen, 2017). If a sixth mass extinction does occur, evidence from the previous events shows us that it could take many millions of years for the biosphere to recover (Kirchner and Weil, 2000). This is perhaps the most powerful lesson from prehistory.

3.2 Extinction of Australian Megafauna

In the Late Quaternary period, Australia was home to a remarkable and diverse megafauna. This included the largest marsupial that ever lived,

3.2 Extinction of Australian Megafauna · 121

Figure 3.4 A reconstruction of Australian megafauna, featuring (left to right, at back) *Genyornis*, *Diprotodon*, *Procoptodon goliah* (front), *Megalania*, *Thylacoleo* and *Thylacine*. Image by Peter Trusler, Monash University, copyright Australia Post, reproduced with permission.

the genus *Diprotodon*, which was a form of wombat the size of a large rhinoceros; *Procoptodon goliah*, the largest-known kangaroo that ever existed, which stood ~2 m tall; *Genyornis newtoni*, a large, flightless bird that reached a similar height; *Palorchestes azael*, a marsupial tapir-like animal the size of a horse; the giant koala *Phascolarctos stirtoni*; and *Thylacoleo carnifex*, the 'drop cat' or marsupial lion, which had the most powerful bite of any mammal that has ever lived and which hunted its prey by dropping from tree branches. There were also some impressive reptiles, such as *Varanus priscus*, a giant monitor lizard that grew to more than 4 m long; *Quinkana*, a genus of terrestrial crocodiles that reached 6 m in length; and *Liasis dubudingala*, a giant python that reached lengths of 10 m. In total, there were around 90 of these large taxa, drawn from a wide variety of different taxonomic groups.

It is difficult to imagine the first encounters of people with this amazing, but perhaps rather intimidating, fauna (Figure 3.4). Humans may have stood eye to eye with giant kangaroos and giant birds and watched wombats the size of a VW Beetle being attacked by giant snakes, lizards and crocodiles. Did people become prey items for 'drop cats' and monitor lizards? It seems likely. Recent evidence indicates that humans were present in northern Australia 65,000 years ago (Clarkson *et al.*, 2017). This colonisation process represents a remarkable achievement by itself; it must have involved numerous long water crossings of more than 65 km and is very likely to have required the construction of boats (Balme, 2013). It has been described as the first great maritime migration undertaken by anatomically modern humans (Norman *et al.*, 2018). But within about 20,000 years of this event, the Australian megafauna was extinct throughout the continent (Roberts *et al.*, 2001).

The causes of this extinction event have been the subject of a lively controversy, which is still ongoing. The principal causes that have been proposed are climate change and human impact, which could have involved overhunting and the use of fire. Initially, climate change was believed to have been the main factor responsible, specifically because of an increase in aridity at the Last Glacial Maximum at around 21 ka (Reeves *et al.*, 2013). This view began to be challenged with the discovery that *Genyornis newtoni* became extinct at around 50 ka (Miller *et al.*, 1999), based on precise dating of eggshells. This was followed by the suggestion, based on dating of remains at 28 sites, that all the megafauna went extinct at around 46.4 ka, a time that predates major climate change (Roberts *et al.*, 2001). When this research was conducted, it was believed that human colonisation of Australia first occurred at about 56 ka, implying that megafaunal extinction occurred within around 10 kyr of human arrival on the continent (Roberts *et al.*, 2001). As noted previously, more recent evidence indicates that initial colonisation occurred around 10 kyr earlier than this. However, dispersal of humans around the continent may have occurred significantly later, although this too is the subject of uncertainty and debate (Bird *et al.*, 2013, 2016).

These analyses were challenged subsequently based on their limited sample sizes and the difficulty of scaling up from a limited number of individual sites to the entire continent. While they provide clear evidence of localised extinctions, a continent-wide mass extinction is another matter entirely (Wroe, 2005). The picture is complicated by the fact that there is evidence of climatic instability from around 45 ka (Hope *et al.*, 2004), which was accompanied by a shift towards a glacial biome in Queensland (Barnosky *et al.*, 2004). Furthermore, some studies have indicated that humans continued to live alongside megafauna for many thousands of years. For example, Trueman *et al.* (2005) present a stratigraphic analysis of Cuddie Springs, an ephemeral lake in southeastern Australia, which indicates the coexistence of humans and now-extinct megafaunal species for a minimum of 15 kyr. Similarly, Westaway *et al.* (2017) examine the occurrence of *Zygomaturus trilobus*, a close relative of *Diprotodon*, in another site in south-eastern Australia, namely the Willandra Lakes Region. This is a unique site, as it provides evidence of continual human occupation from around 50,000 years ago, together with the presence of megafauna. Results indicate a period of coexistence of some 14–17 kyr.

Such evidence contradicts the idea that overhunting by humans led to rapid extinction of megafauna soon after human arrival, as proposed by the overkill, or 'blitzkrieg', hypothesis (Martin, 1984). Instead, these results suggest that a combination of climate change and human impact was responsible for the extinction event, with increased interactions between humans and megafauna occurring during a period of increasing aridity, as both were increasingly restricted to refugia surrounding waterbodies (Westaway *et al.*, 2017). Yet the debate rumbles on. Saltré *et al.* (2016) describe a metadata analysis of 659 Australian megafauna fossil ages, which showed that extinctions were broadly synchronous among genera and independent of climate aridity. On this basis, these authors rejected climate change as the primary driver of megafauna extinctions in Australia. They estimate that the megafauna disappeared from across the continent within around 13,500 years after the arrival of humans, with shorter periods of coexistence occurring in some regions. This same research team subsequently came to a different conclusion, after developing a statistical modeling approach to examine spatial variation in extinction patterns in relation to human appearance in south-eastern Australia (Saltré *et al.*, 2019). Results suggested that both climate change and human persecution were important factors; human populations may themselves have been constrained by the spatial distribution of climate-dependent water sources. Further, Wroe *et al.* (2013) argue in favour of climate change as a significant factor, suggesting that many species were lost before the arrival of people.

The nature of this debate highlights the problem of identifying the causes of extinction from the fossil record, given the uncertainties associated with chronologies, the lack of securely dated fossil material and the biases in the fossilisation process. Wroe *et al.* (2013) point out that there is no direct evidence of human activities as a primary cause of extinction from either the palaeontological or archaeological record, as there is no evidence of predation or consumption of megafauna by people (although Miller *et al.* [2016b] provide evidence of *Genyornis* eggs being cooked). The evidence for human impact is largely circumstantial. It may therefore prove impossible to determine the relative importance of climate change and human activity as extinction drivers. According to Wroe

(2005), extinctions are seldom attributable to a single cause, and the loss of the Australian megafauna is therefore likely to be attributable to both of these factors, which may have acted interactively. On the other hand, after similarly weighing the available evidence, Johnson *et al.* (2016) concluded that humans are primarily to blame.

Whatever caused the extinction event, it may have had a dramatic effect on Australian ecosystems. Miller *et al.* (2005) suggested that human impacts on the landscape resulted in ecosystem collapse. Specifically, systematic burning practiced by human colonisers may have converted a drought-adapted mosaic of trees, shrubs and grasslands to the fire-adapted grasslands and chenopod/desert scrub that still prevails in semi-arid areas of Australia (Miller *et al.*, 2005). This shift in vegetation composition and structure may have contributed significantly to the loss of megafauna. These suggestions were based on isotopic analysis of avian eggshells and marsupial teeth, enabling analysis of dietary changes coincident with the putative ecosystem collapse. Analysis of eggshells was conducted on two flightless bird species, the extant Australian emu *Dromaius novaehollandiae* and the extinct *Genyornis newtoni*. Results indicated that before 50 ka, *Dromaius* consumed a wide variety of food plants, consistent with being an opportunistic feeder living in an environment with high variation in moisture availability. The isotopic data suggested that in relatively moist years, *Dromaius* fed on grass, whereas in drier years, it relied more on shrubs and trees. Most significantly, the diet of this species appears to have become restricted after 45 ka, a pattern that persisted throughout the Holocene. In contrast, the diet of *Genyornis* was more restricted, suggesting that it was a more specialised feeder than *Dromaius*. These food resources appear to have become less available during the period 50 ka to 45 ka, suggesting a major reorganisation of vegetation communities across the Australian semi-arid zone. To test this idea, these authors then measured isotopic values in wombat teeth. Results suggested that before 50 ka, C4 plants accounted for 40% to 100% of the wombat's dietary intake, whereas after 45 ka, their diet was dominated by C3 plants, supporting the conclusions derived from *Dromaius* eggshells (Miller *et al.*, 2005). This evidence for major ecosystem reorganisation is tentatively attributed by these authors to a change in the fire regime resulting from human activity, because other factors, such as overhunting or disease, would not have caused the changes in the food webs that were observed. In the case of *Genyornis* at least, dietary specialisation may have increased its vulnerability to extinction.

These observations were further extended by Miller *et al.* (2016a), who suggested that human activities may also have resulted in a change in the relationships between climate and vegetation across the arid zone of Australia, resulting in increased aridity and reduced ecosystem diversity. For example, after 45 ka, the large lakes of interior Australia that had previously been maintained by summer rainfall, failed to refill. While the mechanisms for these changes are not understood fully, there is some evidence for replacement of trees and shrubs by C4 grasses after 50 ka, as would be expected with an increased frequency of burning. However, sedimentary charcoal records provide no evidence of a change in fire regime following human colonisation (Mooney et al., 2011). This could be attributable to a lack of charcoal records from Australia's arid zone (Miller et al., 2016a), or other limitations of the technique of charcoal analysis (Bird et al., 2013), so the evidence is not entirely clear-cut. If the fire regime did change, this provides a potential mechanism for the observed dietary shifts in *Dromaius* and the expansion of fire-adapted grasses (Miller et al., 2016a).

More recent research provides further evidence of ecosystem collapse linked to this extinction event. Van der Kaars et al. (2017) describe analysis of a sediment core spanning some 150,000 years, obtained 100 km offshore in south-western Australia. Analyses included changes in the number of spores from the dung fungus *Sporormiella*, which was used as a proxy for herbivore biomass. High abundance of *Sporormiella* was observed from 150 to 45 ka, which suggests that megafaunal populations were present throughout this period, under a range of different climatic conditions ranging from warm, wet, interglacials to more arid glacial periods. While changes in vegetation and fire regime were noted around 70 ka associated with an increase in aridity, *Sporormiella* displayed a marked and irreversible decline from 45 to 43.1 ka, which was interpreted as evidence for megafaunal population collapse. These results suggest that human activities such as hunting, rather than increased aridity or habitat change, were probably the main cause of extinction (van der Kaars et al., 2017). However, the potential impacts of the extinction on vegetation composition were not explored.

A similar decline in dung fungi at around 41 ka was observed by Rule et al. (2012) at Lynch's Crater palaeolake in north-east Australia. Before this decline, vegetation was a mixture of rain forest, dry sclerophyll forest and grassland, with little evidence of fire. The decline in dung fungi was followed by a rapid increase in the quantity of charcoal and a gradual replacement of the original rain forest vegetation by a more uniform sclerophyll forest. The increase in fire may be attributable to an accumulation of fuel following the decline in herbivory; the change in vegetation composition may reflect both the shift in the fire regime and release from herbivore pressure. The increase in fire could also have been anthropogenic in origin. According to these authors, this ecosystem shift was as large as any effect of climate change over the last glacial cycle (Rule *et al.*, 2012). Such vegetation changes were not always associated with loss of the megafauna, however, indicating that patterns of ecosystem response varied regionally (Johnson *et al.*, 2016).

The hypothesis that humans may have altered both vegetation and climate following colonisation of Australia is considered further by Bird *et al.* (2013). These ideas actually have a long history. For example, Flannery (1990) proposed that extinction of the megafauna was caused by 'blitzkrieg' overhunting, which would have led to a subsequent increase in fuel availability owing to growth of vegetation following the decline in herbivory. Humans may have responded to this through the innovation of 'firestick farming', the regular burning practice of Indigenous Australians to facilitate hunting and to change the composition of plant and animal species in an area. This may have favoured fire-adapted plant species and some smaller mammals but may have caused an extinction cascade involving loss of those species with life histories incompatible with frequent burning (Bliege Bird *et al.*, 2008). Potential links with climate have also been suggested. For example, Miller and Magee (1992) hypothesised that the introduction of an anthropogenic fire regime could have reduced tree or shrub cover across tropical Australia, which could have reduced the incidence of monsoonal rains into the continental interior, leading to increased aridity and the desiccation of lakes.

Such ideas have stimulated a substantial research effort in the ensuing decades, yet a consensus is yet to emerge regarding their veracity. In reviewing the evidence gathered to date, Bird *et al.* (2013) draw a colourful analogy between megafaunal extinction and Gregori Rasputin, a controversial member of the court of the Russian tsar, whose ultimate demise is still the subject of debate. Was he poisoned, shot or clubbed to death, or all three? Or did he survive these initial attempts on his life, before finally succumbing to drowning after being dumped through a hole in the ice on the Neva River? The fate of the megafauna and the ecological impacts of their demise are similarly mysterious and multifaceted. Bird *et al.* (2013) conclude their very considered review as follows:

- It has been suggested by some modeling studies that forest vegetation can potentially exist over a larger area of tropical Australia than is currently the case, where local soil conditions allow;

- Current climate and fire regime clearly favour the maintenance of open vegetation across much of tropical Australia;
- A number of potential feedbacks, some highly non-linear, have been suggested to operate between vegetation and climate, potentially resulting in alternative states of climate and vegetation;
- A change in fire regime theoretically provides one mechanism by which one state may be advantaged over another;
- There is evidence that fire regime has been manipulated by humans to some degree since their arrival in tropical Australia;
- Ecological processes and vegetation-fire-climate-human feedbacks exist that could have driven a significant shift in ecosystem state at the subcontinental scale through the sustained imposition of an anthropogenic fire regime over tens of millennia;
- These potential feedbacks operate through the inhibition of forest expansion both directly, by targeted burning at established forest edges and newly irrupted forest patches, and indirectly, through lengthening of the dry season because of changes to the timing of burning;
- The impact of any such anthropogenic forcing may have been entirely overshadowed by the effects of natural climate change and variability, as well as the generally low nutrient status of Australian soils;
- Although appropriate mechanisms have therefore been identified, insufficient evidence is currently available to rigorously test whether or not ecosystem collapse was driven by human modification of the fire regime.

3.3 Megafauna Extinctions in the Late Quaternary

One of my favourite films is *Cave of Forgotten Dreams*, a documentary by Werner Herzog about the Chauvet Cave in southern France. Discovered in 1994, this cave contains some of the best preserved and most impressive cave paintings in the world, which date back to the Upper Palaeolithic. Some of the artworks are believed to be at least 28,000 years old, although even greater ages of 30,000–32,000 years have been suggested (Pettitt and Bahn, 2015). As Herzog says in the film, seeing these paintings is like looking back into the abyss of time. The paintings provide an insight into the ecosystems of the last Ice Age and include hundreds of depictions of animals of at least 13 different species, including cave lions, mammoths, aurochs, cave bears, bison, owls, rhinos and hyenas. Some of the paintings are of truly exceptional quality and have

Table 3.5 *Continental extinctions of mammalian megafauna at the generic level*

	Number of globally extinct genera	Number of extinct genera surviving on other continents	Number of Holocene survivors	% extinct
Africa	7	3	38	21
Australia	14	–	2	88
Eurasia	4	5	17	35
North America	28	6	13	72
South America	48	2	10	83

Source: Reproduced from Koch and Barnosky (2006) with permission

been described as great masterpieces of human art despite being among the oldest artworks known. What is most haunting is that many of the species depicted are now extinct. The paintings therefore preserve an eyewitness account of a fauna that has long since vanished.

In the Late Quaternary, between 50,000 and 10,000 years ago, most large terrestrial mammals became extinct in all parts of the world except Africa. In total, around 90 genera of mammals weighing ≥44 kg were lost (Koch and Barnosky, 2006). In addition to the European animals depicted in caves such as Chauvet, there was diverse megafauna in North America, which included giant ground sloths, short-faced bears, tapirs, peccaries, the American lion, sabre-toothed cats such as *Smilodon* and the scimitar cat, camels, cheetahs, dire wolves, mastodons, mammoths and a giant beaver, among others. These animals may have numbered in their tens of millions (Pardi and Smith, 2016). Similarly, South America was home to another distinctive megafauna, including the car-sized glyptodont, the giant ground sloth, short-faced bears, the elephant-like *Stegmastodon* and the three-toed litoptern, which resembled a horse with a camel's neck and a short elephant-like trunk (Koch and Barnosky, 2006). Overall, more than 70% of these genera were lost in both North and South America (Table 3.5).

As with the case of Australia discussed in the previous section, megafauna extinctions in other continents have been attributed to two main causes: climate change and human activity (Figure 3.5). Again, the issue is controversial and has generated a substantial literature. This was comprehensively reviewed by Koch and Barnosky (2006), who summarised their main conclusions as follows:

3.3 Megafauna Extinctions in the Late Quaternary · 129

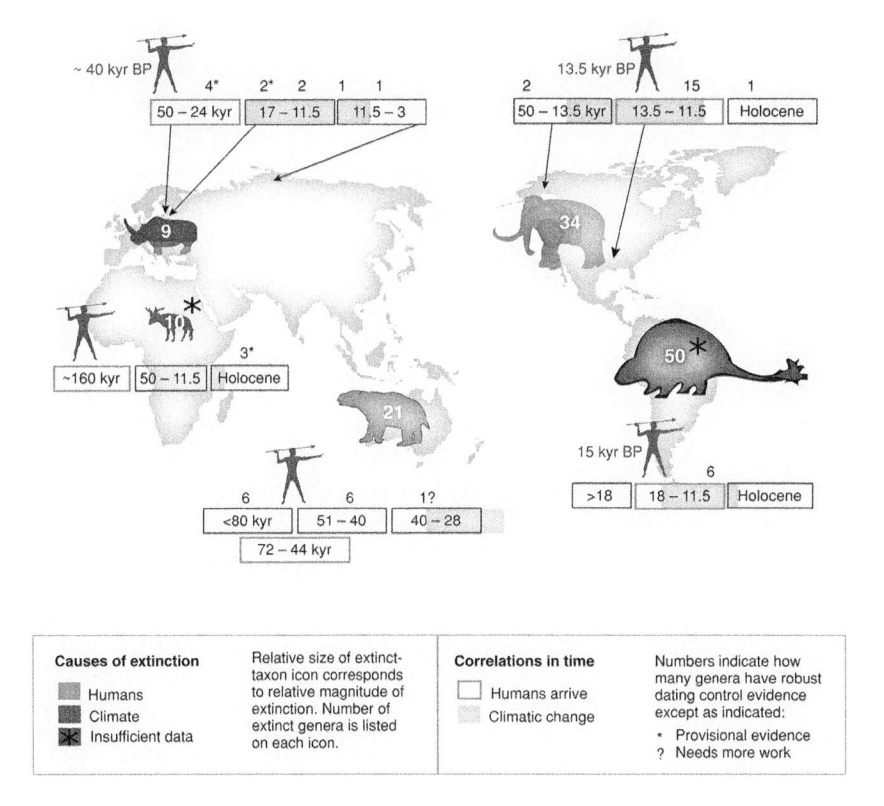

Figure 3.5 The chronology of Late Quaternary extinction, climate change and human arrival on each continent. Reproduced from Koch and Barnosky (2006) with permission.

- The Late Quaternary extinction, with its extreme focus on large and slow-breeding animals, was unusual relative to extinctions earlier in the Cenozoic.
- The unusual body-size selectivity of the extinction and its rough synchrony with the global geographic expansion of modern humans are compelling evidence that the extinction was precipitated by human activities, especially hunting.
- Climate change likely affected the timing, geography and, perhaps, magnitude of this anthropogenically triggered extinction.
- The intersection of rapid climate change with initial human contact seemed especially deadly for megafauna.
- It is an oversimplification to say that an abrupt wave of hunting-induced extinctions swept continents immediately after first human

contact. This 'blitzkrieg' hypothesis can now be firmly rejected in Western Europe, Siberia, Alaska and probably central and eastern North America.
- Taken as a whole, recent studies suggest that humans precipitated the extinction in many parts of the globe through combined direct (hunting) and perhaps indirect (competition, habitat alteration and fragmentation) impacts.

These conclusions have been supported by some subsequent research. For example, Araujo *et al.* (2017) conducted a high-resolution quantitative analysis employing recent advances in fossil dating and climate modeling and concluded that the timing of human arrival in each geographical region was the best explanation for the megafauna extinctions observed. The effects of climate were generally found to be additive, rather than synergistic with human arrival. In South America, Metcalf *et al.* (2016) similarly concluded that it was when human presence and climate warming coincided that the extinction occurred. Sandom *et al.* (2014) provided a global analysis of all mammal extinctions since the previous interglacial and concluded that the severity of extinction was strongly linked to human presence, with a relatively weak association with climate change. Barlett *et al.* (2016) reached very similar conclusions following an analysis of regional climate data. The case is not completely closed, however. Rabanus-Wallace *et al.* (2017) presented some intriguing new research examining the influence of environmental changes on megaherbivores through the analysis of nitrogen isotopes in bone collagen, which provides a proxy for moisture availability in their rangeland habitats. The end of the last Ice Age was associated with an increase in moisture availability, which led to the widespread formation of wetland habitats and the consequent decline of arid grassland and steppe vegetation. This loss of habitat may have contributed to megafauna extinctions and may also explain why proportionally fewer African megafauna became extinct at this time (Table 3.5), as large areas of rangeland persisted in Africa after the end of the Ice Age owing to its trans-equatorial position.

What impact did the extinction of these megafauna have on the structure and functioning of ecosystems? The answer to this fascinating question depends critically on the ecological role of these animals, an issue that has latterly attracted significant interest from researchers. Most of the animals that became extinct were herbivores. Following their extinction, changes in the structure, composition and dynamics of vegetation might therefore have occurred. Furthermore, as argued by Owen-Smith (1988), loss of mega-sized herbivores might be expected to have led to mega-sized effects. In other words, these animals were likely to have been keystone organisms in the communities of which they were a part. Based on analysis of contemporary vegetation, three principal effects might be anticipated (Johnson, 2009):

(i) *Loss of open vegetation and habitat mosaics*, as a result of successional processes occurring that had previously been kept in check by herbivory.
(ii) *Increased fire*, because of an increase in fuel loads resulting from increased availability of plant biomass.
(iii) *Decline of coevolved plants*, such as those species dependent on megaherbivores for dispersal.

Evidence for the first of these impacts is available for Ireland, where the giant deer *Megaloceros giganteus* became extinct around 12 ka because of climatic deterioration. Extensive grasslands appear to have been maintained by this species, which were progressively replaced by shrubs and woodland after it became extinct (Bradshaw and Mitchell, 1999). A larger-scale example is provided by the 'mammoth steppe'. During much of the Pleistocene, extensive areas of northern Eurasia and North America were covered by extensive steppe vegetation composed of grasses, forbs and sedges, which supported a high biomass of large herbivores including woolly mammoths, bison, woolly rhinoceros and horses – hence the name (Johnson, 2009). After about 13 ka, this vegetation was largely replaced by wet tundra, shrubland and forest, although some localised areas are still visible today in places such as the Altai Mountains of Siberia (Chytrý *et al.*, 2018).

The idea that mammoths and other large herbivores actually maintained the 'mammoth steppe' was proposed by Zimov *et al.* (1995), on the basis of both modeling and field evidence of contemporary ecosystems. High grazing intensity would have favoured dominance of forbs and grasses rather than mosses, which increased rates of transpiration and reduced soil moisture, further favouring grass growth over mosses. Conversely, a decline in grazing would have led to increased soil moisture and increased dominance of mosses (Figure 3.6), leading to the expansion of bogs and mires. The evidence presented by Zimov *et al.* (1995) suggests that the grazing impacts of large herbivores were sufficiently large that their extinction could have led to the widespread shift from steppe to tundra observed at the end of the last Ice Age. These ideas

132 · Case Studies from Prehistory

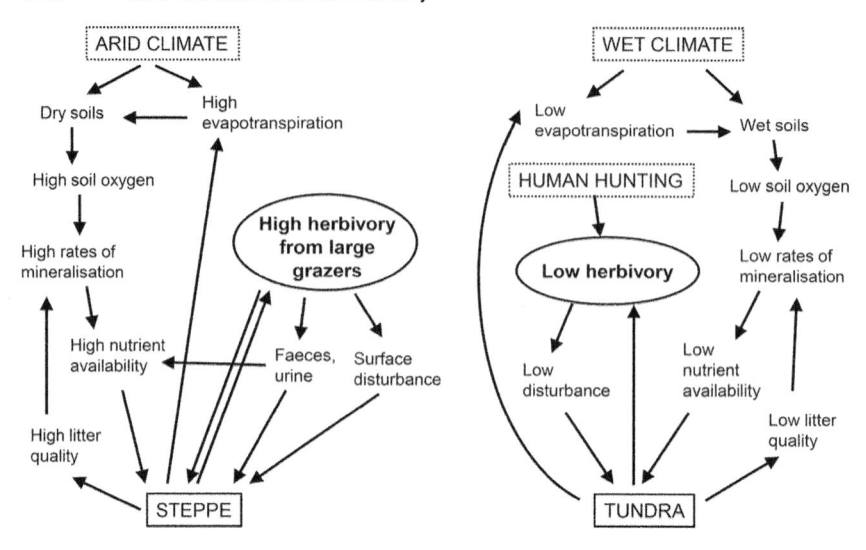

Figure 3.6 Proposed mechanisms for the shift from steppe to tundra ecosystems driven by extinction of large herbivores. Adapted from Zimov *et al.* (1995).

have been supported by some palaeoecological evidence, such as the sudden increase in tree cover coinciding with the loss of mammoths at around 13.6 ka observed in south-western Alaska (Johnson, 2009). However, the effects of climate change and megaherbivore extinction on vegetation are very difficult to disentangle.

North America provides some further examples of how ecosystems may have been affected by megafauna extinction. Analysis of fossil pollen showed that since full-glacial times (21–17 ka), the fastest change in vegetation structure and composition occurred between 13 and 10 ka (Williams *et al.*, 2004), in the immediate aftermath of megafaunal extinction (Johnson, 2009). Relatively open vegetation communities, such as 'spruce parkland' and 'mixed parkland', were relatively widespread in parts of North America in the late glacial but subsequently declined and are uncommon or absent today. Although their loss could be attributable to climate change, it is also possible that megafauna helped to maintain their open parkland character prior to the extinction event (Williams *et al.*, 2001).

Similarly, Robinson *et al.* (2005) presented palynological analysis of four Late Quaternary deposits in New York State, USA, and found evidence of landscape-scale vegetation change that was attributed to human activity. Specifically, a region-wide decline in the dung fungus *Sporormiella* was followed by a rise in charcoal concentrations, indicating

3.3 Megafauna Extinctions in the Late Quaternary · 133

increasing fire frequency following the decline in herbivore populations. This evidence is consistent with the idea that an increase in fire frequency resulted from an increase in plant biomass that accumulated as a result of megaherbivore extinction. These changes in fire and herbivory regimes occurred before the climatic reversal associated with the Younger Dryas (around 12 kPa) and were interpreted as an example of ecological collapse driven by humans that took at least a millennium to complete (Robinson *et al.*, 2005). Although the dominance of conifer forest that had previously prevailed in the area continued after the extinction event, there was some evidence of compositional change in the vegetation, and also a shift in the nature of sediments, from clays to darker organic muds and peats. This may represent an increased input of organic matter to sediments, perhaps reflecting an increase in vegetation density and an accumulation of plant material, which may have been accompanied by a raised water table (Johnson, 2009).

Barnosky *et al.* (2016) extended these analyses through a compilation of palaeoecological evidence from five different regions in the Americas: south-western Patagonia, the Pampas, north-eastern United States, north-western United States, and Beringia. This evidence was explored in relation to persistent ecological state shifts, or regime changes, which are interpreted by some authors as a definition of ecosystem collapse (see Chapter 2). Major ecological state shifts associated with megafaunal extinction were identified in the North American sites but not in the South American ones, highlighting the importance of regional differentiation in the responses of vegetation to these extinction events. Based on this finding, Barnosky *et al.* (2016) identified two factors necessary for defaunation to trigger lasting ecological state shifts: (i) the megafauna that are lost need to have been effective ecosystem engineers, such as proboscideans; and (ii) the ecosystem needs to be characterised by plant species likely to be released by declining herbivore pressure. For example, in North America, it was only after mammoths and/or mastodons declined substantially that denser understory vegetation and deciduous forests began to flourish; in southern Patagonia, no such ecosystem engineers were ever present. In the Pampas, although an ecosystem engineer capable of damaging or killing trees was present, namely *Notiomastodon*, either the soil or climatic regimes remained unsuitable for forest establishment. Consequently, no major change in the structure of the vegetation followed the megafaunal extinction events in South America. Furthermore, these results suggest that the ecosystems most likely to shift into new regimes are those that are simultaneously impacted by both climate change and a major disturbance event, such as defaunation (Barnosky et al., 2016).

With respect to the impact of the extinction event on co-evolved plants, Janzen and Martin (1982) suggested that the large, fleshy fruits of some neotropical plants were an adaptation to dispersal by megaherbivores, examples including the papaya (*Cerica papaya*), avocado (*Persea americana*), honey locust (*Gleditsia triancanthos*), pawpaw (*Asimina triloba*) and Osage orange (*Maclura pomifera*) (Barlow, 2002). In their initial survey, Janzen and Martin (1982) identified about 39 species of woody plant in the lowland deciduous forest of Costa Rica as likely to have been dispersed by megafauna such as gomphotheres, whereas Guimarães et al. (2008) noted 103 neotropical fruit species that appear to demonstrate this same mode of dispersal. According to these authors, megafauna extinction may have had a number of potential consequences for plant species by disrupting dispersal, including the reduction of seed dispersal distances, increasingly clumped spatial patterns, reduced geographic ranges, reduced genetic variation and increased among-population structuring. Plant species of Central and North America with this form of dispersal typically have restricted distributions in lowlands and flood plains, suggesting that they may now be reliant on gravity and water for dispersal of their large seeds (Barlow, 2002; Janzen and Martin, 1982). Modeling-based analyses suggest that long-distance dispersal of South American plant species may have contracted by at least two-thirds after the loss of megafauna (Pires et al., 2018), which may also have caused significant changes to dispersal networks (Pires et al., 2014). In addition, there is evidence that megafauna-dispersed tree species have smaller size ranges than other animal-dispersed trees, perhaps reflecting decreased seed dispersal distance following the megafauna extinctions (Doughty et al., 2016), although this study is based on a lot of guesswork. Furthermore, Campos-Arceiza and Blake (2011) emphasised the important role of extant elephants as seed dispersers, noting that they maintain tree diversity and make a significant contribution to ecosystem function. Their extinct relatives may have fulfiled a similar role in the past.

As noted by Doughty (2013), the extinctions of the Pleistocene megafauna may have led to a cascade of other extinctions of coevolved species, including trees and insects. For example, there is evidence that the loss of megafauna led to the extinction of a number of species of dung beetle, which may have made a significant contribution to a number of ecological processes, including seed dispersal (Doughty, 2013). Such cascading effects may have affected ecosystem processes, including nutrient cycling and interactions with climate (Gill, 2014). It has been suggested that the high productivity of the 'mammoth steppe' may be attributable to the effect of megaherbivores on nutrient cycling (Zimov et al., 1995, 2012), as occurs today in modern African savannas. The extinction of megafauna may therefore have disrupted nutrient cycling by reducing the mobility of nutrients, which are distributed by animals through their bodies and faeces (Doughty, 2013). Recent research has documented the role of animals in translocating nutrients over a variety of scales. For example, in Alaska, a significant proportion of the nitrogen uptake by spruce trees was found to be derived from salmon consumed and translocated by brown bears (Hilderbrand et al., 1999). Similarly, in Amazonia, substantial quantities of phosphate were found to be transported by woolly monkeys through the process of dispersing seeds (Stevenson and Guzman-Caro, 2010).

The possible role of extinct megafauna in such nutrient transfers was examined by Doughty et al. (2013), using a modeling approach based on some rather heroic assumptions. Results suggested that extinction of the Amazonian megafauna may have decreased the transfer of phosphorus by more than 98%, with somewhat lower declines postulated for other geographical regions. The legacy of this impact may still be felt; it is possible that the current phosphorus limitation in the Amazon basin may be partially attributable to the loss of this megafauna. These authors suggested that the Pleistocene megafaunal extinctions resulted in large and ongoing disruptions to terrestrial biogeochemical cycling at continental scales. While this is a possibility, these findings are associated with a very high degree of uncertainty, and palaeoecological evidence for them is currently lacking. While substantial changes in the global nitrogen cycle have been documented at the end of the Pleistocene (McLauchlan et al., 2013), the role of animal extinctions in these changes is unclear.

Given a fertile enough imagination, it is also possible to conceive that the megafaunal extinctions may have influenced global climate. For example, Doughty et al. (2010) argue that the expansion of birch (Betula spp.) in both Alaska and Siberia at around 13.8 ka, which they attribute to the extinction of mammoths, could have increased regional temperatures by up to 1°C through modification of land surface albedo. On the other hand, the spread of birch could have been completely driven by changing climate; once again, the effects of megaherbivore extinction and climate change are difficult to separate. The same issue was examined by Brault et al. (2013) using an integrated climate and vegetation model. Results indicated the potential for a small but 'non-trivial' effect of megafauna extinction on climate, with local-scale warming exceeding 0.3°C in some locations. These authors did however identify a potential feedback mechanism, whereby the replacement of high-latitude tundra with forest led to a release of soil carbon and consequently a small increase in atmospheric CO_2 (Brault et al., 2013). It is also possible that extinction of the megafauna led to a reduction in methane emissions (Smith et al., 2016), which could have affected global climate (Doughty, 2013), although again these calculations are highly uncertain.

Other putative impacts of megafauna extinction include changes in trophic structure and community assembly (Malhi et al., 2016). For example, van Valkenburgh et al. (2015) suggested that the diversity of megacarnivores in the Pleistocene may have had an influence on megaherbivore populations through predation of juveniles, suggesting the possibility of top-down trophic control. This is supported by analysis of the Hall's Cave site in Texas, which showed that a fundamental change occurred during the late Pleistocene in the composition of herbivores, with large grazers being replaced by smaller frugivores or granivores from about 15 ka (Smith et al., 2016). These results also demonstrated significant positive associations between megacarnivores and megaherbivores at this site, suggesting that these species were interacting through tightly linked predator–prey relationships. Considering evidence available at the global scale, the observed changes in body size distribution suggest a fundamental change in energy flow as well as ecological interactions following the extinction event (Malhi et al., 2016). Loss of herbivores also had demonstrable impact on niche separation and competition between carnivores and on overall guild connectivity between trophic levels (Pardi and Smith, 2016).

Although loss of this wondrous fauna has understandably attracted the attention of researchers, we should not let this blind us to the role of more prosaic ecological processes that might also have been influential during the early Holocene. Jeffers et al. (2018) present detailed results from five sites in Britain and Ireland spanning the full period of Late Quaternary megafauna extinction in northern Europe. In these ecosystems, the loss of megaherbivore species and declines in herbivore biomass were coincident with or preceded an increase in both shrub and tree biomass. These authors suggested that a loss of open or grassland vegetation caused by ecological succession to woodland, perhaps facilitated by the presence of shrubs, may have helped drive the herbivores to extinction. Regression analysis indicated that plant growth forms were influential in driving shifts in nitrogen availability and plant community composition across the extinction event, rather than simply the presence or absence of megaherbivores. These results indicate a relatively high level of plant control on ecosystem structure and function and suggest that interactions between plant growth forms and plant–soil feedbacks may have been more important than trophic interactions in determining changes in terrestrial nitrogen availability and above-ground plant biomass in these ecosystems (Jeffers *et al.*, 2018). In other words, the megaherbivores in these ecosystems were not able to prevent the expansion of woody plants at the onset of postglacial warming.

The idea that successional changes in vegetation, driven by climatic warming, may have caused the extinction of at least some megafauna is supported by other research. The woolly rhinoceros, *Coelodonta antiquitatis*, was another impressive animal that, at the height of the last Ice Age, ranged from Spain and Britain in the west, across central Europe and Asia to north-east Siberia and northern China. Stuart and Lister (2012) present an extinction chronology of the species, which charts a sad process of decline in geographic range. From ca. 35 ka, it appears to have contracted towards the east, eventually becoming restricted to north-eastern Siberia, with the last populations becoming extinct in this location at around 14 ka. Significantly, although it is pictured in Palaeolithic art such as at Chauvet Cave, there is little evidence that the species was hunted by humans; there was very limited geographical overlap between the distribution of humans and woolly rhinoceros after 20 ka and little archaeological evidence of its consumption by people. Stuart and Lister (2012) consider that its final extinction is probably attributable to Late Glacial Interstadial warming and increased precipitation resulting in the replacement of low-growing herbaceous vegetation by shrubs and trees. The fact that the species survived longest in north-eastern Siberia may reflect the later persistence of open vegetation in that region. The fact that this species became extinct at the end of the most recent Ice Age, rather than in previous interglacials, is attributed to the fact that the Late Glacial Interstadial was significantly warmer than any other event in the previous 50 millennia (Stuart and Lister, 2012). The rapid climate changes associated with interstadial warming events have also been implicated in the extinction of a number of other species of megafauna (Cooper *et al.*, 2015).

These observations are further supported by the work of Allen *et al.* (2010), who, by using a dynamic vegetation model, noted a marked and rapid decline in the annual net primary productivity of mesophilous herbs shortly after the Last Glacial Maximum (25–15 ka), especially in western Eurasia. This may have contributed to the extinction of a number of megaherbivores, including the woolly rhinoceros (Allen *et al.*, 2010). Similar conclusions were reached by Willerslev *et al.* (2014) based on a DNA metabarcoding study of Arctic plant diversity. For much of the past 50,000 years, Arctic vegetation predominantly consisted of dry steppe-tundra, which was forb rich. Diversity declined markedly during the Last Glacial Maximum, although forbs remained dominant. After 10 ka, the vegetation changed markedly, with the appearance of moist tundra dominated by grasses and woody plants. DNA analysis of the gut contents and dung of extinct megaherbivores indicated that forbs made up a large part of their diet, suggesting that it was the forbs rather than grasses that were key components of the 'mammoth steppe'. It is possible that there may have been a feedback loop between megaherbivores and the vegetation, with activities of the animals such as foraging and trampling providing opportunities for recruitment of the relatively nutritious forbs, which thereby helped to maintain the high herbivore populations. This feedback may have broken down after the end of the Ice Age, with climatic warming and the decline in herbivores acting together to accelerate the expansion of moist tundra (Willerslev *et al.*, 2014). This therefore provides a potential mechanism for the kind of regime shift postulated by alternative stable state theory (Bowman *et al.*, 2015; see Chapter 2) and the collapse of the 'mammoth steppe' ecosystem.

3.4 Holocene Examples of Ecosystem Collapse

The Holocene refers to the current geological epoch, beginning around 11,650 years ago at the end of the last Ice Age. This has been a period of unprecedented change in the global environment, associated with the growth and development of human societies and their major civilisations, and the initiation and spread of agriculture and urbanisation. Human impacts on Earth's ecosystems have been profound and pervasive, as illustrated by the recent debate on the Anthropocene (Crutzen, 2006). This concept is based on the idea that the Earth system has changed so dramatically as a result of human activities that it is detectable as a significant alteration of the stratigraphic record and therefore merits recognition as a new geological epoch. The environmental changes that have excited some stratigraphers include pronounced increases in erosion

3.4 Holocene Examples of Ecosystem Collapse · 139

and sedimentation rates; large-scale perturbations to carbon, nitrogen and phosphorus cycles, together with those of other elements; widespread use and dissemination of plastics; contamination of the environment with radionucleotides; the initiation of significant change in global climate and sea level; biotic changes, including major biodiversity loss and widespread introduction of invasive species. Many of these changes are considered to be geologically long-lasting, and some are essentially irreversible (Zalasiewicz et al., 2017a,b). Despite these facts, and intense lobbying, the stratigraphers who are in charge of formally naming geological epochs have so far ignored the Anthropocene (Walker et al., 2018), much to the bemusement of many commentators and the chagrin of some researchers. Perhaps their decision represents a fit of pique because the concept was not first developed by geologists. Nonetheless, given its popularity both in scientific discourse and in the popular media, the Anthropocene is here to stay – after all, it even has a journal devoted to it now.

Regardless of how it is defined, this era of intensifying human impact has undoubtedly been associated with widespread ecosystem collapse. As this section focuses on prehistory, we are reliant on evidence from palaeoecology and archaeology, coupled with recent advances in the analysis of ancient DNA. Together, these approaches have recently provided new insights into the origins and early spread of agriculture, which should surely be considered the most important driver of ecosystem collapse during the Holocene. The domestication of different plant and animal species occurred in at least 10 different parts of the world, where both natural resources and human populations were concentrated (Price and Bar-Yosef, 2011). Of these, south-west Asia is perhaps the most important of these in terms of its subsequent environmental impact. In the central Fertile Crescent, especially the upper reaches of the Tigris and Euphrates rivers, there was a remarkable coincidence of species with high food value for people, leading to the initial domestication of a number of founder crop plants (einkorn, emmer, pulses) and four livestock animal species (sheep, pigs, cattle and possibly goats) (Zeder, 2011). Recent genetic and archaeobiological evidence has painted an increasingly complex picture, with the emergence of agriculture in the Near East now being seen as having taken place in many locations across the entire region, domestication sometimes having occurred multiple times in the same species (Ibáñez, et al., 2018; Zeder, 2011). The outcome was a suite of domesticated plant and animal species that supported fully agrarian livelihoods, sometimes referred to as the 'Neolithic package' (Çilingiroğlu, 2005).

DNA analysis has proved remarkably useful for understanding the subsequent spread of farming. For example, DNA sequences from early farmers have provided the amazing insight that agriculture spread into Europe not simply through the dissemination of ideas but through the migration of people who took their crops and livestock with them (Hofmanová et al., 2016; Lipson et al., 2017). They also brought other items whose descendants are familiar features of modern culture, including bread (Arranz-Otaegui et al., 2018), wine (McGovern et al., 2017) and beer (Sewell, 2014). This has led to the compelling suggestion that the discovery of beer may have led to the rise of human civilisation, encouraging mankind to adopt an agricultural lifestyle in order to secure a steady supply of beer-making ingredients (Standage, 2005). If you are sceptical of this beer-oriented narrative, don't forget that throughout history, inebriation is one of the defining features of human society (Guerra-Doce, 2015). Regardless of their primary objectives, whether to eat or make merry, farmers created a completely new relationship with the environment, which contrasted markedly with that of hunter-gatherers. Rather than using a diversity of resources over a wide area, farmers use the landscape intensively and modify it profoundly, through the introduction of domesticated crops and livestock and the conversion of natural vegetation into cultivated fields and pastures (Price and Bar-Yosef, 2011).

Kirch (2005) summarises the archaeological evidence regarding the environmental impacts of early farming, noting that after the rise of agriculturally based societies during the early Holocene, humans have had cumulative and often irreversible impacts on natural landscapes and biotic resources worldwide. Substantial archaeological evidence has been accumulated of environmental impacts in prehistory, including deforestation, spread of savannas, increased rates of erosion, permanent rearrangements of landscapes for agriculture and resource depression and depletion. In this review, Kirch (2005) makes the following points:

- The environmental impacts of early agrarian populations were orders of magnitude greater than those of the smaller-scale populations of the Pleistocene. Impacts included forest clearance for fields, agricultural burning, erosion and/or soil nutrient depletion owing to overly intensive cultivation and exploitation of wood resources for fuel and construction. These resulted in cumulative and irreversible modifications of landscapes over increasingly large parts of tropical, subtropical and temperate regions. Impacts progressively expanded from local to landscape and then to regional scales.

- In many areas, the cumulative impacts of agricultural (and pastoral) economies led to widespread degradation of landscapes. An example is provided by the third millennium BC in the Near East, where the cumulative effects of 4,000–6,000 years of farming and herding precipitated social collapse over much of south-western Asia. Archaeobotanical remains from Mesopotamia and elsewhere in the Near East demonstrated major forest declines in the third millennium. Similarly, it has been argued that agriculture-related deforestation and erosion were associated with the demise of the Classic Maya civilisation and with later collapse of agricultural systems in the Tehuacán Valley of Mexico ca. 900 BP.
- Early Holocene settlements in the Levant systematically deforested the landscapes around their villages to provide fuel. Slag deposits from smelting metal ores around the Mediterranean littoral in antiquity represent the clearance of at least 50–70 million acres of trees.
- The islands of Remote Oceania, which were settled by humans between 3,200 and 800 years BP, provide a series of case studies demonstrating how rapidly preindustrial societies can effect massive deforestation through agricultural burning. For example, the interior of Mangaia in the southern Cook Islands was cleared of forest soon after Polynesian arrival. Easter Island lost its forest cover over the course of no more than 600–700 years. In the Hawaiian archipelago, the native lowland vegetation was extensively altered by Polynesians through agricultural clearing, burning and introduced rats. Here, the rapid collapse of dryland forests occurred in advance of the direct effects of human land clearance for agriculture, because of rat predation of tree seeds and seedlings.
- Research in Greece has demonstrated that intermittent pulses of erosion and sediment deposition were tightly correlated with phases of human settlement and land use, and with episodes of farming, grazing and human settlement. Here, a major phase of erosion and alluviation commenced about 4,500 years BP, roughly one millennium after the introduction of agriculture, as the result of increasing clearing of steep, marginal soils. Similar results have been obtained in the Levant.
- In New Caledonia, massive sediment accumulations (up to 6 m deep) occurred in the island's valleys by around 2000 BP, about one millennium after initial human colonisation. This erosion transformed coral reefs and sandy bays into mud flats. In Viti Levu in the Fiji archipelago, significantly increased sediment loads deriving from shifting cultivation resulted in the accumulation of a massive dune field, burying earlier sites of human occupation.

- In New Mexico, soils cultivated prehistorically remain partly degraded nearly 900 years after agriculture ceased, indicated by significant losses of organic matter, nitrogen and phosphorus. In the Hawaiian Islands, elemental soil nutrients in cultivated soils show losses of up to 50% when compared with uncultivated soils in the same areas, after about four centuries of intensive farming practices.
- In southern Mesopotamia, major salinisation occurred from about 2400 to 1700 BC, requiring a shift from wheat to more salt-tolerant barley. The long-term legacy is a saline plain that requires expensive salinity-control methods for modern agricultural production.

In their review of soil erosion in the Anthropocene, Vanwalleghem *et al.* (2017) note that on the basis of multiple case studies, the arrival of the first farmers in many parts of the world resulted in a significant increase in soil erosion, by one or two orders of magnitude. In many locations, the erosion was catastrophic, with significant effects on soil functioning, vegetation patterns and possibly even atmospheric CH_4 and CO_2 concentrations. Given the typical soil erosion rates on cultivated land of 1–4 mm yr^{-1}, the time required to erode through the fertile topsoil is often in the order of 75–300 years, which, in parts of the Mediterranean or Asia, must have occurred many millennia ago (Vanwalleghem *et al.*, 2017). As a result, many regions currently have a significantly lower soil-resource capacity than before the first advent of agriculture. This picture is illustrated further by the situation in central Europe, which was extensively deforested by 1000 BP as a result of intensive, continuous human occupation throughout the Holocene. Likely impacts ranged from effects on regional hydrology to global climate (Kaplan *et al.*, 2009), increased NO_2 emissions from fertiliser use, biodiversity losses and habitat fragmentation (Defries *et al.*, 2004), plus introduction of exotic species and changes to the fire regime (Ellis *et al.*, 2013).

Human impacts were not limited to those associated with agriculture. The wave of animal extinctions that followed the arrival of people as they spread around the world did not cease at the end of the last Ice Age but continued as humans colonised some of the remoter and more isolated parts of the planet. Some of the most powerful examples are provided by islands. On islands of all sizes, in the Mediterranean, Caribbean, Indian Ocean and Pacific, the arrival of humans was typically followed by extinction of species (Kirch, 2005). For example, the islands of Polynesia provide multiple histories of avifaunal extirpation and extinction, associated with the arrival of humans. On Easter Island and

Mangareva, the removal of large populations of nesting seabirds had far-reaching consequences for their island ecosystems because of their role as sources of phosphorus, nitrogen and other nutrients through deposition of guano, which were essential to the maintenance of forest cover (Kirch, 2005). These are now among the most severely degraded islands in the region (Rolett and Diamond, 2004). In Hawaii, 44 of 68 species of landbirds known from fossil-bearing sites became extinct in relatively recent times, a pattern that was repeated in many other Pacific islands (Grayson, 2001). In Mangaia, more than half of the native bird species were lost after human arrival, and native forests were permanently converted to degraded fernland over much of the interior of the island (Kirch, 1997). In total, more than 2,000 bird species were lost from the tropical Pacific, representing around 20% of the global total (Steadman, 1995).

The best evidence for ecosystem collapse following human arrival during the Holocene is provided by another Pacific example, New Zealand, which is considered in detail in the following section. Two other relatively well-documented examples then follow.

New Zealand

New Zealand was one of the last of the world's larger landmasses to be colonised, at around 700 years ago (Wilmshurst *et al.*, 2008). Today, New Zealand ecologists play a rather sombre game: 'spot the native species'. Sighting species that are genuinely native to New Zealand can be challenging; in parts of the country, such as the eastern part of South Island, it is possible to drive for an hour or two without seeing any. This might come as a surprise to those who imagine New Zealand to be a majestic, pristine wilderness, such as the tourists who flocked to the country after watching Peter Jackson's epic series of film adaptations, *The Lord of the Rings*. For the botanically astute, these films actually provide an insight into just how profoundly New Zealand's ecosystems have been altered by people. The first film opens with idyllic scenes of The Shire, which, rather than being some ecologically benign utopia, is in fact a highly modified agroecosystem dominated by exotic species, including introduced pasture grasses, pine trees, poplars and willows. As Frodo and Sam set off on their journey, they wade through dense stands of wandering willie (*Tradescantia* spp.), an invasive species that aggressively outcompetes native plants. No points available here for Frodo or Sam if they were playing 'spot the native species'. Sweeping landscapes that are the scene for epic horseback chases in the film are actually native tussock grasslands being invaded by exotic pines, a process that can have profound ecological consequences (Pawson *et al.*, 2010) – perhaps an example of ecosystem collapse in progress. By the time our heroes arrive in the beautiful forest of Lothlórien, they are finally able to spot native species, as the forests are dominated by southern beech (*Nothofagus* spp.). What the film does not show is the massive populations of invasive, exotic wasps (*Vespula* spp.) that inhabit these forests. These occur at such densities that their biomass is greater than that of birds, rodents and stoats combined (Thomas, 1990); as a consequence, they have a significant impact on carbon and nutrient cycling at the ecosystem scale (Wardle *et al.*, 2010). Ecologically, these wasps are much scarier than orcs.

The story of how New Zealand was colonised by people – another epic tale – is presented by Flannery (1994). At the time of initial human arrival, New Zealand was home to a unique community of some 164 species of birds, which occupied all the major ecological niches that are usually occupied by mammals in other parts of the world. In other words, these mostly flightless birds were the ecological equivalents of giraffes, rhinoceros, kangaroos, sheep and tigers (Flannery, 1994). Most extraordinary among these birds were the 12 species of moa, the largest of which was over 3 m high and weighed up to 250 kg. These were preyed on by the world's largest-ever eagle, Haast's eagle (*Harpagornis moorei*), which had claws as big as a tiger's and attacked at speeds of up to 80 km h^{-1} – a sight that must have terrified the first human colonists. Yet these people encountered such a superabundance of wild food that they ceased raising the domesticated chickens and pigs that they likely brought with them (Flannery, 1994).

Evidence from the thousands of bones collected from archaeological middens indicates that within a few centuries after the initial settlement, a small population (<2,000 individuals) of humans drove the moa to extinction (Holdaway *et al.*, 2014), together with two geese and a duck species, and marine fauna, including sea lion (*Phocarctos* spp.), elephant seal and penguins (*Megadyptes waitaha*). In total, at least 64 species of New Zealand's endemic birds and reptiles became extinct (Seersholm *et al.*, 2018), while other species, including marine mammals and fish, suffered massive declines. Wild resources were so heavily exploited by a growing human population that by the fifteenth century, people were relying on the barely edible bracken fern as a staple food. Subsequently, this formerly peaceful human society became riven by warfare, and cannibalism became a significant source of dietary protein (Flannery, 1994).

3.4 Holocene Examples of Ecosystem Collapse · 145

Figure 3.7 The food web from ancient New Zealand. The upland moa (*M. didiformis*) ate moss, ferns, mycorrhizal fungi that require animal-assisted dispersal, and aquatic vegetation. Dashed arrows indicate interactions lost since the moa extinctions. From Lafferty and Hopkins (2018); see also Boast *et al.* (2018).

Extinction of these bird species may have had a number of effects on New Zealand's ecosystems (Figure 3.7). Johnson (2009) reviewed evidence suggesting that herbivory by moa may have created habitat mosaics in the relatively dry eastern parts of South Island, through the heavy cropping of small trees and shrubs. The creation of an open understorey may have supported the survival of rich assemblages of herb and grass species, perhaps through the maintenance of a patchwork of small glades. Another aspect of the New Zealand flora that has attracted research attention is the fact that many native plant species appear to demonstrate anti-herbivore defences, such as unpalatable chemicals, cryptic colouration, mimicry of dead twigs and especially divarication, a growth form involving branching at a wide angle. Divarication is rare in other parts of the world but is present in 10% of the New Zealand flora. This may have been a defence against avian herbivores such as the moa, as suggested by Diamond (1990) in his wonderfully titled paper 'Biological effects of ghosts'. If this is true, then loss of these herbivores may have influenced subsequent development of vegetation composition, by altering competitive interactions between species. While the idea of avian ghosts haunting vegetation dynamics is an attractive one, evidence for it is limited (McGlone and Clarkson, 1993).

Similarly, loss of the moa and other bird species could have had a major impact on dispersal of plant species (Forsyth *et al.*, 2010), as suggested for other faunal extinctions (see previous section). Some vegetation types may even have been dependent on avian dispersal and herbivory (Lee *et al.*, 2010). However, recent research suggests that the role of moa as dispersers of large seeds may have been overstated (Carpenter *et al.*, 2018). While seeds of a number of large-seeded plants have been found in the remains of moa gizzards, analysis of subfossil faeces found no seeds larger than 3.3 mm in diameter. Experiments indicated that even 4 h in a concrete mixer would have been a poor proxy for the powerful grinding action of a moa gizzard (Carpenter *et al.*, 2018). The only seeds to have survived the moa gut passage intact would therefore have been very small, contradicting the ideas about megafauna dispersal discussed in the previous section.

Human impacts were not solely limited to hunting of animals, however. McWethy *et al.* (2010) reconstructed fire and vegetation history over the last millennium in the South Island of New Zealand from high-resolution charcoal and pollen records obtained from 16 lakes. Results indicated most sites experienced several high-severity fire events within the first two centuries after human arrival. These resulted in major changes in vegetation and slope stability, leading to an increase in erosion. This transfer of sediments and nutrients to water bodies resulted in a change in the water chemistry and biological communities in the lakes, indicated by shifts in diatom and chironomid composition. As native forests had little history of fire and little resilience to burning, this introduction of fire represented a substantial change in disturbance regime. The burning was deliberate and systematic (McWethy *et al.*, 2009). In total, more than 40% of the area of native forests was cleared in only a few decades; this is a remarkable degree of impact given that it was accomplished by a small, non-agricultural population (McWethy *et al.*, 2010). The impacts of this disturbance were very persistent; in most locations, the native forest has still not recovered. Even where fire activity subsequently decreased, recovery of native forests was slow (i.e. timescales of centuries), further highlighting the vulnerability of native forest species to repeated fire. Typically, the forest was replaced by tussock grassland and fern shrubland, which was subsequently transformed by Europeans to pastureland composed of non-native plants. This conversion of forest to non-forested vegetation provides an example of the positive feedbacks that can be initiated with the introduction of a new disturbance regime, involving a shift from fire-sensitive to relatively fire-tolerant and fire-prone plant species (McWethy et al., 2014). This vegetation transition therefore provides a striking example of ecosystem collapse driven by a positive feedback mechanism.

Madagascar

Currently the world's fourth-largest island, Madagascar, has been isolated for about 70 Myr. Its remote geographic position led to the development of a unique endemic biota, featuring a diverse megafauna. This included giant sloth lemurs (*Archaeoindris fontoynonti*), which were the size of male gorillas; koala lemurs (*Megaladapis*), which were the size of orangutans; suspensory sloth lemurs (*Palaeopropithecus* and *Babakotia*), which were about the size of a chimpanzee; and the baboon lemurs (*Hadropithecus* and *Archaeolemur*). Other spectacular animals included the 3-m-high elephant bird (*Aepyornis maximus*), a ratite that at 400 kg was the heaviest bird ever to have existed; the pigmy *Hippopotamus*, of which there may have been as many as four species; and a diverse group of tortoises, two of which were giants (*Aldabrachelys* spp.). Like New Zealand, this was an island where people arrived relatively late. As a result, virtually the entire endemic megafauna was eliminated, in what has been described as an ecological catastrophe (Burney et al., 2003).

Palaeoecological evidence indicates that people were present in Madagascar by around 2,300 years BP (Burney et al., 2004), although genetic evidence suggests that principal colonisation occurred somewhat later, at around 1,200 years BP. Remarkably, this evidence indicates that Madagascar may have been colonised by a single boatload of Indonesian women, who may have made a transoceanic crossing without intending to do so (Cox et al., 2012). Whereas some megafaunal species declined rapidly after human arrival, within the first few centuries (Burney et al., 2004), others appear to have persisted at some sites for a thousand years or more after humans first colonised (Burney et al., 2003). Settlement initially began in the semi-arid south-west of the island, which initiated a process of ecosystem change that spread to other parts of the island and that continues today. Before human arrival, the vegetation of the interior highlands of Madagascar was composed of a species-rich mosaic of wooded grassland, riverine and groundwater forests; today, most of this vegetation has been converted to depauperate steppe grassland (Burney et al., 2003). This is partly attributable to the marked increase in the frequency of fire that is discernible in the stratigraphic record after colonisation, although the extirpation of large herbivores is also likely to have been a contributory factor. The drier south-western part of the island is characterised by a mosaic of dry forest, palm savanna and spiny bushlands; human impact appears to have led to the expansion of the latter two vegetation types at the expense of dry forest, coupled with an increase in grasses and other fire-tolerant species (Burney et al., 2003).

Based on a review of available evidence, the process of ecosystem collapse in Madagascar is summarised by Burney et al. (2003, 2004) as follows:

- The late prehistoric extinctions in Madagascar were not a single, brief event with a simple explanation, but rather a more complex and extended process in which human-perturbed systems underwent a series of drastic phase transitions.
- Prior to human arrival, there is evidence of climate change in late Holocene with a trend of increasing aridity and changing seasonality, which may have caused changes in key megafaunal habitats such as wooded savanna.
- People arrived and hunted the megafauna, which was probably concentrated in woody savannas, a vegetation type that is rare today. Those species that were relatively slow-moving, such as many of the largest lemurs, were hunted to local extirpation, whereas the number of individuals of other species was reduced.
- In the absence of the high herbivore pressure imposed by hippos, tortoises and ratites on the grasslands (and perhaps on woody vegetation by giant lemurs), savanna areas, forest edges and understories would have become increasingly flammable as plant biomass accumulated.
- The change in fire regimes resulting from human activity, including the hunting of herbivores, resulted in preferred megafaunal habitats becoming increasingly fragmented as fire converted vegetation to structurally simpler types with less edible plant biomass (e.g. spiny bushland and steppe). There may also have been feedbacks with local climate.

- Humid forest and high-elevation areas would have provided poor refugia for most of the savanna-adapted megafauna, and the slow-moving giant lemurs would have been vulnerable to hunting even within dense habitats.
- Ecosystem collapse appears to have occurred rapidly, with biomass declining in <200 years locally, at least in the better-documented south-west sites. Some of the larger animal taxa such as *Megaladapis* and *Palaeopropithecus* may have disappeared from all but the most inaccessible parts of the island soon after human arrival, but they were still present more than one millennium later at a relatively inaccessible site. Other species such as *Archaeolemur*, the hippos and perhaps the ratites may have persisted until recent centuries.
- With human presence and few large herbivores, increasingly frequent fires would have transformed more arid savannas to thorny deserts and short grasslands. Wetter savannas in the highlands would lose much of their woody element to increased fire frequency, resulting in the depauperate steppe vegetation that covers ~70% of the land area today.
- This example represents not a single event but a stepwise cascade of extinctions. The synergistic combination of human impacts and non-linear ecological responses created a dynamic mosaic of environmental change, population fragmentation and local extirpation, a process that continues today.

The Sahel-Sahara

The Sahel-Sahara region of northern Africa provides an example of a region where positive feedbacks between precipitation and vegetation are considered to be influential. Consequently, the area has attracted a great deal of interest from theorists and modelers (Scheffer, 2009). Palaeoecological evidence indicates that earlier in the Holocene, around 6,000–9,000 years ago, much of the region was much wetter than today, with extensive vegetation cover, lakes and wetlands. For this reason, it is often referred to as the 'Green Sahara'. Around 5000 BP, there was an abrupt transition to a desert-like state in much of the region, which occurred over a timescale of decades to centuries (de Menocal *et al.*, 2000). The principal factor appears to have been slight shifts in Earth's orbit, which weakened the North African summer monsoon and consequently reduced precipitation. However, Claussen *et al.* (1999) showed, by using a coupled atmosphere-ocean-vegetation model, that orbital forcing acting in isolation was insufficient to account for the observed changes in vegetation cover. Instead, the characteristics of the vegetation itself were found to be influential through an effect on albedo, which strongly amplified the effects of climate change through biogeophysical feedbacks. Specifically, when the Sahara was vegetated, the land surface was darker, reducing the albedo contrast between the land and the sea. This would have helped maintain relatively high precipitation, by decreasing the downwelling of dry air relative to the desert state (Swann et al., 2014).

More recent modeling studies have generally supported this interpretation (e.g. Patricola and Cook, 2007), although the characteristics of vegetation cover further north in Eurasia may also have had an influence (Swann et al., 2014). As a result of such analyses, it has been suggested that vegetation in the Sahel-Sahara region exhibits alternative stable states: a 'desert' equilibrium with low precipitation and absent vegetation and a 'green' equilibrium with moderate precipitation and permanent vegetation cover, with the shift between states controlled by the difference in albedo (Brovkin et al., 1998). This mechanism may account for the collapse in the 'Green Sahara' ecosystem in the mid-Holocene. Specifically, the reflective characteristics of the vegetation cover may have helped maintain the monsoon circulation for some time after the reduction in insolation, but once a threshold was reached, drier conditions and loss of vegetation might have acted together in a positive feedback process, driving a rapid transition towards a desert state (Scheffer, 2009).

This is not necessarily the whole story, however. Recently, Wright (2017) has suggested that humans may have had something to do with it. This argument rests on two main pieces of evidence of human impact, relating to the introduction of pastoralism: when it occurred, and what its ecological impacts might have been. With respect to the first of these, Wright (2017) summarises the available evidence as follows:

- Sediment runoff in the Nile River progressively increased from the early termination phase of the African Humid Period ca. 8,000 years BP, which has been attributed to intensified land use associated with farming and animal pastoralism.
- The arrival of domesticated taxa at Ifri Oudadane, Morocco, by 7,300 years BP was concurrent with a significant reduction in the cover of native trees and grasses with significant increases in shrubs, as indicated by pollen analysis.
- At Tin-Hanakaten, Algeria, a similar pattern of vegetation change has been documented along with evidence of significant aeolian activity at

7,200 years BP, which is stratigraphically associated with the first occurrences of substantial quantities of cattle remains in archaeological deposits.
- A sediment core from Lake Chad indicates an increase in non-tree pollen and shrub vegetation at the expense of tree cover between 6,700 and 5,000 years BP, with an abrupt switch to arid conditions after 5,000 years BP. It has been suggested, but not yet archaeologically proven, that this is when domesticated plants and animals arrived in the region.
- An abrupt reduction in Guinean and Sudanian plant taxa at 3,300 years BP and their replacement by shrublands has been recorded in the Manga Plateau west of Lake Chad, although the role of humans in this transition is not clear.
- At Lake Yoa in the Chad Basin, a transition to a semi-desert plant community on disturbed dunes occurred after 4,600 years BP, and true desert communities were present after 2,700 years BP. Archaeological evidence indicates that pastoralism replaced hunting and gathering as the primary subsistence economy at this time.

The second line of evidence relates to the impacts that introduced livestock may have had on the semi-arid ecosystems of North Africa. This argument rests on observations of livestock impacts made in other parts of the world. For example, in North America, extensive prairie grasslands were present in interior upland regions prior to the arrival of European colonists. The introduction of cattle led an increase in scrubland at the expense of grassland, in what has been described as a regime shift (Wright, 2017). The mechanisms for this change were examined by van Auken (2000), who concluded that the main cause was chronic high herbivore pressure from domestic animals, which reduced the aboveground biomass of grass. This in turn reduced the availability of fuel for grassland fires, which consequently occurred at much lower frequency, or were eliminated altogether. As a result, woody plants were able to colonise the grassland through a process of ecological succession, which may have been assisted by the dispersal of seeds by the livestock. Examples of this phenomenon include southern Texas, USA, where indigenous grasslands have been replaced by mesquite (*Prosopis glandulosa*)-dominated shrubland, and central Oregon, USA, where the introduction of livestock led to expansion of the western juniper (*Juniperus occidentalis*) (Wright, 2017). Experiments also suggest that once the threshold from grassland to shrubland is crossed, it will not return to its pre-disturbance state without significant human intervention (Loeser et al., 2007). These observations provide an interesting contrast to the ideas about herbivore impacts discussed in Chapter 2, such as Vera's theory (Vera, 2000), which postulated a reduction in cover of woody plants owing to the action of large herbivores. Perhaps the impact of herbivores is different in semi-arid systems, owing to the influence of fire and its interactions with herbivory.

Taken together, these pieces of evidence suggest that the activity of humans and their livestock may have accelerated the termination of the 'Green Sahara' and its conversion to desert. If true, it means that people contributed to the most significant, large-scale example of ecosystem collapse that occurred in Africa during the Holocene (Wright, 2017). It may also provide an example of how localised land degradation can trigger cascading effects, which can spread and amplify across landscapes through positive feedback loops, leading to ecological thresholds being crossed (Wright, 2017). However, the evidence is largely circumstantial and is based on correlation, which of course is no evidence of causation. As noted by Wright (2017), it is possible that domesticated livestock spread into areas of expanding shrubland *in response* to ecological changes, rather than actually causing such changes. However, it is worth noting that in tropical western Africa where pastoralism arrived later and was a relatively minor component of subsistence economies, the termination of the African Humid Period occurred significantly later, and proceeded more slowly, compared with areas further north (Shanahan et al., 2015).

Science is most fun when people disagree, and this case study provides a fine example. Clearly somewhat irked by the suggestions made by Wright (2017), which are labelled as an example of the 'outdated doctrine against pastoralists', Brierley et al. (2018) reached precisely the opposite conclusions. Based on use of a climate-vegetation model and observations of contemporary pastoralism, these authors claim that the African Humid Period was most susceptible to collapse between 7,000 and 6,000 ka, which is 500 years before the collapse actually occurred. They therefore propose that the introduction of pastoralism may actually have slowed the deterioration of vegetation caused by climate change, rather than increasing it. However, just because ecosystem collapse did not occur when the system was at its most sensitive does not preclude pastoralism being a significant factor in collapse when it eventually did happen.

Brierley et al. (2018) commit another logical flaw in their assertion that early pastoralists adopted sophisticated land-use practices that maintained grassland vegetation and maximised its regeneration. While this may be true of modern pastoralists, knowledge of such management practices takes time to acquire and may not have been available when pastoralism was first developed. Furthermore, the livestock 5,000 years ago will not have been as domesticated as they are today and, consequently, will have been less easy to manage. Brierley et al. (2018) also suggest that the introduction of livestock essentially fulfiled the same ecological functions as the wild populations of herbivores that they replaced, forgetting that the behaviour of domesticated livestock and wild animals is very different. Not only are livestock largely protected from predators, but they go where they are directed, rather than purely where their behaviour dictates (Wright, 2017), so their ecological impacts are likely to have been quite different. Nevertheless, it is not difficult to identify weaknesses in Wright's suggestions that might be considered legitimate: for example, increased shrub cover is not the same thing as desertification.

The debate doesn't end there, though. Zerboni and Nicoll (2019) provide a survey of zoogeomorphologic evidence from the region, including traces of animal husbandry and transhumance that are left in the North African deserts. Some of these are very ancient and appear to date back to the mid-Holocene. These authors refer to the extensive archaeological evidence (i.e. rock art, pottery, settlements) indicating that people with cattle, goats and sheep were distributed throughout the Sahara and the Sahel by around 5 ka, having dispersed from the Levant between ca. 8.3 and 6 ka. The husbandry of livestock left direct evidence in the form of trails and trampling, which, under conditions of high aridity, have been preserved over long timescales. The herding of animals affected geomorphic processes on slopes, including an increased probability and intensity of debris flows, gully erosion, and surface wash along alluvial fan or escarpments (Zerboni and Nicoll, 2019). These zoogeomorphic processes might also have played an important role in increasing the amount of dust generated from the continental interior from about ~7 ka, which is detectable in oceanic sediment cores. For these authors, this provides clear evidence that early pastoralism was not environmentally sustainable, as suggested by Brierley et al. (2018). Rather, it significantly influenced geomorphic stability in the region and contributed to the collapse of the 'Green Sahara' ecosystem.

Yeakel et al. (2014) provide an account of what happened to wild animal species in Egypt when the 'Green Sahara' ecosystem collapsed. In the early Holocene, 37 large-bodied (>4 kg) mammalian species were present in Egypt, of which only 8 remain today. Many of these extinctions occurred under conditions of increasing aridity at the end of the African Humid Period, including loss of spotted hyenas, warthogs, zebra, wildebeest and water buffalo. This time also marks the beginning of dense human settlement in the region. Analysis of the predator–prey ratio revealed a long-term change in the structure of vertebrate communities: species richness of herbivores began to decline at around 5,000 years BP and was followed by a decline in species richness of predators from around 3,000 years BP. While climate change and resource completion with humans could have contributed to these extinctions, another potential factor is overhunting by people. After the African Humid Period, Egyptian people shifted from mobile pastoralism to agriculture, but likely continued with subsistence hunting of herbivores, which may have increased mortality of these species (Yeakel et al., 2014). The consequent loss of predators therefore provides a possible example of an extinction cascade.

3.5 Evaluation of Propositions

At the end of Chapter 2, some tentative propositions were identified based on existing theories of ecosystem collapse and recovery. These are briefly evaluated below in relation to the evidence from the prehistoric case studies presented in this chapter. In addition, some further propositions are identified based on this empirical evidence.

1. Ecosystem collapse can be caused either by extrinsic or intrinsic factors or by a combination of the two. Extrinsic factors refer to the external disturbance regime affecting an ecosystem, and intrinsic factors to interactions of organisms between each other and with the physical environment.
   - The different prehistoric examples profiled here paint a largely consistent picture: extrinsic factors were identified in every case. Typically, they acted in combination with intrinsic factors, although here the evidence is often less clear. None of the examples involved intrinsic factors acting in isolation. Key extrinsic factors for the 'Big Five' mass extinctions were climate change and volcanism, with some evidence available for intrinsic factors such as extinction cascades and the disruption of food chains caused by a decline in primary production. Late Quaternary ecosystem collapse in Australia and in other regions appears to have been driven by climate change and predation by people, but

3.5 Evaluation of Propositions · 155

again intrinsic factors (such as competition, habitat fragmentation and extinction cascades) appear to have been influential. Holocene examples also highlight the role of humans as extrinsic factors, through hunting of animals and use of agriculture, but, again, intrinsic factors such as extinction cascades are likely to have been influential.

2. Any ecosystem can potentially collapse, if subjected to disturbance of an appropriate type and occurring at sufficient frequency, extent, intensity or duration. Novel disturbances are more likely to cause collapse.
    - The fact that any ecosystem can potentially collapse is best illustrated by the mass extinctions, which were global in scale. Taken together, these prehistoric case studies document collapse in a wide range of ecosystem types, including terrestrial and marine, dryland and wetland, and both lowland and upland types, distributed from the tropics to the poles. These examples clearly highlight the role of novel disturbances, whether these are extraterrestrial bolides, climatic changes, large-scale volcanic activity or humans arriving in an area for the first time. Novel disturbance therefore appears to be a consistent feature of most (if not all) of the examples presented here.
3. Disturbance regimes characterised by interactions between multiple types of disturbance are more likely to cause ecosystem collapse. A particular disturbance event is more likely to result in collapse if it follows another event from which the ecosystem has not yet recovered.
    - Interactions between different types of disturbance appear to be a consistent feature of many of these prehistoric examples. Much of the literature has been characterised by debate about the relative importance of different types of disturbance, whether this be bolides, climate change or volcanism in the case of mass extinctions, or the role of humans versus climate change in Quaternary and Holocene examples. This partly reflects the difficulty of disentangling the relative influence of different causal factors. However, the very existence of these debates is evidence in support of this proposition: more than one possible cause has been implicated in every example of ecosystem collapse presented here. Whether these factors acted additively or in an interactive manner is less clear, although potential interactions and feedbacks

between disturbances have been suggested for a number of the case studies. The second part of this proposition is best supported by the mass extinctions, which often involved a series of different disturbance events that successively led to increasing numbers of species becoming extinct. The loss of megafauna in the Late Quaternary tells a similar story, with declines in many species occurring over many thousands of years in response to successive changes in climate, vegetation and hunting pressure.

4. Anthropogenic pressures are more likely to cause ecosystem collapse than natural forms of disturbance, because they are often continuous ('press') disturbances rather than one-off 'pulse' events. Impacts of press disturbances can be cumulative.
    - This proposition is difficult to sustain in the light of prehistoric evidence. First, 'natural' forms of disturbance responsible for ecosystem collapse are not necessarily of the 'pulse' type; some, such as climate change and some forms of volcanism, can be continuous. Further, anthropogenic pressures can be relatively sudden and short-term, as illustrated by the introduction of hunting and human-set fire to islands such as New Zealand and Madagascar. Might this proposition be refined? Following the suggestions of Arens and West (2008) based on analysis of fossil data, it is the coincidence of both 'press' and 'pulse' disturbance types that is most likely to lead to ecosystem collapse. Although pulse disturbances can cause significant mortality of individuals, the effects of press disturbances can be more insidious, increasing mortality over longer timescales. For example, many of the Late Quaternary megafauna appear to have been vulnerable to human hunting because they were already under pressure from climate change. Evidence from mass extinctions clearly indicates that the effects of press disturbances can be cumulative. But palaeontological evidence also suggests that press disturbances can impede recovery, as in the case of ongoing climate change and volcanism following some mass extinctions. Might this partly explain their role in ecosystem collapse? This proposition could therefore be revised as follows: *The coincidence of both 'press' and 'pulse' disturbance types is most likely to lead to ecosystem collapse. While both 'pulse' and 'press' disturbance events can increase the mortality of individuals, 'press' disturbances can also play a particular role in impeding ecosystem recovery.*

3.5 Evaluation of Propositions · 157

5. Some ecosystems can exist in more than one state or regime, such as different successional stages. Transitions between these states can form part of the natural dynamics of an ecosystem. When such transitions are persistent rather than transient, they can become of concern to human society.
   o The basis of this proposition is the idea that transitions between states can form part of the natural dynamics of ecosystems, and some of these transitions can be persistent. Perhaps because of the limitations of the fossil record in differentiating between ecosystem states, relatively few palaeoecological researchers have chosen to interpret their findings in this way. Examples where this has been proposed include both North America and Australia in the Late Quaternary, where extinction of megaherbivores has been implicated in transitions between vegetation types. In Australia, fire was also identified as a potential mechanism driving ecosystem transitions at this time. Interestingly, fire has also been implicated in persistent ecosystem transitions observed in New Zealand and Madagascar during the Holocene. Herbivory was also identified as a driver of the Holocene transition observed in the Sahara, although here the primary mechanism driving the shift between ecosystem states appears to be albedo and its impact on regional climate. This proposition also suggests that persistent ecosystem transitions can have implications for human society. Although the 'Big Five' mass extinctions obviously predate the evolution of humans, some of the Late Quaternary and Holocene examples clearly did affect people. We will never know what the loss of the mammoth steppe meant to our Palaeolithic ancestors, but major ecosystem transitions such as those that occurred in New Zealand, Madagascar and the Sahara had demonstrable impacts on the people who lived there, and their consequences are still being felt today.
6. A persistent ecosystem transition, or collapse, can arise when ecological recovery is impeded. This can occur if there are stabilising feedback processes that maintain an ecosystem in a degraded state, or when the processes of ecological recovery fail. Understanding these reasons for lack of recovery is key to understanding collapse.
   o Ideas about ecological feedbacks are difficult to test, if all you have access to is palaeoecological evidence. This hasn't prevented researchers from proposing a number of potential feedback

mechanisms that might explain some of the persistent ecosystem transitions recorded in the Late Quaternary and Holocene. For example, in Australia, a number of potential feedbacks between vegetation and climate have been suggested. In the case of the mammoth steppe in North America and Eurasia, a feedback loop may have existed between megaherbivores and the vegetation that enabled the persistence of forbs. In the Sahara, a feedback loop between vegetation and climate could have helped maintain a relatively moist climate. Each of these examples postulates positive feedback mechanisms that helped maintain a particular ecosystem state; collapse occurred when these feedbacks broke down. This implies a need to capture this eventuality in a proposition, as suggested under Proposition 7. What about the reasons for a lack of recovery in these examples? These appear to relate to persistent changes in the disturbance regime, such as fire or climate, a point captured in Proposition 4. However, some negative feedback mechanisms have been identified that might have maintained an ecosystem in a different state. For example, in Australia, tree establishment could have been constrained by the maintenance of a longer dry season, which inhibited canopy closure and thereby affected both climate and the fire regime (Bird, 2013). In general, though, palaeoecological evidence for such negative feedback mechanisms is lacking.

7. Collapse can be caused by feedbacks in the internal ecological processes of an ecosystem, which can drive a system to a different state. As a result of these feedbacks, major ecological shifts can result from minor perturbations. Such shifts can occur when external conditions reach a critical value.

    ○ As noted under Proposition 6, limited palaeoecological evidence is available in support of dynamical systems theory, on which this proposition is based. However, there is evidence that ecosystem collapse can occur when the feedbacks that maintain a particular ecosystem state break down. This proposition could therefore be modified as follows: *Collapse can be caused by breakdown of the stabilising feedback mechanisms maintaining an ecosystem state, or by feedbacks in the internal ecological processes of an ecosystem driving a system to a different state. As a result of these feedbacks, major ecological shifts can result from minor perturbations. Such shifts can occur when external conditions reach a critical value.* Among the prehistoric

3.5 Evaluation of Propositions · 159

examples provided here, evidence for such critical values is perhaps most clearly provided by the modeling work conducted for vegetation change in the Sahara (e.g. Claussen et al., 2017).

8. Collapse may be more likely if disturbance events cause the loss of generalist species, those that are highly connected to other species, top predators and/or trophically unique species. Loss of such species can cause many secondary extinctions.
   - The megafauna extinctions in the Late Quaternary provide evidence in support of this proposition. Loss of megaherbivores as a result of hunting by humans could account for the loss of predators and scavengers that depended on them as a food source. Extinction cascades have also been postulated for the 'Big Five' mass extinctions, although here at least some of the evidence points towards loss of species at the base of trophic pyramids, rather than those nearer the top. This suggests the need for a minor amendment to this proposition, to include *those species at the base of food chains*.
9. The disassembly of ecological networks tends to be characterised by thresholds. In such situations, the consequences of species loss become amplified and self-reinforcing as more and more species are extirpated, leading to network collapse.
   - There is little evidence either for or against this proposition in these prehistoric case studies, perhaps reflecting the limitations of palaeoecological data.
10. Certain types of ecosystem may be especially sensitive to disruption of biotic processes and interactions, namely those with strong top-down trophic regulation or with many mutualistic or facilitation interactions, those that are strongly dependent on mobile links and where positive feedbacks operate between the biota and the disturbance.
    - Again, there is limited evidence for this proposition in these case studies, largely because information on how these ecosystems functioned is lacking. However, these examples do provide some evidence of positive feedbacks between the biota and disturbance, such as the concept of Earth system succession, increasing dominance of fire-adapted plant species in Australia and New Zealand and the positive feedbacks between vegetation and climate proposed for the Sahara.
11. Extinction cascades and community collapses should be less likely to occur in species-rich multi-trophic communities than in species-poor ones, unless the environment is highly variable.

○ Any attempt to use palaeontological data to examine the role of species richness in ecosystem collapse is bound to be fraught with uncertainty, given the limitations of the fossil record. The mass extinctions illustrate that collapse can potentially occur in any kind of ecosystem, regardless of the number of species. The collapses reported in the Late Quaternary and Holocene were also associated with a wide range of different ecosystem types. On balance, there seems to be no evidence in support of this proposition in these case studies, and some evidence against it.
12. The ability of an ecosystem to recover is critically dependent on intrinsic factors, namely, interactions of organisms between each other and with the physical environment. The rate or extent of recovery can be limited by intrinsic and/or by extrinsic factors, such as the disturbance regime and the extent of degradation.
    ○ The mass extinctions provide clear evidence that the rate of recovery can be limited by the disturbance regime. One of the principal lessons of these case studies was that ecological recovery can require very long timescales, in some cases requiring the evolution of new forms in order for ecosystem function to be restored. Such evolutionary processes provide an example of an intrinsic factor affecting both the extent and rate of recovery. A further factor that sometimes affected recovery rate was the extent of degradation, with respect to the number of species lost in particular functional groups such as producers. In the Quaternary and Holocene examples, recovery was essentially slight or non-existent, raising the interesting question of why this was the case, as explored under Proposition 6.

Additional Propositions Based on Prehistoric Case Studies
- Functional and structural change in an ecosystem undergoing collapse may be unrelated to loss of taxonomic diversity; species identity matters.
- Recovery can take a long time, and takes longer than collapse.
- There are feedbacks between ecosystems and the biosphere, or the entire Earth system, which can act as a mechanism for collapse and influence recovery.
- Collapse can be positive by creating new opportunities, for example evolutionary diversification and radiation.

4 · *Contemporary Case Studies*

There has been remarkable growth in ecological science in recent decades, driven by increasing concern about human impacts on the environment. As a result, there is now a rich body of information available on the impacts of contemporary environmental change on many different types of ecosystems. This chapter is therefore highly selective and focuses on those case studies that are relatively well documented or are particularly insightful. Or those that are simply interesting. The aim here is to explore these examples to identify what can be learned about ecosystem collapse and recovery, with a particular focus on evaluating empirical evidence of underlying mechanisms. This evidence is then considered in relation to the propositions presented at the end of Chapter 2 and further revised at the end of Chapter 3.

4.1 Coral Reefs

Coral reefs are very special places. They are one of the most diverse and productive ecosystems on Earth and are home to thousands of species, making them the most species-rich of all marine ecosystems (Mumby and Steneck, 2008). Recent estimates suggest there are around 830,000 multicellular taxa on coral reefs worldwide, which represents about a third of all marine species (Fisher *et al.*, 2015). Coral reefs are also of high value to people, proving a range of benefits including food, recreation, coastal protection and medicines (Moberg and Folke, 1999), which have collectively been valued at hundreds of billions of dollars (Souter and Lindén, 2000). Some 500 million people depend on coral reefs for their livelihoods (Wilkinson, 2008). Yet such bare facts fail to convey why coral reefs are so special. They are incredibly beautiful, combining structural complexity and diversity with a dazzling array of colour and movement. Put simply, they are one of the greatest natural wonders of the world.

Collapse

The degradation and decline of the world's coral reefs has been well documented. In 2008, it was estimated that the world had lost 19% of the original area of coral reefs; a further 15% of reefs were considered to be seriously threatened with loss within the next 10–20 years and 20% in the following 20–40 years (Wilkinson, 2008). By 2011, about 75% of the world's reefs were considered threatened (Burke *et al.*, 2011). The situation has undoubtedly worsened since then. Coral reefs are being subjected to a wide range of intensifying threats, including impacts from overfishing, coastal development, agricultural run-off and shipping (Burke *et al.*, 2011). In addition, global warming has caused widespread damage to reefs. High sea temperatures cause corals to eject their symbiotic algae, resulting in coral bleaching, where corals are reduced to their white skeletons. Mass bleaching events have occurred almost annually since the early 1980s in one or more regions, sometimes resulting in catastrophic loss of coral cover and substantial changes in community structure and composition (Baker *et al.*, 2008). The median return time between severe bleaching events is now only six years, having declined continuously from once every 27 years in the 1980s (Hughes *et al.*, 2018a). In addition, increasing CO_2 emissions are causing the world's oceans to gradually become more acidic, reducing the growth rate of corals and increasing their mortality (Prada *et al.*, 2017). As a result of such factors, reefs are becoming increasingly susceptible to other forms of disturbance such as storms, pollution and diseases (Burke *et al.*, 2011). An assessment of such local-scale threats indicated that more than 60% of the world's reefs are being directly affected by one or more local factors, including overfishing and destructive fishing, coastal development, watershed-based pollution or marine-based pollution and damage. Of these pressures, overfishing is the most pervasive. About 75% of the world's coral reefs are considered to be threatened, when local threats are considered together with thermal stress (Burke *et al.*, 2011).

In a recent overview of the challenges facing the conservation of coral reefs, Hughes *et al.* (2017a) rightly refer to the current situation as a 'crisis'. While stressing that we should not give up hope for coral reefs, these authors highlight the significant challenges that lie ahead if effective conservation is to be achieved. Addressing this coral reef crisis will require immediate action to reduce emission of greenhouse gases, together with the development and implementation of innovative management and policy interventions. These require a clearer understanding of the effects of multiple drivers and ecosystem responses under projected scenarios of global climate change (Hughes et al., 2017a), the provision of which sounds like a very worthwhile career objective for any aspiring ecologist. The task here, though, is to examine what coral reefs can teach us about how and why ecosystems collapse and what influences their ability to recover thereafter.

So the first question is, *have any coral reefs actually collapsed?* To answer this question, we might first usefully remind ourselves of our definition of collapse (Bland et al., 2017a; Chapter 1): a transformation of identity, a loss of defining features and a replacement by a different ecosystem type. Coral bleaching by itself might therefore not be enough to constitute an example of collapse; to qualify, bleaching would need to have led to the death of the corals and the subsequent conversion of the coral reef to another ecosystem type. In their review of the current and future state of coral reefs, Hughes et al. (2017a) highlight the possibility of ecosystem collapse, but do not provide any actual examples. For an assessment of what has happened to coral reefs following bleaching events, we can turn to another review: the one provided by Baker et al. (2008). This examined the longer-term ecological impacts of bleaching and the potential for recovery, rather than bleaching mechanisms and their immediate effects; the review drew upon a number of published meta-analyses and expert knowledge. Some key findings from this paper (Baker et al., 2008) are summarised below:

- *Coral mortality*. Mortality rates vary markedly among bleaching events. Following minor events, when temperature changes are minor and of short duration, coral mortality is low and nearly all corals recover. Severe bleaching events can result in near 100% mortality and local loss of some taxa. No coral species are immune to such events. For example, in the eastern Pacific coral mortality following the 1982–1983 bleaching event reached 90% at Cocos Island and 97% in the Galápagos Islands.
- *Coral reproduction and recruitment*. Although corals can recover from bleaching, they may suffer long-term sublethal effects, such as declines in fertility and reproductive success, or developmental problems. These impacts can lead to reduced recruitment. For example, most flat corals that bleached during the 1998 event at Heron Island, Great Barrier Reef, subsequently demonstrated a reduction in percentage fertile polyps and the number of eggs produced per polyp.
- *Community structure*. Impacts following bleaching can include: (i) changes in the relative abundances of coral species and (ii) changes in the dominance of non-coral taxa occurring on reefs. In general,

scleractinian corals with branching colony morphologies suffer higher rates of mortality than species with other morphologies. For example, in the Caribbean, populations of the branching coral *Acropora* have declined by 97%, leading to persistent changes in the coral community and shifts away from coral dominance.
- *Changes in algal symbiont communities.* Bleaching can also cause changes in symbiont communities, which can have a major bearing on the capacity of corals to tolerate and recover from these events. For example, in the Caribbean, corals that changed their symbiont communities as a result of bleaching experienced less mortality. Bleaching can lead to adaptive shifts in symbiont communities, for example by increased dominance of heat-tolerant taxa.
- *Corallivores.* The structure of coral reef communities can be affected by changes in the relative abundance of species affected by bleaching events. While many species can die during an event, others may increase in abundance; such effects may persist for several years. For example, in Panama, warming in 1982–1983 increased mortality of a gastropod corallivore (*Jenneria pustulata*) but did not alter the mortality of a coral-eating echinoderm (*Acanthaster planci*) or a fish species (*Arothron meleagris*). In such cases, predation of those corals that survive bleaching may intensify.
- *Reef fishes.* Responses of fish to bleaching vary markedly between species, depending on their patterns of resource use and the coral taxa affected. Although obligate corallivores may die within weeks of bleaching, other trophic categories may be less affected. However, fish species that require coral structure for shelter, reproduction or larval settlement sites may decline rapidly. Species that can browse on algae may increase in abundance following bleaching, as the algae colonise coral substrates. Typically, around 50% of herbivorous fish decline after bleaching. Recent evidence has highlighted behavioural changes in fish in response to bleaching, leading to the reorganisation of communities (Keith et al., 2018).
- *Bioerosion.* This refers to the breakdown of limestone skeletons and reef structures as a result of the activity of animals that burrow in, consume or otherwise damage the reef (such as algae, fungi, sponges, polychaete worms, molluscs and crustaceans). Disturbances that reduce the cover of live coral, such as eutrophication, sedimentation and bleaching, can quickly cause a transition to an erosional state. This reduces structural integrity and structural complexity. For example, in the Galápagos Islands, bioerosion following bleaching was found to be mostly

attributable to a sea urchin (*Eucidaris galapagensis*) grazing on algal-coated dead coral skeletons; reef framework structures were subsequently lost over a 10-year period.

This evidence indicates that bleaching can clearly lead to a loss of defining features of coral reef ecosystems and suggests that replacement by a different ecosystem type is also a potential outcome. The best example of such a transition is replacement of corals by algae. This has attracted a great deal of research interest and led to the widely held view that most of the world's reefs are being overrun by seaweed (Bruno *et al.*, 2009; Knowlton, 2008). As one researcher put it, 'algae, algae everywhere, and nowhere a bit to eat!' (Szmant, 2001). Much of this evidence has been interpreted in relation to dynamical systems theory (see Chapter 2), and consequently it is widely referred to as a coral algal 'phase shift' (e.g. Done, 1992; McManus and Polsenberg, 2004). Further, this transition is often cited as evidence of the existence of alternative stable states in ecosystems (e.g. Scheffer, 2009; Scheffer and Carpenter, 2003; Scheffer *et al.*, 2001).

Some of the earliest evidence for this transition came from studies of coral reefs in the Caribbean, such as a classic investigation by Hughes (1994). Here, annual monitoring was undertaken for 17 years at multiple sites, in an area that was intensively fished over a prolonged period. Where fishing pressure was high, the echinoid *Diadema antillarum* was found in abundance, almost certainly as a result of the harvesting of fish species that competed with it for algae. *Diadema* underwent a mass mortality throughout its entire geographic range as the result of a species-specific pathogen, in the period from 1982 to 1984. Following this, the entire reef system of Jamaica underwent a spectacular and persistent algal bloom, forming intensive mats that were 10–15 cm deep. As a result, coral recruitment subsequently failed. Overall, across >300 km of coastline, cover of coral declined from 52% to 3%, whereas cover of macroalgae increased from 4% to 92% (Hughes, 1994). This surely qualifies as a convincing example of ecosystem collapse. Although overfishing was one of the main drivers, other factors that may also have adversely affected the reefs during this period include hurricanes, diseases of corals, eutrophication and coral bleaching events.

Similar transitions from coral- to algae-dominated reefs have been reported elsewhere. The phenomenon has been widely observed in other parts of the Caribbean (McManus and Polsenberg, 2004), including relatively remote reefs such as those of Belize (McClanahan and Muthiga,

1998). In Australia's Great Barrier Reef, McCook (1999) highlighted concerns about abundant macroalgae on inshore fringing reefs, which were attributed to anthropogenic inputs of sediments and nutrients. Cruz et al. (2018) mentioned examples in the Florida Keys and Hawaii and suggested that algal 'phase shifts' could cause the loss of structural complexity. This could then result in loss of habitat heterogeneity, a reduction in the capacity of a reef to maintain its local diversity, changes in trophic structure and a decline in the structural integrity of the reef over the long term (Cruz et al., 2018). Similarly, in Cousin Island in the Seychelles, Ledlie et al. (2007) reported that the reefs suffered extensive coral mortality after a bleaching event in 1998, which was followed by a transition from coral to algal dominance. By 2005, mean coral cover was <1%, macroalgal cover had increased by up to 40% and structural complexity had declined (Ledlie et al., 2007).

The widespread occurrence of this phenomenon, and growing concern about the conservation status of coral reefs worldwide, has led to a substantial research effort to understand its causes. Much of this has been informed by dynamical systems theory, and in particular, the idea that dominance by either corals or algae might represent alternative ecosystem states as postulated by the theory. Some of this research was summarised by Hughes et al. (2017a), who placed particular emphasis on the need to understand the multiple drivers that are currently affecting coral reefs, and how they interact. Towards this, Hughes et al. (2017a) presented an interesting model that illustrates the potential transition between coral- and algal-dominated states, as a function of three different drivers (Figure 4.1). The model incorporated three well-established feedbacks: (1) prevention of coral recruitment by macroalgae, (2) reduced carrying capacity of herbivores when coral cover and structural complexity of the habitat is low and (3) swamping of herbivores by high abundance of macroalgae. The combination of these feedbacks creates a situation in which the modeled ecosystem can either be in a coral-dominated or a macroalgae-dominated state; for some intermediate values of drivers, the system could potentially be in either state (i.e. exhibiting alternate stable states) (Hughes et al., 2017a). Although this presents a useful illustration of how dynamical systems theory might be applied to a real-world situation, it is important to note that the model has not been tested in any way; it is essentially a representation of the purely heuristic model described by Scheffer et al. (2001). It therefore provides no evidence regarding whether these alternative states actually exist in reality.

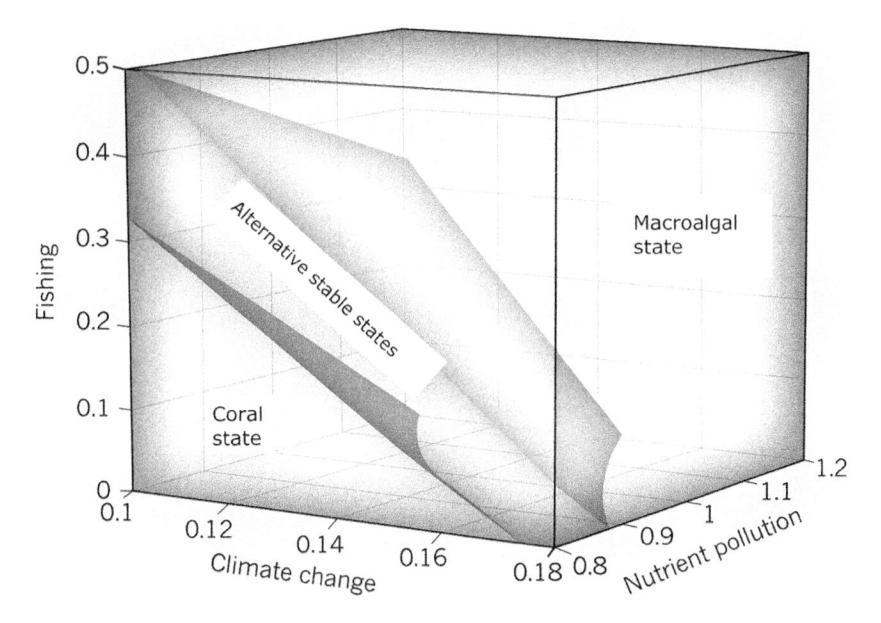

Figure 4.1 Heuristic model of the response of coral reefs to multiple anthropogenic drivers. Reprinted from Hughes *et al.* (2017a), with permission from Springer Nature, from which the following description is reproduced: 'Depending on the strength and interaction between certain anthropogenic drivers (climate change, nutrient pollution and fishing), three outcomes are possible: when the drivers are weak, healthy coral-dominated assemblages form (coral state); when the drivers are strong, a state dominated by macroalgae with few corals forms; and when the drivers are intermediate, alternative states occur. Interfaces between the coral, alternative and macroalgal states represent the tipping points or thresholds for each combination of drivers. The coral state collapses if any single driver is too strong and is eliminated entirely by the cumulative impacts of multiple drivers. The width and shape of the region of alternative stable states depends on the strength of interacting feedbacks'.

The feedback processes incorporated in the model are based on empirical data, however, and these provide an important mechanism by which ecosystem collapse might be brought about. Hughes *et al.* (2017a) go further and indicate that there are in fact 30 types of positive feedback that have been observed on coral reefs, which indicates that there is no shortage of potential collapse mechanisms. This estimate is based on the work of van de Leemput *et al.* (2016), who conducted a literature search to identify feedbacks (of which only 19 are actually listed, although the point still stands). The feedbacks are classified into five broad categories, based on the main process involved: predation;

168 · Contemporary Case Studies

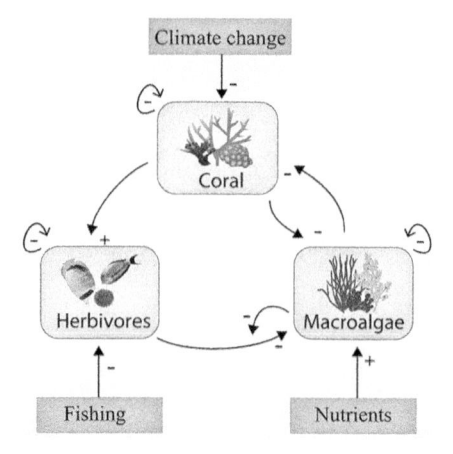

Figure 4.2 An overview of some of the feedbacks affecting coral reefs. The qualitative effect of each feedback route can be determined by multiplying the signs on the arrows of the route taken through the diagram. In this way, two negative feedbacks combine to cause a positive feedback. For instance, herbivores reduce the biomass of macroalgae, but abundant macroalgae can reduce the impact of herbivores if they swamp grazing pressure and become less palatable. A three-phase positive feedback can occur between corals, herbivores and macroalgae. Reproduced from van de Leemput et al. (2016), with permission from Springer.

competition; density-dependent demography; facilitative interactions between species and/or between species and their environment; and social–ecological interactions (Figure 4.2; van de Leemput et al., 2016). These authors note that these feedbacks have rarely been quantified, and their potential role in promoting critical transitions or alternate ecosystem states has generally not been confirmed. They also emphasise that empirical evidence for the existence of any specific positive feedback mechanism does not by itself prove that alternate ecosystem states exist, because the feedback may be too weak or intermittent to shift an ecosystem from one state to another. Further, they highlight the fact that empirical information on spatial and temporal variation in most of these feedbacks is either very limited or nonexistent, making it difficult to understand the response of any coral reef to intensifying disturbance (van de Leemput et al., 2016).

Although the idea that coral reefs switch between alternative states of coral and algal dominance is very widespread in the literature, there is actually very little empirical evidence to support it. According to Dudgeon et al. (2010), this stems partly from misunderstandings of dynamical systems theory and specifically from a widespread confusion regarding the difference between phase or regime shifts, which are changes in the equilibrial community in response to a persistent change in environmental conditions, and alternative stable states, which represent alternative configurations of the same community. In other words, for the latter situation to apply, more than one ecosystem state needs to be able to occur in the same place and under the same environmental conditions at different times (see Chapter 2). Most of the examples in the literature that claim the existence of alternative stable states in coral reefs examine press perturbations, which are inappropriate for this purpose because environmental conditions are different before and after the perturbation event (Dudgeon *et al.*, 2010; see also Chapter 2).

In their review conducted to examine the existence of alternative stable states, Petraitis and Dudgeon (2004) noted a complete lack of experimental evidence to support their existence in coral reefs. Conversely, they did find ample evidence of a transition between coral to algal assemblages resulting from a change in environmental conditions, from one favouring corals to one favouring algae. According to their terminology, these situations would qualify as regime shifts. They also found that such shifts tended to be restricted to specific areas. These results were further corroborated by Dudgeon *et al.* (2010), who accused reef scientists of developing a 'minimalist picture of reefs flipping between the two extremes' and then proposing 'a baroque overlay of mechanisms that lock those reefs into one state or the other'. This reflects the large number of putative feedback mechanisms that have been identified, as noted previously. It is quite an achievement to be both minimalist and baroque at the same time, so reef scientists deserve some recognition for that accomplishment. Lest you be concerned that the distinction between regime shifts and alternative stable states is simply a 'semantic quibble', these authors emphasise that it is a critical issue that has a major bearing on identifying how coral reefs should be managed (Dudgeon *et al.*, 2010), a point considered further in the following sections. Specifically, the former situation implies a need to change external conditions, whereas the latter requires a change within the system itself (van Nes *et al.*, 2016; see also Chapter 2).

Are coral reefs really being overrun by seaweed? Bruno *et al.* (2009) conducted an extensive review to assess how widespread the phenomenon actually is, incorporating an impressive 3,581 surveys of 1,851 reefs, conducted during the period 1996–2006. Results indicated that replacement of corals by macroalgae is less common and less geographically widespread than is often assumed. In the Caribbean, where increases in macroalgal cover have been most often reported, only 4% of reefs were dominated by macroalgae (i.e. >50% cover). Corresponding figures for the Indo-Pacific were 1% of the surveyed reefs. Overall, between 1996 and 2006, the severity of the transition decreased in the Caribbean, did not change in the Florida Keys and Indo-Pacific and increased slightly on the Great Barrier Reef. In conclusion, these authors suggested that: (i) coral reefs are more resistant to macroalgal colonisation than often assumed; (ii) according to analysis of relatively 'pristine' reefs, current values of macroalgal cover may be close to those prevailing prior to human impact across much of the world; (iii) the cover of coral and macroalgae are not strongly correlated, and therefore the abundance of macroalgae is not an effective indicator of either coral loss or ecosystem condition; (iv) the problem of macroalgae has been overstated, and the main form of degradation is the loss of reef-building corals; (v) management efforts should therefore focus on conserving coral populations, as without them, the entire system would collapse (Bruno *et al.*, 2009). Interestingly, these authors also observe that few of the world's reefs fit neatly into either of the alternative stable states that have been proposed, namely coral or macroalgal dominance. Most reefs are currently somewhere between these two extremes, which further undermines the idea that coral reefs switch between coral- and macroalgal-dominated stable states (Bruno *et al.*, 2009).

Recovery

This debate has implications for understanding the recovery of coral reefs following degradation or collapse. If coral reef ecosystems demonstrate alternative stable states, then a change in environmental conditions towards a situation more suitable for corals will not necessarily result in a transition back from algal to coral dominance; a large-scale pulse perturbation may be required to achieve this (Petraitis and Dudgeon, 2004). Conversely, if the transition from coral to algae is a regime shift driven by a change in the external environment, then this could potentially be reversible. Some empirical observations provide possible support for this type of recovery. For example, Edmunds and Carpenter (2001) describe reversal from algae to coral at five sites along 8 km of coastline in Jamaica. Here, by the year 2000, density of the sea urchin *Diadema antillarum* recovered to values similar to those recorded in the late 1970s and early 1980s. This recovery was associated with a decline in the cover of algae and an 11-fold increase in coral density.

The evidence of coral reef recovery after bleaching events was reviewed by Baker *et al.* (2008), who concluded the following (see also Table 4.1):

- Both the severity of bleaching and recovery were highly variable spatially. Despite this, some geographic patterns in recovery have been detected. For example, rates of recovery were higher in the Indian Ocean than in the western Atlantic. No clear trends in coral recovery were evident in the Pacific or the Arabian Gulf, where some reefs declined while others showed some evidence of recovery.
- The recovery rate also varied among sites. In some cases recovery occurred within two years (e.g. in Hithadoo, Maldives), whereas in other locations, recovery was totally absent even after 20 years.
- Recovery rate did not appear to be related to the severity of the bleaching disturbance. Furthermore, the extent of recovery was not related to the amount of coral cover that remained following the disturbance. Many reefs with relatively high coral cover continued to decline after bleaching, whereas other reefs with low cover recovered well (e.g. Dubai in the Arabian Gulf recovered from 0% to 42% coral cover in nine years; Tutuila in the south Pacific recovered from 6% to 40% in four years).
- Key processes for coral recovery include the asexual regeneration of surviving colony tissues and the sexual recruitment of larvae. Important requirements for coral regrowth include light availability and stable, relatively sediment-free substrates. Maintaining reef framework integrity was important in the recovery process. In addition, substrates should be free of other taxa that might outcompete young coral. Heavy fish predation and sea urchin grazing can impede coral recovery.
- Species identity influenced recovery potential. Corals with a broadcasting mode of reproduction, such as species of *Acropora*, *Montipora*, *Porites* and faviids, were the predominant recovering corals in most regions.

What about coral reef recovery following other forms of disturbance? Bellwood *et al.* (2006) provide an account of an interesting experiment conducted in the Great Barrier Reef, where a transition from a coral- to an algal-dominated state was induced artificially by excluding herbivorous fishes, using experimental cages. When the cages were removed, a rapid reversal occurred back to a coral-dominated state, but this was not achieved by increasing the abundance of herbivorous fishes. Rather, the transition was found to be driven by a single batfish species (*Platax pinnatus*), which is an invertebrate feeder. This suggests that the species that caused the

Table 4.1 *Summary of mean per cent change (coral recovery or decline) following bleaching mortality events in five major coral reef regions*

Region	Number of sites recovering/ declining/ no change	Mean (±SD) % change in coral cover Median (range) % change in coral cover			Mean (±SD) number of years after bleaching mortality		
		Recovering	Declining	No change	Recovering	Declining	No change
Indian Ocean	46/12/0	+8.3 ± 9.1 +5.0 (0.2–42.6)	−15.0 ± 13.2 −17.5 (0.5–41.0)	–	4.7 ± 1.8	7.3 ± 5.4	–
Arabian Gulf	2/2/0	+30.0	−7.8 −7.8 (2.0–13.5)	–	9.0	9.0	–
Central/ Southern/ Western Pacific	16/11/2	+30.0 (18–41.9) +25.2 ± 24.1 +17.0 (0.2–89.0)	−12.0 ± 8.4 −11.5 (2.0–28.0)	0	8.9 ± 6.0	6.0 ± 2.6	3.0
Eastern Pacific	9/6/5	+19.0 + 18.1 +13.8 (<0.1–45.5)	−10.8 ± 9.5 −8.5 (0.2–22.0)	0	18.4 ± 1.1	13.2 ± 4.7	16.6 ± 7.6
Western Atlantic	1/16/0	+0.5	−6.6 ± 5.6 −6.0 (0.5–16.5)	–	7.0	4.9 ± 1.8	–

Source: Reproduced from Baker *et al.* (2008), with permission from Elsevier.

ecosystem transition were not those that were able to reverse it, providing an example of hysteresis (see Chapter 2). While this experiment demonstrates that the recovery pathway may be unexpected, and different from the pathway of degradation, it also shows that recovery is possible.

Diaz-Pulido *et al.* (2009) provide further evidence that coral reefs can recover rapidly from seaweed blooms; again working on the Great Barrier Reef, coral recovery was observed within a year. This process of recovery did not involve reestablishment of corals by recruitment of coral larvae, which is often assumed to be the key recovery mechanism. Rather, a number of other mechanisms appeared to be influential, including rapid regeneration rates of remnant coral tissue and high competitive ability of the corals enabling them to outcompete the seaweed (Diaz-Pulido *et al.*, 2009). Yet recovery of the corals does not necessarily result in recovery of other components of the ecosystem. For example, Bellwood *et al.* (2012) showed that 13 years after the 1998 bleaching event in the Great Barrier Reef, the fish assemblage had exhibited no evidence of recovery, even though coral abundances had completely recovered. This finding may reflect changes in the relative abundance of different corals resulting from the bleaching event (Bellwood *et al.*, 2012).

Graham *et al.* (2011) provide a meta-analysis of the recovery rates of coral cover from pulse disturbance events, based on observations made in 48 different reef locations. Disturbances included bleaching events, storms and outbreaks of crown-of-thorns starfish. Results indicated that reefs in the western Pacific Ocean displayed the highest rate of recovery, whereas those in the eastern Pacific Ocean were slowest to recover. These findings were attributed to regional differences in coral composition, the functional diversity of fish communities and the degree of geographic isolation. The type of disturbance was found to have a limited effect on the rates of subsequent reef recovery, although recovery rates were higher following crown-of-thorns starfish outbreaks, and when large areas of benthic space were opened up, allowing coral recruitment to take place. The importance of maintaining reef structural complexity for supporting key reef processes, such as recruitment and herbivory, was also highlighted (Graham *et al.*, 2011).

From the perspective of dynamical systems theory, the potential for coral reef recovery can best be understood in terms of feedback mechanisms. While noting reservations about application of this theory to coral reefs as explored earlier, it is possible that such feedback mechanisms could play a crucial role in the recovery process. For example, there is experimental evidence that macroalgae can prevent coral recruitment (Kuffner et al., 2006) and can cause allelopathic damage to corals (Rasher and Hay, 2010). Such processes could act as stabilising feedbacks that could strengthen the persistence of the macroalgae state (Graham et al., 2013). Conversely, destabilising feedbacks could potentially weaken the macroalgae state. For example, Graham et al. (2013) note that rabbitfishes (*Siganus* spp.) use macroalgae as habitat as juveniles and feed directly on macroalgae as adults; increases in algal cover could therefore lead to increased populations of herbivores that would reduce algal cover. The interplay between stabilising and destabilising feedbacks could therefore affect how readily coral reefs can recover from algal dominance (Graham et al., 2013). If stabilising feedbacks are weak, then it may be easier to achieve such a transition (Nyström et al., 2012).

For a coral reef to recover, it will also typically be necessary to reduce chronic pressures – such as fishing, sedimentation or nutrient loading – beyond the threshold at which the original transition occurred (Graham et al., 2013). Reducing fishing pressure, for example through the creation of protected areas, can demonstrably increase the rate of coral recruitment, although this is not always the case. For example, analysis of long-term ecosystem dynamics in 21 reefs following a bleaching event in the Seychelles indicated that recovery was favoured when reefs were structurally complex, when they were located in relatively deep water, when density of juvenile corals and herbivorous fishes was relatively high and when nutrient loads were low. However, in this case, whether or not reefs were situated inside a no-take marine reserve had no bearing on the recovery trajectory (Graham et al., 2015).

Although it is traditionally believed that coral reef recovery is dependent on propagules arriving from other, undisturbed reefs, Gilmour et al. (2013) showed that relatively isolated reefs can recover successfully from disturbance, despite the low connectivity with other reefs. Here, high growth and survival rates of remnant colonies, followed by an increase in juvenile recruitment from local propagule sources, were key factors in the recovery process. Another interesting question is why reefs in the Caribbean appear to have recovered less well than those in some other regions, such as the Indo-Pacific. It may be that reefs in the Caribbean are particularly predisposed to algal blooms because of the limited scope for top-down community control, owing to relatively low herbivore biomass and diversity. The functional loss of a major urchin and two acroporid coral species may also have had a significant impact on community interactions, perhaps shifting the system from a recovery trajectory to a relatively stable algal-dominated state (Roff and Mumby, 2012). In locations in the Pacific such as Hawaii, experimental evidence indicates that increasing herbivore populations on degraded reefs can be an effective strategy for reducing algal dominance and enabling coral recovery (Smith et al., 2010).

It is also useful to differentiate between the recovery of the coral itself and the reassembly of the entire community. Evidence indicates that these two elements of ecosystem recovery are not necessarily closely related. The processes involved in the reassembly of coral-reef communities include framework building and increases in habitat complexity, diversity of food sources, recruitment and connectivity (Johns et al., 2014). Whereas some studies have shown that coral reef communities can return to their pre-disturbance composition, in other cases the composition shifts; for example, changes in communities of butterflyfishes after disturbance were observed in Mo'orea (Berumen and Pratchett, 2006). In a study of six reefs in the Great Barrier Reef, Johns et al. (2014) found that coral cover recovered in five of them within 7–10 years, whereas four of the six communities reassembled to their pre-disturbance composition in 8–13 years. The trajectories of two communities indicated that they were unlikely to reassemble; these were situated in near-shore locations and had low abundance of the tabulate coral *Acropora* spp., a group of species that are known often to be crucial contributors to the condition of coral reef ecosystems.

The processes of both collapse and recovery will also be influenced by whether coral-reef communities are organised principally by top-down or bottom-up processes. The relative importance of these processes has been the subject of debate. This is illustrated by the issue of algal growth: in isolation, both nutrient enrichment and a reduction in herbivory can lead to an increase in algal cover. Numerous factorial experiments have been conducted to examine the interactions between these factors, with somewhat variable results (Smith et al., 2010). A meta-analysis of some of these studies showed that herbivory generally had larger effects than nutrients on algal biomass (Burkepile and Hay, 2006), a result supported by additional experimentation (Smith et al., 2010). Further evidence of the relative importance of top-down and bottom-up processes is provided by Ruppert et al. (2013), who examined the loss of sharks as top-order predators resulting from fishing (a press disturbance) and the bottom-up processes structuring reef fish communities following cyclones and bleaching (pulse disturbances) on remote atolls in the eastern

Indian Ocean. Monitoring evidence suggested that the loss of sharks can result in impacts that propagate down the food chain, which can contribute to mesopredator release and changing the abundance of primary consumers. The pulse disturbances led to impacts that spread up the food chain through herbivores, planktivores and corallivores, but did not affect carnivores. This suggests that the removal of sharks by fishing could potentially affect the recovery of coral reefs following disturbance, through a trophic cascade. However, as noted by the authors, the precise mechanisms remain elusive (Ruppert et al., 2013).

Despite such results, there is limited evidence for top predators influencing the structure of coral reef communities. This may be because coral reefs' food webs are complex, with large numbers of species. This may lead to functional redundancy within trophic levels, which would dampen the effects of predators (Rasher et al., 2017). As reef sharks are generalists that obtain prey from a variety of different reef habitats, their impacts are diffused across the food web and are unlikely to lead to trophic cascades. However, it is possible that they may have a significant impact on community structure through fear, the effects of which may amplify as they cascade; evidence for such processes has been obtained in Fiji (Rasher et al., 2017). There is also evidence that the abundance of predatory fishes can influence the abundance of coral-eating starfish and thereby affect the structure and function of coral reef ecosystems (Dulvy et al., 2004). This study is also notable for having detected threshold responses in relation to starfish population dynamics.

Observations of recovery provide some hope that coral reef ecosystems will somehow weather the crisis that they are currently experiencing. After all, it is not the first time that reefs have had to cope with rapid environmental change; for example, at the end of the last Ice Age, they similarly faced catastrophic climate and sea-level change (Blanchon and Shaw, 1995; Webster et al., 2018). However, in closing this section, it is pertinent to consider evidence from the most recent bleaching event. Hughes et al. (2018b) summarise observations made on the Great Barrier Reef, after the record-breaking marine heatwave of 2016, as follows:

- Corals began to die immediately where the accumulated heat exposure exceeded a critical threshold of 3–4°C heating weeks.
- After eight months, an exposure $\geq 6°C$ heating weeks caused an unprecedented, regional-scale shift in the composition of coral communities, resulting from contrasting responses to heat stress by different coral taxa.

- Fast-growing staghorn and tabular corals suffered a catastrophic die-off, transforming the three-dimensional structure and ecological functioning of 29% of the 3,863 reefs that make up the world's largest coral reef system.

As the authors state, this study is an important contribution to the theory and practice of assessing ecosystem collapse, as it enabled both initial and collapsed states to be identified, together with the major driver of change and quantitative collapse thresholds. Furthermore, Stuart-Smith *et al.* (2018) reported region-wide ecological changes as a result of this bleaching event, including community-wide trophic restructuring. Some of these changes are likely to affect the process of reef recovery, such as the declines observed in fish species that consume algae from reef surfaces. It is clear that this is a critical moment for the Great Barrier Reef, especially as the area experienced severe bleaching again in early 2017, causing further damage (Hughes *et al.*, 2018b). It is difficult to grasp the enormity of what is happening, as beautiful reefs teeming with fish are being turned into graveyards, but it is enough to reduce experienced researchers to tears (Dr Sally Keith, personal communication).

4.2 Marine Fisheries

As noted in the preceding section, fishing is a major cause of ecological change in coral reef ecosystems and can contribute directly to ecosystem collapse. Given that fishing is widespread throughout the world's oceans, such impacts are not limited solely to coral reefs; many other marine ecosystems must surely be at risk. The issue of marine fisheries has attracted the interest of a large community of ecological researchers, as well as fisheries scientists, who have generated a copious and fascinating literature. While heroically surfing this ocean of information, I have again had to be highly selective. In this task, I acknowledge a debt of gratitude to my principal navigational aid, the very useful blog produced by Branch (2016).

Collapse

Although researchers working on fisheries seem to delight in controversy and debate, there is general agreement that marine fisheries are in crisis. The nearest approximation to an official statement on this topic is that provided by the FAO, whose latest report makes the following points (FAO, 2018):

- The mass of fish harvested from wild populations has remained relatively static since the late 1980s, with ~80 million tonnes harvested from marine environments in 2016.
- The total value of capture fisheries is currently around US$130 billion annually, employing around 40 million people who use about 4.6 million boats.
- Between 1961 and 2016, the mean annual increase in global food fish consumption (3.2%) outpaced human population growth (1.6%) and exceeded that of meat from all terrestrial animals combined (2.8%). Annual food fish consumption is now around 20 kg per capita on average and accounts for about 17% of animal protein consumed by the global population of people.
- According to FAO statistics, the state of marine fishery resources is steadily declining. The fraction of marine fish stocks fished at biologically unsustainable levels increased from 10% in 1974 to more than 33% in 2015, with the largest increases recorded in the late 1970s and 1980s. Nearly 60% of fish stocks are now classified as maximally fished.
- It seems unlikely that the 33% of stocks that are currently overfished can recover in the very near future, because such recovery requires time, usually two to three times the lifespan of the harvested species.

One of the controversies that has exercised fisheries scientists is the reliability of the official statistics used by FAO. The accuracy of these data has implications for estimating the magnitude of the fisheries crisis. Given its position as an intergovernmental organisation, the FAO has to rely on the data provided by its member countries, even if these data are inaccurate. The kind of problems such inaccuracies can create was described by Watson and Pauly (2001), who showed that misreporting by countries with large fisheries, such as China, can lead to erroneous conclusions regarding global trends. This has led to concerns being raised about the value of using official catch data to infer trends (Pauly *et al.*, 2013). To address such concerns, efforts have been made to collect independent fish consumption and trade data to evaluate official catch statistics. Comparison of these data suggests that catches are under-reported by 100–500% in many developing countries and by 30–50% in developed ones (Pauly *et al.*, 2013; Pauly and Zeller, 2016; Zeller *et al.*, 2011). Such results imply that the global fisheries crisis is more serious than official statistics might suggest – a case of 'aquacalypse now' (Pauly, 2009)?

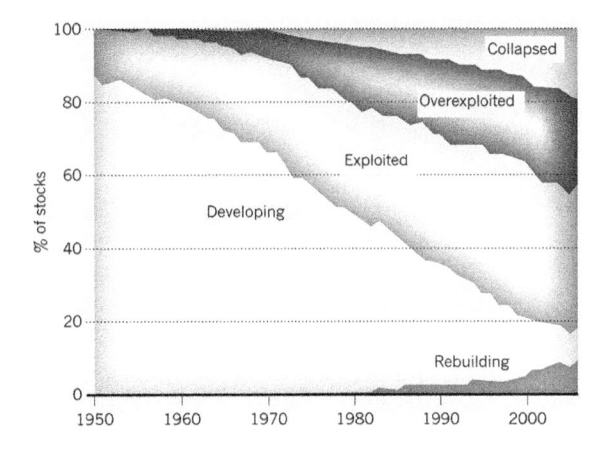

Figure 4.3 Collapse of marine fisheries. Reproduced from Pauly *et al.* (2013), with permission from Springer Nature.

Fishery catch data have been used to assess the incidence of fishery collapse, although here again there has been controversy. Daniel Pauly and colleagues used modified official catch data to produce stock status plots for all the fisheries in the world for which such data were available (Pauly, 2007; Pauly *et al.*, 2013). This suggested that the number of collapsed fish stocks has steadily increased in recent decades (Figure 4.3), such that by the mid-1990s, 20% of the stocks that were exploited in the 1950s had collapsed. Here, fish stocks were classified as collapsed if their annual catch had declined to less than 10% of the highest values ever recorded (Pauly *et al.*, 2013).

The fact that fisheries can collapse is well established. One of the most striking examples is the case of Atlantic cod in Canada, which was once was one of the world's most abundant stocks of this species. In the early 1960s, the northern population of cod numbered nearly 2 billion breeding individuals, but over the next few decades, numbers declined by 97%. Other data, based on catch rates from independent surveys, suggest that northern cod numbers have declined by 99.9% since 1983 because of overexploitation (Hutchings and Reynolds, 2004). Other examples of fisheries collapsing because of overfishing include California sardine in the 1950s, Peruvian anchovy in the early 1970s and several stocks of North Sea herring since the 1960s (Dickey-Collas *et al.*, 2010; Pauly, 2008). In an analysis of time series data for 232 populations of marine fish, Hutchings and Reynolds (2004) reported a median decline of 83% in recent decades, with more than half of the populations exhibiting declines of 80% or more, suggesting that collapse is widespread.

One of the difficulties of interpreting catch statistics is that they do not necessarily provide an accurate indication of the number of fish in the sea (Pauly et al., 2013). For example, recent catch data suggest that 34 large stocks of different species along the west coast of Oregon and Washington in the USA have collapsed, yet this number was reduced to three (anchovies, candlefish and abalone) after detailed stock assessments were conducted (Pauly et al., 2013). Consequently, the economic collapse of a fishery does not necessarily equate directly to ecological collapse of a marine ecosystem. Furthermore, some 'collapsed' stocks can continue to be fished; in fact, this is occurring today in many parts of the world's oceans. Examples include the Swedish west coast, where a long-collapsed Atlantic cod stock continues to be harvested, and the Gulf of Thailand, where a demersal fishery that had collapsed by the 1990s also continues to be exploited (Pauly, 2016).

Such uncertainties help to explain the debate surrounding the publication of Worm et al. (2006), which is a notable piece of research, if only for the controversy that it generated. This paper presented a compilation of long-term trends in 12 coastal and estuarine ecosystems, which showed mean depletion (>50% decline over baseline abundance), collapse (>90% decline), or extinction (100% decline) of 91%, 38% or 7% of species, respectively. Only 14% of species subsequently recovered from collapse. Furthermore, this study analysed FAO catch statistics for 64 marine ecosystems worldwide. Fishery collapses were defined as a decline in catches to below 10% of the recorded maximum. Results indicated an increase in the incidence of collapse over time, with 29% of currently fished species considered collapsed in 2003. Consistent with the results from the coastal ecosystems they examined, Worm et al. (2006) observed that fishery collapses occurred at a higher rate in relatively species-poor ecosystems than in species-rich ones. Perhaps most controversially of all, the regression analyses presented were extrapolated to project the global collapse of all taxa currently fished by the mid-twenty-first century. Or, to be more precise, the year 2048, a date that Pauly (2016) described as both Orwellian and reminiscent of Mayan prophecies of doom.

The idea that all the fisheries in the world would collapse by 2048 was met with widespread scepticism (Pauly et al., 2013). For example, Daan et al. (2011) responded by paraphrasing Mark Twain, suggesting that reports of the death of fish were greatly exaggerated. A scientific debate is surely getting serious when witticisms by literary figures are being co-opted to support the cause. Daan et al. (2011) highlighted the following flaws in the work of Worm *et al.* (2006): (i) use of the underlying algorithm would have showed a similar pattern to that observed in catch statistics if any series of random numbers has been used; and (ii) it is incorrect to interpret the period of maximum harvest in time-series data as reflecting a period during which a stock was fully exploited, as such harvesting intensities might not have been sustainable. This study was one of many that challenged the methods employed in the original investigation. The controversy was further examined by Branch (2013), who examined 664 papers that cited Worm *et al.* (2006). He then used the results to make an interesting contention: that the more knowledgeable the authors of citing papers were about the controversy over the 2048 projection, the less likely they were to refer to it directly. In other words, the projection was seen by those 'in the know' as something of a distraction, perhaps generated by an overzealous press release, and should not be used to undermine broader concerns about biodiversity loss presented in the original paper. Although these conclusions may be sound, one is left wondering whether it really is legitimate to infer the knowledgeability of scientific authors from variables such as the number of co-authors included on a paper, as performed in this study. Discredited papers about homoeopathy (e.g. Davenas *et al.*, 1988) typically have plenty of co-authors, for example.

The controversy surrounding Worm *et al.* (2006) reflected deeper concerns relating to a whole series of publications about marine fisheries in leading journals; Hilborn (2006) alleged that these had been selected for publication more on the basis of their publicity value than their scientific merit. Similar concerns have been raised in other branches of conservation science (e.g. Ladle *et al.*, 2004). But to their credit, and in contrast to some other conservation debates, the fishery science community responded by creating a new collaborative project involving proponents from both side of the argument, including both Hilborn and Worm. This collaboration is doubly impressive, given that Hilborn had previously described Worm's work as 'mind-bogglingly stupid' (Stokstad, 2009). The initiative aimed to compile new data sets to assess the status of marine fisheries. The resulting publication paints a more nuanced picture than the original paper, by recognising the positive achievements that have been made in restoring some marine ecosystems (Worm *et al.*, 2009). In this way, the paper attempts to bridge the divide between ecological researchers and fishery management scientists, which had become rather polarised during this debate. Results noted that the fish exploitation rate has declined in a number of well-studied fisheries and is now at or below the value associated with achieving maximum sustainable yield. Despite this, some 63% of assessed fish stocks worldwide are in need of a reduction in exploitation rates (Worm *et al.*, 2009). This investigation, which was based only on those stocks that are closely monitored by research agencies, therefore suggested that many of the world's fisheries are not in such a perilous state as implied by the 2048 projection (Pauly *et al.*, 2013). However, some 14% of the 166 stocks that were analysed were judged to have collapsed by the mid-2000s (Pauly *et al.*, 2013; Worm *et al.*, 2009).

This wasn't the end of the story, however. A more recent analysis with updated methodology revisited the assessment of global fish stocks and examined where they might be heading in future. Using data from 4,713 fisheries worldwide, Costello *et al.* (2016) showed that 68% of stocks were below the biomass target that supports maximum sustainable yield, and only 35% are currently fished at a level that would allow for recovery. About 118 fisheries were found to be associated with mortality rates higher than 10 times the sustainable yield target. Using a bioeconomic model, the authors showed that by 2050, 88% of stocks would be overfished, suggesting widespread collapse (Worm, 2016). Costello *et al.* (2016) also noted that reforms of fishery management can dramatically increase overall fish abundance, and under such situations, recovery of fish populations can happen quickly, with the median fishery taking under 10 years to reach recovery targets.

Why, exactly, do fish populations collapse if overfished? Hutchings and Reynolds (2004) highlight the importance of the following factors in determining the impact of harvesting on fish populations, in relation to both collapse and recovery:

- *Life-history traits.* According to theory, a high recovery rate should be correlated with low age at maturity, small body size, short lifespan and rapid individual growth. Species with contrasting traits would therefore by at higher risk of collapse. Empirical work on marine fishes has supported these predictions.
- *Population structure.* Fishing always affects size and age distributions of fish populations, as most fisheries target particular age or size classes. Typically, larger, older, faster-growing individuals are more likely to be caught than smaller, younger, slower-growing individuals. This may have implications for reproductive success and therefore population dynamics following harvesting.
- *Community structure.* Changes to community structure and food webs can occur as a result of species in particular trophic levels being

harvested, such as predatory fishes. This can alter predator–prey interactions and interspecific competition, which has been implicated in impeding the recovery of Atlantic cod in Canada.
- *Allee effect*. Low rates of recovery may result from situations when rates of population growth per individual decline when population sizes fall below some threshold level of abundance. This form of positive feedback can be caused by negative density dependence, such as an increased difficulty of finding mates at lower population densities. The Allee effect has been suggested as an explanation for the relatively slow recovery of Atlantic cod and other marine fishes, although empirical data are generally lacking.
- *Evolutionary changes*. Selective harvesting may cause evolutionary changes in some populations. For example, in some areas cod have become smaller at a given age and are reproducing at smaller sizes than they did historically. Reductions in age and size at maturity may negatively affect longevity and mortality and thereby increase risk of collapse. Evolutionary reductions in body size may also lead to reduced fecundity, smaller egg size and reduced offspring survival, all of which would negatively affect population growth.
- *Habitat modification*. Harvesting approaches such as repeated bottom trawling may smooth and flatten the seabed, reducing physical heterogeneity and leading to habitat loss for demersal species.

The idea that large, slow-growing fish species at the top of food chains are most at risk of collapse (Hutchings and Reynolds, 2004) has been challenged by the work of Malin Pinsky and colleagues. Pinsky *et al.* (2011) analysed both stock assessment and landing data of species that have experienced population collapses, in relation to their life history traits. Results showed that up to twice as many fisheries of small, low trophic-level species have collapsed compared with those of large predators. In other words, species with relatively rapid life histories have at least as high a probability (per stock) of undergoing population collapses as larger, slower species. This represents a very interesting contrast to the data available for terrestrial animal species. Possible causes for this difference suggested by Pinsky *et al.* (2011) include that: (i) fishery management often recommends higher exploitation rates for species with more rapid life histories; and (ii) a relatively rapid life history may actually increase vulnerability to collapse. The latter situation could arise because populations of short-lived species can respond rapidly to changes in climate; potentially, changes in fisheries management might lag behind these changes and therefore drive a population to collapse (Pinsky *et al.*,

2011). This idea was explored further by Pinsky and Byler (2015), who examined the influence of both species traits and climate variability on risk of collapse for 154 marine fish populations. While the magnitude of overfishing was the most important factor explaining collapse, rapid growth was the second most important factor, increasing the risk of collapse by more than a factor of three. Relatively fast-growing populations and those occurring in variable environments were most sensitive to overfishing, supporting their original hypothesis (Pinsky and Byler, 2015). Fishery collapses might therefore best be understood as the combined effects of overfishing, life-history traits and climatic variability, with relatively fast-growing species in climatically variable environments to being especially susceptible to overfishing.

A key question, though, is how collapses in the populations of harvested fish species affect other parts of the ecosystem. As Pinsky *et al.* (2011) note, even temporary collapses of relatively small, low trophic-level fish species can have significant ecosystem-wide impacts by reducing food supply to larger fish, marine mammals and seabirds. Although small, short-lived species might typically be able to recover more quickly from population declines than other fishes, collapses in such species can extend from years to decades, the timescale often depending on the duration of the fishing effort. These changes can have substantial impacts on food webs. For example, in the Benguela system off south-western Africa, harvesting of sardine and anchovy has affected seabirds over a period of several decades, illustrated by the 90% and 95% declines in the breeding numbers of penguins and gannets, respectively, that have been observed since the late 1950s (Crawford, 2007). Effects on ecosystem structure and function may be particularly pronounced in 'wasp-waisted' ecosystems, where only a few species occur at intermediate trophic levels, contrasting with the higher diversity occurring at lower and upper trophic levels. In such cases, there may be top-down control of zooplankton abundance by the intermediate small pelagic fish component, as well as bottom-up control exerted by these fish on predators (e.g. marine birds and mammals) higher up the food chain (Cury *et al.*, 2000). An example is provided by sandeels (*Ammodytes marinus*), which support the largest single-species fishery in the North Sea; declining sandeel abundance has been associated with severely reduced breeding success in seabirds such as kittiwakes (Frederiksen *et al.*, 2004).

Although trophic cascades provide a possible mechanism for ecosystem collapse, initial investigations of heavily exploited marine ecosystems failed to find much evidence for them (Frank *et al.*, 2005). While they

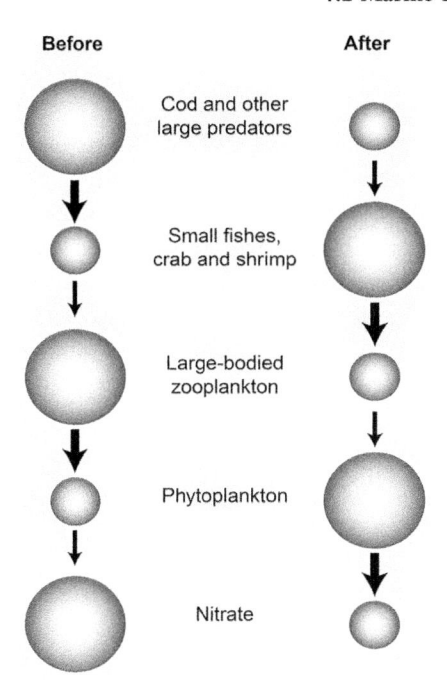

Figure 4.4 Cascading effects of the collapse of cod and other predatory fishes on the coastal shelf ecosystem in Nova Scotia during the 1980s and early 1990s. Reproduced from Scheffer *et al.* (2005), after Frank *et al.* (2005), with permission from Elsevier.

have been demonstrated for limnic and marine benthic systems, it was suggested that trophic cascades might not be widespread in open-ocean pelagic ecosystems (Terborgh and Estes, 2010). However, in a meta-analysis of population interactions between cod and shrimp in the North Atlantic, Worm and Myers (2003) showed that regulation of trophic structure was predominantly top-down. This suggests that in some cases, the effects of overfishing in the ocean can cascade down to lower trophic levels. Further, Frank *et al.* (2005) provide evidence of a trophic cascade in the coastal shelf ecosystem off Nova Scotia, Canada (Figure 4.4) involving four trophic levels and driven by changes in the abundance of large predators (primarily cod). Here, cascading effects were found to influence the entire community (Scheffer *et al.*, 2005). Similarly, in the north-west Atlantic Ocean, the decline of large sharks has led to a rapid increase in populations of cownose rays (*Rhinoptera bonasus*), which resulted in a decline in bay scallop populations (*Argopecten irradians*); this provides an example of one fishery affecting another (Heithaus *et al.*,

2008; Myers et al., 2007). Other examples of trophic cascades in marine fisheries include the Black Sea, the Barents Sea and the Baltic Sea (Andersen and Pedersen, 2010).

There has therefore been a recent shift in our understanding of how marine ecosystems are structured. The traditional view of bottom-up control, where phytoplankton dynamics determines the production and biomass of upper trophic levels, is being replaced by increased recognition of top-down control, where the dynamics of prey species is determined by predator abundance (Frank et al., 2007). However, top-down control is not ubiquitous. This raises the question of what influences the distribution of top-down *versus* bottom-up control of marine ecosystems. Frank et al. (2007) suggest that sea temperatures and species richness may be two influential factors; relatively cold and species-poor areas might be more susceptible to top-down control and recover more slowly (if ever) from collapse, whereas warmer areas with more species might be characterised by either top-down or bottom-up control, depending on exploitation rates and perhaps by changing climate.

Similarly, there is uncertainty regarding when, and under what conditions, the cascading effects of fishing are likely to occur. Salomon et al. (2010) provide a valuable review of evidence relating to this issue and make the following points:

- Response times often differ among trophic levels, and trajectories of change are often non-linear; in addition, trophic cascades are often associated with major changes in community structure. Fishery-induced trophic cascades therefore have the potential to drive shifts in ecosystem states, in which each state is maintained by internal feedback mechanisms. Such transitions could be associated with hysteresis (i.e. system recovery follows a trajectory different from that observed during decline), although there is currently limited empirical evidence for this. However, there are many examples of major changes in ecosystem configuration being triggered by fishing including New Zealand, the Mediterranean, Chile and the Gulf of Maine.
- An example of a state shift associated with hysteresis is provided by the Black Sea, where a cascade across four trophic levels altered nutrient and oxygen concentrations in the surface water, which then led to a second state shift (Daskalov et al., 2007). Excessive exploitation of consumers appears to have been the triggering factor of both state shifts.
- Examples of stabilising mechanisms that enable the persistence of a fishery-induced shift in ecosystem state include nutrient enrichment,

changing competitive relationships and climate, as illustrated in the Baltic Sea (Österblom et al., 2007). Predation is an additional mechanism, illustrated by North Atlantic cod, where recovery following collapse has been slow, partly because of an increase in forage fish that predate juveniles. Such mechanisms can inhibit ecosystem recovery.

- Intense exploitation can affect parasite and disease dynamics throughout food webs. For example, harvesting of predators can lead to increases in sea urchin numbers, which can increase the incidence of disease outbreaks among urchins.
- Removal of predatory fish can lead to significant changes in benthic primary production and carbon flux, showing that trophic cascades can alter ecosystem processes.
- The magnitude of fishery-induced trophic cascades appears to depend on context, with influential factors including species diversity, regional oceanography, local physical disturbance, habitat complexity and the nature of the fishery itself (Table 4.2).
- Trophic cascades can also be caused by multiple stressors, which can potentially interact. For example, reduction of fish stocks to low levels increases their dependence on annual recruitment and therefore their vulnerability to environmental fluctuations. As a result, climatic variability can influence the timing, magnitude and persistence of fishery-induced regime shifts, as illustrated by the Baltic Sea.

Further ecosystem impacts of marine fishing were reported by Maureaud et al. (2017), who used global catch data to examine biomass transfers within food webs. Results indicated a trend towards more rapid and efficient biomass transfers in marine ecosystems, attributable to the over-exploitation of top-predators and increasing dominance of small, fast-growing species. This may have reduced the overall biomass of the ecosystems, particularly at high trophic levels. This study also provided evidence in support of the 'fishing down the food web' phenomenon, which is yet another area of lively debate among fishery researchers. Impressively, Daniel Pauly has once again been at the centre of this controversy, the idea being popularised by his influential *Science* paper from 1998 (Pauly et al., 1998). This highlighted a shift in fish landings from long-lived, high trophic level species towards short-lived species associated with lower trophic levels. In other words, according to this idea marine fisheries tend to first deplete top predators and then successively target species lower and lower down the food chain, as stocks are progressively exhausted.

Table 4.2 *Factors that alter the occurrence and magnitude of fishery-induced trophic cascades by either dissipating or amplifying the transmission of indirect fishing effects throughout marine food webs*

Context-dependent factors	Effect	Explanation
Species diversity and trophic complexity	Dissipate	High species diversity facilitates the replacement of overfished species.
Top-down control	Amplify	Marine ecosystems under strong top-down control are less resilient to the exploitation of top predators and more susceptible to fishery-induced trophic cascades.
Regional oceanography	Dissipate or amplify	Factor mediates rates of recruitment, primary production, growth and maturation, predation and herbivory.
Recruitment limitation and variability	Dissipate	Low or sporadic recruitment at intermediate trophic levels can slow the recovery rate of prey following predator depletion.
Local physical disturbances	Dissipate	Physical disturbance can decouple the trophic links between predators, herbivores and primary producers.
Multi-trophic-level fisheries	Dissipate	Prey release that may have occurred following the exploitation of predators is offset by the harvest of prey.
Disease	Dissipate	A dramatic depletion in predators may cause prey to exceed their host density threshold for epidemics.
Predator-avoidance behaviour by prey	Amplify	Because most organisms behave in ways that moderate their exposure to predation risk, often at the cost of reduced food intake, predator depletion can also lead to increased prey-foraging effort or efficiency.
Habitat complexity and spatial refuges	Dissipate	Predator effectiveness can be dampened by prey-avoidance behaviour and the availability of safe hiding spots.

Source: Reprinted from Salomon *et al.* (2010), with permission from Wiley.

Clearly such a shift could have major implications for the structure and functioning of marine ecosystems, but once again the use of official catch statistics has been questioned, together with the way trophic level was assessed using these data. For example, using different data including those obtained from stock assessments, Branch *et al.* (2010) found no evidence of the trophic shift reported by Pauly *et al.* (1998) – in fact a converse trend was reported. Furthermore, in a study of fishery landings in 48 large marine ecosystems worldwide, Essington *et al.* (2006) found that fishing down the food web occurred in 9 ecosystems, whereas fishing up occurred in 18. However, the most common pattern (21 ecosystems) was fishing through – where catches of top predators remained high or increased, but fishing also expanded to include species of lower trophic levels (Branch, 2015).

Can fishing result in a shift between alternative ecosystem states? Some researchers have interpreted their results in this way. For example, Casini *et al.* (2009) describe a reorganisation of the central Baltic Sea ecosystem caused by a trophic cascade, following harvesting of the top predator, cod. The transition between two ecosystem states was induced by crossing an ecological threshold, represented by a particular abundance of zooplanktivorous fish, namely sprat. The two ecosystem configurations, namely high cod/low sprat *versus* low cod/high sprat abundance, were associated with contrasting zooplankton biomass, zooplankton community composition and abundance of large copepods. Interestingly, the transition between these ecosystem states represented a shift from bottom-up to top-down control of zooplankton dynamics. The transition also had implications for ecosystem recovery: the authors suggested that the current regulation by sprat of the feeding resources for larval cod may restrict recovery of cod populations and also hinder the return of the ecosystem to its prior state (Casini *et al.*, 2009). Such multilevel top-down regulation has rarely been reported from open marine systems; its detection in the Baltic Sea might be attributable to the relatively low complexity of its ecosystem, with its low species richness, simple food web structure and weak omnivory. These factors are believed to increase the susceptibility of ecosystems to top-down regulation (Casini *et al.*, 2008).

In a review of evidence for regime shifts and marine fishing in the northern hemisphere, Möllmann and Diekmann (2012) highlight examples in a number of regions, including the North Pacific, North Sea and the Mediterranean. Overall, climate change was found to be the most influential driver, together with overfishing; eutrophication and introduction of invasive species were also important factors in some cases. The potential for interactions between multiple drivers was noted. In addition, trophic cascades were widely identified as a significant contributor to ecosystem reorganisations. These examples of regime shifts indicate that fishing may often be associated with rapid transitions between different ecosystem states. However, these do not necessarily equate to alternative stable states as defined by dynamical systems theory. For verification of alternative stable states, experimental evidence is ideally required, but as the authors note, this is largely lacking for marine ecosystems. The available evidence for alternative states is largely dependent on analysis of time series data, which can be used to identify multimodality of the frequency distribution of ecosystem states, or to test whether the relationship of state variables to the control variable is described by two different functions, indicating possible hysteresis (Möllmann and Diekmann, 2012). Such analyses have rarely been applied to marine systems, partly owing to the lack of suitable field data. Apart from the Baltic Sea investigation described earlier (Casini *et al.*, 2009), the best example appears to be the Black Sea, where abrupt changes in time series data and bimodality related to climatic and biogeochemical variables have been recorded, associated with overfishing and eutrophication (Daskalov *et al.*, 2007; Oguz and Gilbert, 2007; Oguz and Velikova, 2010).

Bakun and Weeks (2004) provide a further example of a regime shift, focusing on the Northern Benguela Current Large Marine Ecosystem off the coast of Namibia. This is an upwelling zone with very high rates of primary productivity, which formerly supported a large sardine population. Abundance of sardines collapsed as a result of very heavy fishing pressure in the 1970s. This has resulted in a shift to a less desirable ecosystem state characterised by unchecked growth of phytoplankton, on which the sardines feed. After the collapse of sardine, the rate of deposition of dead phytoplankton onto the sea floor would likely have increased, which may have increased the incidence of hypoxia (Bakun and Weeks, 2004). This provides an informative parallel with the widespread oceanic anoxic events that characterised the Mesozoic period and were associated with mass extinction events (see Chapter 3 and in the following text). The incidence of hypoxia and toxic gas eruptions would impact negatively on grazing organisms, including sardines, thereby maintaining the system in its altered state (Bakun and Weeks, 2006). This case study also provides a potential link between fish harvesting and global-scale ecological processes, as decomposition of phytoplankton could lead to an increase in emissions of methane, which is a potent greenhouse gas (Bakun and Weeks, 2004). These authors also highlighted a number of potential feedback mechanisms that might have contributed to this ecosystem collapse, including the *prey-to-predator loop* and the *predator-pit loop* (Bakun and Weeks, 2006). The former is based

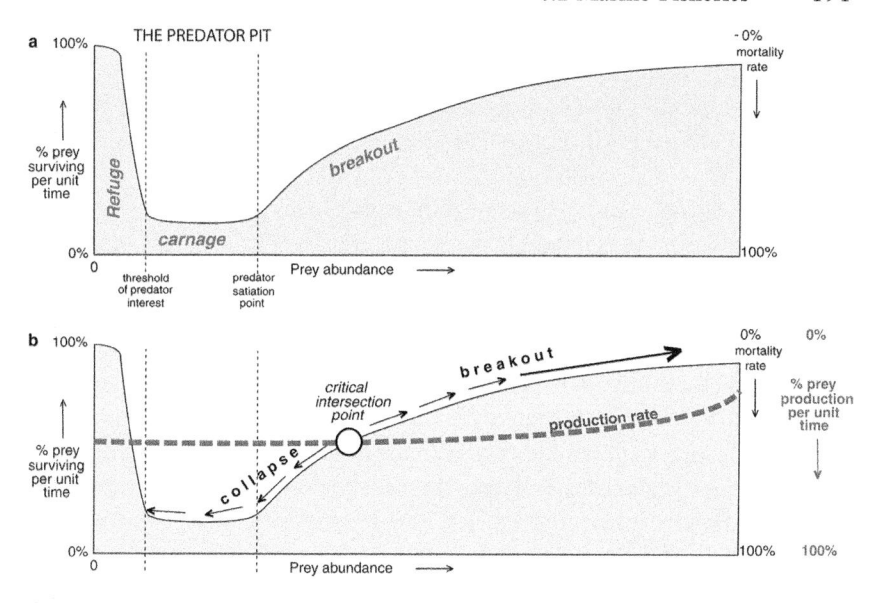

Figure 4.5 Diagram of the predator-pit feedback loop. Below a threshold level of prey abundance, predator interest declines to reach a level of zero mortality as prey abundance approaches zero, thus precluding local extinction and survival of species in 'refuges'. Conversely, above a threshold, the prey population will grow (i.e. 'breakout') in an accelerating fashion, owing to a positive feedback loop. This explosive growth to very high biomasses often includes major expansions in occupied habitat. This mechanism may apply to small oceanic pelagic fish and other small prey organisms, as well as to juvenile stages of larger predatory fishes. Reproduced from Bakun and Weeks (2006), with permission from Wiley.

on the fact that prey fishes often consume the eggs and larvae of the predators that prey on them. A decrease in population sizes of a predatory fish, such as cod, would reduce predation pressure on those species that prey on their early life stages, thereby adversely affecting cod reproductive success still further. This type of feedback loop may partly explain why collapsed cod populations are often very slow to recover (Bakun and Weeks, 2006; Möllmann and Diekmann, 2012). The latter mechanism often applies to relatively small fish, which are often characterised either by relative rarity or superabundance (see Figure 4.5).

In his review of critical transitions, Scheffer (2009) also summarises evidence for regime shifts in marine ecosystems, including kelp forests, sea grass and collapses of oyster fisheries in estuaries. A fascinating example of a potential feedback mechanism is given for the example of sardines and anchovies, populations of which often alternate. Here, individuals of either species have an urge to form schools. When a species is rare, perhaps because of overfishing, individuals may often join schools of the other species, where their fitness is likely to be lower. This 'school trap' could act as a form of Allee effect and may help explain why a gradual change in environmental conditions (such as temperature) could lead to an abrupt change between dominant species. Scheffer (2009) also highlights research undertaken in the North Pacific, where large-scale ecosystem shifts appear to have occurred in the mid-1970s and again in the late 1980s. While changes in physical conditions, such as ocean currents and sea temperatures, are likely to be the main drivers of these shifts, there may also be a biological component. Time-series analysis indicated that whereas key physical variables varied in a stochastic linear manner, biological variables displayed non-linearity. This suggests that large-scale marine ecosystems can be dynamically non-linear owing to biological mechanisms, and as a result they may change abruptly in response to stochastic fluctuations in the physical environment (Hsieh et al., 2005). Ecological dynamics may therefore not simply track environmental change, but may amplify changes in the physical environment, such as climate. Overall, Scheffer (2009) concludes that most of the regime shifts observed in marine ecosystems are likely to be attributable to physical causes, but in some cases, feedbacks in biological communities appear to be large enough to cause threshold-like responses to changing conditions. Although true alternative stable states may exist, these are difficult to prove; evidence is perhaps stronger in coastal ecosystems. In both the coastal areas and the open ocean, fishing pressure and physical conditions have a strongly interactive effect on ecosystem dynamics.

One of the phenomena associated with the ecosystem transitions described by Bakun and Weeks (2006) is the outbreak of jellyfish and 'other undesirable types of pelagic zooplanktivores'. Jellyfish blooms have become a major global environmental issue, owing to the combined effects of overfishing, eutrophication, climate change and habitat modification. Richardson et al. (2009) describe the rapid conversion of pelagic ecosystems from one that is dominated by fish (which keep jellyfish in check by predation and competition), to a less desirable 'gelatinous state' dominated by jellyfish. The idea of being converted into something gelatinous is surely bound to evoke horror in anyone raised on 1950s sci-fi B movies, such as *The Blob*. Jellyfish outbreaks often follow over-exploitation and collapse of a small filter-feeding fish stock. The animals can reach spectacularly high densities, enough to clog fishing nets and even to capsize a trawler. Richardson et al. (2009) suggest that a jellyfish-dominated ecosystem state may persist through a self-enhancing feedback loop, involving the predation of fish eggs and larvae by jellyfish and increased jellyfish reproductive success as their abundance increases. According to these authors, this 'jellyfish joyride' might lead to a more gelatinous future, although intriguingly, this might be more accurately described as a return to the past. Parsons and Lalli (2002) suggest that anthropogenic disturbance threatens to return modern marine ecosystems back to those of the Palaeozoic era, when jellyfish dominated the seas.

Parallels with the deep past are further strengthened by the increasing incidence of dead zones in modern oceans, characterised by hypoxia (Bakun and Weeks, 2006). In coastal waters, such dead zones are typically caused by eutrophication, which results in an increase in abundance of planktonic algae. After these organisms die and are deposited on the seabed, the increase in microbial respiration leads to a decline in dissolved oxygen concentrations. This can subsequently lead to elimination of the benthic fauna, increased fish mortality and the loss of economically important fisheries, such as the collapse of the lobster fishery in Norway (Diaz and Rosenberg, 2008). Hypoxia can also have major impacts on ecosystem processes; for example, energy flows into microbial pathways may increase at the expense of flows into higher trophic levels, which can negatively affect fish recruitment (Diaz and Rosenberg, 2008).

Since the 1960s, dead zones have spread exponentially in the world's oceans and have now been reported from more than 400 ecosystems, affecting more than 245,000 km^2. Recovery has been very limited, with only 4% of these systems showing any signs of improvement (Diaz and Rosenberg, 2008). Where some recovery has been achieved, for example by reduction of nutrient and carbon inputs, it can take a number of years, the timescale depending on the duration and degree of hypoxia. For example, recovery of the benthos after hypoxia in the Black Sea in 1994 was still not complete a decade later (Mee et al., 2005). Furthermore, owing to variation between species in tolerance of oxygen depletion, the pattern of species establishment during recovery will not be the same as during species loss, leading to a hysteresis-like response. Examples of this form of recovery trajectory have been reported from Gullmarsfjord, Sweden and the Black Sea (Diaz and Rosenberg, 2008).

Human disturbances such as eutrophication and hypoxia can interact with other forms of disturbance, such as diseases, storms and climate change. Furthermore, the combined effects of multiple stressors may be increased by the loss of entire trophic levels as a result of fishing.

Examples include the outbreak of toxic blooms and disease following harvesting of oysters, and outbreaks of seagrass wasting disease following removal of seagrass grazers such as green turtles (Jackson et al., 2001). These synergistic effects can display threshold responses, with a time lag occurring before the impacts of fishing become fully manifest, probably attributable to ecological diversity and redundancy within trophic levels (Jackson et al., 2001). Examples include the delay in the collapse of kelp forests in southern California compared with Alaska after the extirpation of sea otters. Other predators and competitors of sea urchins, such as spiny lobsters and abalone, which were present in the more diverse Californian kelp forests, helped keep sea urchin populations in check until these had also been eliminated (Dayton et al., 1998). Similarly in northern California, Rogers-Bennett and Catton (2019) describe the collapse of kelp forest to sea urchin barrens along 350 km of coastline, caused by the combined effects of climate change and the mass mortality of sea stars as a result of disease. A further mechanism for relatively rapid ecosystem collapse is the elimination of refuges that were historically protected from fishing, for example because of difficulty of access. This mechanism has been proposed for the decline in American lobster following loss of larvae from deepwater offshore stocks, as well as collapse in Jamaican reefs (Jackson et al., 2001).

Recovery

Understanding the scope for recovery of marine systems depends first upon appreciating the extent to which they have been degraded. Here the exceptional work presented by Jackson et al. (2001) is of particular value. This investigation employed palaeoecological, archaeological and historical records to assess the likely structure and composition of marine ecosystems prior to the onset of fishing. Results indicated that the historical abundances of many large animals were massively higher than would have been estimated from modern observations. For example, based on estimates of pre-fishing population sizes, dugongs in Moreton Bay, Australia, have declined by more than 200-fold; green turtle has undergone a 30-fold decline in the Caribbean; and white abalone has undergone a >2,000-fold decline in California (Jackson et al., 2001). To these sad trends may be added the extinction of Steller's sea cow, which declined from an initial population of around 2,000 individuals to complete extinction by 1768, within 30 years of first being described (Turvey and Risley, 2006).

In their review, Jackson *et al.* (2001) conclude that overfishing of large vertebrates and shellfish was the first major anthropogenic disturbance to all the coastal ecosystems examined, which was later followed by other forms of disturbance, such as habitat destruction, pollution, species introductions and climate change. Across the different ecosystems considered, ecological changes resulting from overfishing were very consistent, involving simplification of food webs (Figure 4.6). Massive declines in the biomass and abundance of large animals were recorded in all examples; these species are now largely absent from most coastal ecosystems worldwide. The results obtained by Jackson *et al.* (2001) have important implications for establishing an appropriate baseline before evaluating both collapse and recovery. As degradation of ecosystems can continue over many human lifetimes, the prevailing view of the condition of an ecosystem can shift over time. Essentially, people can forget how the ecosystem used to be. Historical data may therefore be needed to assess whether a given ecosystem has collapsed or not. As Jackson *et al.* (2001) point out, the problem of shifting baselines also has implications for assessing the potential for recovery and the identification of appropriate restoration targets.

How long does it take for a depleted marine species to recover? Neubauer *et al.* (2013) note that it doesn't happen very often; only 1% of global fish stocks requiring rebuilding had done so by 2010. Similarly Lotze *et al.* (2006) found that 14% of depleted species in selected estuarine and coastal ecosystems showed some recovery during the twentieth century, with increased abundance mostly recorded among birds, pinnipeds and sea otters. Recovery of large marine animals has been observed in around 15% of populations (Lotze and Worm, 2009). Low recovery rates of fish populations have been attributed variously to Allee effects, adverse environmental conditions and the evolution of life-history traits caused by high fishing mortality (Neubauer *et al.*, 2013). Results of a meta-analysis indicated that the recovery rate of fish populations is driven by their exploitation history, as well as factors such as the extent of depletion and the fishing intensity during recovery. Modeling analysis suggested that most stocks could recover within a decade if fishing pressure were rapidly and substantially reduced, although for many other fisheries, recovery times of several decades were projected (Neubauer *et al.*, 2013). Similarly, Lotze *et al.* (2011) indicate recovery times for finfish and invertebrate stock range from 3 to 30 years, with recovery rates of demersal species generally being lower than pelagic ones.

The extent of recovery of marine species is difficult to evaluate given the lack of historical baselines (Jackson *et al.*, 2001); in their study, Lotze

196 · Contemporary Case Studies

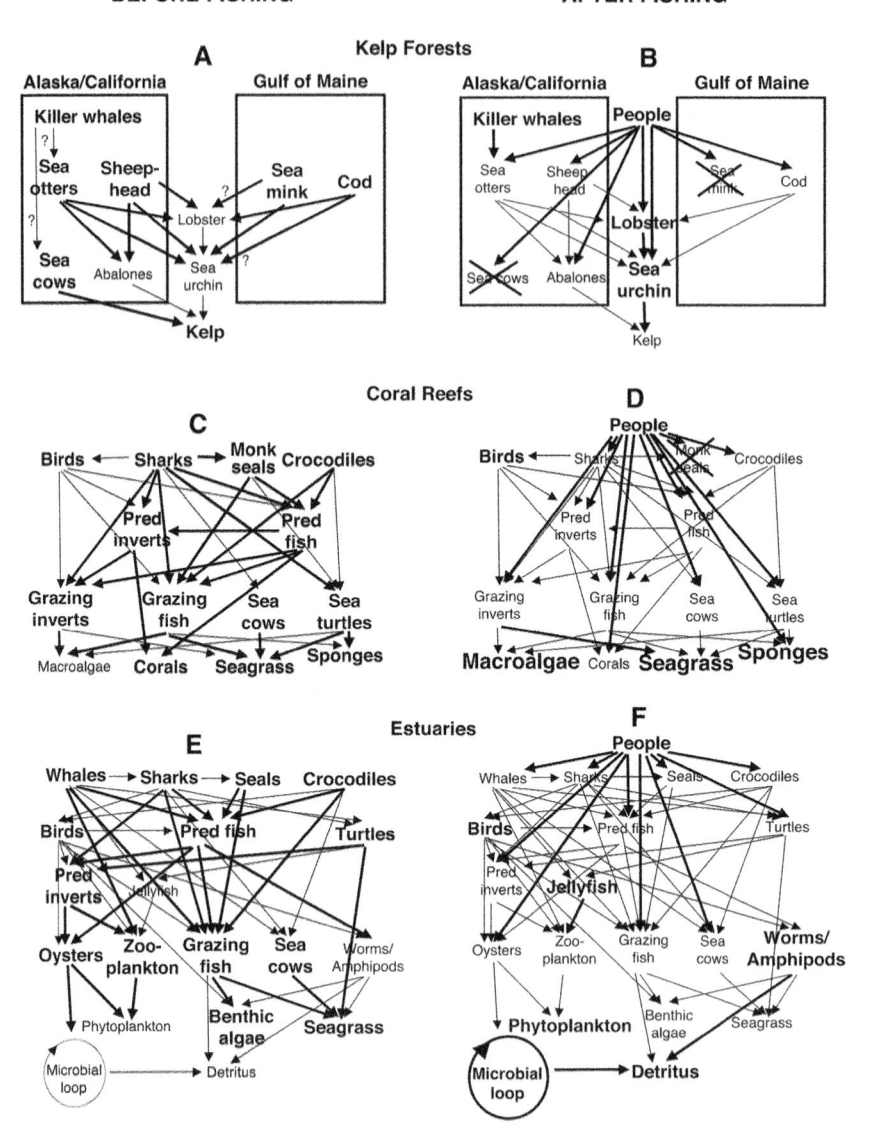

Figure 4.6 Simplified coastal food webs showing changes in some of the important top-down interactions due to overfishing; before (left side) and after (right side) overfishing. (A and B) Kelp forests for Alaska and southern California (left box), and Gulf of Maine (right box). (C and D) Tropical coral reefs and seagrass meadows. (E and F) Temperate estuaries. Bold font represents abundant; normal font represents rare; 'crossed-out' represents extinct. Thick arrows represent strong interactions; thin arrows represent weak interactions. Reprinted from Jackson *et al.* (2001), with permission.

and Worm (2009) reported mean recovery of 13–39% of historical values. Overall, the extent of recovery appears to depend on the duration, magnitude and type of disturbance, with higher depletion associated with lower rates of recovery (Lotze et al., 2011). Frank et al. (2011) provide an interesting account of the fate of a fished ecosystem after it collapsed. Assessing the Scotian Shelf of the east coast of Canada using data spanning several decades, these authors showed that after collapse, forage fish species increased in abundance by 900%. Subsequently, their numbers declined as a result of having outstripped their zooplankton food supply. As a result, their predation of fish species of other trophic levels declined, enabling abundances of the latter to increase. In this way, the system recovered towards its earlier structure, providing evidence that fisheries can recover after collapse. This occurred over a timescale of about 12 years after a fishing moratorium was first introduced.

Hughes et al. (2005) provide an overview of the recovery of damaged marine ecosystems and make the following useful points:

- Chronic human impacts are press disturbances, and therefore a return to original conditions is impossible unless the major ongoing drivers (e.g. fishing pressure, nutrient addition) are reduced.
- Marine ecosystems often exhibit hysteresis; i.e. recovery follows a trajectory different from that observed during decline. Some systems have changed so much that they can no longer return to their original state. It is therefore often easier to sustain an ecosystem in its current state than to enable it to recover after such a transition has occurred.
- Life histories of different species have a major influence on changing patterns of species composition during recovery. Relatively long-lived marine species (e.g. whales, turtles, dugongs, sharks) are likely to recover relatively slowly. For example, recovery of sea cows (*Dugong dugong*) in Queensland, Australia back to the population sizes of the 1970s will take at least 120–160 years, given the limited annual growth rate of sea cow populations of 2–3%.
- All the major fishing grounds worldwide have seen a shift to relatively disturbance-intolerant, fast-growing species that are more prone to environmental fluctuations.
- Alternate ecological states can be maintained by density-dependent mortality (e.g. owing to altered predator–prey ratios) or by density thresholds required for reproductive success.

Table 4.3 *Status and trends of major ocean ecosystems defined by principal symptoms and drivers of degradation, as summarised by Jackson (2008)*

Ecosystem	Status	Symptoms	Drivers
Estuaries and coastal seas	Critically endangered	Marshlands, mangroves, seagrasses and oyster reefs reduced 67–91%; fish and other shellfish populations reduced 50–80%; eutrophication and hypoxia, sometimes of entire estuaries, with mass mortality of fishes and invertebrates; loss of native species; toxic algal blooms; outbreaks of disease; contamination and infection of fish and shellfish; human disease	Overfishing; run-off of nutrients and toxins; warming due to rise of CO_2; invasive species; coastal land use
Continental shelves	Endangered	Loss of complex benthic habitat; fishes and sharks reduced 50–99%; eutrophication and hypoxia in dead zones near river mouths; toxic algal blooms; contamination and infection of fish and shellfish; decreased upwelling of nutrients; changes in plankton communities	Overfishing; trophic cascades; trawling; run-off of nutrients and toxins; warming and acidification due to rise of CO_2; introduced species; escape of aquaculture species

Table 4.3 (*cont.*)

Ecosystem	Status	Symptoms	Drivers
Open ocean pelagic	Threatened	Targeted fishes reduced 50–90%; increase in non-targeted fish; increased stratification; changes in plankton communities	Overfishing; trophic cascades; warming and acidification due to rise of CO_2

Note: Ecosystem status was based on the judgement of the author, not a formal assessment.
Source: Reproduced with permission, copyright (2008) National Academy of Sciences, USA.

I close this section with a quote from Jackson (2008), summarising the current state of ocean ecosystems and the potential for recovery (see also Table 4.3):

Synergistic effects of habitat destruction, overfishing, introduced species, warming, acidification, toxins, and massive runoff of nutrients are transforming once complex ecosystems like coral reefs and kelp forests into monotonous level bottoms, transforming clear and productive coastal seas into anoxic dead zones, and transforming complex food webs topped by big animals into simplified, microbially dominated ecosystems with boom and bust cycles of toxic dinoflagellate blooms, jellyfish, and disease. Rates of change are increasingly fast and nonlinear with sudden phase shifts to novel alternative community states. We can only guess at the kinds of organisms that will benefit from this mayhem that is radically altering the selective seascape far beyond the consequences of fishing or warming alone. Today, most fish and invertebrate stocks are severely depleted globally, and one-half to two-thirds of global wetlands and seagrass beds also have been lost. Nowhere are there any substantial signs of recovery, despite belated conservation efforts, except for nominal increases in some highly protected birds and mammals.

4.3 Freshwater Ecosystems

It has been estimated that freshwater ecosystems contain about 10% of all animal species on the earth and perhaps as much as half of all vertebrate diversity (Geist, 2011). They are also of particular importance to human societies, owing to provision of critical services such as potable water. At the same time, they are arguably the ecosystems that have been most heavily altered by human activity, which has resulted in relatively high losses of biodiversity. Major threats to freshwater ecosystems include overexploitation of species; water pollution; fragmentation, destruction or degradation of habitat; flow modifications; water extraction and invasions by non-native species (Geist, 2011). As illustration of the extent to which human activities have affected rivers, around 2.8 million dams have been constructed, and more than 500,000 km of rivers and canals are now used for navigation and transport (Grill *et al.*, 2019). As a result of infrastructural development such as dams, only 37% of rivers longer than 1,000 km remain free-flowing over their entire length and less than a quarter flow uninterrupted to the ocean (Grill *et al.*, 2019). It has been estimated that 65% of global river discharge, and the aquatic habitat supported by this water, is currently under moderate to high threat, and at least 10,000–20,000 freshwater species are extinct or at risk of extinction (Vörösmarty *et al.*, 2010). This overall pattern of degradation suggests that collapse of freshwater ecosystems is likely to be widespread. It is also worth noting that the global fisheries crisis extends to freshwater ecosystems, although far less is known about its impacts here than in the sea (Allan *et al.*, 2005).

Collapse

Lakes are particularly informative for understanding ecosystem collapse. On geological timescales, all lakes are temporary, as they will eventually fill up with sediment. The oldest lake in the world is Lake Baikal in southern Siberia, with an age of 25 Myr. This is also the world's deepest lake, at 1,700 m; another impressive statistic is that it contains around a fifth of the world's unfrozen reserve of freshwater. Yet even Lake Baikal will eventually collapse, in common with all other lakes. In fact, Lake Baikal appears to have already collapsed at least once in its history; during the last Ice Age, its biota was virtually eliminated owing to environmental changes (Karabanov *et al.*, 2004). Many lakes are currently collapsing because of anthropogenic pressures. A powerful example is the Aral Sea, lying on the borders of Kazakhstan and Uzbekistan. Once the fourth largest lake in the world, its waters were diverted for irrigation from the mid-twentieth century, the volume of water inflow being reduced by 75% since 1960 (Revenga *et al.*, 2000). By 2005, it had lost more than half of its area and nearly 75% of its volume. As the lake

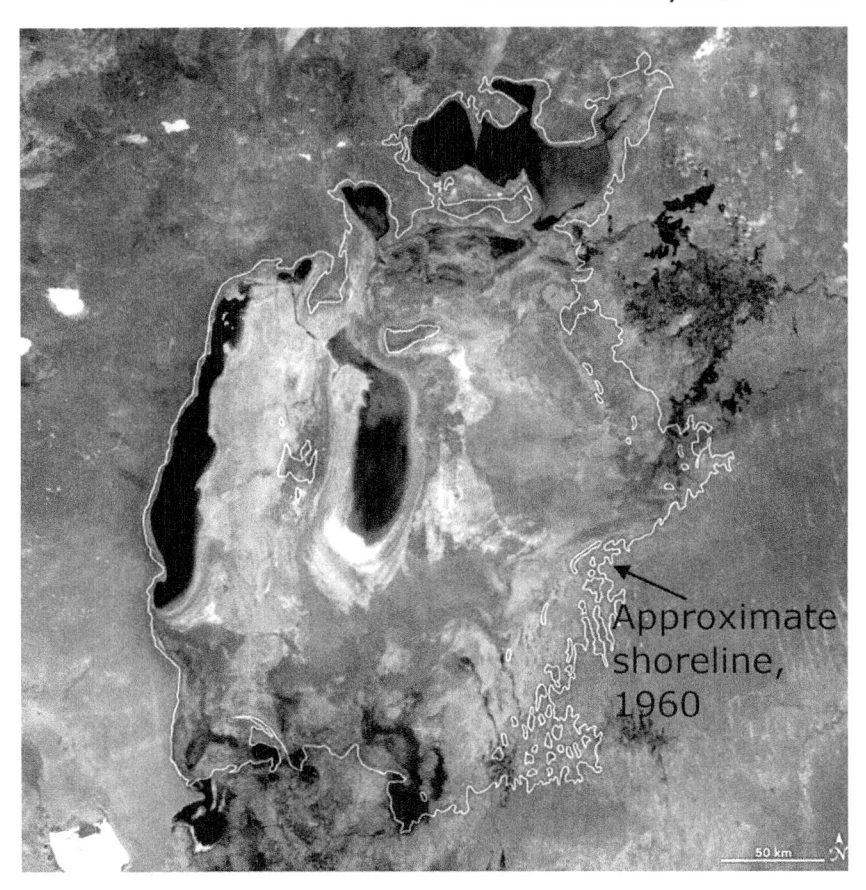

Figure 4.7 Decline of the Aral Sea. *Source*: NASA's Earth Observatory, https://earthobservatory.nasa.gov/world-of-change/aral_sea.php

declined, salinity increased and the fishery collapsed, with the loss of some 60,000 jobs. The eastern basin of the Aral Sea has now completely dried up and is now referred to as the Aralkum Desert (Figure 4.7), providing one of the most extreme examples of collapse of an aquatic system (Revenga *et al.*, 2000). Similarly, the wonderfully named Lake Poopó was once Bolivia's second largest body of water, but by 2015 it had dried up completely, with the loss of millions of fish and thousands of birds. Here the cause again appears to be water extraction, together with climate change and pollution from mine waste (Whitt, 2017).

Lakes have also long been popular among both theoretical and experimental ecologists. As their boundaries are clearly defined, they form relatively discrete systems, providing outstanding opportunities for studying ecological interactions and dynamics. Some lakes are small enough to be manipulated in their entirety, enabling experiments to be conducted (Knowlton, 2004). This helps to explain why limnologists, such as Raymond Lindeman and G. Evelyn Hutchinson, have made such important contributions to the development of ecological science. It also explains why lakes have proved to be so important for the development and application of dynamical systems theory to real-world ecosystems.

Scheffer (2009) provides a summary of his own extensive research on lakes. This has focused on shallow lakes, which often display rapid transitions between states. Many shallow lakes are now affected by eutrophication; high inputs of phosphorus and nitrogen can lead to blooms of phytoplankton, which can convert a transparent lake to a turbid state. After such events, restoration of the original state has often proved surprisingly difficult. However, reducing fish densities for a brief period has often brought the lakes back to a permanently clear condition. Research has demonstrated that both clear and turbid ecosystem states have stabilising feedback mechanisms. For example, as water plants grow, they may reduce turbidity, which can further enhance plant growth by increasing light penetration (Scheffer, 2009). Research has also identified the role of fish, which can have a major influence on the state of lake ecosystems. For example, large bottom-feeding fish such as bream and carp tend to dominate in lakes that are unvegetated. Their method of feeding results in disturbance of the sediment, which can inhibit colonisation by plants, providing another stabilising feedback mechanism (Scheffer, 2009).

Do lakes in the real world display alternative states, as described by dynamical systems theory? A number of different models have been developed that explore the application of dynamical systems theory to lakes, but the focus here is on empirical evidence. This question is examined by Scheffer *et al.* (1993), who mention examples such as the gravel pit complex at Great Linford in England. Here, both turbid and clear lakes are present, and both have apparently been stable for a number of decades. Removal of the fish stock from one of the turbid lakes in 1987 led to a reduction in turbidity and development of vegetation. This new vegetated state also appeared to be persistent. Mention is also made of the Tomahawk Lagoon in New Zealand, which has repeatedly switched between periods of phytoplankton dominance and aquatic vegetation, each lasting one to five years. Similar observations have been made in Lake Tåkern and Lake Krankesjön in Sweden (Scheffer *et al.*, 1993). Further empirical evidence is presented by Scheffer (2009), including the Dutch lake Velue, where time series data are available for periods of both increasing and subsequent decreasing nutrient load. Here, a critical phosphate load for collapse of the system was determined, which was found to be about twice the critical phosphate concentration associated with ecosystem recovery. This therefore provides a potential example of hysteresis (Ibelings *et al.*, 2007).

While time series data might be suggestive of the mechanisms underlying collapse, firmer evidence is available from experimental investigations. An outstanding example of such a study is provided by Carpenter *et al.* (2011), using experimental lakes in Michigan (interestingly named Peter and Paul). The top predator, largemouth bass (*Micropterus salmoides*) was gradually added to a lake dominated by planktivorous fishes, to induce a trophic cascade. A nearby lake, dominated by the same predatory species, was not manipulated. It was hypothesised that predation by the largemouth bass would eventually drive planktivorous fishes to low densities, which would result in an increase in the abundance of zooplankton, leading to cyclical oscillations of zooplankton and phytoplankton biomass. Results were consistent with these predictions; numbers of planktivorous fish declined in the manipulated lake as fish predation increased and approached the values of the reference lake. Zooplankton biomass initially declined then became strongly oscillatory, as did phytoplankton biomass, reflecting the development of predator-prey cycles. These data also indicated that a regime shift occurred in the manipulated lake, which displayed non-linear dynamics. However, the precise mechanisms of this ecosystem transition are still not fully understood (Carpenter *et al.*, 2011).

The experiments on the Peter and Paul lakes in Michigan described by Carpenter *et al.* (2011) are actually part of a long research tradition; the world's first whole-lake experiments (with a control) were conducted in the same area back in the 1950s (Elser *et al.*, 1986). This remarkable experimental resource continues to provide valuable insights. For example, Pace *et al.* (2013) showed that variance in zooplankton biomass could provide early warning of the regime shift, whereas Seekell *et al.* (2012) demonstrated the same for chlorophyll-*a* concentration and the abundance of minnows. In an elegant experiment involving monitoring the response of lake ecosystems to nutrient addition, Wilkinson *et al.* (2018) identified a range of early warning indicators that preceded blooms of cyanobacteria. Further, Seekell *et al.* (2013) examined evidence for the existence of multiple attractors in the same system.

According to dynamical systems theory, abrupt changes in ecosystem state can potentially be caused by the existence of alternate attractors, or different sets of values towards which a system tends to evolve (see Chapter 2). However, it is difficult to test for the existence of multiple ecosystem attractors using empirical data, and consequently evidence for them in real-world ecosystems is still limited (Seekell et al., 2013). In this experimental system, the specific ecological mechanism for the hypothesised alternate attractors is the existence of trophic triangles; these are sets of predator-prey relationships that can potentially drive either predators or prey to dominance, as a result of positive feedback mechanisms (Seekell et al., 2013). In these lakes, the food web includes two size classes of piscivorous fish (adult and juvenile) and a fish species that eats juvenile piscivores as well as herbivorous zooplankton (Carpenter et al., 2008). As we encountered in the ocean ecosystems considered earlier, food webs involving fish are often fascinatingly complex, as prey fishes often consume the juveniles of the predators that eat them, providing scope for trophic triangles.

According to theory, food webs that demonstrate alternate states could undergo abrupt shifts as a result of gradual change in exploitation of a top predator or prey species, reintroduction of a predator or modification of habitat (Carpenter et al., 2008). However, the occurrence of abrupt change from one state to another does not by itself prove the existence of multiple attractors. Similarly, features of empirical data such as shifts in mean values, bimodality, hysteresis and changes in driver–response relationships may be consistent with the theory, but might also be found in systems that are not characterised by multiple attractors. To provide a more robust test of the existence of multiple attractors, Seekell et al. (2013) employed excitingly named statistical approaches, such as generalised autoregressive conditional heteroscedasticity, to the high-frequency time series data available from their experimental setup. The results provided firm evidence that alternate attractors can exist in the real world at the scale of ecosystems, as a result of ecological phenomena (trophic triangles and trophic cascades) that are relatively widespread in lakes (Seekell et al., 2013).

To what extent is dynamical systems theory applicable to understanding ecosystem transitions in lakes? Schröder et al. (2005) provide a valuable review of experimental evidence for alternative stable states. More than half of the 35 studies that they identified were undertaken in freshwater habitats. Of these, eight showed that alternative states were present, four showed they were not present and seven were judged to be methodologically inadequate to determine either way. However, none of the experiments displaying positive results were conducted in the field; the only field investigation listed among the freshwater examples considered failed to find evidence for alternative states. In conclusion, Schröder et al. (2005) note that while there is clear evidence that ecological systems can exhibit alternative stable states, at least in the laboratory, this should not be interpreted as a rule. Instead, each particular system should be experimentally investigated to test for the occurrence of stable states. At present, experimental field evidence is very limited, although this partly reflects the practical difficulties of conducting appropriate experiments. In particular, few field studies have manipulated a natural system over sufficient time to determine whether transitions between ecosystem states follow the same trajectory from perturbation through to recovery (Mittelbach et al., 2006).

Capon et al. (2015) reviewed 135 papers on freshwater ecosystems to assess the evidence for non-linear changes in freshwater ecosystems resulting from environmental change, drawing on both experimental and observational studies. While noting that shifts between turbid and clear-water states have been reported for some shallow lakes in temperate climates, overall these authors found little empirical evidence for transitions between alternative stable states in their review. They also noted a clear geographical bias in the literature, with few case studies reported from Africa, Asia or South America, and none at all from Australasia. Given that the authors were all Australian, this latter discovery must have been particularly galling. Overall, only five papers (3.7%) met Peterson's (1984) criteria for demonstrating a pressure-induced ecological change, in accordance with dynamical systems theory. Many of the studies that did not meet these criteria failed to examine ecological responses to pulsed pressures, by continuing observations beyond the release of the pressures being investigated. While the clear *versus* turbid states of shallow lakes were the most common examples encountered in the literature, often observed in response to nitrogen and phosphorus loads increasing over time, few studies demonstrated the maintenance of an alternative state after nutrient loads were reduced. Rather grudgingly, Capon et al. (2015) admitted that shallow lakes provide the best evidence for alternate states in freshwater ecosystems, with non-linear pressure-response relationships evident in a 'moderate number' of these examples. However, they considered evidence for alternative stable states to be limited overall, because the different ecological states described typically occur under different environmental conditions (e.g. nutrient concentrations) and therefore do not meet the requirements of the theory (Peterson, 1984; Petraitis, 2013; see Chapter 2).

This lack of evidence must partly reflect the difficulty of conducting an experimental study with sufficient duration and temporal resolution, in a way that meets Peterson's (1984) criteria. It may therefore be premature to conclude, as Capon *et al.* (2015) do, that non-linear changes in freshwater ecosystems 'are much rarer than gradual, monotonic changes and high temporal variability'. After all, as Carl Sagan might have said, absence of evidence is not evidence of absence. However, Capon *et al.* (2015) go further by concluding that their analysis revealed 'a problem of falsely associating theoretical constructs with results that do not credibly support them', a problem that they suggest is particularly severe in freshwater ecology. If this sounds a little harsh, they are not alone in suggesting that dynamical theory has been applied somewhat uncritically by ecologists (Montoya *et al.*, 2018; Myers-Smith *et al.*, 2012; see also Chapter 2).

It is well established that a trophic cascade in lakes can produce two different ecosystem states (i.e. few planktivores, large zooplankton, with clear water, *versus* abundant planktivores, small zooplankton and turbid water). However, it is less clear that these equate to the alternative system states consistent with theory. For example, it is not well established whether real-world zooplankton communities show a discontinuous response to varying planktivore density, as would be required by theory (Mittelbach *et al.*, 2006). In some situations, this is definitely not the case. The long-term data for Wintergreen Lake in Michigan, USA, showed dramatic regime shifts associated with changes in planktivore density. However, these transitions occurred smoothly and predictably in response to changes in planktivore numbers, in both forwards and backwards directions (Mittelbach *et al.*, 2006). This again raises the question of how commonly, and under what conditions, alternate ecosystem states occur in nature.

Whether or not dynamical systems theory is applicable to lakes and rivers in the real world, the research undertaken to develop and test the theory has provided many useful insights into the mechanisms of ecosystem collapse. On the basis of experimental and other sources of evidence, Gsell *et al.* (2016) identified the ecological mechanisms that have been most commonly shown to generate transitions between different states of aquatic systems, namely: (i) competition between two or more species; (ii) trophic cascades resulting from the inclusion or exclusion of top predators or parasites; and (iii) intraguild predation, involving species that compete for resources that also prey on one another. In relation to the latter mechanism, Verdy and Amarasekare (2010) note that although the phenomenon is widespread, and theoretical modeling suggests that it may play a role in supporting transitions between alternative ecosystem states, empirical evidence for this is scant. Scharfenberger *et al.* (2013) provide a possible example for the Műggelsee, a shallow eutrophic lake in Germany.

How widespread are trophic cascades in freshwater ecosystems? Ripple *et al.* (2016a) note that much of the early empirical evidence for trophic cascades came from aquatic systems, leading to the widely held idea that they might be commoner in water than on land, because of the relatively simple food webs observed in most aquatic environments. More recent research has contradicted this idea; in fact since the 1990s, many more examples have been reported for terrestrial than for freshwater ecosystems (Ripple *et al.*, 2016a). Carpenter *et al.* (2001) suggest that in lakes, it is now generally accepted that cascades are 'reasonably common'. As illustration of this, trophic cascades are now being widely manipulated as a practical intervention to manage plant biomass (e.g. Meijer *et al.*, 1999).

Taylor *et al.* (2015) provide a valuable review of top-down *versus* bottom-up interactions in freshwater ecosystems and reach the following conclusions:

- Research has established the importance of nutrients, including both nutrient ratios and overall supply, in limiting primary production and in the transfer of energy to herbivores. In this way, nutrient availability plays an important role in influencing the bottom-up processes that regulate freshwater food webs.
- Research has also demonstrated clearly that higher trophic levels can influence primary producer biomass through trophic cascades, providing an example of top-down regulation, which appears to be widespread in both lakes and streams. Factors that influence the strength of trophic cascades include prey edibility and vulnerability, disturbance regimes, water temperature, depth and nutrient enrichment.
- Evidence suggests that both top-down and bottom-up regulation are pervasive in freshwater food webs. Primary producers appear to be more strongly regulated by resources and primary consumers by top-down processes. However, these two processes do not act independently. For example, increasing nutrient availability can intensify the effects of trophic cascades on producer communities and increase the overall contribution of animal-mediated nutrient recycling to the ecosystem.

- *In situ* experiments in streams have shown a range of different results, including (i) independent positive effects of nutrient enrichment and fish-induced trophic cascades on algal biomass, (ii) the weakening of strong trophic cascades after nutrient enrichment and (iii) strengthening of trophic cascades after increasing the supply of resources.
- Ecosystem subsidies (namely landscape-scale flows of energy, nutrients and organisms) can stabilise local food webs, but may also be affected by consumers. Conversely, ecosystem subsidies may modify the strength of trophic interactions. An example is provided by stream ecosystems, where organic inputs from the surrounding landscape are critical to their structure and function.
- Non-trophic interactions can also influence the strength of both top-down and bottom-up processes. For example, consumers in freshwater ecosystems can influence nutrient availability via processes such as excretion. Furthermore, ecosystem engineering can also have significant impacts. Ecosystem engineers are organisms that change the availability of resources to other organisms, by creating, maintaining or modifying habitats. The consequences of these actions can be transmitted to other trophic levels. The classic example is the beaver (*Castor* spp.), which can transform freshwater environments through its dam-building activities.

Overall, after consulting a wide breadth of literature, Taylor *et al.* (2015) concluded that there is little consensus regarding the relative importance of bottom-up *versus* top-down control in freshwater ecosystems, nor the strength of the interaction between the two. As noted by Ripple *et al.* (2016a), it is widely recognised that the two processes often occur simultaneously, but their relative influence is difficult to determine. In a review of trophic cascades in both aquatic and terrestrial systems, Borer *et al.* (2005) concluded that the principal factor determining cascade strength was a combination of herbivore and predator metabolic factors and predator taxonomy. Both within and across all systems, the strongest cascades occurred in association with invertebrate herbivores and endothermic vertebrate predators.

Taylor *et al.* (2015) also considered the process of ecosystem collapse in relation to the loss of biodiversity and its impact on ecosystem function. Declines in biodiversity can negatively affect ecosystem function by altering patterns of nutrient turnover in aquatic systems; loss of species will also influence ecosystem function by affecting top-down and bottom-up controls. There have been a few attempts to measure changes in ecosystem function associated with actual loss of species in the field. For example, Whiles *et al.* (2006) assessed how neotropical stream ecosystems have been affected by rapid extinctions of amphibians owing to attack by the chytrid fungal pathogen. As ecologically important grazers and detritivores have been lost, there have been effects on algal community structure and primary production, sediment dynamics and populations of other grazing species. The energy and nutrient subsidies associated with amphibians have also been reduced. A detailed account of such impacts was described by Whiles *et al.* (2013), working in a wet forest stream in Panama. A 98% decline in tadpole biomass was associated with a doubling of algae and fine detritus biomass in the stream and a >50% reduction in nitrogen uptake rate. Nitrogen turnover rates in organic sediments were also significantly lower after the decline. As a result, the stream cycled nitrogen less quickly, and there was a reduction in downstream exports of particulate nitrogen.

Nutrient cycling provides a potential feedback mechanism that could help drive ecosystem collapse. Increasing evidence indicates that top predators can influence ecosystem nutrient dynamics beyond the trophic level of their prey, which can then influence biological productivity (Schmitz *et al.*, 2010). A number of studies have shown that trophic cascades can affect not only nutrient concentrations but also the distribution of nutrients among different ecosystem pools and the extent and severity of nutrient limitation within an ecosystem (Vanni, 2002). Evidence from research on lakes has identified four general mechanisms of the effect of top predators on nutrient cycling (Schmitz *et al.*, 2010; Vanni, 2002):

(1) predation that leads to spatial and temporal shifts in the composition of communities that graze on zooplankton, and hence the amount of nutrients contained in the zooplankton community; this may arise because nutrient ratios may vary with prey species;
(2) trophic cascades affect phytoplankton size structure owing to size-selective predation on zooplankton, which leads to size-dependent effects on rates of nutrient turnover by phytoplankton;
(3) excretion of nutrients by predators directly into the water column and
(4) the spatial translocation of nutrients via consumption of resources in one location and excretion or egestion into another.

The discovery that salmon may contribute up to 25% of the nitrogen budget of riparian vegetation, following their consumption by bears, provides a particularly interesting example of the latter (Helfield and Naiman, 2006). Potentially, the loss of fish species could therefore have a significant effect on nutrient cycling at the ecosystem scale. McIntyre *et al.* (2007) examined the potential impact of fishing on nutrient cycling using field data from a neotropical river and Lake Tanganyika, Africa, together with a modeling approach. Results indicated that in both of these ecosystems, nutrient cycling was dominated by a relatively small number of species. Loss of the species harvested by fisherfolk led to more rapid declines in nutrient recycling than fish extinctions associated with rarity, body size or trophic position. Overall, the consequences of a decline in fish diversity were found to depend upon the order of extinctions and the compensatory responses of surviving species (McIntyre *et al.*, 2007).

Can exotic species cause ecosystem collapse? Freshwater ecosystems appear to be particularly vulnerable to invasion by exotic species, which can trigger trophic cascades. Gallardo *et al.* (2015) present the results of a global meta-analysis from 733 case studies, which indicated a pronounced negative influence of invasive species on the abundance of aquatic species, particularly macrophytes, zooplankton and fish. Invaded habitats showed increased water turbidity, nitrogen and organic matter concentration, which were attributed to the capacity of invaders to transform habitats and increase eutrophication. Key findings were that: (i) the largest declines in fish abundance were caused by the expansion of invasive macrophytes; (ii) planktonic communities were depleted most by invasive filter feeders; (iii) the greatest decline in macrophyte abundance was associated with the invasion of omnivores; and (iv) benthic invertebrates were most negatively affected by the introduction of new predators (Gallardo *et al.*, 2015).

This review also highlighted the fact that invasive species can affect multiple trophic levels, depending on the trophic position of the invading species and its ability to modify habitats. In some cases, such as filter collectors and predators, impacts propagated up and down the food web, whereas in others, impacts were restricted to one trophic level (Gallardo *et al.*, 2015; Figure 4.8). An example is provided by the introduction of trout to New Zealand streams, which led to major shifts in trophic structure and energy flow, higher annual net production across all trophic levels and greater nutrient uptake and retention (Taylor *et al.*, 2015). A further example is the golden apple snail (*Pomacea canaliculata*), invasion by which can cause aquatic plant communities to collapse, shifting the system from a clear water to a turbid state. This is associated with an

4.3 Freshwater Ecosystems · 211

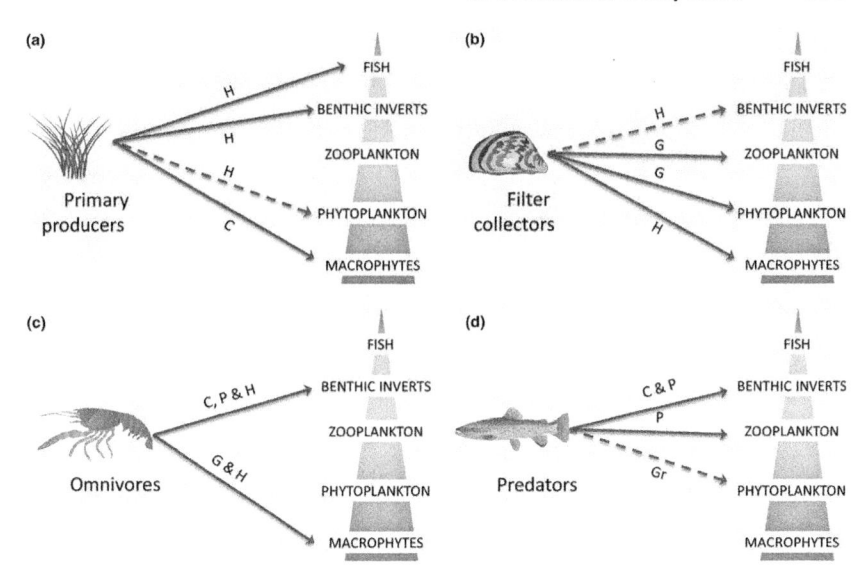

Figure 4.8 Impacts of invasive species on aquatic ecosystems. Arrows reflect the negative (continuous) or positive (dashed) impacts of invasive species on the abundance of five different functional components of resident communities. Impacts are the result of a combination of direct ecological (C: competition, P: predation, G: grazing, Gr: grazer release) and indirect physicochemical impacts of invasive species (H: habitat alteration). Reprinted from Gallardo *et al.* (2015) with permission from Wiley.

increase in translocation of nutrients to the water column via excretion (Carlsson *et al.*, 2004).

The most iconic example of freshwater ecosystem collapse caused by an introduced species is Lake Victoria in East Africa. Before the 1970s, Lake Victoria contained one of the most diverse and unique assemblages of fish in the world, including more than 350 species of cichlid fish, of which 90% were endemic; these accounted for around 80% of the lake's biomass (Revenga *et al.*, 2000). In the 1980s, abundance of the introduced Nile perch (*Lates* spp.) increased rapidly, causing the loss of 65% of the endemic haplochromine cichlid fish species. This eradication of around 200 species in less than a decade has been referred to as the largest vertebrate extinction event in the twentieth century (Goldschmidt *et al.*, 1993). This resulted in profound changes to the food web, with a marked reduction in algal grazing and an associated increased in algal blooms, anoxia and fish kills (Figure 4.9). In addition, detritivorous cichlid fish were partially replaced by atyid prawns (*Caridina nilotica*), which

212 · Contemporary Case Studies

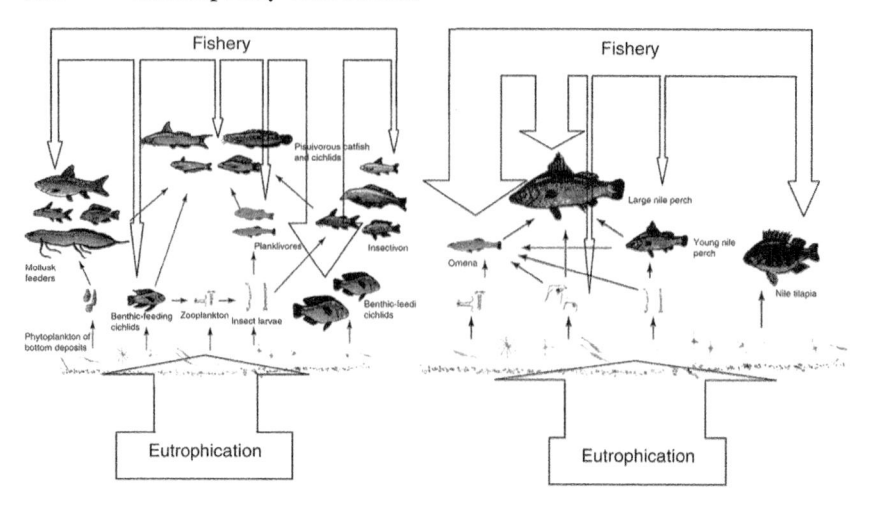

Figure 4.9 Top-down and bottom-up processes in the Lake Victoria ecosystem before and after ecosystem collapse, caused by the explosion of the introduced Nile perch. Fish communities are represented by the major trophic interactions between fish species. Fishing pressure and eutrophication, the main drivers of changes in the fish community, are represented by the open arrows. Reprinted from Kolding *et al.* (2008) with permission.

increased significantly in abundance (Goldschmidt *et al.*, 1993). Although the biomass distribution between trophic levels experienced a major perturbation, it largely recovered its initial state by 2005. In contrast, community structure and composition appear to have undergone a permanent change, despite the fact that some haplochromine species have recently displayed a degree of recovery (Marshall, 2018). The lake has apparently transitioned to a permanently different state, dominated by introduced species of fish and a few native surviving species (Downing *et al.*, 2012). While there has been a debate about the relative importance of top-down *versus* bottom-up processes in driving this example of ecosystem collapse, including a possible role of both eutrophication and climate change, Marshall (2018) concluded in a recent review that Nile perch is indeed primarily to blame. Lake Victoria therefore provides a powerful example of how easily a complex and diverse ecosystem can permanently collapse (Goldschmidt *et al.*, 1993), following introduction of a voracious top predator into an area where none had previously existed (Marshall, 2018).

Although not explicitly explored by Gallardo *et al.* (2015) in their review, invasive species could potentially interact with other stressors to hasten ecosystem collapse. As illustration, climate change might increase the likelihood of introduced species becoming established, for example by increasing water temperatures, and could also enhance their competitive and predatory abilities (Rahel and Olden, 2008). An example is provided by the zebra mussel (*Dreissena polymorpha*), which is projected to increase its geographic range as a result of global warming (Gallardo and Aldridge, 2013). This lowly mollusc, typically about the size of a fingernail, is widely recognised to be one of the most damaging invasive species in freshwater ecosystems. This is not least because of its propensity to invade water-treatment plants and clog up pipes, resulting in costly removal programmes. Given that a female zebra mussel can produce several million eggs within its lifespan and that individuals can attach to most substrates (including other mussels), it is not difficult to see how this species has become such a problem. Remarkably, populations of zebra mussels can be large enough to dominate the biomass of entire ecosystems. For example, after invading the Hudson River in 1991, zebra mussels have often constituted >50% of all heterotrophic biomass in the river and can filter 25–100% of the river's volume of water each day (Strayer, 2010). As a result, phytoplankton biomass in the river has fallen by around 80%, and the pelagic part of the food web has collapsed (Strayer, 2010). This is just one of a number of invasive bivalve species that can severely graze some primary producers, leading to significant increases in concentrations of dissolved nutrients and producing effects that may rival in magnitude those resulting from human activities. After zebra mussels colonised the Seneca River, New York, dissolved oxygen concentrations fell so much that the river was no longer suitable for sewage disposal, forcing the city of Syracuse to change its sewage management plans (Strayer, 2010).

Climate change can interact with a number of other stressors of freshwater ecosystems, including acidification, land cover change and eutrophication (Heino *et al.*, 2009). In Canada, for example, global warming has caused a decline in the phosphorus concentration of lake water, which affected phytoplankton biomass and diversity. Conversely, the thawing of permafrost may lead to increased concentrations of phosphorus in streams and lakes, resulting in increased production of diatoms and mosses (Heino *et al.*, 2009). Ormerod *et al.* (2010) note that many problems relating to freshwater science and management are associated with multiple stressors; this is partly because human activities typically alter more than one environmental factor at a time. For example, urbanisation affects the amount of water run-off quantity, water quality, thermal regimes, habitat availability and the dispersal of invasive species, among other factors. Many of these multiple-stressor issues are now being exacerbated, or at least further complicated, by climate change (Ormerod et al., 2010).

Jackson et al. (2016) performed a meta-analysis to examine the effects of multiple stressors on freshwater ecosystems and found that overall the combined effects of pairs of stressors was often less than the sum of their single effects. In other words, interactions between stressors were often antagonistic rather than synergistic. Whereas climate warming and eutrophication resulted in additive net effects when they occurred together, for many other stressors, warming was found to be antagonistic. This prevalence of antagonistic interactions between multiple stressors provides an interesting contrast with marine systems, where synergies between stressors have been widely reported. This may be attributable to the inherently greater environmental variability of relatively small aquatic ecosystems, which perhaps affords greater potential for acclimation and co-adaptation to multiple stressors. Alternatively, the effects of different freshwater stressors may be highly asymmetric (Jackson et al., 2016).

Which other disturbances can cause the collapse of rivers and streams? Ecologists researching fast-flowing rivers and streams have come to appreciate the pervasive influence of natural disturbance on the structure and functioning of these ecosystems. Flooding and drought are the two most important disturbance types in this context, and these may interact with other forms of disturbance, such as land-use change. For example, storms transport significantly more sediment to streams in landscapes that have been deforested or burned, whereas invertebrate populations can be less vulnerable to repeated flooding in streams that drain forested catchments than those that have been cleared for agriculture (Stanley et al., 2010). In general, stream ecologists consider that such external drivers are of overriding importance in these systems, which appear to lack the internal feedbacks that would be needed to maintain alternative states, or that can maintain ecosystem configuration in the face of external pressures (Stanley et al., 2010). However, there are many examples of streams or rivers being pushed into new states or configurations by one-off disturbance events, or by changes to the disturbance regime, including those with an anthropogenic origin. Ecosystem collapse can often be caused by major changes in the physical environment that affect patterns of water flow through a catchment, resulting from human interventions such as dam construction, water extraction or constraining streams to concrete channels (Stanley et al., 2010).

4.3 Freshwater Ecosystems · 215

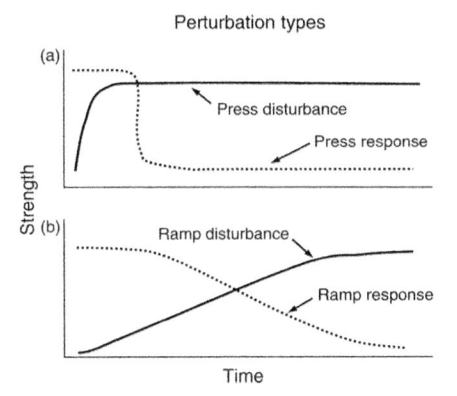

Figure 4.10 Two disturbance types identified for freshwater ecosystems: (a) press disturbance, and the response to a seasonal drought; (b) ramp disturbance, and ramp response to a supra-seasonal drought. Reprinted from Lake (2003) with permission from Wiley.

While both floods and droughts are key disturbances affecting many freshwater ecosystems, they differ in their respective characteristics and impacts. Floods can be considered as a form of pulse disturbance as they arise suddenly. Although their severity is often difficult to predict, flood return times and durations may be more regular. In contrast, the beginning of a drought can often be difficult to define (Humphries and Baldwin, 2003). Seasonal droughts can be considered to act as press disturbances. However, Lake (2003) suggests that supra-seasonal droughts might better be considered as 'ramp' disturbances, which increase in intensity (and often spatial extent) with time (Figure 4.9). Responses of the biota differ between the two types of drought. Species may adapt to relatively predictable seasonal droughts through adaptations such as the use of refugia and life-history changes. Adaptation to less-predictable supra-annual droughts may be more difficult (Lake, 2003).

Ecological impacts of drought include reduction of population densities and species richness, and changes to life-history schedules, species composition, patterns of abundance, type and strength of biotic interactions (predation and competition), food resources, trophic structure and ecosystem processes (Lake, 2003). While such impacts could potentially track the disturbance, creating a ramp response (Figure 4.10), there is another possibility. Boulton (2003) suggests that impacts could involve the crossing of critical thresholds, which would lead to a 'stepped' response with alternating periods of gradual and more rapid change. For example, cessation of water flow causes the abrupt loss of a specific

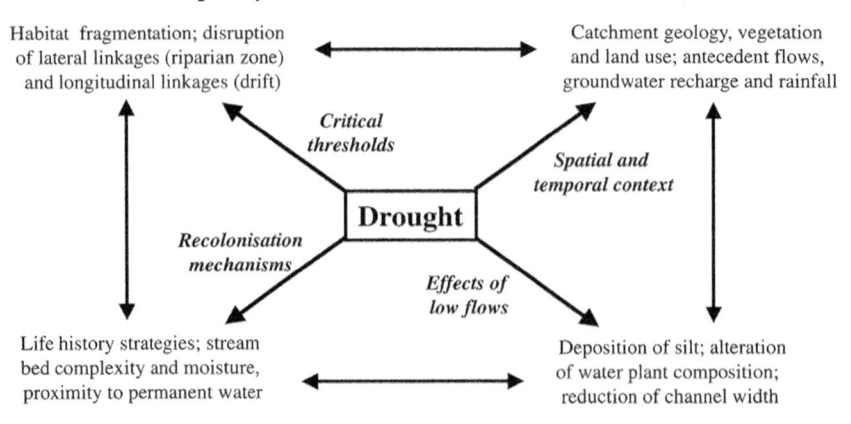

Figure 4.11 Factors influencing the impact of drought on aquatic macroinvertebrates. Reprinted from Boulton (2003), with permission from Wiley.

habitat, alteration of physicochemical conditions in pools downstream and fragmentation of the river ecosystem, each of which could demonstrate threshold responses (Boulton, 2003; Figure 4.11).

Recovery

What are the mechanisms of recovery in lakes? To address this question, let us first return to the nutrient enrichment of shallow lakes, which provides the best-studied example. Carpenter (2005) notes that the eutrophic state of a lake is stabilised by the recycling of nutrients (specifically phosphorus) from lake sediments. In relatively deep lakes, which are thermally stratified, this stabilisation can be caused by a range of factors, including the biogeochemistry of the deep layer of water (hypolimnion), the temperature of the hypolimnion, the shape of the lake basin, the structure of the food web and the abundance of rooted plants (Carpenter, 2005). Regardless of their depth, many lakes have remained in a eutrophic state for long periods of time as a result of such factors. However, the reasons for slow recovery, or a complete lack of recovery from eutrophication, are not fully understood. Carpenter (2005) suggests that persistent eutrophication may often be due to internal recycling from a large pool of phosphorus that builds up in lake sediments, or the chronic release of phosphorus from enriched soils in the surrounding watershed. Modeling of these phosphorus dynamics suggests that eutrophication can potentially last for 1,000 years or more, because of the slow process of removing phosphorus from the system. In other words, eutrophication is often a 'one-way trip' (Carpenter, 2005).

In many areas, significant efforts have been made to reduce anthropogenic nutrient inputs to water bodies and thereby reduce the environmental and economic impacts of eutrophication. Examples include the improvement of farming practices to reduce the amount of nitrates from agricultural sources polluting ground and surface waters, which has reduced nitrate concentrations in parts of Europe, and widespread efforts to improve sewage treatment processes. There is now a substantial literature showing that such nutrient management practices can result in increased water clarity, increased abundance of submerged aquatic vegetation, reduced plankton biomass and reduced nutrient concentrations (McCrackin et al., 2017).

This evidence was examined using a global meta-analysis of 89 studies, including those conducted in both lake and coastal marine ecosystems, to determine the effectiveness of such interventions. Results indicated that nutrient management interventions were sometimes successful in improving phytoplankton, macroalgae, zooplankton, fish and water-column nutrient concentrations over timescales of years to decades (McCrackin et al., 2017). Interestingly, complete cessation of nutrient inputs did not result in a more complete or higher rate of recovery than partial nutrient reductions, probably because of insufficient passage of time, inputs of nutrients from other sources, or shifting baselines. Overall, lakes and coastal marine areas respectively achieved 34% and 24% of baseline conditions within a period of a few decades after the partial reduction or complete cessation of nutrient additions. One third of the response variables considered showed no change or even a decline, suggesting that complete recovery may not always be possible. Recovery times after cessation of nutrient inputs varied widely between studies and between the variables measured, ranging from <1 year to nearly a century. Nutrient concentrations of the water recovered over timescales of one to three decades (McCrackin et al., 2017), providing greater optimism than the results of Carpenter's (2005) modeling study. Similar timescales for recovery, of 10–15 years, were observed by Jeppesen et al. (2005) using long-term data for 35 restored lakes. Somewhat longer recovery durations of several decades were observed by Bennion et al. (2015) in 12 European lakes; this study also provided evidence of hysteresis in some lakes and an apparent failure to recover in others.

Mehner et al. (2008) provide grounds for even-greater optimism by demonstrating how rapid recovery from eutrophication can be achieved in practice. What is particularly informative about this case study is how recovery was achieved: by targeting the feedback mechanisms responsible for maintaining the eutrophic state of a lake. First, the authors considered which mechanisms might be influential in addition to the reloading of phosphorus from lake sediment, as identified by Carpenter (2005). These included (i) increased nutrient transport to the lake owing to the loss of riparian vegetation; (ii) loss of macrophytes from the lake's littoral zones in response to reduced water transparency, resulting in the reduced availability of refuges for smaller organisms, and a higher rate of sediment resuspension; (iii) selective fisheries reducing the abundance of piscivorous predators, thus enabling planktivorous and benthivorous fishes to flourish (this suppresses zooplankton grazers and promotes increased abundance of phytoplankton); and (iv) reduced top-down trophic control owing to reduced periphytic production, owing to low water transparency. This latter mechanism reflects the fact that in addition to macrophytes, the periphyton is an important contributor to the productivity of clear lakes, where it may account for up to 80% of whole-lake primary production (Genkai-Kato et al., 2012). By contrast, in turbid lakes, most primary production is contributed by phytoplankton.

In this particular example, namely Lake Tiefwarensee in Germany, restoration management was designed to simultaneously address the main feedback mechanisms operating in this lake. These were external and internal nutrient loading, lack of keystone predators and lack of structured littoral habitats. As a result, over a five-year period, phosphorus concentrations within the lake declined by more than 80%, and water transparency, zooplankton biomass and fish assemblage structure and biomass all responded positively and linearly to the reduction in phosphorus concentrations (Mehner et al., 2008). The results from this investigation support those from many others indicating that measures aiming to inactivate phosphorus are the most effective for restoration of stratified lakes over the short term (Mehner et al., 2008). However, as noted by Carpenter (2005), complete recovery would only be possible with additional reduction of phosphorus inputs to the system.

What are the mechanisms of recovery in rivers and streams? In comparison with lakes, the recovery of river and stream ecosystems has been relatively little researched. However, ecological restoration initiatives are increasingly being implemented in flowing water ecosystems (Lake et al., 2007), supported by policies such as the Water Framework

Directive in Europe. Verdonschot *et al.* (2013) undertook a systematic review of restoration measures undertaken in lakes and rivers, together with estuaries and coastal waters. Key findings were that:

- River restorations mainly focused on three different kinds of measures: habitat improvement (e.g. additions of wood and boulders), riparian buffer creation and weir removal. This contrasts with the primary focus of lake restoration on addressing eutrophication and acidification.
- In rivers, improvement of the physical environment was reported in 33% of the restoration projects evaluated, while a positive biological response was reported in 50%. Restoration successes for lakes tended to be higher, with 66% of eutrophication projects demonstrating reduced phosphorus and/or nitrogen, while 64% reported positive biological responses.
- Evidence suggests recovery after weir removal may take up to 80 years, whereas recovery after riparian buffer instalment may take more than 30–40 years. In lakes, the recovery time from eutrophication ranges from 10 to 20 years for macroinvertebrates, 2 to >40 years for macrophytes and 2 to >10 years for fish.
- In both rivers and lakes the success rate of restoration measures appears to be much higher for abiotic conditions than for biotic indicators. In lakes internal nutrient loading often delays recovery. In rivers the evidence for macroinvertebrate recovery following hydromorphological restoration is equivocal; some studies have shown recovery while other studies have not, perhaps because nutrient concentrations remained high. Fish removal in shallow eutrophic lakes has repeatedly been shown to have positive short-term effects on water quality and transparency. However, long-term effects (>8 to 10 years) are less evident in the literature, and a return to turbid conditions often occurs unless fish removal is repeated.
- In most restoration projects measures are taken to reduce the primary stressor, but secondary stressors often confound recovery. Confounding factors include climate change, nutrient enrichment, large-scale hydrological change such as floods and droughts and catchment management/land-use practices. Absence of neighbouring source populations can impede recovery.

Another factor that has been shown to be important in many river restorations is improvement of connectivity. Infrastructural developments such as dams can reduce both the longitudinal and lateral connectivity within river systems. This can limit the ability of species to migrate and disperse into new areas and thereby reduce the rate of community recovery. Loss of connectivity can also reduce the flow of subsidies across ecosystem boundaries; for example, decline in organic matter inputs from riparian zones can have a significant impact on stream communities (Jansson *et al.*, 2007). Improvements in the connectivity of rivers can be achieved by reintroducing aspects of natural flow regimes. For example, lateral connectivity can be enhanced by reintroducing floods and vertical connectivity can be restored by increasing the exchange between surface water and groundwater flow (Jansson *et al.*, 2007).

In their review of stream restoration, Lake *et al.* (2007) again highlight the importance of dispersal as a critical process in enabling populations of species to recover. Restoration of refugia is also identified as an important objective, as many organisms survive disturbances through the use of refugia, which may be lost as a result of habitat degradation. Lake *et al.* (2007) also note the importance of food-web structure, which may be altered through human disturbances such as riparian clearing, nutrient enrichment or the introduction of non-native species. While there is evidence of both top-down and bottom-up control of food web structure in streams, manipulations of food web structure have rarely been attempted as a management intervention, in marked contrast to the situation in lakes (Lake *et al.*, 2007). Another feature of streams is that seasonal variation is often very pronounced, and as a result, species abundances and composition are continually changing. The dominance of community structure by exogenous physical factors, generated by water flow and the characteristics of the substrate, may prevent the development of alternative stable states maintained by internal dynamics, even though there have been suggestions that they might occur in streams and rivers (Dent *et al.*, 2002).

Lake *et al.* (2007) describe four models of ecosystem recovery during stream restoration, including the 'rubber band' model, in which recovery is rapid and complete; hysteresis, which is relatively slow (for example because of constraints on species recolonisation); and the wonderfully named 'Humpty Dumpty' model, in which recovery is incomplete (Figure 4.12). This implies that when ecosystems collapse, they can be irretrievably broken, like the unfortunate Humpty Dumpty. Which of these different models are most often encountered in practice? If there is a nearby source of colonists, the disturbance is stopped and the habitat rebuilt, recovery may be relatively rapid and complete, as in the 'rubber band' model. However, as a result of factors such as low dispersal ability

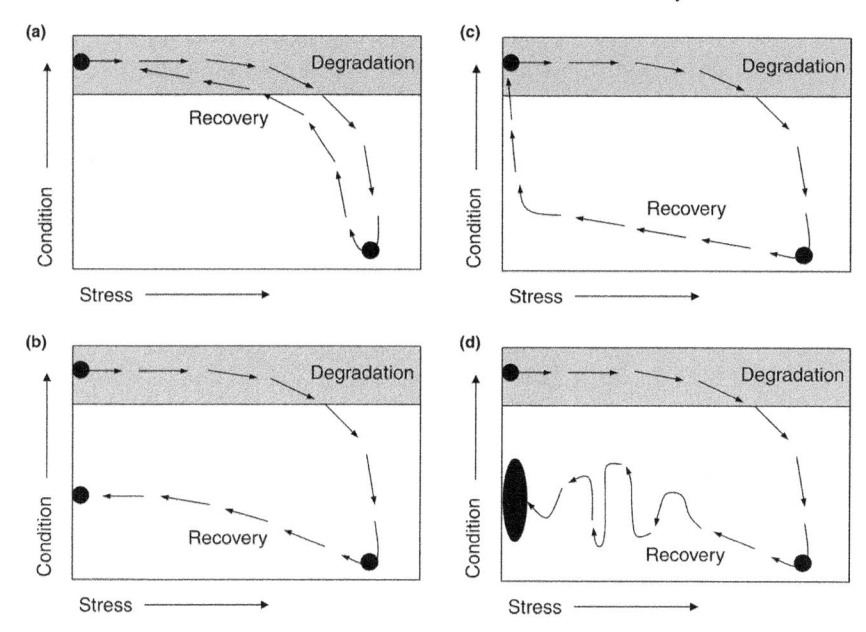

Figure 4.12 Four potential recovery pathways in stream ecosystems: (a) the 'rubber band', model in which recovery follows closely the pathway of degradation; (b) the 'hysteresis' model, in which recovery follows a different, non-linear pathway as stress is reduced; (c) the 'Humpty Dumpty' model, in which recovery could follow various trajectories but the endpoint is distinct from the predegraded condition and (d) the 'shifting target model', an extension of the 'Humpty Dumpty' model in which the recovery pathway itself is unpredictable and the endpoint becomes a shifting target bounded by distinct limits (shaded area). The y axis represents different measures of ecosystem condition, according to the objectives of the restoration action. Reprinted from Lake *et al.* (2007), with permission from Wiley.

and interactions between species, the restoration pathway may be one of hysteresis and may take significantly longer. Restricted dispersal may result in recovery being incomplete, as in the 'Humpty Dumpty' model. In addition, the impact of disturbance on factors such as species presence and resource availability could result in a variety of different recovery trajectories, with uncertain endpoints. Lake *et al.* (2007) provide examples of each of these different recovery patterns in streams, but evidence is lacking regarding their relative frequency.

Do food webs recover after disturbance? Environmental perturbations that change the composition of species, for example by causing extirpations or invasions, can have a significant impact on the structure of food webs. This raises the question of what happens after the perturbation ceases, an issue that has not been widely studied. Research by Mittelbach *et al.* (2006) demonstrated in a lake ecosystem that a community of *Daphnia* spp. returned to the same species composition that was present prior to a temporary decrease in the density of planktivorous fish. Rather different results were obtained by Schröder *et al.* (2012), in an elegant experiment where populations of phantom midges were temporarily suppressed in two small lakes by the introduction of predatory fish, then the impacts on food-web structures were monitored. Results indicated the existence of three distinct food-web states corresponding to the periods before, during and after the perturbation. Interestingly, the post-perturbation food web configuration differed from both the pre- and during-perturbation configurations in phytoplankton biomass and in the species composition of micro- and meso-zooplankton. For example, communities of rotifers shifted from dominance by species with hard shells to those with soft bodies. While the reasons for this shift remain somewhat unclear, the authors suggested that this may provide evidence for the existence of two alternative stable community states (Schröder *et al.*, 2012).

There are even stranger examples of alternative configurations of freshwater food webs. As noted previously, food webs in aquatic systems are often complex. They can even involve cannibalism. Persson *et al.* (2003) describe the occurrence of cannibalism in perch (*Perca fluviatilis*), which can lead to the development of different community states: periods in which the population is dominated by stunted cannibals and other phases in which it is dominated by gigantic cannibals. Shifts between these states can propagate through to lower trophic levels, providing an unusual example of a trophic cascade. Research showed that the existence of just a few giant cannibals dramatically affects the entire lake system down to the lowest level of phytoplankton (Persson *et al.*, 2012). This remarkable story conjures up images of mythic horror, reminiscent of the Laestrygonians, a fictional race of cannibalistic giants that featured in Homer's *Odyssey*. Clearly, a lake can be a scary place for a young perch.

4.4 Forests

It has long been recognised that the widespread occurrence of forest loss and degradation represents a global environmental crisis. The Millennium Ecosystem Assessment provided some striking statements summarising the ecological importance of forests and their current predicament (Shvidenko *et al.*, 2005):

- Forests provide habitat for at least half of the world's known terrestrial plant and animal species. These species play a vital role in the functioning of forest ecosystems and support provision of multiple ecosystem services.
- Forests contain about 50% of the world's terrestrial carbon stocks, about 80% of terrestrial biomass. Consequently, forests play a significant role in the global carbon cycle, and deforestation is a major contributor to global climate change.
- More than three-quarters of the world's accessible freshwater is obtained from forested catchments. As forest cover and condition decline, water quality is reduced, and risks of floods, landslides and soil erosion can increase.
- During the past three centuries, global forest area has declined by around 40%, with three-quarters of this loss occurring during the past two centuries.
- In 25 countries forests have completely disappeared, and another 29 countries have lost more than 90% of their forest cover.
- Although forest cover is increasing in some regions (such as Europe), in the tropics deforestation of natural forests continues at an annual rate of over 10 million ha per year, which is more than four times the size of Belgium.
- Even in regions where forest areas have stabilised, the condition of forest ecosystems is being adversely affected by factors such as air pollution, fire, pest and disease outbreaks, habitat fragmentation and inappropriate management. Climate change threatens all forest ecosystems.
- The main direct causes of deforestation are conversion to agriculture, wood extraction and development of infrastructure such as roads and urban areas. Indirect drivers include economic development and failures in markets, policies and institutions.

Collapse

It is the exceptionally high species richness of many tropical rain forests, coupled with their high rates of loss, which has latterly led to deforestation becoming such a cause célèbre among conservationists. Although there has been some progress during recent decades in reducing forest loss and increasing the area under protection, deforestation is still a major issue. For example, recent analyses suggest that 3.6 million ha of primary rain forest disappeared during 2018 (Weisse and Goldman, 2019), an area the size of Belgium (why is it that campaigners always refer to Belgium?). The widespread reporting of such figures in the international media obscures an informative scientific debate regarding their veracity. This debate focuses on how deforestation is measured.

There are two main sources of information about how forest area is changing over time: (i) the Global Forest Watch (GFW), which is compiled from satellite images by the World Resources Institute, a global non-profit organisation; and (ii) the Global Forest Resources Assessment (FRA), which is compiled by the Rome-based UN Food and Agriculture Organization (FAO) from national statistics provided by member states. GFW employs forest maps developed by Hansen et al. (2013) using Landsat data, which have since been updated regularly. Interestingly, the findings of these two sources differ, and this gap appears to be widening over time. For example, the 2015 FRA reported that the rate of net global deforestation has slowed down by more than 50% over the past 25 years, with a net annual decrease in forest area of 3.3 million ha recorded for 2010–2015. In contrast, GFW reported that 18 million ha of tree cover was lost in 2014 (equivalent to 5.86 Belgiums) and that the rate of loss in the tropics is increasing.

One of the reasons for this disparity is semantic uncertainty surrounding the word 'forest'. On the face of it, deciding what constitutes a forest might appear to be a trivial task, but it is anything but. Lund (2018) found more than 1,700 definitions in a literature review. This confusion reflects the fact that the word 'forest' may refer to an administrative unit, a type of land use, a type of land cover or an ecological unit. Also, there is variation in the amount of tree cover required for an area to qualify as a forest, thresholds varying from zero to 80% depending on the definition adopted (Lund, 2018). FAO uses a standard definition of forest, which establishes minimum thresholds for the height, canopy cover and extent of trees, and also requires that the land is officially or legally designated for 'forest use'. This means that recently logged or burned areas, devoid of trees, could legitimately be classified as forest (Harris et al., 2016). In contrast, GFW uses satellite imagery that measures only biophysical criteria (height, canopy cover and extent of trees), but includes all instances of tree cover loss, whether or not these are temporary (Harris et al., 2016). The capacity of forests to regrow after clearance is therefore not factored into GFW assessments.

Arguments about deforestation rate might appear to have little relevance to understanding ecosystem collapse and recovery, but in fact they are highly instructive. This is illustrated by the issue of plantation forests.

To foresters, plantations are just another kind of forest, and this helps explain why the FAO routinely lumps together plantations and natural forests when reporting changes in forest area. For many years, this has been a bone of contention with conservationists, who recognise that the ecological value of a plantation is almost always far lower than a natural forest that it might have replaced. As illustration, in 2017 nearly 200 organisations signed an open letter to the FAO calling for them to change how they define 'forest', specifically to exclude plantations. This followed a protest march in Durban in 2015, where people held banners stating that 'plantations are not forests!', supported by a petition with 100,000 signatures (WRM, 2016). This is perhaps the only time that an ecological definition has brought protestors on to the streets. The reason why this issue is so emotive is that the FAO definition has been used to legitimise the expansion of industrial plantation forestry, sometimes (such as in Chile) at the expense of threatened native forest ecosystems. Under FAO's definition, replacing native forest with tree plantations is not recorded as deforestation, which enables countries to obscure the loss of native forests when they report forest cover data to the FRA (WRM, 2016). One might assume that the use of detailed satellite imagery to assess forest area would solve this problem, but unfortunately it is often difficult to differentiate natural and plantation forests on such images. Some of the losses of tree cover reported by GFW could, therefore, simply refer to the harvesting of a plantation.

Does conversion to plantation forests represent a form of ecosystem collapse? Despite what some foresters might believe, plantation forests are not equivalent to natural forests. The conversion of a native forest to a plantation actually represents a form of ecosystem collapse, as the identity of the ecosystem is changed and defining features are lost (see Chapter 1). Plantation forests typically represent a very different type of ecosystem to the natural forests that they sometimes replace. They may be entirely different in terms of tree species composition and are often structurally much simpler, with relatively even-aged stands and a lack of understorey vegetation (Brockerhoff *et al.*, 2013). They may also be intensively managed through application of fertilisers, pesticides and other types of intervention. Consequently, it is generally recognised that plantations are of much lower biodiversity value than natural forests. For example, in the Atlantic forests of Brazil, Marsden *et al.* (2001) found that only 7% of the bird species associated with a natural forest reserve were also encountered in a nearby *Eucalyptus* plantation. In a review based on 126 observations drawn from both tropical and temperate regions, Bremer and Farley

(2010) found that primary forests supported an average of 35% more plant species than plantations. Similarly, Stephens and Wagner found in a review of 35 studies that 94% reported lower biodiversity in plantation forests than in native or natural forests.

Barlow *et al.* (2007) describe a detailed study of biodiversity patterns for 15 taxonomic groups along forest gradients in Brazilian Amazonia, comparing primary, secondary and plantation forests. Results again emphasised the importance of primary forest for biodiversity. Across all taxa, many more species (25%) were unique to primary forests than to secondary forests or plantations (8% and 11%, respectively), and secondary forests held more primary forest species (59%) than plantations (47%). However, plantations were associated with relatively high species richness of some groups, such as orchid bees, scavenger flies and fruit flies, where >60% of the species found in primary forest were recorded. These values should be interpreted as something of a best-case scenario, given the intact nature of the forest matrix surrounding the secondary forests and plantations in the area studied (Barlow *et al.*, 2007).

The importance of primary forest for biodiversity was further demonstrated by Gibson *et al.* (2011), who conducted a meta-analysis of the effects of disturbance and land conversion on biodiversity in tropical forests. Overall, human impacts were found to reduce biodiversity, with a median effect size of 0.51 from 138 studies, although values of the effect size varied in response to region, taxonomic group, biodiversity metric and disturbance type (where biodiversity measures included abundance, community structure and species richness). In general, agriculture (e.g. conversion to cropland) had a greater effect than plantations; the effect of selective logging was even smaller, yet still positive (effect size of 0.11). This is consistent with some previous studies that have shown that selectively logged forests can retain a relatively high proportion of forest species (Edwards *et al.*, 2011; Martin *et al.*, 2015). The implication of these results is that not only plantation forestry, but any form of conversion from natural forest to another type of land cover could potentially be considered as constituting ecosystem collapse, as species composition is significantly altered.

While primary forests are clearly irreplaceable for sustaining biodiversity (Gibson *et al.*, 2011), the evidence for ecosystem processes is less clear-cut. Given that China now has more plantation forests than anywhere else, with around 25% of the world's total, where better to examine their ecology? Guo and Ren (2014) used a database of 6,153 Chinese forest stands to compare plantations and natural forests of the same climate zone. Results showed that plantations had similar biomass, but much higher productivity and carbon sequestration rates than natural forests. This finding highlights an interesting point: replacement of a natural forest with a plantation might result in loss of biodiversity, but an increase in carbon sequestration. This supports suggestions that evidence regarding the relationship between biodiversity and carbon storage is inconsistent (Cardinale *et al.*, 2012). Consequently, loss of some features of an ecosystem will not necessarily coincide with loss of others. This means that different dimensions of ecosystem collapse are not necessarily correlated with each other. On the other hand, Liao *et al.* (2010) synthesised 86 experimental studies with a paired-site design and found that total carbon stocks were about 28% lower in plantations than in natural forests. So, perhaps restoring natural forests is the best way to capture atmospheric carbon after all (Lewis *et al.*, 2019).

What are the principal drivers of forest loss? Curtis *et al.* (2018) provide an informative assessment based on analysis of high resolution satellite imagery (Google Earth) together with GFW data (Hansen *et al.*, 2013). This aimed to overcome a limitation of GFW by differentiating between permanent conversion of forest to another land use and other forms of forest disturbance that might be associated with subsequent regrowth (e.g. forestry, wildlife, shifting cultivation). Results indicated that during the period from 2001 to 2015, 27% of global forest loss could be attributed to conversion of land use to commodity production (e.g. agricultural crops and palm oil), representing permanent deforestation. In the remaining areas, where some recovery of forest might be possible in future, losses were attributed to forestry operations (e.g. logging and clear-felling) (26%), shifting agriculture (24%) and wildfire (23%). Drivers of forest loss varied regionally; for example, in temperate and boreal regions, forestry operations and wildfire were the dominant drivers, whereas in tropical regions, shifting agriculture and commodity-driven deforestation were more significant (Table 4.4). This study has a number of limitations: it assessed tree cover, which is not the same thing as forest cover (Harris *et al.*, 2016), and it failed to distinguish between different types of tree cover. Where natural forests have been replaced by commercial plantations, this would simply be recorded as regrowth, even though most of the former biodiversity may have been lost. Nevertheless, this provides the clearest picture of the drivers of forest loss that is currently available at the global scale.

What are the impacts of forest loss and degradation? Deforestation can represent an instantaneous form of ecosystem collapse, if forest cover is

Table 4.4 Drivers of global and regional tree cover loss for the period 2001–2015

	Tree cover loss		Driver				
Region	Tree cover loss (Mha)	Tree cover loss (% of global total)	Deforestation	Shifting agriculture	Forestry	Wildfire	Urbanisation
North America	70	21	2	1	48	48	1
Latin America	78	25	64	24	9	<1	<1
Europe	15	5	0	<1	95	5	None
Africa	39	13	2	93	4	<1	1
Russia/China/ South Asia	64	20	2	1	38	59	<1
Southeast Asia	39	13	61	20	14	2	<1
Australia/Oceania	10	3	8	10	19	62	1
Global	314	100	27	24	26	23	1

Note: Here, 'deforestation' refers to permanent land-use change for commodity production.
Source: Reproduced from Curtis *et al.* (2018), with permission

removed and replaced by another land cover type. How the land is subsequently used will then determine whether or not the forest is able to recover, given the availability of other nearby forest areas that can act as a source of colonists. However, recovery may also be inhibited by feedbacks resulting from deforestation. Runyan *et al.* (2012) provide a valuable an overview of the situations in which this can occur. Potential feedback mechanisms include a decline in precipitation owing to changes in evapotranspiration, documented in tropical rain forests; a change in albedo, observed in boreal forests; a decline in soil moisture, recorded in semiarid ecosystems; and a decline in nutrient and water availability under the vegetation canopy, observed in a wide range of ecosystems. Forest loss on slopes can lead to increased soil erosion and landslides, which can hinder the re-establishment of vegetation. In some cases where these feedbacks are present and of adequate strength, deforestation can lead to persistent state shifts where the forest vegetation cannot recover. Whether such transitions are irreversible is difficult to determine (Runyan *et al.*, 2012).

Deforestation is not the only process leading to forest collapse; forest ecosystems are also being widely degraded. According to Ghazoul and Chazdon (2017), anthropogenic degradation refers to human actions that cause persistent changes in the structure, function or composition of a forest ecosystem, which may reduce some attribute relative to a preferred (non-degraded) condition. Conversely, recovery is the process of returning to the preferred condition through natural succession or active intervention (Ghazoul and Chazdon, 2017). For IPBES (2018), degradation is defined somewhat more precisely as 'the state of land that results from the persistent decline or loss in biodiversity and ecosystem functions and services that cannot fully recover unaided within decadal time scales'. This might work rather well as an operational definition of ecosystem collapse. Forests are being degraded in a wide variety of different ways; examples listed by Watson *et al.* (2018) include forest fragmentation, stand-level damage resulting from logging, overharvesting of particular species (such as overhunting) and changes in fire or flooding regimes. To these may be added factors such as invasive species, introduction of browsing livestock, drought and incidence of pests and diseases.

Watson *et al.* (2018) provide a valuable overview of forest degradation, while highlighting the societal importance of intact forest ecosystems. Key points made by these authors include:

- Less than a quarter of the world's forests are now considered to be structurally intact or free of degradation. Forests are continuing to be degraded; for example, Tyukavina *et al.* (2015) recorded an 18%

decline in the area of structurally intact forests from 2007 to 2013, based on satellite imagery. Similar declines were reported from analysis of the global road network (Ibisch et al., 2016).
- Anthropogenic pressures have been shown to alter a range of forest characteristics, including physical structure, species composition, diversity, abundance and functional organisation. These pressures can also interact with natural disturbance regimes such as fire to perturb forests beyond their capacity to regenerate.
- Assessments of degradation using satellite imagery fail to detect many forms of degradation, such as altered fire and flood regimes and overhunting. For example, large areas of Central Africa are mapped as 'intact' but have lost populations of the forest elephant (*Loxodonta cyclotis*), which is a keystone species. Therefore, intensity of forest degradation may often be higher than assessments suggest.
- Loss of fauna through overhunting can significantly affect vegetation composition and structure, as well as carbon storage potential. In some areas, overhunting is pervasive wherever human access is facilitated by new infrastructure, such as roads. Impacts can extend over very extensive areas (for example, at least 36% of the Amazon) and can include major disruptions of ecological processes such as pollination, seed dispersal, nutrient cycling, decomposition and tree regeneration, and can cause significant loss of biomass (Peres et al., 2016). This shows that human impact can collapse an ecosystem from within, in a way that can be difficult to detect from assessments of forest cover or tree density.
- Degradation can reduce the capacity of forests to act as carbon sinks. Collectively, global forests remove 25% (2.4 Pg C yr^{-1}) of anthropogenic carbon emissions and hence greatly slow the pace of climate change. Large-scale forest degradation will likely result in a major reduction in this critical ecosystem service (Griscom et al., 2017).
- Logging is the most pervasive threat facing many forest species. Impacts can increase with the intensity of logging and with the number of times a forest is logged (Edwards et al., 2014).
- Degraded forests also have increased risk of, and susceptibility to, natural disturbances such as fire. For example, logging can increase the risks of high-intensity fires.
- Degraded forests are also at higher risk from invasion by exotic invasive species when compared with non-degraded forests. Degraded forests may also burn more frequently, leading to a cascade of damage or 'landscape trap' where repeated disturbances lock a forest into early successional states (Lindenmayer et al., 2011).

Watson et al. (2018) also note that forest degradation often intensifies over time, through the action of a sequence of potentially interacting factors: (1) construction of new roads and other infrastructure, (2) the entry of new extractive development projects such as mining and logging, (3) successive cycles of logging of often progressively lower value trees, (4) improved human accessibility enables entry of farmers leading to increased forest clearance and fragmentation and (5) increased hunting pressure. For example, in the Brazilian Amazon, Asner et al. (2006) found that nearly all logging occurred within 25 km of main roads; within that area, logging increased the risk of subsequent deforestation by up to a factor of four.

What are the impacts of forest fragmentation? Forest fragmentation has received particular attention from researchers. Taubert et al. (2018) provide an informative analysis of tropical forest fragmentation, showing that South America's largest forest fragment spans around 45% of the total forest area of the Amazon, whereas the largest fragment in Asia covers only 18% of the area of Borneo. Analyses showed the intriguing result that fragmentation displays criticality (see Chapter 2), meaning that its large-scale behaviour is independent of the underlying small-scale mechanisms. Brinck et al. (2017) show that 19% of the remaining area of tropical forests lies within 100 m of a forest edge. And there's more! Haddad et al. (2015) conducted a further analysis of global satellite imagery (MODIS) and concluded that 70% of remaining forest area is located within 1 km of a fragment edge (although again the imagery actually assessed tree cover rather than forest cover).

Some outstanding experimental investigations have been undertaken to assess the ecological impacts of forest fragmentation, notably the Biological Dynamics of Forest Fragments project in Brazil, the SRS Corridor Experiment in the USA and more recently the S.A.F.E. project in Borneo. Laurance et al. (2002) summarise results of the former obtained after 22 years, noting that fragmentation alters species richness and abundances, species invasions, forest dynamics, the trophic structure of communities and a variety of ecosystem processes. A particularly important finding of this project was that fragmentation disproportionately kills big trees (Laurance et al., 2000). In their subsequent update, Laurance et al. (2011) emphasised the role of edge effects as a dominant driver of fragment dynamics, strongly affecting forest microclimate, tree mortality, carbon storage, fauna and other aspects of fragment ecology. Overall, populations and communities of species in fragments were found to be hyperdynamic compared with nearby intact forest.

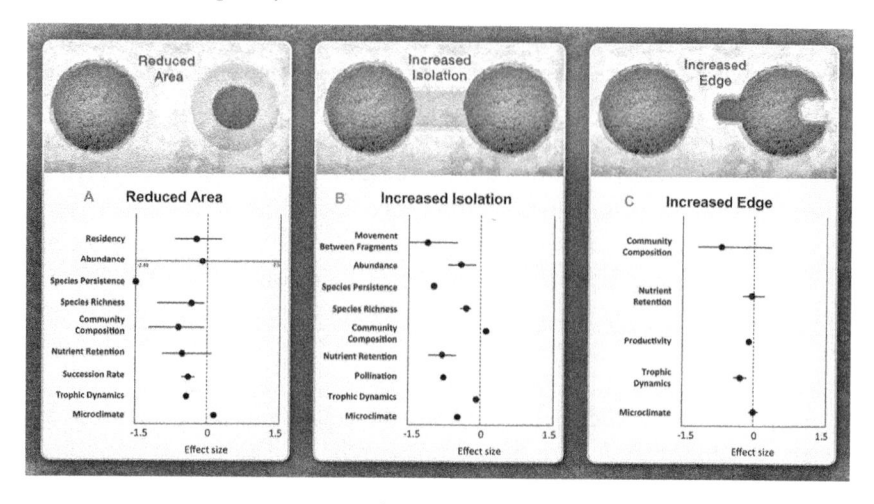

Figure 4.13 Summary of the impact of habitat fragmentation on ecological characteristics and processes. These results summarise the findings of a series of five field experiments undertaken in Brazil, the USA, Australia and UK/Canada, three of which focused on forests and two on grassland or moss. Each dot represents the mean effect size. Either negative or positive values could represent degradation. Horizontal bars are the range when a dot is represented by more than one study. Reproduced from Haddad *et al.* (2015).

Collectively these impacts were referred to as a process of 'ecosystem decay' (Laurance *et al.*, 2002), but they could equally be described as a process of ecosystem collapse. This investigation provides the most detailed account of what the collapse of a forest ecosystem actually entails, when driven by fragmentation.

Haddad *et al.* (2015) summarised the results of these experimental studies, together with some others, which rather bizarrely included experiments on mosses and grasslands as well as forests. These provide evidence of a wide range of fragmentation impacts on different ecological characteristics and processes (Figure 4.13). Overall these authors concluded that across a range of ecosystem types, fragmentation consistently reduced species persistence, species richness, nutrient retention, trophic dynamics and, in more isolated fragments, species movement. These findings are supported by a large number of observational studies. For example, in the Chaco Serrano forest of Argentina, Valladares *et al.* (2012) found that fragmentation has led to contraction of both plant–herbivore and host–parasitoid food webs around a central core of highly connected species. Magrach *et al.* (2014) performed a meta-analysis of forest fragmentation studies that investigated interspecific interactions and found that the effects of fragmentation on mutualisms were primarily driven by habitat degradation, edge effects and fragment isolation. They also found the fascinating result that impacts were consistently more negative on mutualisms such as pollination and seed dispersal, than on antagonistic interactions such as predation or herbivory. Further, Pfeifer *et al.* (2017) assessed the impact of edge effects for 1,673 vertebrate species in a compiled data set with wide geographic coverage and found that abundances of 85% of species were affected by forest edges, 46% positively and 39% negatively. Forest-core species, particularly smaller-bodied amphibians, larger reptiles and medium-sized non-flying mammals, appeared to be more sensitive to edges.

Given this substantive body of evidence, one might conclude that the case against fragmentation is cut and dried, but in fact this turns out to be another area of active debate. Lenora Fahrig recently described fragmentation as a 'zombie idea' – an idea that should be dead but isn't (Fahrig, 2017). There are many examples of zombie ideas in ecology – I encourage you to search for some others in this book. As noted by Fox (2011), 'in some cases they've survived decades of attacks from the theoretical and experimental equivalents of chainsaws and shotguns, only to return to feed on the brains of new generations of students'. So, you have been warned! Nevertheless, it is rather startling to see Professor Fahrig place habitat fragmentation among the zombies, given that – with her >37,000 citations – she has played a significant role in popularising habitat fragmentation as a research topic.

Specifically, Fahrig (2017) highlights the fact that it is very difficult to separate fragmentation from habitat loss; many studies confound them. Other zombie ideas that she nominates, and challenges in her review, are that edge effects are generally negative; fragmentation reduces connectivity; habitat specialists show particularly negative responses to fragmentation; and negative fragmentation effects are particularly strong with low habitat areas and in the tropics. Overall she concludes that the widespread occurrence of non-significant fragmentation effects, and the fact that both positive and negative significant effects have been reported, indicates that the overall effect of habitat fragmentation on ecological responses is likely to be slight (Fahrig, 2017). In other words, when only those studies are considered that empirically estimated fragmentation effects at the landscape scale independently of changes in habitat area, most either reported no significant effects, or found fragmentation to be positive. Unsurprisingly, some researchers who have dedicated their careers to studying habitat fragmentation reacted rather negatively to these findings. Fletcher *et al.* (2018) suggest that Fahrig's conclusions are drawn from a narrow and potentially biased subset of available evidence, which ignores many of the studies that have demonstrated a negative effect of fragmentation. Fahrig (2017) makes no mention of recent meta-analyses of fragmentation impacts, for example.

Like any good academic spat, it doesn't end there, of course. Fahrig *et al.* (2019) counter by noting that the arguments presented by Fletcher *et al.* (2018) are based on extrapolation from patch-scale patterns and mechanisms to the landscape scale. They suggest that this approach is unreliable because it ignores landscape-scale mechanisms (e.g. increased habitat diversity) that can counteract the effects of patch-scale processes. Further, it fails to provide a test of whether impacts occur at the landscape scale – which Fahrig's (2017) review attempted to do. It might be tempting to dismiss this debate as just another example of academics simply doing what they do – having an argument. But as Miller-Rushing *et al.* (2019) point out, this issue really does cut to the core of conservation biology and has major implications for conservation policy and practice, much of which is currently aiming to address the negative impacts of both habitat loss and fragmentation. It might even reverse the policy drive to protecting relatively few, large protected areas, in favour of conserving many small habitat fragments (e.g. see Wintle *et al.*, 2019). In attempting to throw some water on the fire, Miller-Rushing *et al.* (2019) reach the conclusion that scientists always like best: we need more research. Specifically, in this case, research on the impacts of fragmentation at the landscape scale.

How might multiple stressors interact? Whether or not fragmentation *per se* is a driver of forest collapse, there is a widespread belief that it can interact with other anthropogenic pressures. For example, Brook *et al.* (2008) suggest that forest clearance and fragmentation can cause localised drying and regional rainfall shifts, thereby increasing fire risk; fragmentation can also increase overharvesting by improving human access (Figure 4.14). Similarly, in their review of tropical forests, Malhi *et al.* (2014) suggest that synergistic effects among drivers are commonplace. Fragmented forests are generally more prone to fire, logging and hunting. Selective logging and fragmentation can increase forest flammability owing to increased fuel loads resulting from elevated tree mortality, ignition

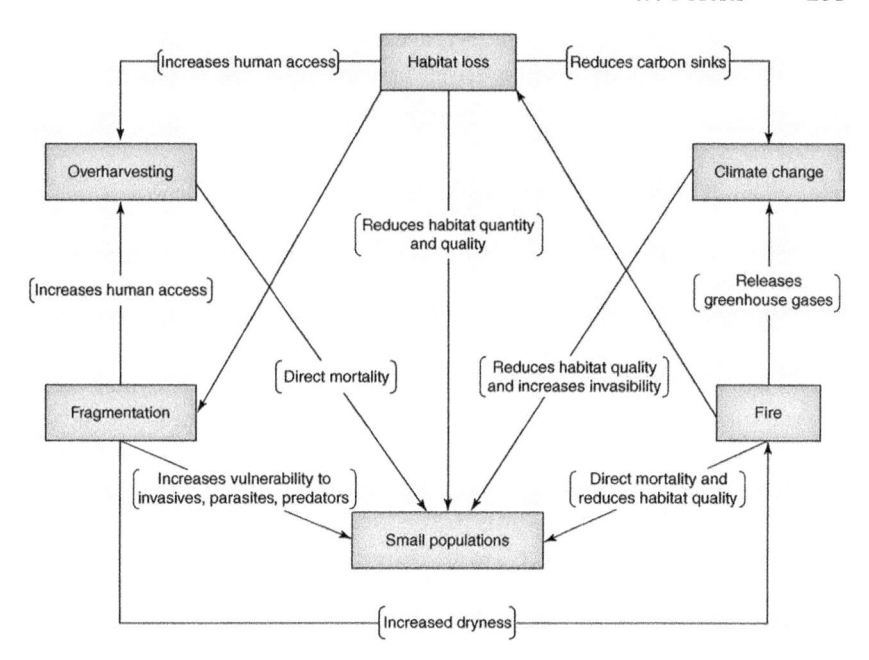

Figure 4.14 Synergistic interactions between multiple stressors that can increase the risk of species extirpation in disturbed tropical rain forests. Reprinted from Brook *et al.* (2008), with permission from Elsevier.

sources can spread from agricultural expansion and road building, and fire risk is particularly high during droughts.

Barlow *et al.* (2016) provide a valuable study of the impacts of anthropogenic disturbance on biodiversity in the Brazilian Amazon, using a large data set of plants, birds and dung beetles. Here, forest disturbances included wildfire and selective logging, which were assessed separately from the effects of forest loss and fragmentation. Results indicated that the impacts of disturbance can double the biodiversity losses from deforestation and fragmentation alone. Despite these observations, there have been relatively few systematic studies of interactions between multiple stressors affecting forests. Darling and Côté (2008) highlight the need for caution: evidence from meta-analysis of experimental investigations on animals suggests that synergies between stressors may be somewhat over-reported in the literature. In their study, only a minority (35%) of the experiments conducted (which included aquatic as well as terrestrial systems) showed synergistic interactions.

Interactive impacts of multiple stressors on forests have received particular attention in relation to climate change and the occurrence of large-scale dieback of forest stands. Allen *et al.* (2010) provided a striking account of large-scale tree mortality in many different parts of the world that have been attributed to drought and heat stress, while noting the occurrence of interactions with other climate-mediated processes such as insect outbreaks and wildfire. In their review, 88 examples of such forest mortality were identified from the literature since 1970, the case studies ranging from local increases in tree mortality rates to regional-scale forest die-off. A range of interactions were identified that have driven these examples of forest mortality, including physiological impacts of drought and heat stress, and the incidence of biotic agents (pests and diseases). Under drought conditions, the latter can increase rapidly in abundance and then overwhelm the trees because they are already physiologically stressed. Examples cited of significant tree death include the severe loss of Atlas cedar (*Cedrus atlantica*) in the region from Morocco to Algeria, mortality of *Pinus tabulaeformia* across 0.5 million ha in east-central China and extensive mortality of *Nothofagus dombeyi* in Patagonia, South America. Many of the best-documented examples are from North America, including death of >1 million ha of multiple spruce species in Alaska, >10 million ha of *Pinus contorta* in British Columbia, 1 million ha of *Populus tremuloides* in Saskatchewan and Alberta and >1 million ha of *Pinus edulis* in the south-western USA (Allen *et al.*, 2010).

Allen *et al.* (2015) provide a further account of this phenomenon, highlighting the following findings associated with global warming: droughts are projected to occur everywhere, eventually; climatic warming produces hotter droughts; tree mortality can occur faster in hotter drought; the frequency of lethal drought is likely to increase non-linearly; and tree mortality occurs relatively rapidly compared with the intervals needed for forest recovery. Again, evidence is summarised of interactions between drought, insect pest outbreaks and more frequent and severe fires. A suggestion is also made that post-mortality forest recovery could also be limited, delayed for long time periods or prevented entirely because of hotter droughts, in combination with other forms of disturbance. Overall, Allen *et al.* (2015) conclude the future vulnerability of forests globally is being widely underestimated, including those located in regions that currently have relatively wet climates.

There are many more examples. For example, van Mantgem *et al.* (2009) showed that mortality rates in unmanaged old forests in the western United States have increased rapidly in recent decades, with doubling periods ranging from 17 to 29 years, owing to global warming and increasing water deficits. Similarly, Peng *et al.* (2011) reported that tree mortality rates in boreal forests in Canada increased by an overall average of 4.7% per year from 1963 to 2008, again because of regional droughts. In southern Europe, Carnicer *et al.* (2011) provide evidence of similar drought-induced tree mortality increasing during recent decades, which was associated with changes in the structure of tree crowns. Observations suggested that drought can lead to sudden increases in defoliation by insects and fungi, which can potentially disrupt food webs. In south-western Australia, Matusick *et al.* (2013) describe a sudden and unprecedented forest collapse in a Mediterranean-type ecosystem as a result of record heat and drought in 2010–2011. This indicates that this forest type, which was thought to be resilient to climate change, is in fact susceptible to sudden forest collapse when thresholds of tolerance have been crossed.

Returning to the south-western USA, Park Williams *et al.* (2012) suggest that owing to projected climate change, the mean forest drought-stress by the 2050s will exceed that of the most severe droughts in the past 1,000 years. This will lead to forests in the region changing 'towards distributions unfamiliar to modern civilization'. Taking these examples together, it is difficult not to reach the same conclusion as McDowell and Allen (2015), who suggest that this century will witness 'massive disruption' of today's forests, resulting in substantial reorganisation of their structure and carbon storage. Such changes in forest ecosystems could also affect the exchange of carbon, water and energy between forests and the atmosphere, providing a feedback mechanism driving further change in global climate (Weed *et al.*, 2013).

Seidl *et al.* (2017) provide a global review of the interactions between climate and other forms of disturbance affecting forest ecosystems. Results indicated that disturbances from fire, insects and pathogens in particular are likely to increase as a result of global heating. Disturbances caused by other agents, such as drought, wind and snow, will be dependent on changes in water availability, which can be expected to display greater variability as a result of changing climate. Overall, biotic forms of disturbance (i.e. insect attack) showed the highest propensity to display interactions with climate; however, all the six forms of

238 · Contemporary Case Studies

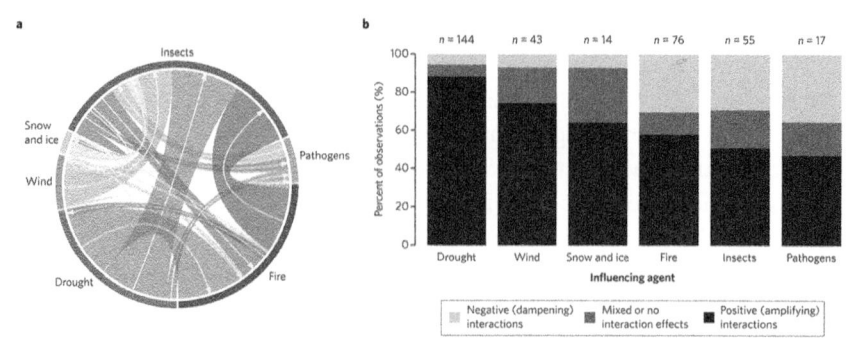

Figure 4.15 Interactions between different types of forest disturbance. (a) The outer circle indicates the distribution of interactions between disturbance types, while the flows through the centre of the circle illustrate the relative importance of interactions between individual disturbance types (as measured by the number of observations reported in the literature). Arrows indicate the direction of influence. (b) The direction of the interaction effects observed in the literature. *n*: number of observations. Reprinted from Seidl *et al.* (2017), with permission from Springer Nature.

disturbance considered displayed some evidence of interactions (Figure 4.15). As might be expected, Seidl *et al.* (2017) found that warmer and drier conditions increase the risk of fire, drought and insect disturbances, while warmer and wetter conditions increase the likelihood of disturbances from wind and pathogens. Most of the interaction effects reported in the literature were found to be positive; in other words, climate change amplifies the disturbance. However, climate-mediated vegetation changes can also potentially dampen disturbance interactions, for example when trees susceptible to an insect pest are outcompeted by individuals or species that are better adapted to warmer climates, resulting in a system that is less susceptible to disturbances (Seidl *et al.*, 2017).

How widespread are trophic cascades in forest ecosystems? As noted by Ripple *et al.* (2016a), many trophic cascades have now been documented in terrestrial ecosystems, including forests. One of the most iconic examples is that of grey wolves in Yellowstone National Park. After wolves were reintroduced to the park in the mid-1990s, elk populations declined, and changes in elk behaviour were observed following the re-establishment of a 'landscape of fear' (Laundré *et al.*, 2001). As a result of declining herbivore pressure, regeneration of tree species such as aspen has increased, providing evidence of a trophic cascade (Beschta *et al.*,

2018). While this is just the kind of ecological story that captures the imagination, and has therefore found its way into ecology textbooks and the popular media, that does not mean the evidence is necessarily robust. Fleming (2019) criticised the methods used by Beschta *et al.*, notably the lack of replication, dependence on observation rather than experimentation and a lack of measurement of the underlying processes. Some other field-based tests of the hypothesis in Yellowstone have failed to support it (e.g. Kaufman *et al.*, 2013; Winnie, 2012). Of course Beschta *et al.* (2018) may be right, and wolves may indeed be saving the Yellowstone aspen – there is a considerable body of evidence to support the idea (Beschta *et al.*, 2019). Nevertheless I can't resist quoting from Fleming's (2019) conclusion to his tirade, which reads as a kind of existential howl on behalf of ecological science:

My critique here exemplifies a bigger problem. The credence of the claims about ecological processes and mechanisms should not rely on the confidence with which they are asserted, but on the veracity, applicability and repeatability of their supporting evidence. The publication of papers that do not clearly articulate and acknowledge caveats around the interpretation of conclusions has detrimental consequences for the reputation and credibility of ecology as a discipline.

Amen to that.

Another classic example is Lago Guri in Venezuela, where a series of forested islands were created in a hydroelectric scheme in the 1980s. Fortuitously, some of these excluded predators. Predator-free islands were characterised by hyperabundant howler monkeys, iguanas and leafcutter ants, which then significantly increased the mortality of young trees as a result of herbivory. This threatened to reduce the species-richness of these forests, leaving only those plant species that are resistant to herbivory (Terborgh *et al.*, 2001). Hyperabundant herbivores have caused major transformations to many other forest ecosystems; a powerful example is provided by deer, especially when introduced to countries such as New Zealand where there are no predators. High deer densities modify forest composition and structure, leading to cascading effects on other species, including insects, birds and other mammals; they can also change successional trajectories and alter nutrient and carbon cycling (Côté *et al.*, 2004; Ripple *et al.*, 2010). In this way, herbivores can shift forest ecosystems to a different state, which may be very persistent (Côté *et al.*, 2004).

Terborgh (2015) suggests that many terrestrial ecosystems are likely to be vulnerable to such catastrophic regime shifts in the absence of predators. Such transitions often involve reorganisation of the community and loss of diversity, suggesting that top-down forcing is important for maintenance of species richness. While noting the potential for a state change in entire ecosystem following the removal (or addition) of a single trophic level, Terborgh (2015) notes the lack of experimental evidence for this in terrestrial ecosystems, particularly for wide-ranging top carnivores such as wolves and big cats. This largely reflects the difficulty of conducting experiments with top predators at the landscape scale, given the large areas involved. There are some examples, however, such as the replicated experimental removal of coyotes in Texas (Henke and Bryant, 1999) and isolation of predator-free forest fragments in a reservoir in Thailand (Gibson *et al.*, 2013). In the latter example, a near-total loss of native small mammals was recorded within 5 years in smaller fragments (<10 ha) and within 25 years from larger fragments (10–56 ha), probably because of predation by invasive rats. Terborgh (2015) notes that such increases in mesopredators can have catastrophic impacts on smaller animals such as birds, lizards, frogs and arthropods, leading to either positive or negative impacts on plants, depending on the prey preferences of the smaller predators. This shows that trophic cascades can have complex effects, depending on the response of species at intermediate levels of food webs.

Both top-down and bottom-up processes are now widely recognised to operate simultaneously in many ecosystems (Ripple *et al.*, 2016a). However, their relative importance in forests does not appear to have been systematically evaluated. Jia *et al.* (2018) describe a meta-analysis of results from 123 experimental exclusions of native animals in terrestrial ecosystems, including forests. Overall results indicate that top-down regulation by herbivores is widespread, as herbivores significantly reduced plant abundance, biomass, survival and reproduction, but not species richness. However, the strength of top-down effects was found to be more site-specific and difficult to predict than in aquatic systems.

What are the impacts of harvesting wildlife? Consideration of what happens to a forest ecosystem when animals are removed is relevant to another threatening process affecting forests – what has been termed the 'empty forest' syndrome (Redford, 1992). This refers to the extirpation of wildlife populations (i.e. defaunation) as a result of hunting by humans

Table 4.5 *Selected research findings assessing the impact of defaunation on forest ecosystems*

Geographic region	Research finding	Reference
Atlantic Forest, Brazil	Defaunation has the potential to significantly reduce carbon storage capacity even if only a small proportion of large-seeded trees are extirpated as a result.	Bello et al. (2015)
Pan-tropical	Carbon losses of up to 12% projected for some regions. Regional variation influenced by distinct combinations of dispersal mode, seed size and size of adults among tree species.	Osuri et al. (2016)
Amazonia	Projected losses in above-ground biomass between 2.5–5.8% on average, with some losses as high as 26.5–37.8%.	Peres et al. (2016)
Central Africa	Trophic webs significantly disrupted, with knock-on effects for other ecological functions, including seed dispersal and forest regeneration.	Abernethy et al. (2013)
North-west Borneo	Pervasive changes in tree population spatial structure and dynamics, leading to a consistent decline in local tree diversity over time.	Harrison et al. (2013)
Central Africa	Reduction in the mean dispersal distances of nine tree species by 22%. Significantly reduced herbivory owing to the lower abundance of mesoherbivores. Significantly lower above-ground biomass than in logged and undisturbed forest types.	Poulsen et al. (2013)

for food. Recent estimates suggest that around half of all tropical forests have been at least partially defaunated (Benítez-López et al., 2019); as a result of overhunting, together with other threats, many terrestrial mammals are experiencing a 'massive collapse' in their population sizes (Ripple et al., 2016b). The scale of the problem is hard to comprehend: some 89,000 tonnes of wild meat are harvested annually in the Brazilian Amazon, and exploitation rates in the Congo Basin are estimated to be five times higher (Fa et al., 2002). According to Ripple et al. (2016b), the loss of large-bodied animals can lead to rapid, potentially irrevocable changes in forest ecosystems, as a result of their role in seed dispersal and other ecological processes. Similarly the loss of small mammals can result in significant changes in forest regeneration, composition and structure.

242 · Contemporary Case Studies

Bats, for example, are the largest group of small mammals threatened by hunting, but many of these play a specialised role in pollination or seed dispersal (Ripple et al., 2016b).

Bushmeat hunting clearly represents a very significant environmental problem, which has justifiably attracted the attention of many researchers. A variety of different impacts on forest ecosystems have been documented, some of which are briefly summarised in Table 4.5. Based on a review of 42 studies, Kurten (2013) concluded that as a result of defaunation, larger-seeded tree species consistently experience reduced seed dispersal, increased seedling aggregation around parent trees and changes in recruitment, consistently leading to lower species richness, higher species dominance and lower diversity in tropical forest communities. Jorge et al. (2013) suggest that as a result of defaunation, 88–96% of the remaining Atlantic Forest of Brazil is likely to be suffering from trophic cascade effects, such as changes in patterns of seed dispersal, herbivory and plant recruitment, changes in community structure) and mesopredator release. It has even been suggested that the impacts of defaunation may become manifest at the global scale, by affecting forest structure and composition, carbon storage and potentially climate (Galetti and Dirzo, 2013).

Given the magnitude of these effects, could a forest that has been stripped of its larger animals be considered to have collapsed? Referring back to the definition of 'collapse' adopted in this text (see Chapter 1), defaunation could certainly be associated with a loss of defining biotic features, a shift in composition outside the natural range of variation and even a transformation in the identity of a system. Does, though, a defaunated forest represent a different type of ecosystem? In the short term, floristic composition might be unaffected, but over the longer term, the changes could be profound (Table 4.5). This raises the fascinating question of how much ecosystem change needs to occur before it can be considered to have collapsed, an issue I will return to in Chapter 5.

Are forests affected by anthropogenic disturbance characterised by thresholds? In their review of global tipping points, Lenton et al. (2008) identify two forest examples: boreal forest and the Amazon rain forest. Large-scale dieback of boreal forest is projected as a possible consequence of climate change, as a result of increased heat stress and drought, which could lead to transitions to open woodland or grassland. The example of the Amazon is based on the fact that much of the precipitation in the region is derived from evapo-transpiration from rain forest in the Amazon basin; projections of climate change suggest that widespread dieback of the rain forest may occur as a result of reduced precipitation and lengthening of the dry season. According to Lenton *et al.* (2008), this process could potentially interact with deforestation to cause a tipping point, although the precise feedback mechanisms involved in either this or the boreal forest example were not presented by these authors. Nonetheless, this study raised the prospect of non-linear changes occurring in both ecosystems.

Field observations provide some evidence that rapid changes are occurring in both of these ecosystems. Much of the boreal forest has recently experienced mean annual temperature increases of 1.5°C or more, which has been associated with severe outbreaks of pest insects such as the mountain pine beetle in Canada and the Siberian silk moth in Siberia, in addition to the drought-induced dieback described earlier (Gauthier *et al.*, 2015). There is also some evidence of a slow northward migration of temperate deciduous tree species into the boreal zone of eastern North America and spread of evergreen coniferous species into deciduous larch forests in Siberia (Gauthier *et al.*, 2015). In North America, the area of boreal forest that has burned each year has increased in recent decades, associated with an increase in lightning ignitions (Veraverbeke *et al.*, 2017); similar observations have been made in Siberia (Soja *et al.*, 2006). Some of the changes recorded in boreal forest ecosystems appear to be happening more rapidly than predicted, suggesting the occurrence of rapid, non-linear change and ecological thresholds (Chapin *et al.*, 2004; Mann *et al.*, 2012; Soja *et al.*, 2006).

In the Amazon, severe droughts were observed in 2005, 2010 and 2015–2016, and severe floods were experienced in 2009 and 2012, suggesting an instability in the entire system. There is also evidence that the length and intensity of the dry season is increasing in southern Amazonia (Fu *et al.*, 2013; Lovejoy and Nobre, 2018). The possibility of climate-driven forest dieback in the Amazon has consequently attracted a great deal of research attention. However, findings have been associated with a degree of uncertainty associated with the use and interpretation of different climate models. For example, whereas some modeling studies have indicated a high risk of dieback (e.g. Cox *et al.*, 2004; Good *et al.*, 2011; Malhi *et al.*, 2009), others have suggested that the forest may remain largely intact, or may even increase in biomass (e.g.

244 · Contemporary Case Studies

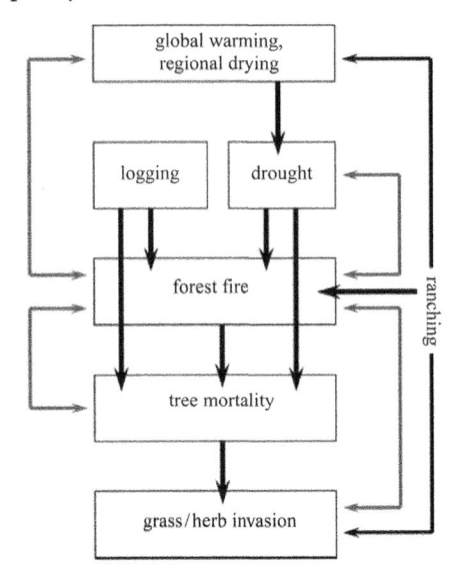

Figure 4.16 Interactions and feedbacks that could potentially lead to a near-term Amazonian forest dieback. Reproduced from Nepstad *et al.* (2008).

Cox *et al.*, 2013; Huntingford *et al.*, 2013; Thompson *et al.*, 2004; Zhang *et al.*, 2015). Other researchers have highlighted the potential for interactions between climate-induced droughts, fire and logging, which could provide a mechanism for abrupt loss of forest over large areas (Malhi *et al.*, 2008). For example, Nepstad *et al.* (2008) suggest that if climate change and associated droughts continue as projected, ~55% of the forests of the Amazon will be lost owing to feedbacks between these factors (Figure 4.16), which collectively provide the mechanism for a region-wide tipping point. It has been suggested that such a tipping point might occur if the total deforested area exceeds a threshold of 40% (Nobre and Borma, 2009).

As noted by Malhi *et al.* (2008), the Amazonian forest is one of Earth's greatest biological treasures and is a major component of the Earth system, so any potential collapse of this ecosystem is deeply concerning. But what is the empirical evidence that this is actually happening? In an important investigation, Phillips *et al.* (2009) showed through an assessment of field plots that Amazonian forest is vulnerable to drought stress; following the 2005 drought, forest subjected to a 100-mm increase in water deficit lost 5.3 Mg of above-ground carbon per hectare, partly as a result of increased tree mortality. While Amazonian rain forest may be somewhat tolerant of seasonal droughts, long-term field experiments have shown that multi-year droughts can lead to increased tree mortality; after four to seven years of rainfall exclusion, tree mortality rates increased by a factor of nearly three (Davidson et al., 2012).

There is also clear empirical evidence highlighting a link between drought and an increased risk of fire; for example, fire incidence increased by 36% during the 2015 drought compared with the preceding 12 years (Aragão et al., 2018). With respect to the impacts of fire, Barlow and Peres (2008) demonstrated that wildfires can lead to substantial changes in forest structure and composition, with additional fire events leading to cascading shifts in forest composition. After a forest is burned, it is more likely to burn again, because it dries out more readily; conversion of fire to pasture can also promote drought by increasing albedo and decreasing the flow of water vapour to the atmosphere (Nepstad et al., 2001). In summary, Davidson et al. (2012) suggest that changes in precipitation and river discharge associated with deforestation already observed in the southern and eastern parts of the Amazon could cause significant vegetation shifts and further feedbacks to climate. However, the idea that the entire basin could transition to a relatively dry, alternative state is more difficult to support, based on current evidence. The possibility of such a basin-wide tipping point might event distract researchers from the large-scale regional changes that are already taking place, such as lengthening of the dry season and increases in river discharge (Davidson et al., 2012).

Examples of ecological thresholds in boreal and Amazonian forests relate to a broader literature examining the occurrence of such thresholds in other terrestrial ecosystems (Groffman et al., 2006; Huggett, 2005). A number of investigations have demonstrated threshold relationships between habitat area or connectivity and abundance of a range of species, such as forest-dependent birds (Radford et al., 2005) and jaguars (Zemanova et al., 2017). These studies were stimulated by the classic review by Andrén (1994), who proposed a threshold value of 70–90% habitat loss before bird or mammal species were affected by isolation of habitat patches. Subsequent research has demonstrated that such thresholds are highly species- and context-specific (Johnson, 2013). For example, in their study of 39 small mammal species in the Atlantic forest of Brazil, Pardini et al. (2010) found that the occurrence of threshold effects varied among species and according to the degree of habitat fragmentation. As a rule of thumb, though, the thresholds proposed by

Andrén (1994) seem to work quite well. For example, in their review of evidence from a range of different ecosystems (not limited to forests), Swift and Hannon (2010) found that most of the thresholds determined empirically fell within a similar range. Overall, threshold responses were found to be commonly encountered in the literature, although interestingly, they were observed more frequently in modeling studies than in field-based investigations (Swift and Hannon, 2010). Ecological thresholds relating to habitat loss have been attributed to a number of different mechanisms, including increasing Allee effects with declines in habitat area, the breakdown of mutualisms and habitat change feedbacks such as an increasing abundance of fire-promoting plant species with an increase in fire frequency (de Oliveira Roque et al., 2018).

What is the impact of secondary extinctions? While ecological thresholds associated with forest loss and degradation appear to be widespread (Evans et al., 2017; Martin et al., 2017; Swift and Hannon, 2010), there is rather less evidence for extinction cascades or secondary extinctions driving the process of ecosystem collapse. As noted by Brodie et al. (2014), this partly reflects the difficulty of studying such phenomena. Secondary extinctions are difficult to document and many may occur un-noticed; alternatively, such extinctions might not have happened yet but may be inevitable in the future (referred to as 'secondary extinction debt'). The best examples are probably the experimental examinations of forest fragmentation and trophic cascades cited earlier, although even here most studies have documented changes in relative abundance of species rather than actual extirpation. Evidence suggests that those species that are critically dependent on specialised interactions with specific partners may be particularly at risk of secondary extinctions, such as specialist parasites and predators (Brodie et al., 2014). Disruption of mutualisms has received particular attention from researchers; many studies have demonstrated that the loss of pollinators or seed dispersers can negatively affect plant populations. Few studies have documented subsequent extinctions, however (Brodie et al., 2014). For example, Anderson et al. (2011) showed that the functional extinction of bird pollinators reduced pollination, seed production and plant density in the New Zealand shrub *Rhabdothamnus solandri*. Seed production was reduced by 84% and the number of juvenile plants by 55%, yet the species has not (yet) been driven to extinction.

Brodie et al. (2014) make the important point that species can sometimes tolerate the loss of a mutualistic partner. For example, introduced pigs and rats can act as seed dispersers for native plants, replacing lost avian dispersers; some species of plant can respond to loss of animal pollinators by switching to wind pollination or self-fertilisation; and species may also display evolutionary changes. As illustration of the latter, Galetti *et al.* (2013) found that populations of palm species in Brazilian Atlantic forest that had been deprived of large avian frugivores for several decades displayed smaller seed sizes than those in non-defaunated forests, enabling dispersal by smaller frugivores. While removal of seed-dispersing vertebrates has repeatedly been shown to change the species composition of tree seedlings and saplings, the impacts of disperser loss on abundance of adult plants or community composition are less well established (Brodie *et al.*, 2014). In a Bornean forest monitored over a period of 15 years, Harrison *et al.* (2013) found that overhunting led to pervasive changes in tree population spatial structure and dynamics, and a decline in local tree diversity over time. However, no effect of hunting was observed on above-ground biomass or tree growth rates, regardless of whether the associated seed dispersers had been extirpated or not. Overall, Brodie *et al.* (2014) concluded that secondary extinction risk is likely to be highest when there are synergistic interactions among multiple anthropogenic stressors. For example, plant species deprived of vertebrate seed dispersers or exposed to elevated herbivory as a result of the loss of top carnivores are most likely to become extinct when they are also subject to habitat loss or fragmentation.

The fact that forest loss and degradation can lead to high rates of species loss is illustrated by the example of Singapore. Here, habitat loss over the past 183 years exceeds 95%; less than 10% of the remaining forest is primary (most being secondary regrowth) (Brook *et al.*, 2003). Comparison of historical and modern checklists indicated that large numbers of species were lost during this period, especially forest specialists, with the highest proportion of extinct taxa (34–87%) recorded for butterflies, fish, birds and mammals. In other groups, such as vascular plants, amphibians and reptiles, observed extinction rates were generally lower, but inferred losses were often higher (5–80%) (Brook *et al.*, 2003). With respect to plant species, epiphytes were found to be particularly prone to extinction (62% loss) (Turner *et al.*, 1994), perhaps reflecting the loss of specialist pollinators and microhabitats. Evidence for secondary extinctions is also available for the Brazilian Atlantic forest, where dramatic non-linear changes in frugivory networks were observed when forest cover was reduced to less than 40% (Mariana Morais *et al.*, 2019).

How useful is dynamical systems theory for understanding forest collapse? Application of this body of theory to forests is less well advanced than for other ecosystem types, such as coral reefs or shallow lakes. However, there have been a number of recent attempts to explore forests as complex dynamical systems. For example, Messier *et al.* (2013) presented an overview of how the theory could potentially be applied to a range of different forest types, with a particular focus on implications for forest management. A number of studies have provided potential evidence for the existence of alternative stable states (Table 4.6), drawn from a wide range of different forest types. A widely cited example is the case of spruce budworm in the boreal forests of North America, which sometimes destroys large areas of conifer forests dominated by spruce and fir. After defoliation, the regenerating forest is often dominated by aspen and birch, although the forest can return back to a state of conifer-dominance following selective browsing by moose (Folke *et al.*, 2010; Holling, 1978).

It seems clear that many forest ecosystems can exist in different states. It is much less clear, however, that these are consistent with the alternative states described by dynamical systems theory. Much of the evidence is anecdotal, or based on models that are difficult to test (Newton and Cantarello, 2015). There is also a complete lack of experimental evidence (Schröder *et al.*, 2005), which partly reflects the difficulty of conducting experiments with forest ecosystems. According to Petraitis (2013) (citing Peterson, 1984), to meet the requirements of the theory, a given site should be shown to have the potential to be occupied by two or more distinct communities, and the environment must not differ between the two putative states. In addition, the disturbance that induces the state transition should be a pulse perturbation (see Chapter 2). Many of the examples given on Table 4.6 fail to meet these criteria (Newton and Cantarello, 2015). Some of them might be better considered as regime shifts, namely transitions in the equilibrial community resulting from a change in environmental conditions or disturbance regime, rather than alternative stable states, which represent alternative configurations of the same community (Dudgeon *et al.*, 2010). It should also be noted that some researchers have explicitly sought evidence for alternative states and have failed to find it (Knox and Clarke, 2012).

Table 4.6 *Suggested examples of alternative stable states in forest ecosystems*

Ecosystem type	Alternate state 1	Alternate state 2	References
Boreal forests	Tundra	Boreal forest	Bonan et al. (1992), Higgins et al. (2002)
	Spruce-fir dominance	Aspen-birch dominance	Holling (1978), Paine et al. (1998)
	Spruce-moss	Spruce-lichen	Jasinski and Payette (2005)
Temperate forests	Pine dominance	Hardwood dominance	Peterson (2002)
	Hardwood-hemlock	Aspen-birch	Frelich and Reich (1999)
	Birch-spruce succession	Pine dominance	Danell et al. (2003)
	Old-growth mountain ash forest	Regrowth mountain ash forest	Lindenmayer et al. (2011)
	Sitka spruce	Alder-dominated deciduous forest	Haeussler et al. (2013)
	Oak dominance	Maple dominance	Nowacki and Abrams (2008)
	Beech dominance	Shrub dominance	Busby and Canham (2011)
	Dominance of species palatable to deer	Dominance of species unpalatable to deer	Coomes et al. (2003)
	Forested wetland	Non-forested wetland	Fletcher et al. (2014)
	Pyrophobic vegetation	Pyrophytic vegetation	Kitzberger et al. (2016)
Tropical forests	Rain forest	Sclerophyll vegetation	Warman and Moles (2009)
	Lowland rain forest	Lowland rain forest, different composition	Vandermeer et al. (2004)
	Subtropical wet forest	Grassland	Henderson et al. (2016)
	Semi-deciduous forest	Shrub- and grass-dominated	Dwomoh and Wimberly (2017)

Source: Adapted from Newton and Cantarello (2015), with input from Folke et al. (2010), Messier et al. (2013), Petraitis (2013) and Resilience Alliance and Santa Fe Institute (2004). Note that the original authors of some of these studies may not have interpreted their research as evidence for alternative stable states, but other researchers have subsequently done so.

Nonetheless, these examples provide some valuable insights into the dynamics of forest ecosystems and the mechanisms of both collapse and recovery. For example:

- Feedback loops operating in tropical forests include the widespread population regulation of plants by density-dependent, distance-dependent or frequency-dependent processes, including the influence of species-specific herbivores and pathogens. Widespread biotic interactions, including mutualisms between plants and fungi, insects, mammals and birds, can create feedback loops between species, leading to self-organising interaction networks. Plant-animal mutualistic networks are often highly nested, with a core set of interacting generalist species, providing functional redundancy and a measure of resilience to species loss (Chazdon and Arroyo, 2013).
- Feedbacks maintaining alder-dominated deciduous forest in degraded stands of Sitka spruce included changes in soil nutrient status induced by the nitrogen-fixing alder, together with other factors such as rodent and insect predation (Haeussler *et al.*, 2013).
- Conversion of spruce-moss forest to spruce-lichen woodland can be caused by the combined effects of fire and outbreaks of spruce budworm, which reduce the stand density and regeneration potential of black spruce. Long-term maintenance of spruce-lichen woodland may result from inhibition of black spruce regeneration by the lichens. This provides an informative example of a persistent transition between ecosystem states attributable to natural forms of disturbance (Jasinski and Payette, 2005).
- A transition from a forested to a non-forested wetland in Tasmania that was persistent over the long term was attributed to a catastrophic fire, which opened the forest canopy and reduced water loss through transpiration. This resulted in local waterlogging, which prevented re-establishment of the forest for over 7,000 years (Fletcher *et al.*, 2014). Similar mechanisms have been identified in high rainfall areas of southern Chile (Díaz and Armesto, 2007), although here a direct effect of fire on soil structure (i.e. loss of organic matter) may also have been influential.
- Post-fire spread of pyrophilic grasses and shrubs produced a positive feedback driving the conversion of semi-deciduous forest in Ghana (Dwomoh and Wimberly, 2017).
- In the temperate forests of southern South America and New Zealand (Kitzberger *et al.*, 2016; Figure 4.17), an increase in fire frequency can

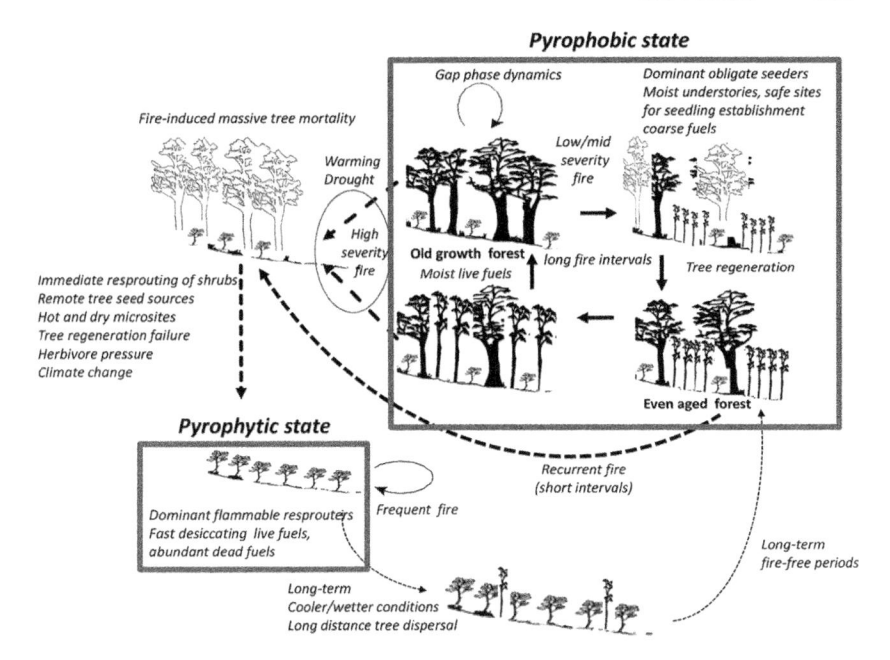

Figure 4.17 Conceptual model of factors influencing transitions between fire-driven alternative states in the forests of southern South America and New Zealand. Undisturbed old-growth forest is dominated by pyrophobic tree species, the shade of which excludes or suppresses pyrophytic plant species, such as shrubs, grasses and bamboo. These pyrophytic species dominate after fire and may persist through repeated fires (which they may promote) or by suppressing regeneration of pyrophobic species. Reproduced from Kitzberger *et al.* (2016), with permission.

result in a transition from pyrophobic old-growth forest to pyrophytic species such as shrubs (e.g. *Leptospermum* and *Kunzea* spp.) and bamboo (*Chusquea* spp.). Forest recovery may be inhibited by an increased fire frequency, promoted by the pyrophytic species, providing a positive feedback mechanism (e.g. see Tepley *et al.*, 2016). Other influential factors include fire-induced edaphic changes (Kitzberger *et al.*, 2005), herbivory, dispersal failure and seed predation. As a result of these factors, pyrophytic states can be stable over the long term (Kitzberger *et al.*, 2016). These processes help explain why there has been so little recovery from the historical fires that occurred in New Zealand (Perry *et al.*, 2012; see Chapter 3).

These latter examples highlight the unique characteristics of fire as a type of ecosystem disturbance. As pointed out by Moritz *et al.* (2005), plants, in part, create their own fire regime. As a result, self-reinforcing alternative states are particularly likely in fire-prone vegetation. This is because under the same environmental conditions, different plant communities can develop with contrasting traits determining their relative flammability and/or resistance to fire. As a consequence, self-promoting pyrogenic vegetation types such as grasslands and shrublands predominate in many parts of the world, which under a different fire regime would naturally develop into forest (Bond et al., 2005; Bond and Midgley, 2012; Kitzberger et al., 2012). Fire-adapted vegetation therefore provides a remarkable example of an ecosystem type that can actually promote certain types of disturbance event, for example through the accumulation of combustible fuel loads. Arguably, this phenomenon isn't adequately captured by any of the theoretical ideas presented in Chapter 2. While it might be tempting to view pyrophobic and pyrophytic vegetation as clear examples of alternative stable states (e.g. see Kitzberger et al., 2016), fire can cause significant edaphic impacts, and changes in the fire regime also represent a form of environmental change. Following Dudgeon et al. (2010), such changes might therefore be more appropriately referred to as regime shifts.

Another example is provided by recent outbreaks of bark beetles (Curculionidae: Scolytinae) in North America, which have caused mortality of up to 90% among larger trees over millions of hectares, with consequent impacts on forest structure, composition and function. Raffa et al. (2008) reported that some 47 million ha of forest had been affected in the previous decade. Similarly, Kurz et al. (2008) assessed a bark beetle outbreak in British Columbia, Canada, which was an order of magnitude larger and more severe than any previous occurrence. These authors estimated that cumulative impact of the beetle outbreak on carbon storage would amount to 270 Mt of carbon during 2000–2020, converting the forest from a carbon sink to a major carbon source. Given the fact that the beetle outbreaks were attributable to climate change, this provides a potential positive feedback mechanism with global climate (Kurz et al., 2008). Further, Raffa et al. (2008) noted that the conifer-beetle system includes key elements often associated with regime shifts by theorists (Folke et al., 2004; Scheffer and Carpenter, 2003), namely cross-scale interactions, positive feedback, multiple causalities, critical thresholds and sensitivity to external drivers.

I close this section with an all-too-brief overview of the work of David Lindenmayer, who has perhaps done more than any other forest ecologist to examine the collapse of forest ecosystems in the context of dynamical systems theory. Lindenmayer and Sato (2018) provide a

4.4 Forests · 253

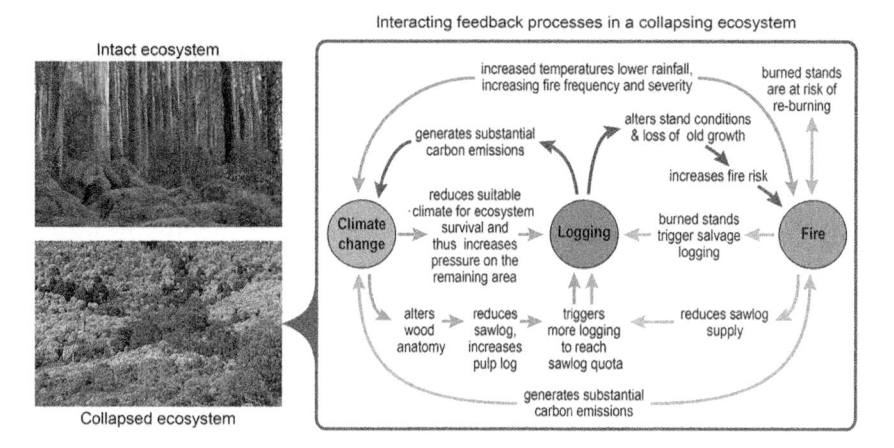

Figure 4.18 Interacting drivers of decline and collapse in the Mountain Ash ecosystem, south-eastern Australia. The arrows indicate whether an interacting driver is initially driven by fire, logging or climate change. Reprinted from Lindenmayer and Sato (2018), with permission.

particularly valuable account of a forest undergoing collapse. Multi-decadal monitoring of Mountain Ash (*Eucalyptus regnans*) forests in south-eastern Australia revealed a number of important insights: (i) ecosystem collapse is occurring, as indicated by changes in ecosystem condition, including the rapid decline of keystone ecosystem structures and associated biodiversity and ecological processes; (ii) the collapse is 'hidden', in that superficially the ecosystem appears to be relatively intact, but prolonged decline coupled with long recovery times result in collapse being inevitable; (iii) different drivers produce different pathways to collapse, and these drivers can interact in ways that hasten and perpetuate collapse (Figure 4.18). In relation to dynamical systems theory, the authors of this study were cautious in suggesting that intrinsic feedbacks might have contributed to the process of collapse. Rather, the main causes appear to have been interactions between multiple drivers (Figure 4.18). However, a positive feedback loop was identified between fire frequency/severity and a reduction in forest age at the stand and landscape scales, leading to an increased risk of young regenerating stands repeatedly reburning before they reach a more mature state (Lindenmayer *et al.*, 2011). The endpoint of this process was referred to as a regime shift.

The concept of 'hidden collapse' is a particularly important one and may be widespread in forest ecosystems. Put another way, an ecosystem could be committed to collapse in a similar way to a species being committed to extinction (Fordham *et al.*, 2016). For example, if regeneration of canopy-dominant tree species is prevented for a period, this will have long-term impacts on the structure and dynamics of the forest, even if the factors preventing regeneration are subsequently removed. In the case of the Mountain Ash ecosystem, extensive historical logging and repeated past wildfires have led to losses of specific tree cohorts, owing to periods of recruitment failure. Consequently as large, old trees die, they are not immediately replaced (Lindenmayer and Sato, 2018). It is well known that in forest ecosystems throughout the world, large, old trees are of exceptional importance as habitat for a wide range of other species, as well as for ecological processes such as carbon sequestration and nutrient cycling (Lindenmayer and Laurance, 2016a,b). In the Mountain Ash forest, significant declines in arboreal marsupials and bird species that require tree cavities were documented over the past two decades, associated with the progressive loss of large trees (Lindenmayer and Sato, 2018).

Recovery

What are the mechanisms of forest recovery? Consideration of this question provides an opportunity to pay a brief tribute to another outstanding forest ecologist, Robin Chazdon, who has been studying recovery processes in tropical forests for many years. Her recent reviews (Chazdon, 2003, 2008, 2014; Chazdon and Arroyo, 2013, Chazdon and Guariguata, 2016) make the following points, among many others. Although these refer explicitly to tropical forests, many are equally applicable to other forest types.

- Natural disturbances (such as tree-fall gaps, landslides and flooding) are integral to forest dynamics; forests will typically recover rapidly from such perturbations. These 'pulse' disturbances contrast with anthropogenic impacts, which are often chronic or 'press' disturbances. The latter tend to be relatively long-lasting and have a homogenising effect on forest composition. Anthropogenic disturbances can also interact with natural events to reduce the regenerative capacity of forests.
- Many tropical forests that were once thought to have been 'pristine' are now known to have recovered from major human disturbances in the past, such as logging and clearance for agriculture. For example, forests in many parts of Mesoamerica have developed on abandoned

agricultural land. Palaeoecological investigations provide evidence for forest recovery from both natural and anthropogenic disturbances, showing that biomass and species richness can recover within a few decades, but recovery of species composition can often take centuries.
- Successional processes are key in determining forest recovery. Recovery from large-scale disturbances often involves patch-dynamics with significant spatial heterogeneity in disturbance producing a spatial mosaic of post-disturbance vegetation. The survival of residual organisms or their propagules also plays a critical role in forest recovery. Many tree species are capable of resprouting after disturbance, and some form soil seed banks.
- The route and rate of recovery will depend on the severity of damage together with a variety of abiotic and biotic factors, including previous land-use history. For example, following abandonment of farmland, rates of recovery are higher if prior land-use intensity is low, recovering areas are relatively small, soils are fertile and there are remnant forest areas nearby. Pastures that have been subjected to overgrazing, repeated burning or that have been bulldozed have slow or no recovery.
- The early establishment of pioneer tree species can be an important contributor to rapid forest recovery. These species function as critical nutrient and carbon sinks that can drive the successional process forward. However, secondary succession following pioneer tree establishment is inevitably slower and less predictable than recovery of residual vegetation.
- Forests are often faced with multiple, overlapping disturbances, which can interact to influence damage and recovery processes. For example, fires that follow cyclones in tropical monsoon forests can be particularly devastating. Effects of prior land use and other human impacts, such as logging, can interact with natural disturbances in complex ways to influence disturbance intensity and recovery processes.
- Many factors can potentially limit forest recovery. Initial tree establishment can be limited by soil conditions, competition from herbaceous vegetation, frequent burning, grazing and other factors. In addition, seed production and dispersal are often a particularly important limitation to tree recruitment following human disturbances. For example, after abandonment of agricultural land on degraded, infertile soils with no residual vegetation and no local sources of seed dispersal, forest recovery can fail to initiate within the expected time frame of 5–10 years. Forest recovery following logging can also fail owing to

insufficient seedling regeneration of commercially exploited species, or if exotic species become competitively or numerically superior.
- Overall, severe disturbances that impact forest canopies, such as large-scale wind disturbances, often have relatively short-lived effects on forest structure and species composition. In contrast, disturbances that affect soils as well as above-ground vegetation, such as logging machinery, heavy grazing and fires, can significantly reduce the rate of forest structural recovery and can have long-lasting effects on species composition.
- The proximity of disturbed areas to remnant forest patches is a key factor influencing the rate of recovery, particularly of species composition. An area of heavily disturbed forest embedded in a matrix of structurally and compositionally intact forest is likely to recover much more rapidly than a similar area surrounded by a matrix of agricultural land or heavily degraded forest.
- Forests often display multiple successional pathways. For example, within the same landscape, some forest stands may show rapid recovery rates and relatively smooth successional pathways, while others exhibit erratic pathways and/or slow recovery rates. This reflects the relative roles of different factors influencing forest recovery, which may vary with spatial scale and landscape composition (Arroyo-Rodríguez et al., 2017; figure 4.19).

Ghazoul and Chazdon (2017) mention a series of variables that underpin forest recovery, including accumulation of below-ground nutrients; rapid fluxes of nutrients, biomass and populations; biotic control of ecosystem functions; species richness and functional redundancy; and critical mutualisms. Similarly, in their review of the application of complex systems science to forests, Filotas et al. (2014) highlight the following additional factors that can influence the recovery capacity of forest ecosystems:

- *Heterogeneity*, in vertical, horizontal and temporal dimensions of stand composition and structure.
- *Cross-scale interactions*, including the ranking structure within food webs, and spatial dimensions such as the size structure of disturbances.
- *Regulating mechanisms that enhance species coexistence*, including density-, distance- and frequency-dependent processes regulating populations, and other self-organising processes such as gap dynamics and the role of mycorrhizal networks. These mechanisms can include many non-linear processes and feedbacks.

4.4 Forests

Ultimate variables

Regional
- Climate
- Topography
- Natural disturbances

Landscape
- Land-use history
- Forest cover
- Matrix characteristics
- Landscape connectivity
- Total forest edge
- Total forest core

Local
- Disturbance regime
- Soil characteristics
- Patch size and shape
- Patch isolation

- Population dynamics of seed dispersers, pollinators and herbivores
- Mesoclimate
- Biological invasion

- Pool of native and exotic plant and animal species
- Speciation and extinction
- Migration

Secondary succession
- Seed/seedling availability
- Soil properties
- Microclimate
- Biotic interactions

(Spatial scale / Proximate variables)

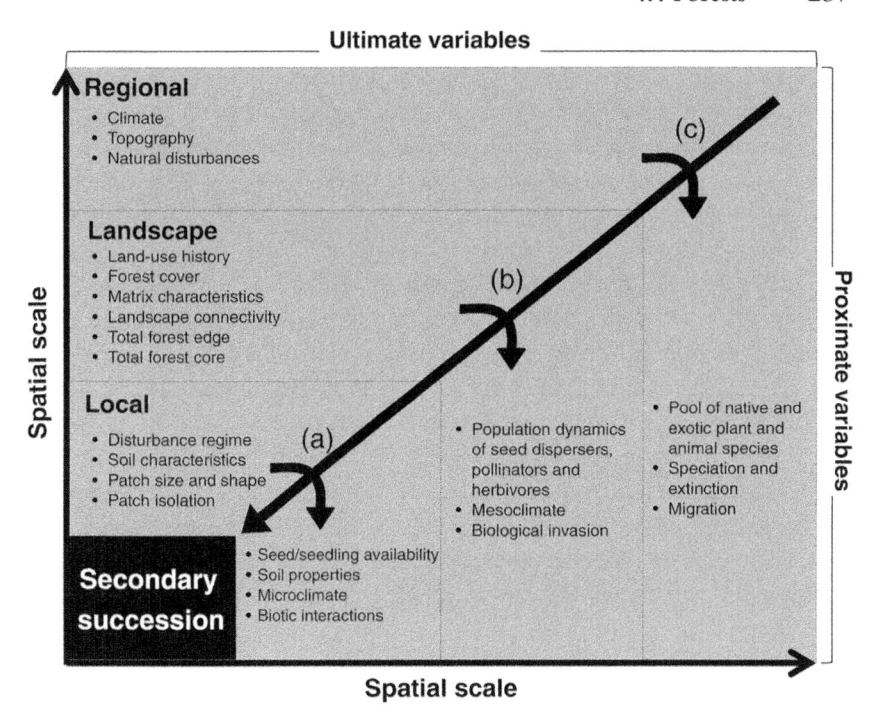

Figure 4.19 Influence of different factors on forest succession, according to spatial scale. Variables can interact among scales (represented by arrows), potentially resulting in synergistic outcomes that can affect forest recovery. Reproduced from Arroyo-Rodríguez et al. (2017), with permission from Wiley.

- *Historical legacies.* These include the legacies of previous disturbance events, which can persist over many centuries owing to the longevity of many tree species. Biological legacies such as seed banks, nurse logs, coarse woody debris and old or dying trees provide continuity of ecological functions and can facilitate forest regeneration.

Overall, the capacity of forest ecosystems to recover from human disturbances is considered to be higher and more predictable in landscapes that: (i) have been modified only recently, (ii) possess a higher proportion of remaining forest cover, (iii) possess remnant trees, seed and seedling banks composed of native species and (iv) possess some areas of well-preserved biodiversity-rich native forests (Arroyo-Rodríguez et al., 2017). Increasing human disturbance or intensification of land use can reduce the ability of forests to recover, by driving the system towards a state with slow or arrested succession (Arroyo-Rodríguez et al., 2017).

Examples presented by Lamb et al. (2005) include situations where degradation leads to loss of topsoil and a reduction in soil fertility, or when land becomes dominated by grasses; similar problems may be encountered following invasion by lianas or invasive ferns (Chazdon, 2014). In such cases, recolonisation by many of the tree species that were originally present can be difficult, if not impossible, leading to persistence of a degraded forest state (Arroyo-Rodríguez et al., 2017).

Further insights into the mechanisms of forest recovery can be gained from experience with forest restoration approaches. The restoration of degraded forest has become a major global activity supported by recent policy initiatives such as the 'Bonn Challenge', which seeks to restore 150 million ha of degraded forest by 2020; the 2014 New York Declaration, which is aiming for 350 million ha by 2030; and the IPCC, which has proposed a forest restoration target of 1 billion ha (Bullock et al., 2011; Crouzeilles et al., 2016; IPCC, 2018; Meli et al., 2017). This partly reflects increasing recognition that forest restoration is the most effective approach to addressing climate change. Recent analyses suggest that worldwide, there is room for an extra 0.9 billion ha of forest cover; if this area were to be completely covered in trees, it would increase global forest area by about a third (Bastin et al., 2019). Most of this area is located within only six countries, namely Russia, the USA, Canada, Australia, Brazil and China. Interestingly, this same analysis indicates that under the Bonn Challenge, 10% of countries have committed to restoring an area of land that is significantly larger than what is actually available for restoration (Bastin et al., 2019). This provides an insight into either the problems associated with developing environmental policy commitments, or the problems associated with meaningfully mapping forest restoration potential (Veldman et al., 2019) – or both.

Crouzeilles et al. (2016) present a global meta-analysis of the factors influencing forest restoration success. Based on data from 221 study landscapes, results indicated that forest restoration enhances biodiversity by 15–84% and vegetation structure by 36–77%, compared with degraded ecosystems. This study also identified some conditions under which forest restoration is most likely to be successful, namely when: (i) there is sufficient time for ecological succession, (ii) previous disturbance is of low intensity and (iii) the forest habitat is not severely fragmented at the landscape scale. This set of conclusions is perhaps not altogether surprising. A further meta-analysis of 196 study landscapes was reported by Crouzeilles and Curran (2016). Results indicated that the degree of recovery of plant diversity was influenced by overall forest cover, while recovery of plant diversity, mammals and invertebrates, as well as forest height, cover and leaf litter, were influenced by the degree of contiguous forest cover. This shows that the likelihood of forest restoration being successful increases as the area of contiguous forest cover increases, another finding in line with expectations. Not stopping there, Crouzeilles *et al.* (2017) conducted an additional meta-analysis of 133 studies, which showed that natural regeneration is more likely to be successful than active restoration approaches for recovery of biodiversity and vegetation structure in tropical forest. No surprises there either, perhaps.

A further global meta-analysis is presented by Meli *et al.* (2017), featuring 166 studies of forests undergoing restoration. Across these studies, species diversity reached a mean of 83% of reference values, whereas biogeochemical functions reached a mean of 81%, showing that many forests had not achieved full recovery during the period of assessment. Time since the inception of restoration was again found to be a strong predictor of the degree of recovery, both in terms of diversity and ecosystem function. Past land-use types were also found to be an important factor, especially for recovery of biogeochemical functions; this suggests that past land use can have strong legacy effects on recovery rate. Overall, recovery was found to occur relatively rapidly in this investigation, namely within the first few decades after the restoration commenced (Meli *et al.*, 2017). This supports results from a previous meta-analysis, which showed forest recovery within a mean of 42 years (Jones and Schmitz, 2009). However, other meta-analyses have reported much longer recovery times for forest ecosystems, ranging from decades to centuries or even thousands of years (Curran *et al.*, 2014; Martin *et al.*, 2013). These contrasting results must partly reflect the different variables being measured. For example, in a meta-analysis of more than 600 tropical forest sites, Martin *et al.* (2013) found that on average tree species richness recovered after about 50 years and above-ground biomass within 80 years, whereas epiphyte species richness appeared to require much longer timescales. Palaeoecological evidence suggests that such longer timescales may be more typical, at least for tropical forests. Based on the results of 71 studies, Cole *et al.* (2014) observed recovery of tree abundance within 42 years in only 20% of studies; median recovery time was 210 years and the mean value was 503 years.

Unfortunately, these reviews do not provide sufficient information to determine the relative importance of all the factors that can limit forest recovery. Chazdon (2014) highlights the need to identify the ecological barriers to tree-seedling establishment for restoration to be successful and suggests that these barriers fall into four main categories: (1) poor soil conditions owing to soil erosion and loss of topsoil, (2) inadequate colonisation of species owing to dispersal limitation, (3) dominance by weedy or invasive species of grasses or ferns or (4) altered microclimatic conditions. There is an abundant literature providing examples of these different barriers, which has not so far been assessed using meta-analysis. For example, Griscom and Ashton (2011) note that exotic grasses have been shown to represent a major barrier to forest succession throughout the tropics, owing to their ability to outcompete tree seedlings. Removal of non-native grasses has been shown to increase the growth rates of planted tree seedlings and to increase the diversity of natural regeneration (e.g. Holl, 1998; Holl et al., 2000). Dominance by grasses provides an example of feedbacks maintaining an ecosystem in a degraded state: either fire and/or grazing will act to increase the dominance of grass, which may increase the incidence of further burning or grazing. However, it is unclear to what extent such feedbacks account for failures in forest recovery.

Observations of forest recovery also provide insight into the process of ecological succession. Norden et al. (2009) highlight a debate regarding the mechanisms of succession in the ecological literature. Some researchers claim that succession is a predictable process governed by niche-assembly rules (see Chapter 2), which enable forests to recover after disturbance. Conversely, others believe that human-impacted forests are doomed because their original functioning has been disrupted to such an extent that species composition will never return to its original state (Brook et al., 2006). Norden et al. (2009) provided an elegant test of whether secondary tropical forests follow either equilibrium or non-equilibrium dynamics by evaluating community reassembly over time in long-term plots in Costa Rica. Results showed that secondary forests are undergoing reassembly of canopy tree and palm species composition through the successful regeneration of mature forest species. Secondary forests showed a clear convergence with mature forest in terms of community composition, supporting an equilibrium model. This pattern of recovery was attributed to three key factors: high abundance of generalist species in the regional flora, high levels of seed dispersal and local presence of old-growth forest remnants (Norden et al., 2009). Conceivably different mechanisms of recovery might prevail in situations where these factors are not met, for example in forest that has suffered greater degradation.

The relationships between degradation, recovery and succession are further explored by Ghazoul and Chazdon (2017). These authors usefully highlight the difference between forest disturbance and degradation: a severely disturbed state that is able to recover quickly might be considered to be less degraded than one that cannot recover. In other words, a highly disturbed state is not degraded if it retains the biological legacies and ecological processes that enable forest regeneration and succession to occur, even if these recovery processes are slow. This leads to the concept of recovery debt, which refers to the interim reduction of biodiversity and ecosystem functions occurring during the recovery process (Moreno-Mateos *et al.*, 2017). In an analysis of data from more than 3,000 field survey plots from a range of different ecosystem types, Moreno-Mateos *et al.* (2017) found that recovering ecosystems have less organism abundance, species diversity and cycling of carbon and nitrogen than 'undisturbed' ecosystems and that complete recovery may not be achievable even after several decades. The amount of recovery debt can be viewed as a measure of ecosystem degradation, which is a product of both the degree of disturbance and the rate and trajectory of recovery (Ghazoul and Chazdon, 2017; Figure 4.20). Furthermore, an ecosystem can be considered to be degraded if it falls into a condition of arrested succession, for example if a shrubland is prevented from developing into forest. A degraded system might therefore be described as an alternative persistent state, where a return to its prior ecosystem dynamics or recognizable successional trajectories is not possible (or even likely) by natural processes (Ghazoul and Chazdon, 2017). As noted earlier, this situation could also be considered as ecosystem collapse.

Forest ecosystems may display a variety of different states, and, as noted in Chapter 2, they may be characterised by more than one endpoint of ecosystem recovery. For example, the cold temperate forests of northeastern United States contain stands of white pine, birch, or mixed maple and hemlock, depending on the frequency and severity of fires (Frelich and Reich, 1995). Each of these states can be considered to be stable given appropriate fire return intervals, but if the frequency of fires declines, birch forests gradually undergo successional development towards pine and then to maple or hemlock (Frelich and Reich, 1995). None of these states is degraded, as all are natural states maintained by natural fire regimes; this demonstrates how degradation and recovery should be assessed with respect to the set of persistent states that together characterise a particular ecosystem (Ghazoul and Chazdon, 2017). This also highlights the risk of considering an ecosystem state as degraded

262 · Contemporary Case Studies

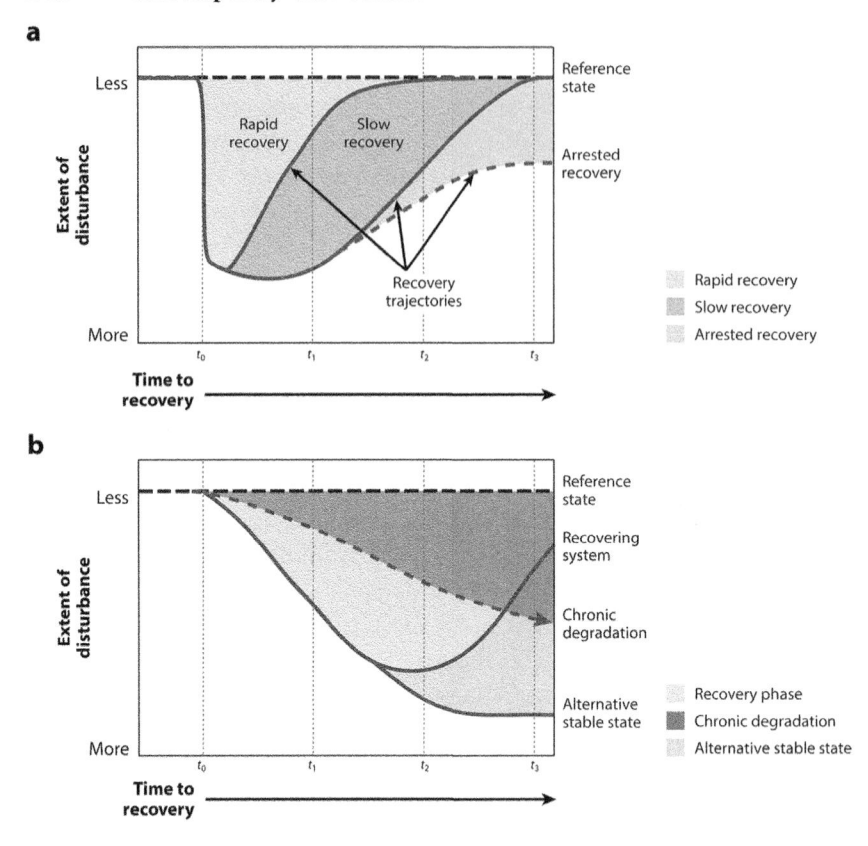

Figure 4.20 The relationship between forest recovery, disturbance and degradation. The recovery debt, or the decline in ecosystem function or composition in relation to an undisturbed reference state, can be viewed as a measure of ecosystem degradation. (a) Shaded areas represent recovery debt under scenarios of rapid recovery (light shading), slower recovery (dark shading) and arrested recovery (intermediate shading). (b) Recovery debt accumulates during prolonged degradation (light shading); under chronic degradation, recovery debt continues to increase over time (dark shading). A disturbance event leading to an alternative ecosystem state incurs substantial recovery debt (intermediate shading). Reproduced from Ghazoul and Chazdon (2017), with permission.

without taking into account the natural dynamics of the system; this could potentially lead to incorrect inferences about the present and future condition of the ecosystem (Ghazoul *et al.*, 2015).

In closing this section, I couldn't resist including the diagram presented by Malhi *et al.* (2014) (Figure 4.21) illustrating current and future changes in tropical forests, partly because it is such a beautiful work of art. The

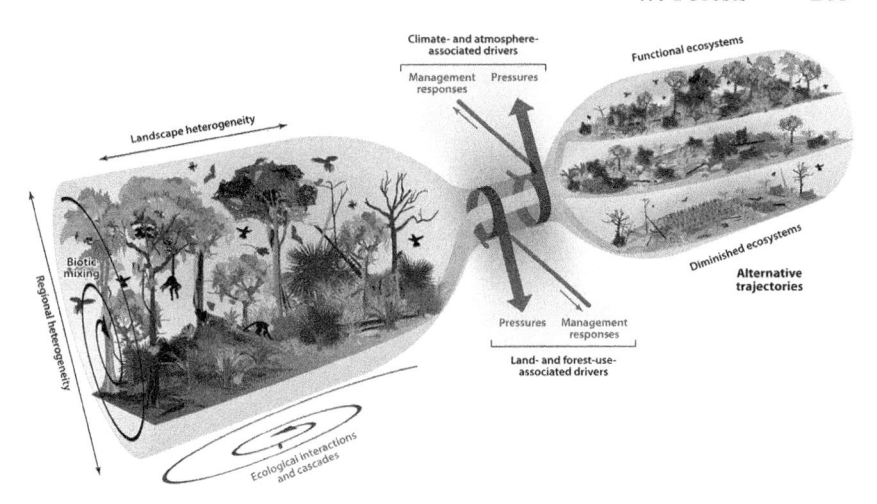

Figure 4.21 The 'early Anthropocene bottleneck' and the future of tropical forests. This suggests that tropical forests are currently experiencing a bottleneck, influenced by the current state and heterogeneity of tropical forests and ongoing pressures and management responses. These factors will influence the likely trajectories of environmental change in tropical forests around the world. Pressures (large arrows) include both interacting land- and forest-use-associated drivers (including forest clearance and fragmentation, overexploitation of natural resources, fire), exacerbated by the effects of biotic mixing and cascading species interactions, and coupled with atmosphere- and climate-associated drivers. These could be offset by management interventions (illustrated as hooks) that act to avoid and mitigate ongoing pressures and to restore degraded areas. Note the alternative trajectories that might arise following the bottleneck. Reproduced from Malhi *et al.* (2014), with permission.

diagram suggests that tropical forests are currently passing through a bottleneck, as a result of both climate change and land use, plus associated drivers. So far so good – the bottleneck arguably provides a useful metaphor for ecosystem collapse. The drivers act to reduce different ecosystem characteristics, illustrated as dimensions of an actual bottle – including landscape heterogeneity, ecological interactions, regional heterogeneity and biotic mixing. But why does this framework focus more strongly on measurements of pattern rather than ecological processes? And what about forest structure, composition and function? Once the pressures are removed, alternative trajectories are illustrated in a second inverted bottle, illustrating the potential for recovery. The diagram nicely conveys some possible ecosystem states that might arise in future, ranging from 'functional' to 'diminished' ecosystems. These have some parallels to the recovery pathways illustrated in Figure 4.20. However, this diagram implies that the different pressures currently affecting tropical forests will somehow be alleviated, enabling degraded forests to recover. Unfortunately, whatever action is taken, climate change is not going away anytime soon. Future trajectories will not take place under conditions of relaxed drivers; more likely the pressures will be chronic and intensifying. The bottleneck may extend indefinitely.

4.5 Other Ecosystems

In addition to those considered previously, many other terrestrial ecosystems are being significantly affected by environmental change and appear to be at risk of collapse. Examples include many wetlands, drylands and peatlands, especially those in the high Arctic (Farquharson *et al.*, 2019; Harris *et al.*, 2018; Nauta *et al.*, 2014; Schmidt *et al.*, 2019). Without wishing to belittle these important phenomena, I am going to briefly consider here just two additional case studies: savannas and temperate agroecosystems. These have recently been the focus of particular research attention and debate.

Savanna

Savanna ecosystems are typically characterised by mosaics of grassland and woodland patches, often marked by sharp transitions, a pattern that has long intrigued ecologists. They are also sometimes home to spectacularly interesting fauna, which makes them even more deserving of study. At first glance, savannas appear to provide compelling evidence for the existence of alternative stable states. This brief overview therefore focuses on application of dynamical systems theory to savannas and its relationship to ecosystem collapse. Note that savanna dynamics have been extensively explored using models (e.g. Murphy and Bowman, 2012; Staal *et al.*, 2018; Touboul *et al.*, 2018; van Nes *et al.*, 2014, 2018). While these studies are surely insightful, they have not been considered further here, as the focus is on empirical evidence.

Hirota *et al.* (2011) used remote sensing data (MODIS) data to analyse the influence of precipitation on the distribution of tree cover in tropical and subtropical regions of Africa, Australia and South America. The relationship between tree cover and precipitation was found to be trimodal, suggesting the existence of three alternative states – forest, savanna and treeless grassland. Forest was found to be more likely to occur at higher precipitation values and treeless vegetation at the lowest,

4.5 Other Ecosystems · 265

with savanna associated with intermediate values. Interestingly, relatively few locations were characterised by values of 60% or 5% tree cover, in any of the different regions. Hirota *et al.* (2011) speculated that this might be because these situations are unstable, because of positive feedbacks involving fire. The authors suggested that the likelihood of fire increases if grasses produce sufficient flammable fuel, but once tree cover becomes sufficiently dense, grass growth would be inhibited, leading to a decline in fire frequency. This feedback process could lead to development of closed forest. Another positive feedback might occur at low values of tree cover, for example if the establishment of some young trees modified the environment to facilitate further tree establishment. However, no field evidence was presented evaluating these feedback mechanisms.

A similar study was presented by Staver *et al.* (2011), who again used the MODIS data but with a wider range of explanatory variables, including soil texture and fire frequency as well as rainfall. Although herbivory is known to influence tree cover and fire regimes, this variable was omitted from the analysis owing to the lack of a suitable global data set. Mean annual rainfall was again found to be strongly related to tree cover on all continents, but at intermediate rainfall (1,000–2,500 mm) with mild seasonality (<7 months), tree cover was found to be bimodal. On all three continents, fire was found to be ubiquitous in savanna, and forest occurred more commonly where fires were absent (although cause and effect are difficult to disentangle here). The authors interpreted these results as evidence for alternative states prevailing over large areas, including parts of Amazonia and the Congo, with the fire regime determining whether savanna or forest prevail. Results were also interpreted as evidence for a positive feedback involving fire, as fire suppresses tree cover but low tree cover promotes fire spread. Similar conclusions were reached by Favier *et al.* (2012), again based on analysis of MODIS data.

Although such results are suggestive of the existence of alternative states, these investigations essentially employed correlational approaches. As we all know, correlations are no proof of causes, and therefore alternative explanations for the results obtained may be equally valid. However, the problems go deeper than that. For example, Hanan *et al.* (2014) neatly demonstrated that discontinuities in tree-cover estimates can emerge simply as an artefact of analysis, when satellite data are examined using particular statistical approaches. They showed this by analysing an artificially generated data set, which despite being entirely random in origin, generated results very similar to those obtained by

Hirota et al. (2011) and Staver et al. (2011). This rather damning result elicited a very considered response. While defending their overall results, Staver and Hansen (2015) admitted that MODIS is not well resolved below 20–30% tree cover and should not be used to infer multimodality in tree cover in that range. Conversely, in a further study using MODIS, Kumar et al. (2019) conclude that multimodal distribution patterns are not necessarily attributable to local-scale feedbacks, but to other factors including edaphic, disturbance and/or anthropogenic processes.

Oliveras and Malhi (2016) summarise other evidence suggesting that soil and strong climate control play a role in determining the distributions of both forest and savanna, and the legacies of historical disturbance may also be highly influential. The relative importance of different drivers in shaping savanna vegetation also appears to vary from place to place (Lehmann et al., 2014). Overall, therefore, it is doubtful whether these observations meet the requirements of dynamical systems theory described by Petraitis (2013), namely that a given site should be shown to have the potential to be occupied by two or more distinct communities, and the environment must not differ between the two putative states. So a transition between these putative alternative stable states might be more appropriately referred to as a regime shift (Dudgeon et al., 2010).

Whether or not the different states of savanna ecosystems are consistent with those of dynamical systems theory, the hypothesised feedback processes (Figure 4.22) provide a potential mechanism for abrupt transitions between forest and grassland. While there is an extensive literature on the role of both fire and herbivory in savannas, there is relatively little rigorous field evidence indicating whether these feedbacks can actually lead to transitions between ecosystem states. With respect to herbivory, Staver and Bond (2014) showed experimentally that release from browsing overrides grazer–grass–fire interactions to strongly promote tree growth, the effects of which persist even after browsers are reintroduced. In relation to the effects of fire, Dantas et al. (2013) examined a gradient of field plots from open savanna to closed forest in the Brazilian Cerrado. Results indicated that many of the community variables measured displayed threshold responses along the gradient, which separated two community states: (1) open environments with low-diversity communities growing on poor soils and dominated by plants that are highly fire-resistant; and (2) closed environments on more fertile soils with plant species that were less fire-resistant. Fire regimes also differed between these two communities, with shorter fire return intervals observed in the

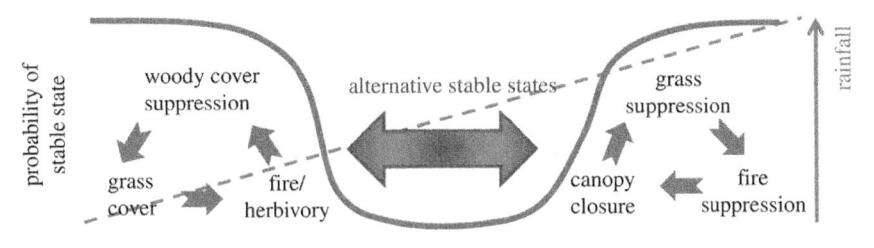

Figure 4.22 Conceptual illustration of the feedback mechanisms associated with forest–savanna transitions. Under low rainfall, grasslands with limited woody cover dominate. Grasses produce fuel that enhances the probability of fires and/or herbivory, which prevents establishment of juvenile woody vegetation. Under high rainfall, the tree canopy closes, moistening the microclimate, shading out light-demanding grasses and therefore suppressing fire. At intermediate levels of rainfall, if fire/herbivore disturbances are suppressed, the tree canopy will eventually be closed enough to suppress the herbaceous layer and therefore reduce fire frequency. On the other hand, if fire frequency is high or herbivore pressure is heavy, an open environment is maintained by continuously preventing tree saplings from establishing. The presence of these two positive feedbacks acting in different directions results in a tendency for relatively abrupt transitions between tree and grass cover. Reproduced from Oliveras and Malhi (2016), with permission.

more open communities. These results were interpreted by the authors as evidence for fire feedbacks driving transitions between the ecosystem states (Dantas *et al.*, 2013). However, conversion from one state to another was not directly observed, and the apparent influence of soil properties on the fire regime was not fully elucidated. The fact that soil characteristics varied along the gradient, and differed between the two putative states, raises the possibility that different ecosystem states are actually driven by edaphic factors.

A further investigation was presented by Dantas *et al.* (2016), involving analysis of tree basal area data compiled from field studies across tropical America and Africa together with environmental data. Regression analyses between tree basal area and resource availability (based on climate and soil data) suggested the existence of three different ecosystem states in Afrotropical savannas and two in neotropical examples. These results were again interpreted as evidence for fire feedbacks maintaining savannas and forests as alternative ecosystem states (Dantas *et al.*, 2016), although no direct observations of either fire feedbacks or ecosystem transitions were made in this study.

Many other studies have examined the empirical evidence for feedbacks driving forest–savanna transitions, including the following:

- Pausas and Dantas (2016) stated that long-term manipulative experiments in tropical ecosystems consistently show that fire exclusion leads to pyrophobic forests while recurrent burns support the development of pyrophylic savannas (e.g. Bond, 2008; Louppe *et al.*, 1995; Woinarski *et al.*, 2004). However, some fire-exclusion experiments have failed to produce shifts from savanna to forest after several decades (Murphy and Bowman, 2012).
- Negative feedback processes have been experimentally demonstrated in Amazonian transitional forest that can contribute to the maintenance of forest cover. In this case, an annual burning regime led to a decline in subsequent forest flammability owing to a reduction in fuel availability (Balch *et al.*, 2008).
- Based on field data from central Brazil, Hoffmann *et al.* (2012) found that saplings of savanna trees accumulate bark thickness more rapidly than forest trees and are therefore more likely to become fire resistant during fire-free intervals. Forest trees accumulate leaf area more rapidly than savanna trees, thereby accelerating the transition to forest.
- Veenendaal *et al.* (2018) assessed data from 11 experiments examining the effects of the timing and/or frequency of fire on tropical forest and/or savanna vegetation structure over one decade or more. Results indicated that the magnitude of fire effects was strongly dependent on both the frequency and timing of fire, the extent of prior disturbance and the canopy cover in the absence of fire. The authors concluded that the effects of fire on tropical vegetation have been overestimated, because of the use of high-intensity burning regimes in experiments that do not reflect those occurring in natural ecosystems. In addition, Veenendaal *et al.* (2018) suggested that is unlikely that self-sustaining non-anthropogenic fire regimes could ever be sufficiently intense as to maintain open savanna-type vegetation in areas that would otherwise be forest.

If by this point you are beginning to wonder whether savanna–forest transitions really are driven by fire feedbacks, you are not alone. In a provocative but wonderfully titled paper, Lloyd and Veenendaal (2016) ask: Are fire-mediated feedbacks burning out of control? In other words, has the savanna research community been carried away by its enthusiasm for dynamical systems theory? These authors suggest that logical fallacies are pervasive in the literature on this topic. These are grouped into five categories: fallacies of 'confirmation bias', 'misplaced concreteness', 'suppressed evidence', 'wishful thinking' and 'hasty generalisation'. This is quite a list of charges. Essentially, researchers who have interpreted savanna dynamics as evidence for dynamical systems theory are accused of being biased in their approach. Lloyd and Veenendaal (2016) consider 'all arguments presented to date in support of the widespread existence of alternative stable states in the tropical regions to be flawed'. Instead, these authors suggest that forest–savanna transitions may be better understood as 'reflecting the effects of soil physical and chemical properties on tropical vegetation structure and function with fire-effected feedbacks simply serving to reinforce these patterns'. Case closed?

So, are savannas at risk of collapse? The difficulty in answering this question partly reflects the challenge of demonstrating feedback processes at the landscape scale, such as those between fire and vegetation (Bowman *et al.*, 2015). Another issue is the 'image problem' that savannas have among scientists, policymakers and land managers. Often, grassy biomes are considered to represent the result of deforestation followed by arrested succession; in other words, they are frequently viewed as degraded forest ecosystems rather than ecosystems in their own right (Veldman *et al.*, 2015). In fact, many grassy biomes originated millions of years ago, long before there were people around to cut down trees or set fire to vegetation. A potential solution to this problem is the recognition of 'old growth' or ancient grasslands and savannas, which possess particularly high conservation value and can usefully be differentiated from short-term vegetation arising from human land uses (Veldman *et al.*, 2015). Frequent fires and herbivory are part of the internal ecological dynamics of 'old growth' savannas, rather than externally imposed disturbances; in this they differ from most forest ecosystems (Veldman, 2016). In savannas, when fires are excluded and/or herbivores are not managed appropriately, woody plants may rapidly increase in abundance. Under such conditions, savannas may be replaced by forests. Evidence suggests that a number of grassland ecosystems are being threatened by woody plant encroachment (e.g. Parr *et al.*, 2012; Ratajczak *et al.*, 2014).

However, in large parts of the tropics, mosaics of savannas and forests are relatively stable, owing to differences in the flammability of savanna and forest fuels, functional differences between savanna and forest trees and the seasonality of fire (Veldman, 2016). It is also important to note the marked difference in fire–vegetation dynamics at the boundaries between 'old growth' savannas and forests and those associated with forest edges created by human-caused deforestation (Veldman, 2016). For example, Silvério *et al.* (2013) demonstrated experimentally that positive feedbacks can occur between the build-up of grass fuel and fire intensity in tropical forests that have not previously been subjected to burning. Working in a transitional forest in the south-western Brazilian Amazon, experimental burning increased the mortality of trees, which resulted in canopy opening and a consequent increase in radiation at the forest floor. This enabled grasses to spread from the forest edge. The authors speculated that recovery from this disturbance might require many fire-free decades (Silvério et al., 2013). Similar findings were reported by Brando et al. (2014) in an experiment conducted in the seasonally dry forests of south-eastern Amazonia. Together these results suggest that feedbacks between extreme drought events, forest fragmentation and anthropogenic ignition sources might increase the likelihood of ecosystem collapse in these drier parts of the Amazon basin.

Temperate Agroecosystems

In 2017, Hallmann et al. (2017) published evidence suggesting that flying insect biomass had declined by more than 75% over a 27-year period in protected areas in Germany. These striking results generated a great deal of media coverage; the British newspaper the *Guardian* (18 October 2017) described them as evidence of 'ecological Armageddon', which sounds like ecosystem collapse on steroids. When interpreting this finding, it is pertinent to consider some features of the methods employed. The study was unusual in focusing on insect biomass, rather than species richness or functional diversity. The sampling procedure was also unusual; while data were collected according to a standard protocol, most of the 63 locations were sampled in only one year. These data do not represent a compilation of time-series monitoring data collected at individual sites, as might be expected. Unsurprisingly, the data are therefore characterised by a high degree of uncertainty. The headline figure of 75% was derived from a regression model, in which parameters values were obtained using statistical simulation as a way of dealing with this uncertainty.

Hallmann et al. (2017) suggested that this decline might be attributable to the agricultural intensification that has occurred in recent decades (e.g. increased pesticide usage, year-round tillage, increased use of fertilisers, etc.). The reserves in which the traps were placed were relatively small and were mostly surrounded by intensive agricultural landscapes. This implies that the insect populations of such protected areas might be being affected by surrounding land use. I found this aspect of the study to be particularly worrying, as it suggests that the ecological impacts of farming are not limited to agricultural land, but can affect sites of high conservation value situated nearby. As the authors note, it is now well established that agricultural intensification is associated with an overall decline in the abundance of many species of plants, insects, birds and other groups, but a potential decline in insect biomass is particularly concerning as it would likely result in cascading effects across trophic levels and throughout the ecosystem.

A decline in insects in temperate agricultural landscapes such as those of Germany can be viewed as part of a global decline in biodiversity. As noted by Dirzo et al. (2014), we are currently experiencing a global wave of anthropogenically driven biodiversity loss, which involves widespread species and population extirpations, together with many declines in local species abundance. For example, Dirzo et al. (2014) notes that among terrestrial invertebrates at a global scale, 67% of monitored populations show 45% mean abundance decline since 1970. However, these figures mask the major shortfall that exists in our current state of knowledge about trends in invertebrate populations. These estimates were based on only 29 studies, most of which were located in Europe. This shows that the number of invertebrate populations that are being systematically monitored is very low indeed. However, evidence of rapid declines in insect biomass is not limited to temperate regions. For example, in the Luquillo rain forest of Puerto Rico, Lister and Garcia (2018) found biomass losses of between 98% and 78% for ground-foraging and canopy-dwelling arthropods since 1976. This represents a similar rate of annual decline (2.2–2.7%) to that reported by Hallmann et al. (2017) (2.8%) but was attributed to climate change rather than agricultural intensification.

Are such declines typical? Sánchez-Bayo and Wyckhuys (2019a) examined the evidence for insect declines through a literature review and meta-analysis of long-term survey data. Analyses were based on the results of 73 studies, which although drawn from throughout the world, were concentrated in Europe and North America. Results indicated that the current proportion of insect species in decline (41%) is about twice as high as that of vertebrates, and the median annual biomass decline is around 2.5% (based on the results of only five studies). The review noted high rates of decline in particular groups of insects, such as butterflies, moths and bees, supporting current concerns about global declines in pollinators (Ollerton et al., 2014; Potts et al., 2010). Insects in freshwater ecosystems also appear to be declining relatively rapidly. Habitat loss and pollution were identified as the principal causes of insect declines, especially the intensification of agriculture in recent decades, with the associated widespread and relentless use of synthetic pesticides. When considering these findings, it is again important to note the limited data on which this review was based. Nevertheless, Sánchez-Bayo and Wyckhuys (2019a) are surely right to highlight the potential implications of these trends, as insects play a crucial role in the functioning of all ecosystems, particularly in ecological processes such as pollination, nutrient recycling and decomposition. Insects also are an important element in many food webs, providing food for many vertebrates including reptiles, amphibians, most bats, many birds and fish. Given the structural and functional importance of insects for many of the world's ecosystems, Sánchez-Bayo and Wyckhuys (2019a) concluded that the consequences of insect declines are likely to be 'catastrophic' for the planet's ecosystems.

Perhaps unsurprisingly, given the limitations of the underlying data, these conclusions have drawn a substantial amount of criticism. For example:

- Cardoso *et al.* (2019) suggested that the methods used by Sánchez-Bayo and Wyckhuys (2019a) were biased and the analyses flawed and that factors other than agricultural intensification (such as habitat loss, fragmentation, invasive species and climate change) were downplayed.
- Komonen *et al.* (2019) also highlighted methodological flaws in the study, such as biases in the search terms used, and accused its authors of being alarmist through the use of dramatic, non-scientific wording.
- Wagner (2019) similarly highlighted biases in the search terms used and suggested that the most important error was the suggestion that extinction of 40% of the world's insect species might occur in the next few decades, based on a rather heroic extrapolation. Sánchez-Bayo and Wyckhuys (2019b) provided a response to these points.
- Thomas *et al.* (2019) suggested that reports of imminent 'insectageddon' may be exaggerated owing to the considerable uncertainties and biases in the available data. (It is somewhat ironic to see Thomas *et al.* accuse others of hyperbole, when the lead author has himself been accused of similar practices in the past! See Ladle *et al.*, 2004.)

So, is insectageddon happening or not? Do plummeting insect numbers 'threaten the collapse of nature', as further reported in the *Guardian* (10 February 2019)? There are a number of other pieces of evidence, which together paint what is perhaps a slightly more nuanced picture. For example:

4.5 Other Ecosystems · 273

- Van Strien *et al.* (2019) used opportunistic butterfly records to estimate trends in occurrence of 71 species in the Netherlands. Results suggested a decline of 84% over the period from 1890 to 2017, although the authors suggested that this might be an underestimate. Farming intensification and expansion, together with aerial pollution and habitat loss, were suggested as potential causes.
- Shortall *et al.* (2009) described findings from 12-m-tall suction traps used by the Rothamsted Insect Survey in the United Kingdom. Only one of the four traps analysed (Hereford) showed downward trends in insect biomass over the 30 years (1973–2002) examined.
- Ewald *et al.* (2015) analysed 42 years of monitoring data collected in cereal fields in southern England. Of the 26 invertebrate taxa studied, only 9 showed a significant decrease in abundance. As noted by Leather (2018), cereal fields are of course not a natural habitat and are intensely managed, including application of pesticides, so they might not be representative of other habitats.
- Fox *et al.* (2012) reviewed evidence for moth declines in Great Britain, which indicated a 31% decrease in moth abundance over 35 years. Agricultural intensification and climate change were suggested as major drivers.
- Conversely, Macgregor *et al.* (2019), using data from the Rothamsted Insect Survey, found that annual moth biomass estimates increased between 1967 and 1982, then declined gradually from 1982 to 2017. Overall there was a net gain in mean biomass between the first and last decades of the survey, although high variability between years was noted.
- Powney *et al.* (2019) reported that a third of wild pollinators (353 hoverfly and bee species) have decreased in extent of spatial distribution in Britain between 1980 and 2013, while approximately a tenth have increased.

Clearly, these studies are all limited in their taxonomic and geographic coverage; they also vary in their temporal span. It is therefore difficult to know whether the results suggesting rapid insect decline are representative. Yet anecdotal evidence also suggests that something truly devastating has happened to insect populations in recent decades. As entomologist Simon Leather describes it, 'people of my age will all tell you that insects are less abundant than when we were children' (Leather, 2018). His anecdote about earning pocket money as a child by cleaning his parents' car, laboriously scraping off the smeared bodies of insects that smothered the front of the vehicle (Leather, 2016), certainly resonated with me. This is an experience that is difficult to duplicate today, at least in most of the United Kingdom. Recent research has suggested that collision with road vehicles might itself represent a significant cause of death for invertebrates, perhaps killing hundreds of billions of individuals each year, at a continental scale (Baxter-Gilbert et al., 2015).

If insectageddon is happening, might this then lead to collapse of whole ecosystems? Given the importance of insects as food for other animals such as birds and bats, one might predict declines in these other ecosystem components. There is indeed evidence that bats have experienced major declines throughout Europe during the last century, although owing to the lack of systematic monitoring throughout this period, this evidence is uncertain and often anecdotal. In the United Kingdom, for example, Haysom et al. (2010) report declines of 60–70% in abundance of some bat species since the early 1960s, although in the past 20 years, populations of most species have stabilised or recovered owing to conservation action (BCT, 2017). The impact of agriculture on bats is illustrated by the lower diversity and abundance of bat species in intensive agricultural landscapes than on organic farms owing to the effects of pesticide use in the former (Wickramasinghe et al., 2003). Similarly, declines in farmland birds are well established, and these have again been linked to intensification of agriculture. Of the 28 farmland bird species studied in Britain, 24 showed a contraction in range between 1970 and 1990, with 7 species estimated to have undergone population decreases of at least 50% (Fuller et al., 1995).

Arguably, agricultural intensification might therefore be considered as an important driver of ecosystem collapse. Agroecosystems have been transformed through the repeated application of chemical pesticides, herbicides and fertilisers and defining features (such as insects, birds and bats) have been lost. Relatively low-input agricultural systems that were strongly dependent on manual labour have been replaced by high-input, mechanised systems based on the extensive use of heavy machinery and other technologies. In the United Kingdom and other European countries, this process of intensification has led to a dramatic reduction in landscape diversity, with farms becoming larger and more homogeneous. At the scale of individual farms, habitat diversity has been reduced through the loss of non-cropped areas, hedgerows, field margins and ponds (Robinson and Sutherland, 2002).

These processes can be illustrated by what has occurred in the area where I live, namely the county of Dorset in southern England. Today, around 75% of Dorset's land area is farmed, of which about a third is arable; this is fairly typical of England as a whole. Analysis of historical land cover maps shows that the total proportion of the land area comprising agricultural land has remained roughly constant over the past 80 years. However, there have been significant changes in the extent and distribution of different agricultural land-use practices. In assessing the impacts of these changes, we are fortunate to be able to access the pioneering research of Ronald Good, who undertook a vegetation survey in the 1930s at 7,575 sites across the county. His field notes were detailed enough to enable a resurvey of these same sites around 70 years later. Using these data enabled us to accurately determine the changes in vegetation that have occurred throughout Dorset during a period of agricultural intensification. Results indicated the substantial loss of some habitats, particularly neutral grassland and calcareous grassland, with respective declines of 97% and 70% of their initial area. These losses were primarily attributable to ploughing and conversion to arable land, or the conversion to improved grassland through the application of fertilisers. Significant declines were also recorded in wetland, acid grassland and heathland sites, with losses recorded of 63%, 54% and 57%, respectively (Newton *et al.*, 2019). Analysis of the vegetation of remaining heathland, calcareous grassland and woodland sites indicates that significant changes have occurred in plant community composition, attributable primarily to the combined effects of eutrophication and climate change (Diaz *et al.*, 2013; Keith *et al.*, 2009, 2011; Newton *et al.*, 2012b).

While we have strong evidence that the vegetation composition has changed dramatically in Dorset as a result of land-use change, we have much less information on the impacts of intensification on other components of the ecosystem. This highlights an important lesson: ecosystem collapse can be cryptic and difficult to detect. There are very few systematic monitoring data for relatively inconspicuous organisms such as insects. The situation for other megadiverse but cryptic groups, such as fungi and bacteria, is immeasurably worse, despite their importance for ecosystem function. It may be that agricultural intensification in the United Kingdom has led to widespread losses of the insect fauna of a magnitude similar to those reported by Hallmann *et al.* (2017) and Sánchez-Bayo and Wyckhuys (2019a), which may have led to a cascade of secondary extinctions at other trophic levels, coupled with profound changes in ecosystem function. However, we have no way of knowing.

The example of temperate agroecosystems offers a number of further lessons. This form of collapse appears to be less amenable to dynamical systems theory; extrinsic rather than intrinsic factors appear to be of overriding importance. There is little evidence of the internal feedback processes that are critical to the theory, or the occurrence of alternative stable states. Furthermore, agroecosystems themselves are a form of degraded ecosystem. In Dorset, as in many other areas of northern and central Europe, Holocene vegetation was principally dominated by forest prior to the introduction of farming (Binney et al., 2017; Fyfe et al., 2013; Whitehouse and Smith, 2010). As forest areas were cleared for agriculture, they were replaced by a mosaic of habitats, including grassland, heathland and wetland, as well as areas of crop cultivation. A key point is that low-input agricultural land can be of high value for conservation; in the United Kingdom, distinctive communities of plants and animals are associated with different types of unimproved grassland and heathland, as well as arable land. Indeed, much of the conservation management undertaken in the United Kingdom is designed to prevent such habitats reverting to forest through succession, reflecting the fact that many species require open conditions (Sutherland, 2000). The loss of these communities following the introduction of intensive farming practices could therefore represent the collapse of an ecosystem that has already collapsed previously.

4.6 Evaluation of Propositions

The propositions listed at the end of Chapter 3, based on the theoretical ideas presented in Chapter 2, are here further evaluated in the light of evidence presented in the current chapter. A robust evaluation of these propositions would require a systematic survey of the literature, rather than the discursive exploration presented here, so these comments represent only a tentative first step.

1. Ecosystem collapse can be caused either by extrinsic or intrinsic factors, or by a combination of the two. Extrinsic factors refer to the external disturbance regime affecting an ecosystem, and intrinsic factors to interactions of organisms between each other and with the physical environment.
    o There is clear evidence for extrinsic factors inducing collapse in all the ecosystem types considered. With respect to intrinsic factors, the evidence is more equivocal and appears to differ among ecosystem types. For example, there is good evidence for intrinsic factors contributing to collapse of some freshwater ecosystems,

4.6 Evaluation of Propositions · 277

particularly shallow lakes. Conversely, the principal cause of collapse in many forest ecosystems is deforestation or land cover change, which is an extrinsic factor. The proposition, as stated here, is in fact neutral regarding the relative importance of extrinsic and intrinsic factors in driving collapse, but according to dynamical systems theory, intrinsic factors play a leading role in causing collapse, once a threshold value of an extrinsic factor is met. This issue will therefore be explored further in Chapter 5. As it stands, this proposition is supported.

2. Any ecosystem can potentially collapse, if subjected to disturbance of an appropriate type and occurring at sufficient frequency, extent, intensity or duration. Novel disturbances are more likely to cause collapse.
    o Evidence suggests that collapse is widespread; it has been observed in all the ecosystem types considered here. The first part of the proposition is therefore tentatively supported. However, a more robust assessment would involve a systematic survey of all ecosystem types to evaluate the occurrence of collapse. It is conceivable that there are ecosystems so remote from human impact that collapse is unlikely, such as the deep sea. However, even the deep sea is not immune; fisheries resources, hydrocarbons and minerals are now being exploited at depths greater than 2,000 m, and the effects of increases in atmospheric CO_2 associated climatic warming are likely to be felt here too (Ramirez-Llodra *et al.*, 2011). With respect to the second part of the proposition, there are clearly examples where novel disturbances have contributed to collapse, such as global warming and the spread of disease in coral reefs; harvesting of fish, which now extends through much of the world's oceans (Watson, 2017); invasive species, for example in lakes; outbreaks of novel pests and diseases in forests. This is therefore also supported. However, it is not necessarily the novelty of the disturbance that is most important, but its intensity.

3. Disturbance regimes characterised by interactions between multiple types of disturbance are more likely to cause ecosystem collapse. A particular disturbance event is more likely to result in collapse if it follows another event from which the ecosystem has not yet recovered.
    o There is evidence for interactions between multiple stressors driving ecosystem collapse in each of the ecosystem types considered. For example, in coral reefs, interactions were recorded between nutrient loading, overfishing and global warming

(Hughes *et al.*, 2017a); in marine fisheries, between global warming and nutrient pollution (FAO, 2018); in forests, between climate change, fire, drought and insect attack (Seidl *et al.*, 2017). Interestingly, a potential difference between ecosystem types emerged in relation to whether interactions between stressors were typically synergistic or antagonistic. Jackson *et al.* (2016) found that in freshwater ecosystems, interactions between stressors were often antagonistic rather than synergistic, whereas in marine systems, the converse seems to be true (although see Crain *et al.*, 2008). In coral reefs, although synergistic interactions have been widely reported (particularly among irradiance and temperature), results from meta-analysis suggest that the prevailing interaction type is additive rather than synergistic (Ban *et al.*, 2014). In forest ecosystems, evidence suggests that interactions may tend to be synergistic (Seidl *et al.*, 2017). This may mean that this proposition is less likely to be true in freshwater ecosystems and perhaps coral reefs than in other ecosystem types, although clearly this merits further investigation. To date there have been no systematic comparisons of stressor interactions across all these ecosystem types.
    - There is also some evidence for the second part of the proposition, such as repeated bleaching events of coral reefs, repeated fishing of marine ecosystems and repeated logging of forests. However, the precise role of repeated disturbance events in driving collapse awaits investigation. There is also the possibility of cascading effects, where a disturbance interaction can extend the impacts of a driver for one disturbance into another disturbance type (Buma, 2015); this phenomenon has not yet been examined in detail, in relation to ecosystem collapse.
4. The coincidence of both 'press' and 'pulse' disturbance types is most likely to lead to ecosystem collapse. While both 'pulse' and 'press' disturbance events can increase the mortality of individuals, 'press' disturbances can also play a particular role in impeding ecosystem recovery.
    - Generally, this proposition was not strongly supported. Most anthropogenic disturbances leading to ecosystem collapse are of the 'press' type, for example climate change and eutrophication affecting coral reefs; overfishing of marine ecosystems; eutrophication, climate change and invasive species affecting freshwater ecosystems; climate change, invasive species, repeated burning,

4.6 Evaluation of Propositions · 279

grazing and logging of forests. However, there are also examples of 'pulse' disturbances being important in some cases, such as storms on coral reefs or floods in freshwater ecosystems. In the case of forests, deforestation or land-cover change is the principal driver of ecosystem loss, which might be viewed as a 'pulse' disturbance. However, lack of recovery did appear to be associated with press disturbances in all ecosystem types.
- It is also worth noting that is not always easy to classify disturbances as either 'pulse' or 'press'. For example, a one-off fire event might be considered as a 'pulse' event, but frequent fire as a 'press' disturbance, highlighting the fact that variation in disturbances can be continuous. There has certainly been great inconsistency in the way these terms have been used in the literature (Glasby and Underwood, 1996), which limits their application. This is partly because both the cause and effect of the disturbance were conflated in the original definition of these disturbance types (Glasby and Underwood, 1996). So perhaps there are better ways of considering disturbance? This point will be examined further in Chapter 5.

5. Some ecosystems can exist in more than one state or regime, such as different successional stages. Transitions between these states can form part of the natural dynamics of an ecosystem. When such transitions are persistent rather than transient, they can become of concern to human society.
   - There is good evidence for the first and second parts of this proposition: all the ecosystems considered demonstrated different states, some of which were successional in nature. For example, coral reefs demonstrate alga *versus* coral dominance; marine ecosystems demonstrate different communities of fish depending on fishing pressure; plus the turbid and clear states of shallow lakes; and the different successional states of forest ecosystems. There is also strong evidence of persistent transitions (i.e. collapse) that are of human concern in each of these ecosystem types.
   - A key issue relates to whether transitions between states are transient or persistent. As noted by Fukami and Nakajima (2011), and as illustrated by the case studies presented here, many communities are maintained in a transient state by disturbance; such transient states are not considered by dynamical systems theory. However, they are very relevant to understanding ecosystem collapse. The proposition might be usefully reworded to

reflect this, by inserting: *Many ecosystems are kept in a transient state by disturbance.*

6. A persistent ecosystem transition, or collapse, can arise when ecological recovery is impeded. This can occur if there are stabilising feedback processes that maintain an ecosystem in a degraded state, or when the processes of ecological recovery fail. Understanding these reasons for lack of recovery is key to understanding collapse.
    ○ Overall, evidence supports this proposition. If some factor impedes recovery processes, then a degraded ecosystem state may become persistent rather than transient, thereby contributing to collapse. However, in many ecosystems, the reasons for lack of recovery are often not well understood, despite a wide variety of different mechanisms having been proposed. Stabilising feedback processes have been identified in all the ecosystem types considered here, although again, empirical evidence for their role in preventing recovery is often limited. Examples include the negative impact of algae on coral recruitment and allelopathic interactions between algae and corals; phytoplankton blooms leading to hypoxia in lakes; and the role of fire and herbivory in forests. Some mechanisms appear to unique to particular ecosystem types; for example, food-web structure appears to be different in marine ecosystems than on land, as species at a relatively low positions on the food chain can control populations of top predators by eating their juveniles.
    ○ One issue that emerged from these case studies was the fact that a lack of recovery can be the result of ongoing, chronic disturbance. For example, ongoing nutrient addition can maintain a lake in a degraded state. It is therefore suggested that the wording of this proposition be amended to reflect this.
7. Collapse can be caused by breakdown of the stabilising feedback mechanisms maintaining an ecosystem state, or by feedbacks in the internal ecological processes of an ecosystem driving a system to a different state. As a result of these feedbacks, major ecological shifts can result from minor perturbations. Such shifts can occur when external conditions reach a critical value.
    ○ This proposition describes some of the basic elements of dynamical systems theory. There have been attempts to apply this theory to all the ecosystem types considered here, particularly in freshwater ecosystems such as shallow lakes. In each case, a number of different feedback mechanisms have been proposed,

4.6 Evaluation of Propositions · 281

both in terms of stabilising a particular state and driving a transition between states. However, the empirical evidence for these processes operating in nature is somewhat limited. Stronger evidence is often available for a change in environmental conditions or disturbance regime causing a change in community structure or composition.

- There are some examples of critical values of environmental pressures, such as critical seawater temperatures bleaching and killing corals, which can lead to ecosystem transitions even in the absence of feedback mechanisms. Other examples include values of nutrient concentrations in lakes, flow rates in rivers and insect pest outbreaks in forests. It is less clear how widespread such critical values are in nature; relatively few examples were encountered in the case studies examined here, although this might simply reflect our current state of knowledge.

8. Collapse may be more likely if disturbance events cause the loss of generalist species, those that are highly connected to other species, top predators and/or trophically unique species, and those at the base of food chains. The loss of such species can cause many secondary extinctions.

- This proposition refers to four types of species whose loss might contribute to ecosystem collapse. Evidence for each of these four types was provided by the case studies, although their relative importance appears to differ between ecosystem types. For example, in coral reefs, there is strong evidence that collapse occurs when species of coral that are highly connected are lost, such as those species that provide habitat complexity and heterogeneity. The loss of canopy dominant tree species from forest ecosystems has similar repercussions. Both corals and tree species can also be the base of food chains, emphasising their importance. Less evidence was available for the importance of generalists in coral reefs, although in forest ecosystems, it was noted that the presence of generalists in mutualistic networks can provide a measure of resilience to species loss.
- With respect to top predators and/or trophically unique species, their importance depends on whether community composition is regulated from the top down or bottom up. Evidence suggests that both bottom-up and top-down trophic regulation are widespread in each of the ecosystem types considered here, although their relative importance can clearly vary from place to place. This

suggests that loss of top predators can often contribute to ecosystem collapse, although this will depend on the specific context.
- o Evidence also indicated that loss of species with particular functional roles can contribute to ecosystem collapse, as demonstrated for example in freshwater ecosystems (Whiles *et al.*, 2006). This suggests a need to amend the proposition, as explored further in Chapter 5.
9. The disassembly of ecological networks tends to be characterised by thresholds. In such situations, the consequences of species loss become amplified and self-reinforcing as more and more species are extirpated, leading to network collapse.
    - o Threshold responses appear to be widespread in ecosystems affected by anthropogenic disturbance. Examples include seawater temperature for coral reefs and nutrient concentrations for shallow lakes. Less evidence is available for thresholds specifically in ecological networks rather than in ecosystems more broadly, although some examples were found (e.g. Dulvy *et al.*, 2004; Marshall, 2018).
    - o There is limited evidence for extinction cascades or secondary extinctions driving the process of ecosystem collapse. This partly reflects the difficulty of studying such phenomena (Brodie *et al.*, 2014). Trophic cascades have been reported in all the ecosystem types considered here, although most studies have documented changes in relative abundance of species rather than actual extirpation (Brodie *et al.*, 2014).
10. Certain types of ecosystems may be especially sensitive to disruption of biotic processes and interactions, namely those with strong top-down trophic regulation or with many mutualistic or facilitation interactions, those that are strongly dependent on mobile links and where positive feedbacks operate between the biota and the disturbance.
    - o It is difficult to evaluate this proposition without a systematic comparison of different ecosystem types, which was not performed here. Nevertheless, both coral reefs and some forest ecosystems are characterised by large numbers of mutualistic and facilitation interactions; this is perhaps less the case in many freshwater ecosystems. As noted previously, top-down trophic regulation has been reported in each of the ecosystem types considered here, although its relative importance appears to vary. Less evidence is available for positive feedbacks between biota and

disturbance; examples include the effects of fire and herbivory in some terrestrial ecosystems, but few examples were encountered in freshwater and marine ecosystems.
- Similarly, few references to mobile links were encountered in this literature. Chazdon and Arroyo (2013) note that in many tropical forests, many seed dispersers and pollinators move large distances and provide spatial links between different parts of the ecosystem. This can effectively buffer populations at larger spatial scales, which would reduce vulnerability of the forest to disturbance. Loss of these species might therefore increase sensitivity to disturbance, supporting the proposition.

11. Extinction cascades and community collapses should be less likely to occur in species-rich multi-trophic communities than in species-poor ones, unless the environment is highly variable.
    - Evidence does not support this proposition. Many examples of species-rich ecosystems such as coral reefs and tropical forests have demonstrated collapse. Lake Victoria provides a powerful freshwater example. The contradiction between empirical evidence and this proposition is supported by the lack of consistency in both theoretical and empirical studies, regarding the effect of environmental variability on species richness (Kaneryd et al., 2012).

12. The ability of ecosystem to recover is critically dependent on intrinsic factors, namely interactions of organisms between each other and with the physical environment. The rate or extent of recovery can be limited by intrinsic and/or by extrinsic factors, such as the disturbance regime and the extent of degradation.
    - Extrinsic factors that limit recovery were found in all ecosystem types, such as nutrient enrichment and fishing pressure in coral reefs, marine and freshwater ecosystems, and both fire and herbivory in forest ecosystems. A wide variety of different intrinsic factors that influence recovery ability were also noted in each ecosystem type. This proposition was broadly supported by the evidence. However, evidence indicates that typically, some elements of an ecosystem might recover while others do not, indicating that recovery does not have a single dimension.
    - While many ecosystems display a remarkable capacity for recovery, often recovery is relatively slow and only partial. In some cases, evidence suggests that disturbance has caused a permanent change in the ecosystem, so complete recovery might be

impossible. Key questions that emerge include: (i) What is the relative importance of intrinsic and extrinsic factors in limiting the rate or extent of recovery? This appears to vary from case to case. (ii) Is ecosystem recovery possible if the anthropogenic disturbance is removed? This will depend on the extent of degradation that has occurred. There may be thresholds of degradation beyond which recovery is not possible, but this has not been determined for most ecosystem types. (iii) Can recovery even occur if some forms of press disturbance, such as climate change, continue?

13. Functional and structural change in an ecosystem undergoing collapse may be unrelated to loss of taxonomic diversity; species identity matters.
    - There is clear evidence that species identity matters. For example, in coral reefs, the loss of key coral species with structural functions is critical to collapse. Similarly in marine, freshwater and forest ecosystems, harvesting may target species with particular functional roles in the ecosystem. There is ample evidence for the existence of 'keystone' species (Chazdon and Arroyo, 2013; Mills et al., 1993).

14. Recovery can take a long time and takes longer than collapse.
    - This proposition is strongly supported. In all ecosystem types, decadal timescales of recovery are typical. Longer durations of centuries or even millennia may be required for recovery of some characteristics of some ecosystems, such as species composition of tropical forests. Collapse typically occurs over much shorter timescales, such as a few years (or less), although often the timescales of collapse are not well defined.
    - Again, evidence indicates that recovery is multidimensional, and recovery rate will differ depending on which ecosystem attribute or ecological process is being considered. Species composition, for example, may take much longer to recover than species richness or measures of ecosystem function. Nevertheless, the durations needed for ecosystem recovery seem to be consistently long relative to the timescales of collapse.

15. There are feedbacks between ecosystems and the biosphere, or the entire Earth system, which can as a mechanism for collapse and influence recovery.
    - The evidence for this is limited, although a number of potential feedback mechanisms have been identified. The most compelling are those associated with global climate, for example the release of

4.6 Evaluation of Propositions · 285

CO_2 as a result of forest dieback or deforestation, which could contribute to global warming. In the oceans, the development of hypoxia following phytoplankton blooms could lead to an increase in emissions of methane, which is a potent greenhouse gas (Bakun and Weeks, 2004). Other feedbacks may await discovery.

16. Collapse can be positive by creating new opportunities, for example evolutionary diversification and radiation.
    o Ecosystem collapse consistently provides opportunities for some species. For example, the collapse of coral reefs is often associated with increased colonisation by algae; many invasive species are flourishing in freshwater ecosystems; and forest collapse may encourage species associated with grassland or cropland. In the oceans, jellyfish may be doing better than at any time since the Palaeozoic. The characteristics of species that benefit from human disturbance are well known: such 'weedy species' are phenotypically plastic and physiologically tolerant, with high growth rates and reproductive output (Doubleday and Connell, 2018).

5 · *Synthesis*

At the start of this book, as we set out on our peripatetic voyage through the scientific literature, my aim was to arrive at some kind of tangible destination – ideally, a place where ecosystem collapse and recovery are understood well enough that we might be able to do something practical about them. So, where have we actually arrived, more than a thousand scientific publications later? Have we achieved our ultimate goal? Or have we taken, at least, some significant steps towards it?

As an attempt to answer these questions, this chapter first evaluates the empirical evidence presented in Chapters 3 and 4 in relation to the theoretical ideas described in Chapter 2. This is achieved through the evaluation of some tentative propositions. Selected issues that emerged during our journey are also considered, and some initial questions are addressed that still await an answer. A second section then examines the implications of our current understanding for conservation policy and practice. Specifically, how can we avert the collapse of ecosystems and support their recovery?

Before exploring the terrain of our destination, I would like to pay tribute to all the researchers whose work we have encountered along the way. While not wanting to belittle the efforts of theoreticians and modelers – ecology would not be a robust science without them – I especially acknowledge the contribution made by field researchers. While reading this literature, I have been struck by just how much hard work has been involved in collecting the evidence I've summarised. It isn't easy to collect insightful data about the natural world; many of the studies cited here (and the investigations that they themselves cite) involved long hours working under challenging conditions. It is thanks to the dedication of field ecologists that we have some idea about how the world's ecosystems are currently changing. Despite these magnificent efforts, it is clear from this scientific mystery tour that there is still much to learn. Therefore, one of the objectives of this chapter is to identify some priorities for future research.

5.1 Understanding Ecosystem Collapse and Recovery

In Chapter 1 (Table 1.3), I listed a number of initial questions that I hoped this book would help to answer. Many of these have been addressed in the subsequent chapters, as illustrated by the list of propositions that emerged (Table 5.1). These propositions represent an attempt to identify a set of generalisations based on the theoretical ideas presented in Chapter 2, which were then evaluated using the empirical evidence presented in Chapters 3 and 4. It should be noted that this empirical evidence was not systematically identified or collated; a robust testing of these propositions would require systematic review procedures and formal meta-analysis (Gurevitch *et al.*, 2018; Pullin and Stewart, 2006; Vetter *et al.*, 2013). Yet, at least we have here a set of initial ideas that might usefully form the basis of future research, and who knows, they might eventually develop into an empirical theory of ecosystem collapse and recovery (Ford, 2000; Pickett *et al.*, 2013). They might also help to inform conservation policy and practice (see Section 5.2).

However, these propositions are clearly not the whole story. Some of the initial questions remain unanswered. Other issues have emerged during development of the propositions that merit further consideration. Furthermore, one of the striking features of the literature considered in this book is the lack of consensus: many of the ideas presented are still being contested and debated. Some of these questions, issues and controversies are therefore considered further in the following sections.

Extrinsic *versus* Intrinsic Factors

It is well established that populations and communities of species are continuously fluctuating; ecosystems are therefore dynamic. Scheffer and Carpenter (2003) note that these fluctuations will always be generated by an 'intricate mix' of internal processes and external forces. Furthermore, the relative influence of intrinsically generated dynamics and external forces is difficult to unravel, which is perhaps unsurprising given the historical difficulties in demonstrating virtually all important ecological mechanisms (e.g. competition, chaos or density dependence) (Scheffer and Carpenter, 2003). These points apply equally to ecosystem collapse. Evidence suggests that extrinsic factors are likely to be involved always; although intrinsic processes are also often likely contributors, their role is often more difficult to determine (Table 5.1). Identification of the relative influence of extrinsic and intrinsic factors in driving ecosystem

288 · Synthesis

Table 5.1 *Final set of propositions relating to ecosystem collapse and recovery, refined throughout the book (see preceding chapters)*

No.	Proposition	Comments*
1	Ecosystem collapse can be caused either by extrinsic or by intrinsic factors, or by a combination of the two. Extrinsic factors refer to the external disturbance regime affecting an ecosystem, and intrinsic factors to interactions of organisms between each other and with the physical environment.	Supported. Extrinsic factors are always likely to be involved. The role of intrinsic factors is less clear and may vary with location, context and ecosystem type.
2	(a) Any ecosystem can potentially collapse, if subjected to disturbance of an appropriate type and occurring at sufficient frequency, extent, intensity or duration.	Supported, as stated. However, a systematic assessment of all ecosystem types has not been conducted. This proposition raises the question of whether some ecosystem types are more vulnerable to collapse than others; this is currently unknown.
	(b) Novel disturbances are more likely to cause collapse.	Tentatively supported. There is evidence of novel disturbances causing collapse in all ecosystem types considered. However, the relative importance of novelty *versus* other aspects of disturbance, such as intensity, has not been explicitly tested.
3	(a) Disturbance regimes characterised by interactions between multiple types of disturbance are more likely to cause ecosystem collapse.	Strongly supported. Interactions have been documented in all ecosystem types considered, although some evidence suggests that they may be less likely to drive collapse of freshwater ecosystems than other ecosystem types.
	(b) A particular disturbance event is more likely to result in collapse if it follows another event from which the ecosystem has not yet recovered.	Tentatively supported. There is evidence for repeated disturbances negatively impacting ecosystems, but their role in driving ecosystem collapse merits further investigation.

Table 5.1 (cont.)

No.	Proposition	Comments*
4	(a) The coincidence of both 'press' and 'pulse' disturbance types is most likely to lead to ecosystem collapse.	This was not strongly supported by the literature. Most examples of collapse appear to be driven solely by 'press' disturbances, although 'pulse' disturbances are also important in some situations. There was little evidence that collapse was most likely when both types of disturbance were present. There is also some difficulty about applying this classification in practice, suggesting that the 'pulse-press' framework might need revising (see text).
	(b) While both 'pulse' and 'press' disturbance events can increase the mortality of individuals, 'press' disturbances can also play a particular role in impeding ecosystem recovery.	Strongly supported in all ecosystem types.
5	Some ecosystems can exist in more than one state or regime, such as different successional stages. Transitions between these states can form part of the natural dynamics of an ecosystem. Many ecosystems can be kept in a transient state by disturbance. When such transitions are persistent rather than transient, they can become of concern to human society.	Strongly supported in all ecosystem types.
6	A persistent ecosystem transition, or collapse, can arise when ecological recovery is impeded. This can occur if there are stabilising feedback processes that maintain an ecosystem in a degraded state, if there is chronic disturbance, or when the processes of ecological recovery fail.	Supported; a limited capacity for recovery is typically associated with collapse. Information is needed on the relative importance of these different reasons for a lack of recovery.

(cont.)

Table 5.1 (cont.)

No.	Proposition	Comments*
	Understanding these reasons for lack of recovery is key to understanding collapse.	
7	Collapse can be caused by breakdown of the stabilising feedback mechanisms maintaining an ecosystem state, or by feedbacks in the internal ecological processes of an ecosystem driving a system to a different state. As a result of these feedbacks, major ecological shifts can result from minor perturbations. Such shifts can occur when external conditions reach a critical value.	This proposition is best supported by the classic research on eutrophication of shallow lakes. Research efforts have also been directed at the other ecosystem types considered. This has led to a wide variety of different feedback mechanisms being identified. However, empirical evidence of their role in driving ecosystem collapse is often lacking. Evidence of critical values has also been obtained from all ecosystem types, although these do not necessarily need to be coupled to feedback mechanisms to cause collapse (e.g. if organisms are heated to a lethal temperature).
8	Collapse may be more likely if disturbance events cause the loss of: (a) generalist species, (b) those that are highly connected to other species, (c) top predators and/or trophically unique species and (d) those at the base of food chains. Loss of such species can cause many secondary extinctions.	Generally supported. Relative importance of the four different types of species may differ between ecosystem types. However, this has not systematically been investigated. Tentatively, it appears that loss of those species that are highly connected to others may be the most important contributor to collapse.
9	(a) The disassembly of ecological networks tends to be characterised by thresholds.	Strong evidence for threshold responses in ecosystems affected by anthropogenic disturbance. However, less evidence is available specifically in relation to ecological networks.
	(b) In such situations, the consequences of species loss become amplified and self-reinforcing as more and more	There is limited field evidence for this. Such observations have been made in some studies, but these phenomena are difficult to study

Table 5.1 (cont.)

No.	Proposition	Comments*
	species are extirpated, leading to network collapse.	and may be more widespread than we are aware of.
10	Certain types of ecosystems may be especially sensitive to disruption of biotic processes and interactions, namely: (a) those with strong top-down trophic regulation, (b) those with many mutualistic or facilitation interactions, (c) those that are strongly dependent on mobile links and (d) where positive feedbacks operate between the biota and the disturbance.	While disruption of any of these ecosystem attributes could potentially increase risk of collapse, evidence for this is limited; these attributes may also differ in frequency. Although mutualistic and facilitation interactions are widespread in many ecosystems, strong top-down trophic regulation appears to be less widespread, and a strong dependence on mobile links is even rarer. There are also relatively few documented examples of positive feedbacks between the biota and disturbance.
11	Extinction cascades and community collapses should be less likely to occur in species-rich multi-trophic communities than in species-poor ones, unless the environment is highly variable.	This proposition was not supported, highlighting the lack of consensus in both theoretical and empirical studies regarding the relationship between species richness and the stability of communities.
12	Ecosystem recovery is dependent on intrinsic factors, namely interactions of organisms between each other and with the physical environment. The rate or extent of recovery can be limited by intrinsic factors, and/or by extrinsic factors, such as the disturbance regime and the extent of degradation.	Proposition broadly supported, but key emerging questions relate to the relative importance of intrinsic and extrinsic factors limiting recovery; whether there are thresholds of degradation beyond which recovery is not possible; and whether recovery can occur in the presence of press disturbance (e.g. climate change).
13	Functional and structural change in an ecosystem undergoing collapse may be unrelated to loss of taxonomic diversity but may be affected by loss of species with particular functional traits; species identity matters.	Strongly supported. In recovering ecosystems, recovery of ecosystem function may similarly be related to the functional characteristics of species, rather than taxonomic diversity.

(cont.)

292 · Synthesis

Table 5.1 (cont.)

No.	Proposition	Comments*
14	Recovery can take a long time and takes longer than collapse.	Strongly supported. While recovery rates vary depending on which ecosystem attribute is being considered, the timescales of recovery are consistently longer than those of collapse.
15	There are feedbacks between ecosystems and the biosphere, or the entire Earth system, which can act as a mechanism for collapse and influence recovery.	Limited evidence for this, although some mechanisms have been proposed. This issue merits further research.
16	Collapse can be positive by creating new opportunities, for example evolutionary diversification and radiation.	Supported. 'Weedy' species, in particular, are expected to benefit.

* Note: Brief comments are also provided regarding the evidence considered in Chapters 3 and 4.

collapse awaits further systematic research; however, it seems likely that their relative contributions vary between locations and perhaps between ecosystem types.

It is important to note that ecosystem collapse can occur solely or principally because of extrinsic factors. A powerful example is provided by deforestation, where a forest ecosystem may be permanently converted to some other form of land cover, such as agricultural or urban land. Such collapse can be very abrupt, perhaps occurring within a few days once the bulldozers have rolled in. Even in the case of coral reefs, where so many intrinsic processes have been identified as potential contributors (see Section 4.1), ecosystem collapse associated with bleaching events is principally driven by an extrinsic factor, namely a rise in sea temperature. This point is worth making, because some authors explicitly link ecosystem collapse to dynamical systems theory and the concept of alternative stable states (e.g. Keith et al., 2015; Lindenmayer et al., 2016). According to these theoretical ideas, transitions between ecosystem states are driven by intrinsic feedback processes, after a critical threshold of an extrinsic factor is reached. However, forests can be clear-felled, lakes and wetlands drained, coral reefs dynamited, rivers poisoned and grasslands paved over without the need to invoke any contribution of intrinsic

5.1 Understanding Ecosystem Collapse and Recovery · 293

processes to collapse. Perhaps intrinsic processes and the theories that relate to them are therefore only relevant in situations where the ecosystem has not been completely destroyed by human activity. Sadly, as we all know, such destruction is very widespread and perhaps accounts for most current examples of ecosystem collapse.

The Characteristics of Disturbance

The likelihood of an ecosystem collapsing and its subsequent ability to recover depend strongly on the characteristics of the prevailing disturbance, such as its intensity, frequency and type. For example, we learned from the prehistoric case studies (Chapter 3) that climate change associated with elevated CO_2 concentrations can have a particularly profound effect on global ecosystems, being consistently associated with mass extinction events. The importance of the disturbance regime for understanding both collapse and recovery is illustrated by its explicit inclusion in most of the propositions (10/16) developed here (Table 5.1). Based on this survey of available evidence, further research is needed to determine whether ecosystem collapse is more likely if the disturbances are novel or if disturbances are repeated (e.g. see Villnäs et al., 2013), although tentatively both propositions are supported (Table 5.1). Similarly, information is needed on the extent to which ecosystems can recover in the presence of ongoing disturbance. This is a highly topical issue, given that any ecosystem recovery taking place in the current era will need to occur under conditions of climate change.

Two emerging issues merit further consideration: interactions between different types of disturbance and the relative impact of 'press' and 'pulse' forms of disturbance. Empirical evidence provides strong support for ecosystem collapse often being associated with interactions between different types of disturbance (Table 5.1) despite such interactions not being well represented in available theory (see Chapter 2). This suggests a need for further theoretical development. Buma (2015) notes the need for a better understanding of the potential for interactions between different disturbances, as well as the occurrence of cross-scale interactions and for comparisons between different ecosystem types (e.g. see Shackelford et al., 2017). In this context, the suggestion that interactions may be less likely to drive collapse of freshwater ecosystems than other ecosystem types is of particular interest (Table 5.1). Buma (2015) also provides a conceptual framework for understanding disturbance interactions, which differentiates between 'linked interactions' (involving alteration of the likelihood, extent or severity of a subsequent disturbance) and 'compound interactions' (involving alterations of the recovery time or trajectory). To understand and predict disturbance interactions, further research is required that disaggregates disturbances into their constituent legacies, identifies the mechanisms underlying the interactions and determines when and where these mechanisms might be altered by the legacies of prior disturbances (Buma, 2015).

Some uncertainty also surrounds the relative impact of 'pulse' and 'press' forms of disturbance on both collapse and recovery, although there seems to be consistent evidence that 'press' (or chronic, continuous) disturbance is the form most often associated with limited ecosystem recovery (Table 5.1). This issue is particularly pertinent given that anthropogenic disturbances tend to be 'press' perturbations, while natural disturbance regimes are often characterised by relatively discrete 'pulse' events (Smith *et al.*, 2009). Furthermore, according to Petraitis (2013), tests of alternative stable states should employ the latter rather than the former, as it is 'pulse' perturbations that are incorporated in the underlying mathematics of dynamical systems theory (see Chapter 2). Overall, there was relatively little evidence for collapse being driven by the combined effects of both 'press' and 'pulse' disturbances, as is often explored by theoretical models (Table 5.1). Rather, most examples of collapse appeared to be primarily driven by chronic forms of anthropogenic disturbance. This suggests another possible disconnect between empirical evidence and underlying theory, which again merits further research attention.

Researchers are still grappling with how best to characterise and understand disturbance regimes. Jentsch and White (2019) propose a general theory of pulse dynamics and disturbance, which comprises four fundamental postulates and seven generalisations (Table 5.2). This represents a fascinating attempt to develop a body of theory in what is a complex area of ecology, in which theoretical developments have been rather limited to date (see Chapter 2). By proposing law-like postulates, the authors aim to identify a set of first principles that might provide a basis for further theoretical development, following Marquet *et al.* (2014). The postulates are essentially a set of propositions, which are conceptually similar to those listed in Table 5.1. However, the approach used to develop them was different to that employed here. Rather than attempting to induce generalisations from empirical evidence, Jentsch and White (2019) employed deductive logic. As these authors note, deduction and empiricism are complementary approaches in the

Table 5.2 *Elements of a theory of pulse dynamics and disturbance, proposed by Jentsch and White (2019)*

Postulates	
1. *Resource dynamics*	Pulse events initiate a series of predictable changes in resource ratios, storage and availability that are controlled by abiotic and biotic processes, including the stoichiometric requirements and resource accumulation rates of organisms.
2. *Energy flux*	Energy flux determines the rate of ecological processes and responses to pulse events.
3. *Patch dynamics*	The distribution of patches in space and time determines resource flows, variation in ecosystem structure, the availability of biota and, thus, the nature of future pulse events.
4. *Biotic trait diversity*	Pulse dynamics produce evolutionary forces that generate trade-offs, leading to predictable patterns of trait diversification, and, in turn, the diversity, complementarity and redundancy of biotic traits determine how ecosystems respond to pulse events.
Generalisations	
1. *Disturbance magnitude*	Disturbance magnitude, expressed as changes in resource availability, resource ratios, limiting resources and biotic legacy, are the product of the force (or intensity) of the disturbance and ecosystem resistance.
2. *Recovery trajectory*	Resource trajectories after disturbance are determined by the stoichiometric requirements of organisms, energetic constraints, physiological capacity, luxury uptake and loss through spatial transfers (e.g. via water, wind and gravity). Resource uptake changes the hierarchy of limiting factors.
3. *Rate of change*	Energy flux, resource availability and biotic traits determine the rate of resource change after disturbance. In primary successions characterised by no biological legacy and multiple limiting resources, rates of change are slow at first, increase to a maximum and then decrease again as resources are accumulated in biomass or lost through biotic processes, abiotic processes or export. In secondary successions, rates of change are initially high and similarly decrease through time as available resources are accumulated in biomass or are lost from circulation.

(*cont.*)

Table 5.2 (*cont.*)

Postulates	
4. *Disturbance probability*	Feedbacks between ecosystem structure and disturbance frequency change the probability and magnitude of future disturbance events as a function of time since disturbance and the legacy of those past disturbances. A major consequence is that feedbacks and disturbance interactions can increase or decrease the probability of future disturbances and can lead to cascading disturbances (e.g. Buma, 2015; Seidl *et al.*, 2017). Examples of ecosystem feedbacks include increasing fuel levels through time, which increases fire risk. Examples of disturbance interactions include droughts that increase susceptibility to insect infestations, or flooding, landslides and fragmentation that reduce fuel connectivity and thus fire occurrence. A case of special interest in disturbance ecology occurs when, for systems and processes under human management (e.g. fire and flood), the decrease in disturbance frequency owing to suppression leads to rarer but higher-magnitude disturbances (the suppression hypothesis).
5. *Biotic trait diversification in evolutionary time*	Trait diversification is a saturating, evolutionary process that results in predictable trait distributions for such traits as dispersal, resource accumulation, longevity and competitive interaction. The evolutionary pressure for trait diversification increases with energy flux, resource availability, abruptness of change and resource heterogeneity.
6. *Species and trait diversity in ecological time*	Species and trait diversity at the landscape scale increase with biotic trait differentiation over time (Postulate 4), patch distribution in time and space (Postulate 3), energy flux (Postulate 2) and resource heterogeneity (Postulate 1).
7. *Resilience*	Resilience, defined as recovery after pulse events, can be quantified by the degree of return to reference level, rate of return and time for reaching the former or a new steady state (e.g. Ingrisch and Bahn, 2018). Resilience is a function of disturbance magnitude and frequency relative to trait diversity under the assumption that the higher the biotic trait differentiation, the greater the asynchrony, redundancy and complementarity of traits and thus the capacity for functional recovery (Oliver *et al.*, 2015).

5.1 Understanding Ecosystem Collapse and Recovery · 297

development of ecological theory, and typically there will be a dialogue between these two approaches. For example, deduction is ultimately based on analysis of empirical patterns and their potential causes (Jentsch and White, 2019).

Given that the theory presented by Jentsch and White (2019) focuses on pulse dynamics, where do 'press' disturbances fit in? Interestingly, they hardly receive a mention. Instead, it is suggested that the difference between 'pulse' and 'press' is essentially one of scale. What is perceived as a 'press' disturbance when aggregated at the scale of an entire forest, for example, might be experienced as a series of pulse events at the scale of an individual tree or leaf. All disturbances can essentially be considered as 'pulse' events, which become continuous ('press') as their duration or frequency increase, or their magnitude decreases (Jentsch and White, 2019). This has some fascinating implications. Does it mean that concerns about disturbance type in relation to dynamical systems theory raised by some authors (e.g. Dudgeon *et al.*, 2010; Petraitis, 2013) essentially disappear? Also, given that 'pulsedness' is itself a continuous variable (Yang *et al.*, 2008), might there be some critical value of this variable that is associated with an increased risk of ecosystem collapse?

As Jentsch and White (2019) put it, the most important focus of their theory is the 'prediction of conditions for stable and unstable ecosystem dynamics'. Specifically, they identify the need for further development of the theory to 'define the variable combinations that result in mechanisms of stability, comprising resistance, recovery and adaptation *versus* those conditions that create unstable dynamics and regime shifts'. In other words, further work is needed to extend these ideas to the mechanisms underpinning ecosystem collapse and recovery. Perhaps some linkage with the propositions identified here (Table 5.1) might help that endeavour.

Even if the distinction between 'press' and 'pulse' disturbances is difficult to maintain, these categories are still being widely used by researchers. For example, MacDougall *et al.* (2013) describe an elegant experiment involving both 'press' and 'pulse' treatments, which showed how loss of species richness in a degraded grassland ecosystem can increase the risk of ecosystem collapse. Harris *et al.* (2018) provide a valuable account of recent collapse of a series of Australian ecosystems, including riverine forest, kelp forest, mangroves and arid zone ecosystems (more than a third of the continent having burned in 2011–2012). The pulse-press framework was found to be useful in this context, because collapse appeared to be driven by the combination of pulse events (such as drought, heatwaves, extreme rainfall, fire and storms) and press disturbance (climate change). This study therefore provides evidence in support of Proposition 4 (Table 5.1). It also highlights the particular role of climate change in driving ecosystem collapse. As the authors note, there are many other recent examples of extreme biotic responses to climate change. For example, the mass mortality of sea fans in the Mediterranean reported by Cerrano and Bavestrello (2008) was again attributed to the interaction between global warming and other pressures, including pollution, eutrophication, habitat destruction, overfishing and spread of invasive species.

Harris *et al.* (2018) suggest that future responses to extreme events will likely be taxonomically and geographically idiosyncratic and therefore difficult to predict. However, analysis of how pulse and press events interact could help understand how climate change affects ecosystems. This issue is explored further by McDowell *et al.* (2018), who make the important point that future climate change coupled with extreme events could force ecosystems beyond their historical range of variability, providing another useful way of visualising the process of ecosystem collapse (Figure 5.1).

Transient and Persistent Ecosystem States

The case studies described in Chapter 4 clearly indicate that different states or regimes can be recognised in all the different ecosystem types considered. One of the issues that arises is whether such different ecosystem states are equivalent to the alternative stable states postulated by dynamical system theory, as widely proposed in the literature. However, in each of the ecosystem types considered, this suggestion has been challenged, as key assumptions of the theory have often not been met (e.g. Bruno *et al.*, 2009; Capon *et al.*, 2015; Möllmann and Diekmann, 2012; Newton and Cantarello, 2015). For example, transitions between ecosystem states are often associated with a change in environmental conditions, which is not consistent with theory relating to alternative stable states (Dudgeon *et al.*, 2010; Petraitis and Dudgeon, 2004). Proposition 5 (Table 5.1) has therefore been worded in such a way that its validity does not depend on the validity (or otherwise) of dynamical systems theory.

One of the questions raised at the start of this book (see Chapter 1) was whether a transition between the different successional stages of a community might be an example of ecosystem collapse. When evaluating the

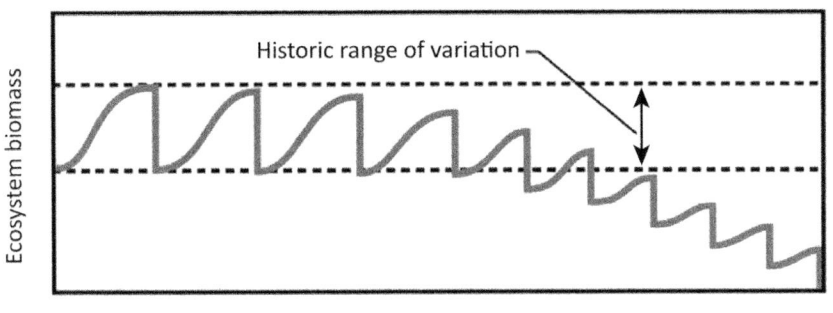

5.1 Understanding Ecosystem Collapse and Recovery · 299

Figure 5.1 Figurative representation of the impact of climate change and extreme events on ecosystems. Here, biomass is used as a proxy for ecosystem function, although other ecological characteristics and processes might respond in a similar way. Historically, discrete disturbance events would reduce ecosystem functioning, but these would eventually dissipate, and the ecosystem would recover. Under current and future climate, the post-disturbance system will be subject to chronic warming. Additional extreme events will become more likely to occur sooner and with greater severity, preventing the ecosystem from reacquiring its prior structure and function. The progressive increase in disturbance will accelerate biomass decline (and alter other ecosystem processes) if recovery time is insufficient. As a result, the ecosystem will experience novel conditions beyond the historical range of variability, leading to a 'chronic disequilibrium'. Adapted from McDowell *et al.* (2018), with permission from Elsevier.

risk of collapse, Keith *et al.* (2015) highlighted the need to differentiate between transitions among ecosystem states that occur as part of the natural variability within an ecosystem type, *versus* a process of collapse and replacement by a different ecosystem type. At first glance, successional transitions would seem to form part of the natural variation occurring within an ecosystem. Yet, as illustrated by many of the case studies considered in Chapter 4, ecosystems can often be maintained indefinitely in a successional state by chronic disturbance (Fukami and Nakajima, 2011). For example, disturbances such as fire or herbivory can prevent the successional transition from grassland to forest (see Chapter 4). Does this mean that the forest has collapsed? According to Keith *et al.* (2015), this partly depends on whether or not transitions between states involve the loss of defining features (characteristic biota and ecological processes) that explicitly describe the ecosystem type. Certainly, the biota and ecological processes of a grassland and a forest can be very different, and in the literature, they are consequently often considered as different types of ecosystem.

On this basis, persistent transitions between successional states might be considered as a form of ecosystem collapse, even though these transitions form part of the natural dynamics of an ecosystem. This raises the question of what is meant by 'persistent'. Obviously, the definition of what constitutes 'persistent' is essentially arbitrary; the wording of Proposition 5 (Table 5.1) implies that timescales relevant to human lifespans might be a pragmatic choice (i.e. durations of decades). An interesting corollary of this argument is that in some situations, succession itself could be considered to be a driver of collapse. For example, if a grassland were maintained by herbivory over long timescales, then allowed to develop into forest through succession after removal of the herbivores, this would constitute collapse. This situation is perhaps illustrated by the loss of the mammoth steppe described in Chapter 3. Another interesting corollary is that different ecosystem states do not need to be equilibrial in order to qualify as examples of collapse when transitions between them occur (e.g. the non-equilibrial 'alternative transient states' of Fukami and Nakajima, 2011). Ecosystem collapse is therefore not limited to transitions between the alternative stable states associated with dynamical systems theory. Neither is it limited to regime (or phase) shifts, which reflect transitions between equilibrial communities associated with different environmental conditions (Dudgeon et al., 2010).

Feedback Mechanisms

Feedbacks are a fascinating property of dynamical systems, which form a central element of dynamical system theory. As Scheffer (2009) puts it, threshold behaviour, runaway processes, hysteresis, cycles and chaos all have positive feedbacks as a key mechanism. But Scheffer (2009) also offers some words of caution: the identification of a feedback process does not automatically mean that a particular theoretical situation applies in practice. For example, positive feedbacks will not lead to transitions between alternative stable states unless the feedbacks are sufficiently strong. An example is provided by the positive feedback between water clarity and the growth of submerged plants. While widespread in many lakes, the effect of this feedback is stronger in shallow lakes, as plants can cover the complete bottom of the lake in such situations, and their effect is stronger on shallow water columns. This positive feedback mechanism is therefore only likely to lead to transitions between ecosystem states in shallow lakes and not deep ones (Scheffer, 2009).

5.1 Understanding Ecosystem Collapse and Recovery

Feedbacks were an element of five of the propositions listed here (Table 5.1), namely numbers 6, 7, 9, 10 and 15. Empirical investigations have led to a wide variety of different feedback mechanisms being identified (Chapter 4), some of which might stabilise an ecosystem in a particular state, while others could potentially cause an ecosystem to shift rapidly to another state. Ecosystem transitions are sometimes attributed to a shift from negative (stabilising) to positive feedback processes (Briske *et al.*, 2010). As a result of positive feedback mechanisms, minor perturbations resulting from human activity can potentially lead to major ecological changes (Scheffer, 2009), and that is why they are currently the focus of such societal concern.

There are several challenges associated with feedbacks. While the ingenuity of ecologists has enabled a wide variety of potential feedback mechanisms to be identified, for example in coral reefs (Section 4.1) and seagrasses (Maxwell *et al.*, 2017), it is often difficult to demonstrate that these processes actually operate in the real world and are responsible for driving the ecological changes observed. In some situations it can even be difficult to surmise what the feedbacks might be, as we found in our analysis of ecological change in agricultural landscapes (Newton *et al.*, 2019). These issues are considered by Bowman *et al.* (2015) specifically in relation to terrestrial landscapes, although their findings are equally relevant to other ecosystem types and indeed to both ecological and Earth sciences more broadly. While in engineering systems there are well-developed methods to examine feedbacks, this is less the case in ecological systems, because here dynamics involve numerous interactions, many of which are poorly characterised and may even be unknown. These interactions occur across a large range of spatial (m^2 – 1,000 km^2) and temporal (<1 s – >1,000 years) scales. Ecological feedbacks are also difficult to study because they present a form of 'chicken and egg' problem, in which simple cause and effect relations are complicated by circular causality (Bowman *et al.*, 2015).

To address this problem, Bowman *et al.* (2015) recommend an integrated approach that combines long-term monitoring, experimentation, conceptual models, simulation and synthesis (Figure 5.2). The application of approaches such as this is needed to determine the role of feedbacks in driving ecological change. Surely it is not good enough to simply infer the role of feedbacks from a study of ecological pattern, as we encountered in research on savannas (e.g. Hirota *et al.*, 2011; Staver *et al.*, 2011; Section 4.5). On the other hand, just because the role of a feedback process is difficult to verify doesn't mean that it isn't influential. Also,

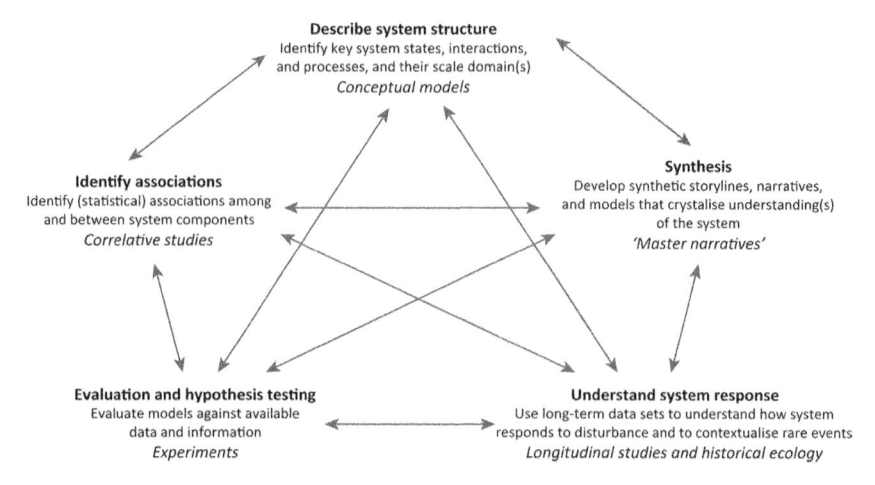

Figure 5.2 An integrated approach to detect feedbacks in ecosystems. Reprinted from Bowman *et al.* (2015), with permission from Elsevier.

feedback mechanisms may often be hidden from view. For example, in marine ecosystems, as Bakun and Weeks (2006) beautifully put it, there may be important feedback processes operating that may not be immediately obvious to our terrestrially based human intuitions. This implies that caution is needed whenever invoking feedbacks, but also when invoking their absence.

Further research is particularly needed on the feedbacks referred to in Propositions 10 and 15 (Table 5.1). The idea that certain elements of the biota may promote further disturbance, causing a positive feedback loop, is well established for fire-adapted vegetation and for grazing-adapted plant species such as grasses (Chapters 3 and 4). However, much less information is available for other forms of disturbance. Other comparable feedbacks must surely exist, but their detection perhaps requires ecologists to display some more of that ingenuity referred to earlier. Examples include positive feedbacks identified between invasive plant species and mycorrhizal colonisation (Zhang *et al.*, 2010) and other plant-soil feedbacks favouring the spread of invasive plant species (van der Putten *et al.*, 2013). There may also be positive feedbacks involving the human component of socioecological systems (Hull *et al.*, 2015), for example when development of forested land encourages the spread of further development by opening up markets and changing the attitudes of landowners (BenDor *et al.*, 2014).

It is important to remember that feedbacks can also play an important role in ecosystem recovery. For example, seagrasses can alter sedimentation patterns to benefit their further spread (Maxwell *et al.*, 2017). A number of different feedback mechanisms limiting recovery are considered by Suding *et al.* (2004), including:

- Species associated with degraded ecosystems can change ecosystem processes, leading to positive feedbacks that can reduce scope for recovery. For example, introduced grasses in Hawaii alter nutrient cycling that further benefits introduced grasses, a situation that has proved very difficult to change.
- Patterns of herbivory and other trophic interactions can alter in degraded systems, which can reduce recovery potential. For example, the redistribution of herbivores in rangelands can cause positive feedbacks with plant cover.
- Lack of landscape connectivity and propagule production can limit the regeneration of native species in degraded communities. Consequently, invasive species can dominate seed banks and the seed rain. Management interventions can exacerbate the problem through feedbacks.

Proposition 15 suggests that there might also be feedbacks between ecosystems and the Earth system that could potentially influence ecosystem collapse and recovery. Although evidence for these is limited at present, this is surely another area where further research is justified. An example is provided by the Amazon forest (see Section 4.4), which has been listed as one of the tipping elements in the Earth system (Lenton *et al.*, 2008). Climate change, together with increasing deforestation, logging and fire, could lead to major vegetation shifts. Potentially these stressors could trigger self-amplified forest loss in the Amazon basin owing to the coupling of vegetation and climate at the regional scale (Zemp *et al.*, 2017), an issue considered in further sections.

Ecological Networks and Secondary Extinctions

Propositions 8–10 are especially grounded in community ecology, that subdiscipline of ecological science that is concerned with understanding the composition and structure of biotic assemblages and their dynamics in changing environments (Jackson and Blois, 2015). As mentioned in Section 2.8, a substantial body of research has examined community dynamics from the perspective of food webs and other ecological networks, including analysis of the vulnerability of ecosystems to species loss and the possibility of collapse (McDonald-Madden *et al.*, 2016). The propositions distilled from relevant theory attracted varying degrees of empirical support (Table 5.1). Here, I further consider Proposition 9b, owing to its particular importance and the uncertainty associated with it. Specifically, is there evidence from the real world that ecosystems are collapsing in a non-linear manner owing to cascading secondary extinctions?

The suggestion that ecological networks might collapse in this way was made by Bascompte and Stouffer (2009) in their valuable review of ecological networks. As these authors point out, much of the research into the disassembly of ecological networks has been conducted using simulation models. The focus of this book has been on empirical evidence rather than that derived from models, so the latter has not been examined in depth. I know from experience how valuable modeling approaches can be, both for exploring the implications of theory relating to ecological networks and for helping to make sense of data gathered from the field (Newton et al., 2018). So now seems like a good moment to pay tribute to the community of researchers who have begun to use such approaches to explore how ecological networks might collapse. Simulation of network dynamics is surely an important way forward, given the enormous challenges of collecting field data that accurately capture how ecological networks are changing in response to human activity. Furthermore, some of the simulation studies employ empirical data and are not solely based on theoretical ideas.

Bascompte and Stouffer (2009) summarise some of the key findings from this research, including:

- Network structure influences the vulnerability of networks to disassembly; for example, ecological networks are relatively robust to the loss of the most specialised species, but more vulnerable if more generalised species are the ones that are extirpated (Dunne et al., 2002a; Solé and Montoya, 2001). Similar results have been obtained in pollination networks (Memmott et al., 2004).
- If species are randomly removed, secondary extinctions increase gradually. However, removal of the most connected species produces a sharp increase in secondary extinctions after an initial lag period (Dunne et al., 2002a). In both cases, there is an increased sensitivity with greater removal level (Dunne et al., 2002a). Therefore, as the intensity of the perturbation increases, the consequences of the perturbations also increase. This suggests that the initial impact of human activity on ecological networks might provide a poor indication of what might happen after greater disturbance.

5.1 Understanding Ecosystem Collapse and Recovery · 305

- As a rule of thumb, based on analysis of empirical data, Dunne *et al.* (2002b) showed that food webs characterised by higher connectance were better able to tolerate loss of species. Also, using simulation of empirical data, Memmott *et al.* (2004) showed that the nested structure of pollination networks was related to their tolerance of species extinction.
- Some species have particularly important roles in the network of interactions among species, and their disappearance may induce structural changes in the remaining network. For example, in pollination networks species of Hymenoptera and Diptera are network hubs. Loss of these species, and the other species that depend upon them, could change the structural properties of a network and reduce its ability to tolerate further perturbation (Olesen *et al.*, 2007).

Together, these observations led Bascompte and Stouffer (2009) to propose a general rule captured here in Proposition 9 (Table 5.1), suggesting that network collapse is non-linear as secondary extinctions cascade through the network. This conclusion is primarily based on the results presented by Dunne *et al.* (2002a), who suggested that when undergoing selective removal of species, food webs demonstrate thresholds. If species are removed at random, this initially leads to a low rate of secondary extinctions. However, once highly connected species begin to be removed, a threshold is exceeded, after which networks collapse much more rapidly. A non-linear collapse trajectory could therefore emerge from the structure of ecological networks. Positive feedback might also exist between a decline in the number of species and the number of extirpations, which could contribute to a similar result. However, it should be noted that not all studies have reported this pattern; for example, in their study of pollination networks, Memmott *et al.* (2004) reported a linear decline in plant species diversity with simulated species loss.

So, is the collapse of ecological networks in the real world a non-linear process or not? Interestingly, in their otherwise instructive review of secondary extinctions, Brodie *et al.* (2014) do not consider this possibility. Instead, they suggest that synergistic interactions among multiple stressors will determine secondary extinction risk, rather than the magnitude of any single stressor. In their conceptual model, multiple human impacts occur concurrently, which directly affect species and the interactions between them, at different positions within the ecological network (Figure 5.3). No consideration is given here to the idea that secondary

306 · Synthesis

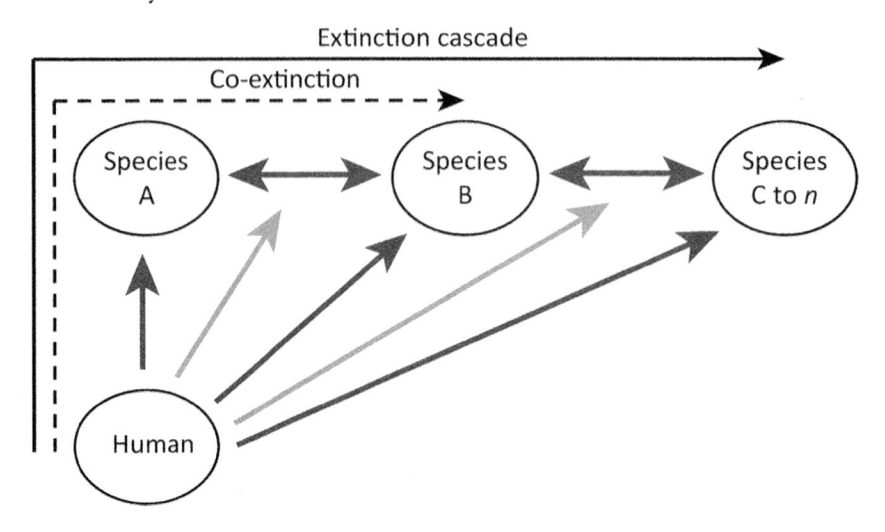

Figure 5.3 Conceptual model illustrating extinction cascades caused by human impacts. Grey arrows show the direction of demographic impacts among interacting species. In co-extinction, direct impacts on species A lead to its extirpation, which in turn causes the secondary loss of species B (broken black arrow). In extinction cascades, the secondary extinction(s) can also occur farther along the food chain (species C through *n*; unbroken black arrow). Multiple human impacts can occur concurrently, which affect species (dark arrows) and modify the interactions among them (light arrows). The direction and strength of these cumulative impacts will determine whether secondary extinctions occur. Reprinted from Brodie *et al.* (2014), with permission from Elsevier.

extinctions might amplify as they progress through a network, leading to non-linear collapse. This might be because they excluded modeling studies, such as those of Dunne *et al.* (2002a), from their review.

In fact, there appears to be insufficient evidence available to evaluate whether or not the collapse of ecological networks in the real world is typically non-linear. As noted by Rodriguez-Cabal *et al.* (2013), field-based empirical studies that demonstrate the disassembly of mutualistic webs are extremely rare; the same could be said of other forms of ecological network. The example described by Rodriguez-Cabal *et al.* (2013) conducted in Argentine Patagonia is notable because of the methods used; the study employed field experiments to trace the impacts of introduced species along an 'interaction web' that included both trophic and mutualistic interactions. Results showed that introduction of a suite of exotic vertebrate and invertebrate species led to the disassembly of a mutualistic plant–animal interaction network (Figure 5.4).

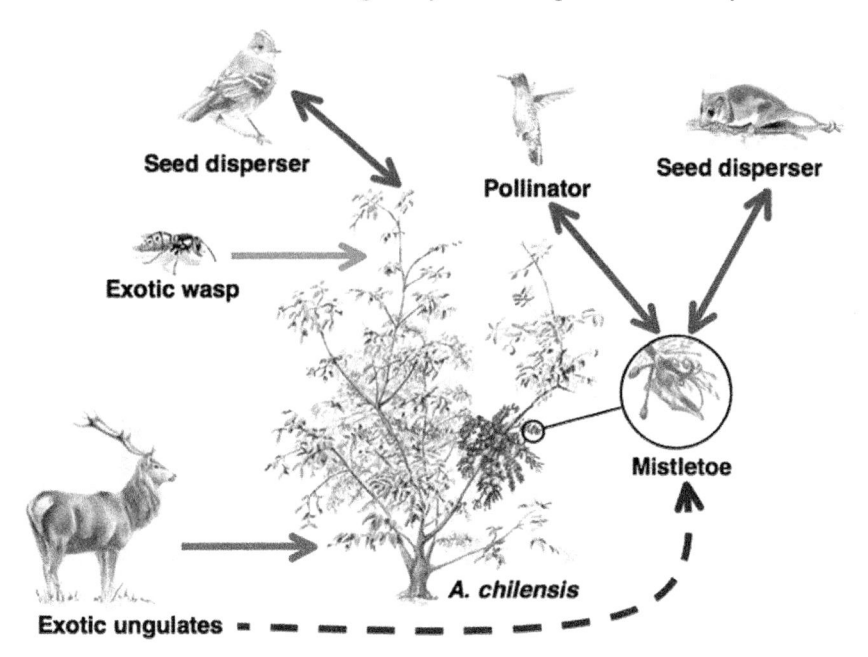

Figure 5.4 Interaction web in Argentine Patagonia. Browsing by exotic ungulates on the shrub *Aristotelia chilensis* reduced the number of available hosts for the mistletoe species (*Tristerix corymbosus*), thereby indirectly affecting populations of a marsupial species (*Dromiciops gliroides*), which is its only known disperser and the hummingbird pollinator (*Sephanoides sephaniodes*). Furthermore, fruit predation by introduced wasps reduced the abundance of *Elaenia albiceps*, the most common seed-predating bird. The impact of exotic ungulates and the German wasp cascaded through the interaction web to influence the number of seeds removed by both the marsupial and *E. albiceps*. Reprinted from Rodriguez-Cabal *et al.* (2013), with permission.

Because of the importance of the hummingbird as a key pollinator, and marsupial and bird species as key seed dispersers, the decline or loss of these vertebrates could lead to a dramatic decline in the species richness of the forests in which they live (Rodriguez-Cabal *et al.*, 2013). However, no evidence was presented of the potential shape of this trajectory.

A large research community has been busily exploring the collapse of ecological networks using models, which have generated many fascinating and sometimes conflicting insights (Table 5.3; see also Section 2.8). The lack of consensus regarding some of these findings is illustrated by the relationship between species richness and the likelihood of network collapse, where contradictory results have been obtained (e.g. see Dunne

Table 5.3 *Some key findings from selected recent literature examining the collapse of ecological networks using modeling approaches*

Key finding	Type of network	Reference
The structure of ecological communities could be more vulnerable to realistic extinction sequences than previously believed.	Food webs	Berg et al. (2015)
Food webs are least robust to the loss of species that have many trophic links or that occupy low trophic levels. Secondary extinctions can be triggered by both bottom-up and top-down cascades.	Food webs	Curtsdotter et al. (2011)
Increased species richness and connectance are associated with decreased likelihood of network collapse.	Food webs	Dunne and Williams (2009)
There is a critical value of habitat destruction beyond which interactions are lost more rapidly.	Plant-animal mutualisms	Fortuna et al. (2013)
The early removal of a species that would eventually become extinct can significantly reduce the number of cascading extinctions.	Food webs	Sahasrabudhe and Motter (2011)
Even a relatively small increase in the mortality rate of large-bodied species can lead to the extinction of another species in the food web.	Food webs	Säterberg et al. (2013)
Regional habitat loss contributes directly to species loss and indirectly by reorganising interactions in a local community. Networks became more highly connected and more modular with habitat loss.	Plant pollinator	Spiesman and Inouye (2013)
The abundance of the rarest species is linked to the resilience of the community.	Mutualistic	Suweis et al. (2013)
A highly connected and nested structure promotes community stability in mutualistic networks, whereas the stability of trophic networks is enhanced in compartmented and weakly connected structures.	Mutualistic and trophic	Thébault and Fontaine (2010)
In the absence of intraspecific competition for consumers, the risk of cascading extinctions increases with species richness, whereas it generally decreases in the opposite case.	Food webs	Thébault et al. (2007)

Table 5.3 (cont.)

Key finding	Type of network	Reference
Extinction cascades may be more likely in highly connected mutualistic communities.	Mutualistic	Vieira and Almeida-Neto (2014)
Highly connected and species rich food webs were more likely to lose species at a low level of disturbance than sparsely connected food webs with few species.	Food webs	Wootton and Stouffer (2016)

and Williams, 2009; Thébault et al., 2007; Wootton and Stouffer, 2016). This reflects the uncertainty associated with Proposition 11 (Table 5.1), and more broadly with the relationship between the stability and diversity of ecological systems, which has been the focus of a long-running debate in ecology (McCann, 2000). It is also worth noting the focus of these studies, which typically simulate the sequential loss of species. For example, models are often used to analyse how secondary extinctions occur after a primary species is removed from the network. Cascading secondary extinctions therefore provide an example of how intrinsic processes can lead to community collapse. This is relevant to specific situations in nature, such as when an individual species is being harvested, but has limited relevance to many other real-world situations where extrinsic factors predominate (see the section titled *Extrinsic versus Intrinsic Factors*). For example, when a forest is clear-felled and converted to a field of crops, an entire community is destroyed, and ecosystem collapse is essentially instantaneous. In such a situation, there is limited scope for secondary extinctions to occur. Nevertheless, it is worth noting that extinction cascades provide a mechanism for the kind of intrinsic feedback postulated by dynamical systems theory, so they are likely to have played a role in some of the transitions between ecosystem states documented in Chapters 3 and 4. Also, food webs can themselves be modeled as dynamical systems, leading to some unexpected insights (Allesina and Tang, 2012; Tang et al., 2014).

Despite the undoubted value of this modeling research, there have been relatively few attempts to verify whether the findings obtained apply in the field. While many of these studies have examined the role of network structure, other factors may also influence secondary extinctions, including spatial interactions and modification of the abiotic environment. Few field-based investigations have examined the relative influence of these different factors. De Visser et al. (2011) described food web structure along a gradient of human impact in the African savanna and examined the effects of a realistic species extinction sequence. Results showed that human impacts on network structures were non-linear, species loss typically beginning with large herbivores and top predators. As a result, poorly connected species tended to be lost first, while more highly connected species were lost as human impact progressed. In this latter situation, the food web showed high sensitivity to decreased network connectance. Results also suggested that a failure to consider non-trophic interactions, such as relationships with the abiotic environment, may lead to an underestimation of human impacts on wildlife communities. A further empirical example is provided by Valiente-Banuet and Verdú (2013), which focused on the ecological impact of harvesting plant species in the Tehuacán-Cuicatlán Biosphere Reserve in Mexico. Under one of the scenarios explored, feedbacks were incorporated that represented facilitation of natural regeneration by different plant species. In this case, interaction networks collapsed after a threshold value was reached (i.e. when the habitat availability of nurse species was reduced to below 76% of its original extent).

In contrast, a substantial amount of empirical evidence is available for trophic cascades, which were widely encountered in the contemporary case studies considered in Chapter 4 (see also Propositions 8 and 10, Table 5.1) and perhaps even in some prehistoric examples (Pires et al., 2015). This supports suggestions that loss of large apex consumers can result in extensive cascading effects in all types of ecosystem, including impacts on disturbance regimes and biogeochemical cycles (Estes et al., 2011; Ripple et al., 2016a). For example, Ripple et al. (2014) documented trophic cascades in 7 of the 31 largest mammalian carnivores, demonstrating their role in supporting the abundance of a variety of different groups of species. Although trophic cascades (and their 'bottom-up' analogues) could potentially lead to cascading secondary extinctions, few examples have been recorded; most documented effects are on species abundance (Brodie et al., 2014).

Cascading secondary extinctions may be particularly likely in mutualistic and parasitic interactions. Even though few co-extinction events have actually been recorded, some models suggest that co-extinction may be the most common form of biodiversity loss (Dunn et al., 2009). This discrepancy may reflect the fact that most mutualists and parasites are small and understudied. Co-extinction is also notable in that it

5.1 Understanding Ecosystem Collapse and Recovery · 311

interacts with other causes of extinction, or the other 'horsemen of the extinction apocalypse' (Diamond, 1984; Diamond et al., 1989), such as habitat loss, species invasion and overharvesting (Dunn et al., 2009). Impacts of these factors and other forms of global environmental change on species interactions are widespread. In a review of 688 published studies, Tylianakis et al. (2008) found that mutualisms involving plants were generally negatively affected by such factors, whereas pathogen infection of plants and animals generally increased; impacts on soil food webs were less consistent and highly context-dependent. Every species can be considered as a node in a complex web of interactions; these interactions may be lost before the species themselves are extirpated (O'Gorman and Emmerson, 2009; Valiente-Banuet et al., 2015). Re-establishment of functional interactions between species may consequently be an important contributor to ecosystem recovery (Colwell et al., 2012).

Ecosystem Function

One of the purposes of this chapter was to paper over the cracks that have appeared in this book while I was writing it. The largest of these lacunae relates to ecosystem function. Rather than address it directly as a distinct theme, I have let ecosystem function thread its own way through rest of the text, making mention of it whenever prompted to do so by the literature consulted. This is perhaps appropriate, given that function is a property that emerges from consideration of an entire ecosystem. Hopefully some understanding of the topic might similarly emerge from consideration of this entire text. On the other hand, no explicit consideration of theory relating to ecosystem function was presented in Chapter 2. This might wrongly imply that a gap exists in the literature, whereas in fact this is an area that is theoretically rich. It has also exercised some of the best brains in the business. So why leave it until the end? Partly this reflects a personal bias; as a conservation ecologist, I'm more concerned about the potential loss of species and communities than changes in their contributions to biogeochemical cycling. Many functions of natural ecosystems can be replaced, at least partially, by ecosystems created artificially by people, such as plantation forests, agroecosystems or artificial lakes (e.g. Clifford and Heffernan, 2018; Paquette and Messier, 2010). If you consider these two statements to be contentious, you are in good company, as this is an area that has generated some of the most acrimonious debate in ecology (Kaiser, 2000). This was another reason for leaving it until the end of the book.

One area of this debate has focused on the relationship between species diversity and ecosystem function. Whereas some proponents believe that there are causative relationships between diversity and ecosystem functioning, others maintain that the principal drivers of ecosystem properties are not species diversity, but the functional attributes or traits of the species present (Wardle et al., 2000). This might seem like a rather limited reason for such a major controversy, but the implications are profound. In the latter view, ecosystem function would be predicted to be much less affected by species loss than in the former view, unless those species of particular functional importance – such as keystone species – were eliminated. In other words, in the latter view, species identity matters: the consequences of species loss will depend on which species actually become extinct and what their functional roles are.

Despite labelling their opponents' views with colourful terms such as 'harmful', 'dangerous' and 'a house of cards', which you might think would have precluded any possibility of future collaboration, a brave attempt was made to forge a consensus between these warring parties. In their review of both sides of the debate, Hooper et al. (2005) identified areas of agreement, which included:

- Species' functional characteristics strongly influence ecosystem properties.
- The effects of species loss or changes in composition can differ among ecosystem properties and ecosystem types.
- Some ecosystem properties are initially insensitive to species loss because: (a) ecosystems may have multiple species that carry out similar functional roles, (b) some species may contribute relatively little to ecosystem properties or (c) properties may be primarily controlled by abiotic environmental conditions.
- Presence of a range of species that respond differently to different environmental perturbations can reduce the impact of disturbances on ecosystem processes.

Here, the term 'ecosystem properties' was used to refer to characteristics such as productivity, carbon storage, hydrology and nutrient cycling, which are often referred to as 'ecosystem functions'. Biases in previous research were also highlighted, including little emphasis on interactions between multiple trophic levels, a lack of long-term field experiments, limited understanding of feedbacks and a lack of studies in freshwater and marine ecosystems (Hooper et al., 2005).

5.1 Understanding Ecosystem Collapse and Recovery

This continues to be a very active area of research; some recent findings are summarised on Table 5.4. Sharp-eyed readers will notice that the debate highlighted previously hasn't really gone away; instead, both sides appear to be accumulating ever more evidence in support of

Table 5.4 *Summary of some recent research findings on the relationship between biodiversity and ecosystem function*

Key findings	Ecosystem type	Reference
Meta-analysis of experimental evidence shows that the effects of biodiversity change on processes are weaker at the ecosystem compared with the community level; and productivity-related effects decline with increasing number of trophic links between those elements manipulated and those measured.	Global	Balvanera et al. (2006)
Experiments suggest that the impact of biodiversity on any single ecosystem process is non-linear and saturating, such that change accelerates as biodiversity loss increases. However, relationships in natural ecosystems sometimes show a different pattern.	Global	Cardinale et al. (2012)
Loss of diversity across trophic levels has the potential to influence ecosystem functions more strongly than loss of diversity within trophic levels, because of trophic cascades affecting plant biomass.	Global	Cardinale et al. (2012)
Deep-sea ecosystem functioning and ecosystem efficiency is exponentially related to deep-sea biodiversity. A higher biodiversity supports higher rates of ecosystem processes and an increased efficiency with which these processes are performed.	Deep-sea benthos	Danovaro et al. (2008)
Meta-analysis of experimental results indicates that mixtures of species tend to enhance levels of ecosystem function relative to monocultures. A power function describes the relationship between consumption and species richness, but the relationship between richness and production is linear.	Marine	Gamfeldt et al. (2015)

(cont.)

Table 5.4 (*cont.*)

Key findings	Ecosystem type	Reference
Meta-analyses of published results show that the effects of species loss on ecosystem productivity and decomposition are of comparable magnitude to the effects of many other global environmental changes. The identity of species lost is also significant.	Global	Hooper *et al.* (2012)
In a field experiment, human disturbance created a negative relationship between diversity and function, contrary to theoretical predictions.	Grassland in western North America	MacDougall *et al.* (2013)
Analysis of empirical data showed that multifunctionality is positively and significantly related to plant species richness.	Drylands	Maestre *et al.* (2012)
Recent evidence suggests that the general findings of early biodiversity and ecosystem functioning studies are robust and may even underestimate diversity's role in nature. Yet frontiers remain in linking this research to the complexity of wild nature.	Global	Naeem *et al.* (2012)
The effects of diversity-dependent ecosystem feedbacks and interspecific complementarity accumulate over time. Consequently, simplification of diverse ecosystems will likely have greater negative impacts on ecosystem functioning than has been suggested by previous short-term experiments.	Grassland experiments, Minnesota, USA	Reich *et al.* (2012)
Hundreds of experimental studies spanning terrestrial, aquatic and marine ecosystems show that high-diversity mixtures are approximately twice as productive as monocultures of the same species. These results reflect trade-off-based mechanisms that allow long-term coexistence of many different competing species. Diversity loss has a significant effect on ecosystem functions such as productivity and nutrient dynamics. These effects are as great as those attributable to other drivers of environmental change.	Global	Tilman *et al.* (2014)

Table 5.4 (*cont.*)

Key findings	Ecosystem type	Reference
The functioning of many ecosystems is being altered both by the loss and addition of species. Invading species with novel traits can alter biotic interactions and ecosystem processes. Less is known about the impacts of species loss on function. Integrated above-ground–below-ground interventions may be needed to facilitate recovery of ecosystem function.	Global	Wardle *et al.* (2011)

their respective positions (e.g. Tilman *et al.*, 2014). The relevance of diversity-ecosystem functioning experiments to real-world ecosystems continues to be questioned (e.g. Wardle, 2016), and these criticisms continue to be rebuffed (Eisenhauer *et al.*, 2016; Naeem, 2008; Naeem *et al.*, 2012). Further theoretical developments continue to be made. For example, Loreau (2010a,b) proposed a theoretical framework that unifies community- and ecosystem-based approaches to ecology. This recognises that whereas communities are structured by networks of interactions, ecosystems are structured by networks of biogeochemical pathways. As organisms are pools of elements in biogeochemical pathways, these two types of network are inextricably linked (Naeem *et al.*, 2012).

Recent research on ecosystem function has also examined the issue of stability. This is another topic that I approach with great trepidation, if only because of the cloud of semantic uncertainty that surrounds it (see Chapter 1). Grimm and Wissel (1997) suggested that this 'confusion of tongues' has biblical proportions, with more than 163 different definitions of stability having been proposed in the literature. This uncertainty has significant implications for environmental policy, which often refers to stability without defining it clearly, thereby creating problems for implementation. According to Donohue *et al.* (2016), the fault lies partly with ecologists, who are not only inconsistent in their use of the term, but also one-dimensional in their thinking. Many different components of stability can be identified; for example, Pimm (1984) identified five, namely asymptotic stability (i.e. an asymptotic return to equilibrium after disturbance), variability, persistence, resistance and resilience. Other researchers have sliced the stability cake rather differently. Yet the important point is that different components of stability, regardless of how one defines and assesses them, each need to be considered individually so that their underlying mechanisms can be fully elucidated. According to Donohue et al. (2016), we currently have a very poor understanding of how environmental change influences different stability components. Theoretical studies are generally based on dynamics of a system close to an equilibrial state, which has limited applicability to the real world, because of its highly dynamic nature and the strong directionality of many elements of environmental change. Furthermore, the relationships between different components of stability can change over time, depending on the nature of the disturbance regime (Donohue et al., 2013).

In contrast to this rather bleak prognosis, other researchers have claimed that significant progress is being made in understanding stability in ecological systems. For example, Wang and Loreau (2016) refer to recent 'major advances', such as a general consensus that a higher biodiversity can enhance the temporal stability of aggregate ecosystem properties (e.g. biomass, productivity, etc.). This is supported by the results of a number of experimental investigations, such as those of Isbell et al. (2009, 2011) and Tilman et al. (2006). Similarly, Cardinale et al. (2012) cite five other syntheses that broadly support the same conclusion. This is consistent with the suggestion that this relationship is statistically inevitable, because of the averaging of fluctuations in species' abundances (Doak et al., 1998). A closely related idea is the 'insurance hypothesis', which assumes that species respond differently to environmental change because of niche differentiation, thereby buffering the impact of environmental changes (Griffin et al., 2009). This depends upon functional redundancy, or the idea that species within the same functional group can replace each other with little impact on ecosystem function. The insurance hypothesis has proved popular among theoreticians and provides a clear mechanism for reducing the risk of ecosystem collapse, but does it apply in the real world? Perhaps not. In their elegant study using long-term field data, Valone and Barber (2008) conclude that the insurance hypothesis does not appear to be a strong mechanism stabilising fluctuations in natural terrestrial communities.

Loreau (2010a) beautifully summarises our current understanding of the diversity–stability relationship: it is complex and multifaceted and 'does not lend itself to sweeping statements'. The same could be said of the biodiversity-ecosystem function relationship more generally. This

5.1 Understanding Ecosystem Collapse and Recovery · 317

helpfully justifies my reticence to suggest propositions relating to ecosystem function (see Table 5.1). Whether or not there are useful generalisations that one can make about these relationships seems to depend on which group of researchers one agrees with and which terminology one adopts.

Personally, I confess that I have a lot of sympathy for the views of David Wardle, who has repeatedly highlighted the limitations of much research in this area. Specifically, most experiments have been conducted under highly controlled conditions using synthetic communities. These experiments arguably have little relevance to natural ecosystems, as they fail to mimic what happens when extirpations occur or when new species arrive (Díaz et al., 2003; Wardle, 2016). The limitations of this approach are demonstrated by Vellend et al. (2013), who used monitoring data from >16,000 local-scale vegetation plots to show that mean species diversity has not changed over the timescales studied (5–261 years). This suggests that controlled experiments that have artificially manipulated species richness have little relevance to most field situations. Wardle and his collaborators have conducted a series of ingenious field-based investigations using natural and removal experiments, which have provided numerous valuable insights into the relationships between species richness and ecosystem properties (e.g. Wardle et al., 1997, 1999, 2004a; Wardle and Zackrisson, 2005). These provide an inspirational example of the kind of research approaches that are needed to determine what happens to function as ecosystems collapse or recover.

Here, only a single proposition (no. 13) was suggested that explicitly links to ecosystem function, which referred to the idea that species differ in their functional importance: in other words, species identity matters. Based on the case studies profiled in Chapter 4, the final version of the proposition (Table 5.1) was amended to include explicit reference to functional traits, which determine the contribution of individual species to overall functioning of the ecosystem. This proposition was strongly supported by the empirical evidence reviewed in Chapters 3 and 4 and by reviews of experimental evidence (Cardinale et al., 2012; Hooper et al., 2005). Many further propositions could potentially be developed in future. One of the consequences of variation between species in their functional traits is that relationships between species richness and ecosystem function are likely to be non-linear. Such non-linear relationships are strongly supported by experimental evidence (Cardinale et al., 2012) and might be widespread in natural ecosystems, although they clearly are not encountered everywhere (Mora et al., 2011). On this basis, one might predict that ecosystem function would tend to display a non-linear pattern of decline during ecosystem collapse and a non-linear pattern of increase during ecosystem recovery. As noted in Chapter 4, meta-analyses of recovery patterns in forest ecosystems support this latter suggestion. Less evidence is available for examples of collapse, highlighting the need both for further meta-analyses and collection of monitoring data in ecosystems that are either collapsing or recovering. Following Hooper *et al.* (2005), one might further expect patterns of change in ecosystem function to differ among types of ecosystem, types of function and types of degradation.

Non-linear patterns of change in ecosystem function could also be driven by feedback mechanisms. Although the possibility of feedbacks is mentioned by Hooper *et al.* (2005), few details are given; the issue is not considered at all by Cardinale *et al.* (2012). Yet the potential for such feedbacks clearly exists. Some of the best documented examples are those between plant communities and the soil biota (Bardgett and Wardle, 2010; Wardle *et al.*, 2004b, 2012). For example, in Arctic tundra, there are positive feedback loops that are driving the spread of woody shrubs such as *Betula nana*, a process that is superimposed on the effects of a warming climate. As shrubs trap and hold snow, the soil beneath them is better insulated, which leads to increases in temperature and consequently microbial activity. This in turn increases rates of N mineralisation, which can then facilitate greater shrub encroachment (Wookey *et al.*, 2009). Although increased growth of a shrub as small as *Betula nana* might sound trivial, its expansion can drastically alter the structure and function of tundra ecosystems by changing energy fluxes and exchange of water, carbon and nutrients, and affecting regional climate, as well as causing the decline of other plant species (Myers-Smith *et al.*, 2011). This feedback could therefore contribute significantly to ecosystem collapse.

Another issue that merits further research attention is the rate and extent of change in ecosystem function during either collapse or recovery. For example, Martin *et al.* (2013) showed that carbon pools recover more quickly than plant species richness in tropical forests. This can be understood as a consequence of Proposition 13: in this ecosystem, many tree species can make similar contributions to carbon storage (cf. Hooper *et al.*, 2005). Furthermore, it is important to note that a decline in ecosystem function is not an inevitable feature of collapse. This is illustrated by the conversion of natural forest to tree plantations, which can sometimes lead to maintenance – or even an increase – in functional processes such as carbon storage (see Chapter 4). Conversely, recovery of ecosystem structure and composition is not necessarily associated with recovery of function (Cortina *et al.*, 2006). So Loreau (2010a) was right; the dynamics of ecosystem function are indeed complex and multifaceted.

Collapse and Recovery Cascades

One of the questions posed at the outset of this book (see Chapter 1) was this: if one ecosystem collapses, might others follow? The case studies considered in Chapters 3 and 4 provided little evidence of this phenomenon, so it is briefly considered here.

Rocha *et al.* (2018) provide a review of cascading regime shifts. As noted earlier in this book, ecosystem collapse can potentially be considered as a type of regime shift, although Rocha *et al.* (2018) cast the net wider to include changes in the global climate system. Nevertheless, one could substitute the term 'collapse' for 'regime shift' in the definition of a cascade provided by these authors. This would provide the following definition: *an ecosystem collapse could be considered as cascading if its occurrence might affect the occurrence of another collapse.* Such a situation requires some form of interconnection (or teleconnection) between ecosystems. Rocha *et al.* (2018) considered two types of interconnections: 'domino effects' and 'hidden feedbacks'. 'Domino effects' refer to a situation when the feedback processes of one regime shift affect the drivers of another regime shift, creating a one-way dependency. 'Hidden feedbacks' involve two-way interactions. These could arise when two regime shifts combine to generate new feedbacks, which if strong enough, could potentially amplify or dampen the coupled dynamics. These feedbacks are considered 'hidden' because they would only emerge when both situations are considered together. These two situations are contrasted with a third possibility, namely when two different regime shifts are caused by the same drivers. If drivers are shared, then the two regime shifts would be correlated in some way, even if they are not otherwise interdependent (Rocha *et al.*, 2018).

These authors used a database of regime shifts (including examples of both ecosystems and the climate system) to examine the occurrence of cascading effects, using a network approach. Hypothesised feedbacks were represented by networks, which were then subjected to a pairwise comparison to identify linkages. The occurrence of shared drivers was similarly identified using analysis of descriptive variables of each case study included in the database. Based on analysis of 30 regime shifts, results suggested that ~45% of these examples showed some evidence of structural interdependence, suggesting that cascades might be widespread.

As the authors recognise, the study has a number of limitations, including biases in the selection of case studies, and the fact that feedback mechanisms were hypothesised and not necessarily tested empirically (Rocha et al., 2018). It should also be noted that regime shifts were equated to transitions between alternative stable states, in accordance with dynamical systems theory. In other words, it was assumed that the regime shifts were driven by intrinsic feedback mechanisms. In fact, the term 'regime shift' refers to any sudden change, regardless of mechanism (Scheffer, 2009), so these transitions might have been caused solely by changes in external drivers (see Chapter 4).

Nevertheless, the investigation identified a number of potential linkages between different examples of ecosystem collapse, most of which are familiar from Chapter 4. These linkages are primarily of the 'domino' type, for example the knock-on effect of a fishery collapse on marine food webs (Table 5.5). It is possible to envisage other mechanisms that might cause collapse cascades to occur. An ecosystem might collapse owing to an external factor, which might then create a new driver that could cause the collapse of another ecosystem, without the need to invoke any feedbacks. For example, if a forest is clear-felled, or a river catchment degraded, increased soil erosion could destroy coastal ecosystems such as coral reefs. McCulloch et al. (2003) provide an example of this process for the Australian Barrier Reef. This suggests that the definition of a 'domino' cascade should be broadened to include situations not involving feedbacks.

Atmospheric moisture recycling provides a particularly important mechanism that could connect the collapse of one ecosystem to another. It has been estimated that as much as 25–50% of total rainfall in the Amazon basin is derived from evapotranspiration from vegetation; much of the rainfall on the eastern Andes is dependent on the presence of lowland vegetation further east. This provides a mechanism for a potent feedback that could cause self-amplified forest loss in the Amazon (Betts et al., 2004; Davidson et al., 2012). Reduced rainfall could increase the risk of forest dieback, which in return could intensify regional droughts, leading to further tree mortality (Eltahir and Bras, 1994; van der Ent et al., 2010; Zemp et al., 2014, 2017). The Amazon therefore provides an informative example of how linkages between ecosystems can potentially

5.1 Understanding Ecosystem Collapse and Recovery · 321

Table 5.5 *Selected examples of ecosystem collapse cascades, based on information presented by Rocha et al. (2018), which are referred to as regime shifts*

Regime shift 1	Regime shift 2	Cascade type	Reference
Marine eutrophication	Hypoxia	Driver sharing, domino effect, hidden feedback	Diaz and Rosenberg (2008)
Hypoxia	Coral transitions	Domino effect	Altieri et al. (2017)
Forest to savanna	Mountain forest transitions	Domino effect	Clark et al. (2015), Morueta-Holme et al. (2015), Weng et al. (2018)
Hypoxia	Marine food webs, fisheries collapse	Domino effect	Weng et al. (2018)
Forest to savanna	Dryland degradation	Domino effect	Bakun (2017)
Fisheries collapse	Marine food webs	Driver sharing, domino effect	Branch et al. (2010), Carlson et al. (2018), Moore et al. (2018), Pauly et al. (1998), Springer et al. (2018), Tickler et al. (2018)
Forest to savanna	Lake food web assemblage	Domino effect	Röpke et al. (2017)
Lake eutrophication	Submerged to floating plants	Domino effect	Van Gerven et al. (2016)

Notes: The examples referring to regime shifts in the global climate system (e.g. loss of Arctic or Antarctic sea ice, weakening of thermohaline circulation, ENSO, North Atlantic oscillation, etc.) have been omitted. For details of cascade types, see text.

occur across different scales (Figure 5.5). However, moisture recycling is not limited to the Amazon; evidence suggests that 10–40% of precipitation over India and 50% over the Congo may be recycled from vegetation (Spracklen *et al.*, 2018).

Examples of 'hidden feedbacks' (perhaps better referred to as 'reciprocal cascades') are harder to identify in nature (Table 5.5). As Rocha *et al.*

322 · Synthesis

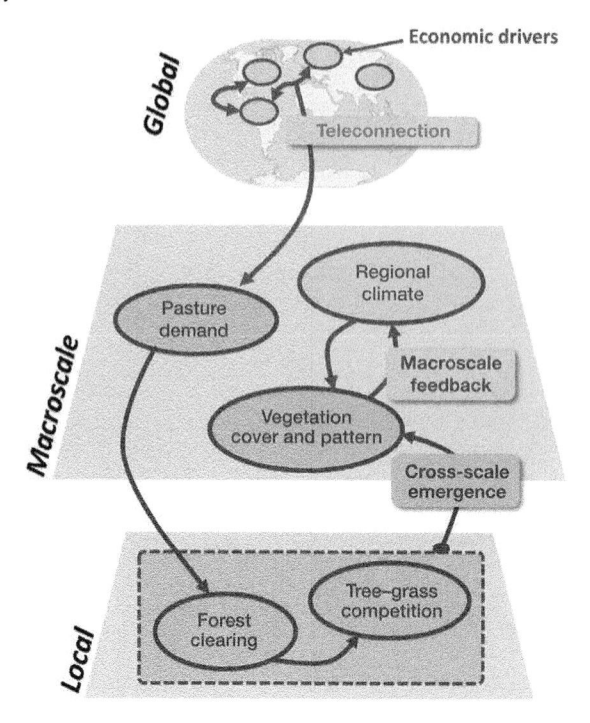

Figure 5.5 Some key interactions within and across scales in the Amazon rain forest, including a feedback at the macroscale, cross-scale emergence (namely, interactions at local scales that produce patterns and processes at the macroscale) and a teleconnection. Reproduced from Heffernan *et al.* (2014) with permission from Wiley. © The Ecological Society of America.

(2018) point out, relatively few studies have been conducted with coupled systems. However, in Chapter 4 we already encountered the remarkable flow of marine nitrogen to terrestrial riparian ecosystems mediated by the consumption of salmon by bears. For example, Holtgrieve *et al.* (2009) found that bears increased soil concentrations of some nitrogen fractions by a factor of 32. It is not difficult to imagine how this could develop into a feedback leading to collapse: forest dieback could affect both bear and salmon populations, which would reduce nitrogen input into the forest ecosystem, perhaps reducing scope for forest recovery. There may be many other such linkages between distant ecosystems, which could potentially develop into reciprocal feedbacks. Most research on teleconnections has focused on the role of climate. For example, structural changes in Arctic tundra vegetation have been linked to the extent of Arctic sea ice, as well as to weather systems at lower latitudes; the result is a shift towards structurally novel ecosystems similar to those described for Beringia in the early Holocene (Macias-Fauria et al., 2012). Examples of teleconnections are also known from the fossil record, where they have been implicated in the collapse of marine ecosystems (Algeo et al., 2011).

Ecosystem collapse cascades present another clear research priority for the future, especially given that cascades that are also envisaged in the global climate system (Steffen et al., 2018). Conversely, if there are cascades of ecosystem collapse, might there also be cascades of ecosystem recovery? In other words, might the recovery of one ecosystem support the recovery of another? This notion appears to be completely unexplored.

5.2 Living with Ecosystem Collapse and Recovery

Given our current understanding of ecosystem collapse and recovery, what are the implications for conservation policy and practice? This is a topic that might justifiably merit a book by itself, so the brief consideration provided here should be viewed as highly preliminary. A conservation manager confronted by the possibility of ecosystem collapse might want to know what the risks are of it actually happening, how to detect when it might happen, how to prevent it from happening and what to do about it if it does happen. Alternatively he or she may be confronted by an ecosystem that has already collapsed and might wish to understand the prospects for recovery and how to bring this about. In either case, there is also a need to understand what the implications of different conservation actions (or inactions) might be, for both wildlife and people. Before addressing these information needs and how they might be met, let us first consider again what ecosystem collapse actually is.

Defining Ecosystem Collapse

As we have already learned (Grimm and Wissel, 1997), one thing that ecologists love to do is to come up with new definitions of terms, even if perfectly acceptable definitions are already available. Nothing I can say here will prevent future researchers from proposing new definitions of ecosystem collapse as they see fit; after all, they see this as part of their scientific duty. However, the IUCN Red List of Ecosystems (RLE) is different, as this is a formal assessment process that needs to be rigorous and consistent. Let us briefly recall what the purposes of the RLE are (see Chapter 1): to provide a global standard for assessing the status of ecosystems and specifically to determine whether or not an ecosystem is facing imminent risk of collapse. It is anticipated that the RLE will help prioritise conservation action and support land-use planning, improve governance and livelihoods and even support macroeconomic planning (IUCN, 2019).

As a reminder, the RLE considers collapse to be the endpoint of ecosystem decline and defines it as 'a transformation of identity, a loss of defining features, and a replacement by a different ecosystem type' (IUCN, 2019). 'Collapsed' is also one of the categories of collapse risk, which is defined as (IUCN, 2019):

An ecosystem is Collapsed [*sic*] when it is virtually certain that its defining biotic or abiotic features are lost from all occurrences, and the characteristic native biota are no longer sustained. Collapse may occur when most of the diagnostic components of the characteristic native biota are lost from the system, or when functional components (biota that perform key roles in ecosystem organisation) are greatly reduced in abundance and lose the ability to recruit.

The RLE recognises that, unlike species becoming extinct, ecosystems do not disappear when they collapse; rather they transform into 'novel ecosystems' (Bland *et al.*, 2017a). In fact, as we have seen elsewhere in this book, such transformations are not limited to 'novel ecosystems', but could include conversion to another ecosystem type that was previously present on the same site (e.g. a transition from forest to grassland in a savanna landscape). Furthermore, Bland *et al.* (2017a) do not consider what happens after the ecosystem has collapsed. Potentially the ecosystem could recover to the type that was originally present, if the causes of collapse were removed. Alternatively, it could transform into something different, which could indeed be new to that area. A key point is that this novel ecosystem type could itself be of high conservation value. This is perhaps one of the most significant shortcomings of the RLE approach; the entire assessment is predicated on the assumption that ecosystem collapse represents a form of biodiversity loss. This is not necessarily the case; in some situations, it could even potentially lead to net conservation gains. Ecosystem collapse should therefore perhaps be viewed as a form of biodiversity change, rather than loss.

Consider the place where I live, situated on the south coast of England. My house overlooks a river valley that around 12,000 years

5.2 Living with Ecosystem Collapse and Recovery · 325

ago was home to a large community of Palaeolithic people who hunted reindeer in a tundra ecosystem. Part of that valley now lies under the sea. By 8,000 years ago, this area was covered in dense deciduous forest (Grant *et al.*, 2014). From around 6,000 years ago, the landscape was transformed by its progressive conversion to extensive agriculture, following the arrival of Neolithic people. By around 3,000 years ago, much of the landscape was dominated by extensive pastureland, which over time was largely converted to cropland as human populations increased (Gale, 2003). In the past century, agricultural land use has itself changed through intensification and industrialisation, causing widespread loss of species and habitats (Newton *et al.*, 2019). This sequence of events represents at least three phases of ecosystem collapse that have happened since people first arrived here. This successive collapse of a series of ecosystem types, each one transforming into another, might be usefully referred to as *serial collapse*. In other words, there is no single endpoint, but a succession of different endpoints. This phenomenon is undoubtedly widespread, at least in areas with a long history of human habitation such as southern England.

In conservation terms, what happened in my local area during the early Holocene could be viewed as the complete loss of glacial ecosystems owing to climate change, which facilitated colonisation by a large number of non-native species. A Palaeolithic conservation manager would have viewed this as a disaster: the reindeer departed and never returned. Relics of the forest ecosystem that subsequently developed are today highly valued for conservation, but so are species-rich habitats that were created by centuries of agricultural land use, such as heathland and calcareous grassland. Conservation metrics such as species richness and landscape heterogeneity probably reached a peak around 200 years ago, before the advent of industrialisation. Whether one views these successive ecosystem transformations as representing conservation gains or losses depends on the relative values accorded to different species and communities and other metrics such as species richness.

This example shows that the collapsed state of one ecosystem type could be a non-collapsed state of another ecosystem type. This second ecosystem type might subsequently collapse itself. The problem this presents to conservation management is illustrated by transformations between two ecosystem types that both have high conservation value. For example, in a savanna landscape, both forest and grassland elements might be associated with characteristic species, worthy of conservation. Conversion of forest to grassland might be considered as an example of biodiversity loss, but the same might also be true if the grassland develops into forest. In this situation, recovery of forest could represent a driver of grassland ecosystem collapse, and vice versa. This shows how ecosystem collapse and recovery are inextricably linked.

Surprisingly, the RLE gives little consideration to the possibility of ecosystem recovery. However, it is important to remember that collapse in the context of the RLE refers to loss from throughout the entire geographical range of an ecosystem, and in this case, the potential for recovery might indeed appear to be limited. The fact that ecosystems can recover after global-scale disturbance is illustrated by the aftermath of the 'Big Five' mass extinctions considered in Chapter 3, although it is worth reflecting on the nature of that recovery. Case studies profiled here (Chapters 3 and 4) illustrate the fact that there are many different dimensions of ecosystem recovery, including structure, function, composition and ecological interactions, which are not necessarily correlated. After the mass extinctions, for example, recovery of ecosystem functions was often more rapid than recovery of species richness and composition. The idea that different dimensions of both collapse and recovery might change at different rates is not explicitly considered by the RLE.

Although not included as part of the formal definition, Bland *et al.* (2017a) note that ecosystem collapse is judged to occur when native biota have 'moved outside their natural range of spatial and temporal variability in composition, structure and/or function'. As observed in some of the case studies profiled here, collapse could form part of the natural dynamics of an ecosystem, at least at the local scale. Collapse might only be only temporary, subsequently being followed by recovery. This presents a challenge to conservation managers, who will typically be making decisions at the local scale, where ecosystems may be described as 'locally collapsed' (Bland *et al.*, 2017a). At such scales it may be difficult to determine what constitutes the natural variability of the system and therefore whether the ecosystem has collapsed locally or not. In this context, the notion of persistent transitions is useful (Propositions 5 and 6, Table 5.1). In future it might be useful to develop definitions of ecosystem collapse specifically for implementation at the local scale, to support practical conservation management. Such definitions might usefully differentiate between transient and persistent changes and refer to the recovery ability of the ecosystems in question.

One of the problems with the word 'collapse' is that it is simultaneously both a noun and a verb, a characteristic it shares with a number of other words in the English language. This can create confusion regarding its intended meaning, an issue of which Dumbo was well aware (check out the lyrics to 'When I See An Elephant Fly' for a masterclass on exploiting the comedic potential of this ambiguity). In the current context, 'collapse' could refer either to an ecological process or the endpoint of that process. To illustrate, consider the ambiguity inherent in the statement 'I observed an example of ecosystem collapse' – what has been observed, an ecosystem in decline or one that has already collapsed? The RLE is explicit in limiting use of the word collapse to the state of an ecosystem (i.e. a noun), namely the endpoint of ecosystem decline. However, many authors also use the word to describe the process leading towards that endpoint. I have done this myself elsewhere in this book, for example when referring to 'collapse trajectories'. There is a reason for this: the word collapse, according to standard dictionary definitions (e.g. Cambridge Dictionary, 2019), implies an abrupt or rapid change. This is therefore an emotive term, which could generate fear or concern about impending ecological catastrophe. Increasing fear is a recognised approach in environmental campaigning (Chen, 2015), and it is no coincidence that reference to environmental 'collapse' has become increasingly popular in the international media in the past few years.

Presentation of the RLE as a process to evaluate ecosystem 'collapse risk' is therefore arguably misleading, as it implies that there is a risk of sudden change occurring. Yet the RLE clearly indicates that transitions to collapse can be gradual and linear, as well as sudden and non-linear (Bland et al., 2017a). The RLE criteria make little reference to the rate of decline, apart from defining a 50-year window over which a specified degree of change needs to have occurred, in order for an ecosystem to qualify as threatened with collapse (Criteria A, C, D and E; Bland et al., 2017a). At the same time, we know that some ecosystems, such as the Great Barrier Reef, really are declining rapidly, a situation that genuinely merits concern. This suggests that the RLE may be a rather coarse and insensitive tool for assessing the pace of change currently occurring in global ecosystems, as documented elsewhere in this book. In a conservation management context, addressing ecosystem decline at the local scale, the term 'collapse' can surely be of value in communicating both the pace of decline and its potential endpoint. Again, specific definitions of collapse might need to be developed for use in this context, to ensure consistency.

The RLE also makes interesting use of the term 'degradation', without ever defining it. Bland et al. (2017a) suggest that an ecosystem may be driven to collapse through multiple pathways, including trophic cascades, loss of foundation species and climatic forcing, as well as degradation. Other observers might classify these alternative pathways simply as different forms of degradation. Furthermore Criterion C of the RLE, which relates specifically to environmental degradation, limits this consideration only to abiotic components of the system, such as water and mineral nutrients. Criterion D, which relates to biotic aspects of degradation, refers instead to 'disruption of biotic processes and interactions'. As noted in Chapter 4, IPBES (2018) have produced an authoritative overview of land degradation that defines it both in terms of a persistent decline in biodiversity and ecosystem functions and a lack of full unaided recovery within decadal timescales. (The report also makes repeated reference to collapse, without ever defining the term.) In my mind this provides an excellent basis for developing an operational definition of ecosystem collapse for use at the local scale, including as it does reference to both persistence and lack of recovery. Perhaps an abrupt decline could be referred to as a process of collapse, and a system that has been degraded according to the IPBES (2018) definition could be considered to have collapsed.

Assessing the Risk of Ecosystem Collapse

As noted in the previous section, the IUCN RLE provides a framework for assessing the risk of extinction collapse, which could be of value in informing conservation policy and helping to prioritise ecosystems for conservation action. Let's now consider the assessment process in a little more detail. Classification of an ecosystem into one of the categories of collapse risk is achieved by application of five criteria (IUCN, 2016; see also Chapter 1), namely:

- *Criterion A. Reduction in geographic distribution*, over a period of 50 years either into the past or the future. Different categories are associated with different thresholds of loss, namely $\geq 80\%$ for Critically Endangered (CR), $\geq 50\%$ for Endangered (EN) and $\geq 30\%$ for Vulnerable (VU). Higher thresholds are defined for losses since 1750.
- *Criterion B. Restricted geographic distribution*. This employs two measures of geographic distribution also used by the IUCN Red List of Threatened Species, namely the extent of occurrence (EOO) and the area of occupancy (AOO). The latter is the number of 10×10 km grid squares occupied by the ecosystem, whereas EOO is the area of a minimum convex polygon enclosing all occurrences. Different

thresholds of these variables are defined for different categories, together with other subcriteria relating to declines in spatial extent, environmental quality and disruption to biotic interactions. For example, the thresholds of EOO for CR, EN and VU are 2,000 km^2, 20,000 km^2 and 50,000 km^2, respectively. The corresponding thresholds of AOO are 2, 20 and 50 grid cells.

- *Criterion C. Environmental degradation.* This focuses on change in an abiotic variable affecting a fraction of the extent of the ecosystem, over a 50-year period into the past or future. Thresholds of this fraction are again $\geq 80\%$ for CR, $\geq 50\%$ for EN and $\geq 30\%$ for VU. Higher thresholds are again defined for the extents affected since 1750.
- *Criterion D. Disruption of biotic processes or interactions.* This is precisely equivalent to Criterion C, except it focuses on biotic variables instead of abiotic variables. The same thresholds apply.
- *Criterion E.* Quantitative analysis estimating the probability of ecosystem collapse, over a period of 50 years. Here the thresholds of probability are $\geq 50\%$ for CR, $\geq 20\%$ for EN and $\geq 10\%$ for VU.

It is worth reiterating that the thresholds designed to separate the different categories are essentially arbitrary. Bland *et al.* (2017a) suggest that these values are based 'partly on theoretical considerations and partly on utilitarian considerations' but also note that available theory offers 'limited guidance' for setting their actual values. This is supported by the overview of available theory presented in Chapter 2. In fact, the threshold values were set at relatively even intervals 'to achieve an informative rather than highly skewed ranking of ecosystems among categories' – a 'utilitarian consideration', in other words. Clearly the thresholds, as well as the overall design of the criteria, were strongly influenced by those employed in the IUCN Red List of Threatened Species (RLTS) (IUCN, 2012).

Implementation of the RLE is still in its early stages, but a number of assessments have been conducted (e.g. see Bland *et al.*, 2017b, 2018b). At the global scale, the Aral Sea has already been designated as Collapsed, whereas another 10 ecosystems have been identified as threatened with collapse (Table 5.6). At the time of writing this book, 23 assessments had been conducted at the regional scale, which have identified eight ecosystems to be Critically Endangered at this scale, including raised bogs in Germany, reefs in Mesoamerica, Cape Flats and fynbos in South Africa, and five different ecosystems in Australia: Cumberland Plain Woodland, montane heath and thicket communities in the Eastern Stirling Range,

330 · Synthesis

Table 5.6 *Global assessments of the risk of ecosystem collapse, conducted as part of the IUCN Red List of Ecosystems*

Ecosystem assessed	Year	Risk category	Country
Aral Sea	2013	CO	Uzbekistan and Kazakhstan
Coorong lagoons and Murray Mouth inverse estuary, south Australia	2013	CR	Australia
Gnarled mossy cloud forest, Lord Howe Island	2015	CR	Australia
Gonakier forests of Senegal River floodplain	2013	CR	Senegal and Mauritania
Lake Burullus	2019	EN	Egypt
Southern Benguela upwelling ecosystem	2018	EN	South Africa
Coolibah – Black Box Woodlands	2013	EN	Australia
Giant kelp forests, Alaska	2013	EN	United States of America
Tapia forest	2013	EN	Madagascar
Tidal flats of the Yellow Sea	2015	EN	China, North Korea and South Korea
European reedbeds	2013	VU	Various
Antarctic shallow invertebrate-dominated ecosystems	2015	NT	Various
Tepui shrublands	2013	LC	Venezuela

Note: Categories: CO, Collapsed; CR, Critically Endangered; EN, Endangered; VU, Vulnerable; NT, Near Threatened; LC, Least Concern.
Source: From IUCN (2019).

Ironstone shrubland, the karst rising-spring wetland of SE Australia, and Mountain Ash forest (IUCN, 2019).

The IUCN Red List has not been without criticism. Assessments of the extinction risk of species have occasionally been controversial and overly political, and there have also been accusations that the results have sometimes been misused (Possingham, et al., 2002; Rodrigues et al., 2006). One might expect these arguments to continue, now that the Red List process has been extended to include ecosystems. However, I believe that the results speak for themselves. Until these assessments were conducted, many of us (including me) will not have heard of some of these ecosystems, let alone the fact that they are at risk (Table 5.6). Despite any concerns one might have about the rigour or objectivity of

5.2 Living with Ecosystem Collapse and Recovery · 331

the assessment process, it is surely important that these ecosystems are placed on the conservation map. Hopefully the RLE will continue to extend its coverage globally, as further assessments are urgently needed. Let's not forget though that its success will depend on individuals throughout the world being prepared to contribute their time to the assessment process; a lack of trained volunteers has limited the number of species listed on the RLTS (Newton and Oldfield, 2008).

The most significant criticisms of the RLE are those levied by Boitani et al. (2015). Some of these were considered in Chapter 1, where the following points were examined:

- there is no single way that ecosystems can be defined consistently;
- there are multiple possible endpoints for an ecosystem that is being degraded, and there may be no consensus on what is desirable or undesirable;
- it is difficult to quantify the natural range of variability in an ecosystem;
- the concept of collapse is both vague and ecosystem-specific; in practice it will often be necessary to define collapse separately for each ecosystem, using a variety of attributes and thresholds.

The empirical evidence presented in Chapters 3 and 4 support all these points. The last of them is perhaps the most problematic for the RLE, as it implies that comparisons of the risk levels between different ecosystems will be difficult, undermining the basis of a comparative assessment (Boitani et al., 2015). The criticisms don't stop there, though; Boitani et al. (2015) also evaluated the assessment criteria and reached the following conclusions:

- the theoretical basis for most of the criteria proposed is generally unsatisfactory or absent;
- there is no clear basis for applying EOO or AOO to ecosystems;
- the proposed quantitative thresholds are not supported by ecological theory in a way that can be applied consistently at all scales, and they have no justification other than a suggested analogy with the RLTS;
- the choice of variables to apply Criteria C and D is subjective, hindering comparison of risk level among ecosystems;
- the even intervals of thresholds used in the criteria assume a linear relationship between probability of collapse and values of the criteria, which has no empirical or theoretical support.

Overall, Boitani et al. (2015) conclude that there is a lack of a clear endpoint for collapse and a lack of a general theory linking ecosystem reduction (of area, components and functions) to collapse that is applicable at all scales and to all ecological systems. Consequently, the identification of scientifically robust thresholds for ecosystem decline is likely to remain very difficult. Again, it is difficult to contradict these conclusions; they are consistent with the evidence presented in this book. One might hope that the propositions listed on Table 5.1 might support development of relevant theory, but this endeavour clearly still has some way to go.

So, where does that leave the RLE? Boitani et al. (2015) suggest a number of potential ways forward, such as the use of vegetation types as a proxy for ecosystem types. Unfortunately, this would be of little help to the assessment of freshwater and marine ecosystems. A rather better suggestion was to clearly identify undesirable endpoints of ecosystem decline and clearly identify why these matter for conservation purposes. In other words, there is a need to identify the relationship between the trajectory of ecosystem decline and its endpoint(s). Furthermore, they recommend that the concept of collapse be standardised for the chosen units of assessment. Boitani et al. (2015) suggest that this might be achieved by focusing more on the loss of species and less on the loss of ecological processes. Yet for me, the real value of the RLE is that it moves beyond the traditional conservation focus on species to include loss of ecological processes and interactions. There is a powerful reason for conducting assessments of ecosystems: when an entire ecosystem collapses, for example when a coral reef bleaches then dies, something more than individual species is being lost. It is the ecological networks and interactions that are also important, and the potential loss of entire communities, which highlights the overriding importance of Criterion D to the overall assessment.

The fundamental problem identified by Boitani et al. (2015) is that the RLE has essentially duplicated the approach of the RLTS, without being able to base it on the kind of scientific foundations that underpinned the latter. On the plus side, this provides a clear rationale for future research on the decline and collapse of ecosystems, to help put those missing foundations in place; on the downside, time is running out. In the meantime, implementing the RLE as it currently stands is undoubtedly worthwhile, if only to learn about the practicalities of conducting an ecosystem-based conservation assessment. Hopefully the RLE will be able to evolve and improve over time through the engagement of a large community of researchers, a process that has been critical to the success of the RLTS.

There are other approaches that could potentially be used to evaluate the risks of ecosystem collapse. Ebenman et al. (2004) describe a technique referred to as community viability analysis, which involves a measure called the 'quasi-collapse risk'. This represents the probability that the number of species in a community falls below a particular value within a fixed period of time following the loss of a species (Ebenman et al., 2004). There are three major steps in a community viability analysis (Ebenman and Jonsson, 2005): (i) construct models of communities; (ii) investigate the response of the model communities to species deletions; and (iii) quantify the risk and extent of secondary extinctions. The analysis is typically conducted using an individual-based stochastic model (Ebenman et al., 2004), which may be relevant to conducting RLE assessments involving Criterion E.

Following the concerns raised by Boitani et al. (2015) about the RLE, assessments of collapse risk might be more usefully conducted at the local or landscape scales, where they might directly inform conservation management. At this scale, the methods described by Wilson et al. (2005) might be of value. These are approaches that have been developed to assess the vulnerability of ecosystems to environmental change, for incorporation in conservation planning. Mostly these methods focus on various forms of spatial analysis, which can be employed to estimate the spatial spread and potential impacts of different threatening processes. Other relevant approaches include spatially explicit models of ecosystem dynamics, dynamic species distribution and population models (Newton, 2007a) and the state-and-transition models considered in Chapter 1. Future research might usefully examine how such methods might be applied to provide a quantitative estimate of collapse risk at such scales. Spatially explicit models of ecosystem dynamics can also be used to estimate rates and locations of ecosystem recovery (Birch et al., 2010; Cantarello et al., 2011).

Early Warning of Collapse

Despite the extensive literature on ecosystem collapse, there are very few examples of field-based investigations that have actually documented collapse while it was happening, especially in terrestrial environments. Consequently, evidence of collapse is most often compiled after the event. Lindenmayer and Sato (2018) suggest that this is one of the reasons why it is so difficult to accurately predict if and when collapse might occur. These authors also point out the importance of developing robust early-warning indicators of collapse, to help detect it at an early stage and thereby prevent it from happening.

In a strikingly titled paper – 'Regime shifts in ecological systems can occur with no warning' – Hastings and Wysham (2010) provide an overview of the development of such early warning indicators. Typically, efforts have involved use of dynamic ecosystem models that include both slowly changing underlying conditions (i.e. slowly changing parameters) and some form of stochasticity. It has been suggested that there are three features of such models that might provide advance warning of a regime shift (Hastings and Wysham, 2010): (i) an increase in variance around the mean population size or some other measure, or (ii) an increase in skew, namely a change in the third statistical moment (which is a measure of the lopsidedness of a statistical distribution), or (iii) critical slowing down, which is a decreasing rate of recovery from small perturbations. Hastings and Wysham (2010) go further and suggest that these approaches are unlikely to succeed because ecological systems often exhibit complex dynamics. In other words, there may be a sudden change in system behaviour as a parameter is varied; under such conditions, the assumptions on which these three approaches are based are not valid. In their paper, Hastings and Wysham (2010) present models that they claim more realistically represent ecological systems than those typically used in indicator development, and in none of their simulations were the proposed early warning indicators actually found. For this reason, they conclude that drastic changes can appear in nature without warning. Rather than base management decisions on early warning indicators, it may therefore often be necessary to act extremely quickly in response to observations of any shifts occurring (Hastings and Wysham, 2010).

This paints a rather bleak picture regarding the prospects for developing early warning indicators. Fortunately, a ready antidote is provided by Scheffer et al. (2015) in their detailed review of the topic (although here they are referred to as indicators of ecological resilience, loss of which implies an increasing risk of a critical transition to a contrasting ecosystem state). Scheffer et al. (2015) conclude that there are many limitations of available indicators, but rather than finding the 'glass half empty', they take the optimistic view of finding it 'half full with many unexplored possibilities of filling it further'. This provides an interesting contrast to that of Hastings and Wysham (2010), for whom the glass was not so much half empty, as completely drained. One is left wondering which of these eminent scientists would make the most interesting drinking companions.

5.2 Living with Ecosystem Collapse and Recovery · 335

I have to confess, I found the optimistic slant adopted by Scheffer *et al.* (2015) to be highly infectious. Their review demonstrates that this is a very active area of research, and why not? If offers the prospect of linking a rich body of ecological theory to some of the most pressing challenges in practical conservation; it might even be considered as a Holy Grail of ecology (Burnett, 2019). It is therefore understandable that so many researchers have joined the quest. But is their optimism well founded? First, let us briefly highlight some key points made by Scheffer *et al.* (2015) in their review:

- Critical slowing down refers to situations close to the equilibrium state. In practice, ecological systems are mostly in transient states owing to high environmental stochasticity and/or intrinsic processes, so critical slowing down may often not apply.
- If stochasticity is high enough, the system may repeatedly visit different basins of attraction, a process termed as flickering (Scheffer *et al.*, 2009). This may be assessed by analysing the probability distribution of states or by estimating the mean exit time, which is the mean time taken for a system to leave the basin of attraction. Use of this latter measure has been limited in ecology to date.
- It might be possible to develop early warning indicators of cascading collapse by examining spatial tipping points. Spatial boundaries between two alternative stable states will usually be unstable; the boundary will tend to move towards the state with the smallest basin of attraction in a travelling wave. This can lead to the spatial spread of ecosystem transitions.
- Changes in resilience may be monitored using time series data by measuring indicators within a sliding window over time. However this requires long-term records sampled at temporal and spatial scales relevant to the dynamics of the studied system. Biases in the data, such as seasonal variation or longer-term trends, should be removed before analysis.
- Early warning signals of regime shifts can be difficult to obtain using time series data because of: (i) strong environmental noise; (ii) changes in the magnitude and type of environmental stochasticity; (iii) changes in the system from other external causes, such as temperature-driven changes in growth rates; (iv) rapid changes in environmental conditions relative to ecosystem response rates; (v) measuring the wrong variable, as some variables might provide weak signals of slowing down or no signal at all.

- Probabilistic estimates of the mean exit time, or the probabilities of switching to another ecosystem state in a given time span, can potentially be achieved by examining temporal dynamics. However the results can be biased because of: (i) heterogeneity of unobserved driving variables, which can lead to an overestimation of the range of hysteresis (e.g. the role of soil conditions influencing forest-savanna transitions, see Chapter 4); (ii) directionality in historical change in conditions.
- As there is no silver bullet, it is important to explore more than one indicator, observe how consistent results are and potentially combine different indicators to characterise resilience.

While the review provided by Scheffer et al. (2009) is full of suggestions and ideas, the main problem is that to date, only a small number of the theoretically predicted indicators have actually been tested using field data. To address this limitation, the authors point out the need for large-scale field studies with a good understanding of underlying mechanisms. They also highlight the potential offered by large data sets that are increasingly becoming available through technological developments in remote sensing and environmental monitoring. We noticed in Chapter 4 how remote sensing data have been used to assess tree cover (Hansen et al., 2013), which has been used to identify different ecosystem states in savanna. The probability of different states can then be related to potential drivers (e.g. Hirota et al., 2011) to produce a model that could be used to evaluate risks of potential collapse. Similarly, satellite data have been used to assess state transitions in large numbers of lakes (Kosten et al., 2012). However, different ecosystem states are not always easy to identify using remote sensing data (Nijp et al., 2019).

Practical guides to analysing resilience indicators using time series or spatial data are becoming available (e.g. Dakos et al., 2012; Kéfi et al., 2014; see also www.early-warning-signals.org/). However, it is important to remember that all these methods assume that dynamical systems theory applies to the field situation; in other words, that ecosystem transitions are driven by intrinsic feedbacks. As noted earlier (see also Chapter 4), there are many field situations where ecosystem collapses do not accord with this theory and where these indicators would therefore be unlikely to work. This point is examined by Dakos et al. (2015), who make it clear that early-warning indicators are not a panacea, citing cases from drylands (Bestelmeyer et al., 2013) and marine ecosystems (Lindegren et al., 2012) where indicators were found to be ineffective. Dakos et al. (2015) refer to a range of situations where a regime shift may not be associated with a critical transition driven by positive feedbacks, including abrupt changes in patterns of environmental drivers, or by combinations of slowly changing environmental and anthropogenic pressures, climatic events and complex internal biological processes. The latter situation may account for the majority of regime shifts observed in the North Atlantic and Baltic Sea (Dakos et al., 2015).

So when these indicators are applied to field data, are they typically effective or not? This is not so easy to answer directly, as many studies have used empirical data to parameterise or explore theoretical models. In other words, field data have not necessarily been used to test theoretical predictions; rather, a more blended approach has been adopted. Such studies do not necessarily attempt to test whether the underlying theory is relevant to the field situation; instead, it is often assumed to be valid. Some examples of recent investigations that have incorporated field data in their analyses are listed in Table 5.7.

Field evidence therefore suggests that early warning indicators are sometimes effective, but at othertimes not (Table 5.7). Evidence from lakes with reported regime shifts, for example, suggests that they are successful about half of the time. A number of recent reviews have examined the reasons for this variation in performance. Clements and Ozgul (2018) suggest that failures may often be attributed to the inherent complexity and low signal-to-noise ratios of ecosystems. They also note that previous research has tended to be biased towards those systems that are known to exhibit critical transitions, referred to as the 'prosecutor's fallacy' (Boettiger and Hastings, 2012), which may lead to their value to real-world conditions being overestimated. Furthermore, use of early-warning indicators to support practical conservation management is likely to be challenging, because of the trade-off between methodological simplicity and predictive accuracy (Clements and Ozgul, 2018).

In a further review, Spears et al. (2017) highlight another fallacy, namely the 'conjunction fallacy', which refers to statements that the co-occurrence of two events is more likely than either event alone. Apparently researchers have been guilty of this too, by assuming that sudden ecosystem-scale change has occurred in response to changes in an environmental stressor, without verifying that this has actually happened. Other suggested problems with previous research include a lack of consideration of temporal, spatial and ecological scales, and a failure to base the indicators on a mechanistic understanding of ecosystem function (Spears et al., 2017). These authors conclude that confidence in early-warning indicators is currently too low to support their wide-scale

Table 5.7 *Selected examples of recent research into early-warning indicators, which have employed the use of field data*

Finding	Study focus	Reference
Relatively simple characteristics of ecosystems that were observed directly (dissolved oxygen, pH and chlorophyll-a concentration) were better indicators of an approaching threshold than were the estimates of rates (gross primary production, respiration and net ecosystem production).	Temperate lake	Batt et al. (2013)
The utility of early warning indicators is often limited and can depend on the abruptness of a transition relative to the lifespan of responsive organisms and observation intervals.	Marine and semi-arid grassland	Bestelmeyer et al. (2011)
Variance and autocorrelation in abundance data consistently fail to predict non-linear change.	Lakes in northern Britain and the North Sea	Burthe et al. (2016)
Results demonstrate that spatial signals of approaching thresholds can be detected at the ecosystem scale.	Temperate lake	Cline et al. (2014)
Critical slowing-down indicators derived from time series data of biomass can signal the proximity of community collapse.	Plant-pollinator and seed-dispersal networks	Dakos and Bascompte (2014)
Changes in the coefficients of a correlation between compositional disorder and species richness anticipated regime shifts in fossil diatom and chironomid communities. Compositional disorder measures unpredictability in the composition of a community.	Chinese lakes	Doncaster et al. (2016)
Early warning indicators do not provide reliable and consistent signals of impending critical transitions even when using some of the best monitoring data available.	Freshwater ecosystems	Gsell et al. (2016)
A model based on two dynamical variables accurately predicted the occurrence of a tipping point, even in the presence of stochastic disturbances.	Plant-pollination networks	Jiang et al. (2018)

5.2 Living with Ecosystem Collapse and Recovery · 339

Table 5.7 (cont.)

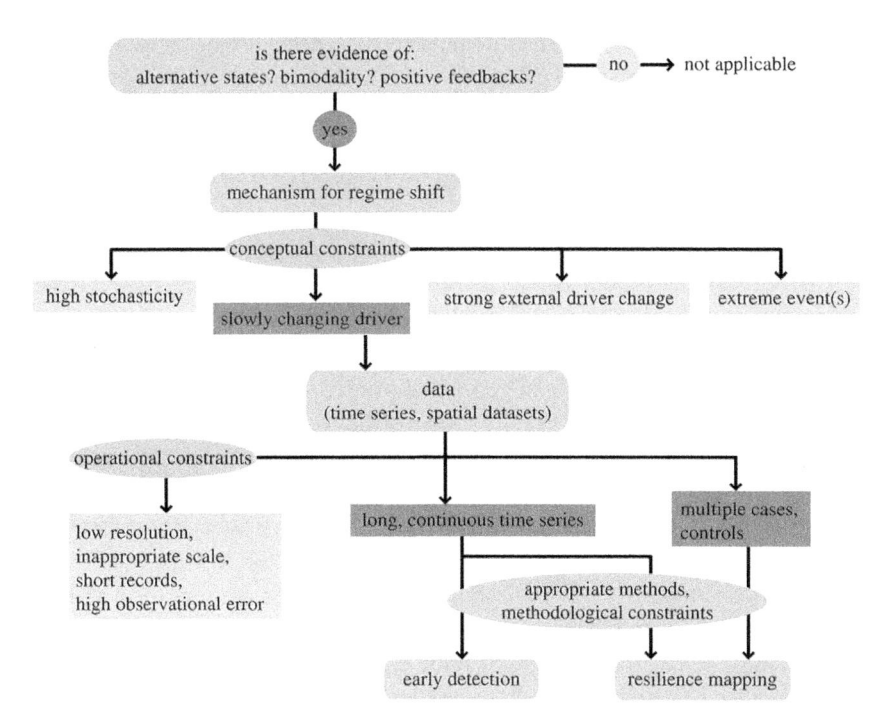

Finding	Study focus	Reference
Patch-size distributions may be a warning signal for the onset of desertification.	Semi-arid vegetation	Kéfi et al. (2007)
Flickering provided an effective indicator of a transition to eutrophic lake conditions.	Chinese lakes	Wang et al. (2012)

Figure 5.6 A flow chart indicating different considerations when applying early-warning indicators for the detection of critical transitions. Reproduced from Dakos et al. (2015), with permission.

practical application, which sounds as though the glass is half empty again.

Might there be alternative approaches to developing early-warning indicators, which are based primarily on analysis of empirical data rather than on theory and models? As noted by Boettiger and Hastings (2013), one way forward might be to focus on making predictions about real systems using readily available data. For example, Burthe *et al.* (2016) suggest that multivariate analyses of a suite of potential indicators might usefully be conducted, incorporating data relating to anthropogenic drivers and relevant ecological processes. Potentially, long-term monitoring data could provide insights into which ecological variables best capture the process of ecosystem decline and the relationship to its eventual endpoint. Lindenmayer and Sato (2018) propose a set of early-warning indicators for Mountain Ash forests in Australia, based on the results of their monitoring and research. These include: (i) rates of decline of key ecosystem structures (e.g. large, old trees), (ii) rates of decline of shorter-lived species dependent on these key ecosystem structures (e.g. arboreal marsupials) and (iii) the spatial extent of key ecosystem structures (e.g. stands of old-growth forest). Similar results were obtained by Evans *et al.* (2019) along gradients of forest collapse in the United Kingdom, where structural variables such as basal area were found to correlate strongly with ecosystem condition.

Although early-warning indicators of ecosystem collapse have received a great deal of research attention, indicators of recovery have been almost completely overlooked. Clements *et al.* (2019) examined the occurrence of indicators prior to the recovery of overexploited marine ecosystems using a trait-based ecological model together with analysis of real-world fisheries data. Results showed that both abundance and trait-based variables were effective as indicators, but when the two were combined they provided the best predictions of recovery. This research suggests that when aiming to support the recovery of ecosystems that have collapsed, indicators could be used to project the impact of conservation interventions, with potential implications for conservation policy and practice. Surely this is a topic worthy of further research.

Dakos *et al.* (2015) perhaps provide the best summary of the current state-of-the-art regarding early-warning indicators, which is so nicely worded that I have reproduced it here:

While the growing number of successful empirical examples highlights their potential, we still need to come up with innovative tools, take advantage of multiple sources of information, design novel monitoring programs based on new technologies, and most importantly, understand for which ecosystems or fields of research in general these approaches may prove most promising for application.

Even if they are shown to be effective,

early-warning indicators can tell us that 'something' important may be about to happen, but they do not tell us what precisely that 'something' may be and when exactly it will happen. Thus, next to the warnings, knowledge of the underlying ecosystem behaviour is important to put the signals in the right context. For this there is no substitute for site-specific knowledge and experiments combined with models.

Managing Collapsing Ecosystems

If collapse is the endpoint of ecosystem degradation or decline, then prevention of collapse will largely need to focus on addressing the causes of this decline. In this respect, the management approaches that are required are essentially the same as those that comprise effective conservation practice in any other context. My personal favourite contribution to the conservation literature is that of Salafsky *et al.* (2002), who set out to improve the practice of conservation through application of a framework, which can be considered as a series of steps: (i) identify the conservation target, in other words the specific area or population the conservation initiative is trying to influence; (ii) identify the threats affecting the target; (iii) identify the conservation actions needed to counter these threats, which might include approaches, strategies and specific tools; (iv) implement these actions through an adaptive management process. The concept of adaptive management involves the integration of design, management and monitoring approaches; the development and implementation of a monitoring plan is a crucial part of this process. The results obtained by monitoring enable conservation managers to learn what happened as a result of their actions so that they can then adapt and improve their management approach through a process of learning from both their successes and their failures (Salafsky *et al.*, 2002).

The framework developed by Salafsky *et al.* (2002) was actually part of a broader initiative launched by a community of nongovernmental organisations (NGO) called the Conservation Measures Partnership (CMP). Their aim was to establish standards of effective conservation practice and encourage the uptake of adaptive management approaches. So how have they fared so far? Redford *et al.* (2018) provide an update, which paints a rather mixed picture. While the value of adaptive management and the supporting framework appear to be widely recognised, the goal of establishing it as the standard operating procedure for the largest NGOs has not been fully achieved, partly reflecting the practical challenges of achieving fully adaptive management. Reasons for this failure cited in a survey of practitioners included the time it takes to move projects fully around the adaptive cycle, and a lack of funding, particularly over the timescales involved for full implementation (Redford *et al.*, 2018). Survey results also indicated a lack of funding specifically for monitoring and a failure to use the results of monitoring to inform management actions. These problems are widespread in conservation; it is well established that the monitoring of most conservation programmes is inadequate (Legg and Nagy, 2006; Newton, 2007a; Sanchirico et al., 2014; Yoccoz et al., 2001). As noted earlier, monitoring is of particular importance for providing early warning of ecosystem collapse; if an ecosystem is deteriorating rapidly, this will only be detected if frequent monitoring is being carried out.

Other initiatives designed to strengthen conservation practice have been developed in recent years (Schwartz et al., 2017), including evidence-based approaches to help identify effective actions (Sutherland et al., 2004), systematic conservation planning approaches to help prioritise locations for action (Margules and Pressey, 2000) and structured decision making to help choose between different management options (Gregory et al., 2012). While ecosystem collapse is not explicitly addressed by any of these initiatives, they all have a potential role to play in identifying appropriate management responses. CMP is distinct in having originated from within the conservation practice community rather than academia, so the approaches they have developed (referred to as Open Standards) are perhaps more likely to achieve buy-in from practitioners. Researchers and practitioners need to work together better, though, if the full potential of this approach is to be realised (Schwartz et al., 2012).

The CMP approach places identification of threats as a key step in identifying appropriate conservation responses. Clearly a wide range of different threats can contribute to ecosystem collapse. Salafsky et al. (2008) list 11 main categories of threat (see also Chapter 2), which have now been incorporated in the threat classification scheme used by IUCN and by the RLE. These are:

- residential and commercial development
- agriculture and aquaculture
- energy production and mining
- transportation and service corridors
- biological resource use
- human intrusions and disturbance
- natural system modifications
- invasive and other problematic species and genes
- pollution

- geological events
- climate change and severe weather

What is the relative importance of different threats, in terms of their impact on ecosystems? The most authoritative current statement is probably that presented by IPBES (2019), based on a systematic literature review of thousands of case studies, but using a somewhat different threat classification. Results indicated that land-use/sea-use change is the most important direct anthropogenic driver of environmental change, accounting for 30% of cases, followed by direct exploitation (23%), climate change (14%), pollution (14%) and invasive alien species (11%). Other threats (e.g. fire, human disturbances) account for the remaining 9%. When broken down by region, some differences were observed; for example, direct exploitation (30%) exceeded land-use/sea-use change (25.5%) in Africa, whereas in the Americas, these two drivers had similar values (23.5% and 25%, respectively). In the other regions, land-use/sea-use change was the most important driver. (Note that habitat fragmentation was not mentioned – have the zombie ideas we encountered in Chapter 4 been killed off?)

There has been no systematic assessment of the association between different threatening processes and the risk of collapse; this clearly merits further research. However, we noted in Section 5.1 that certain threats (notably fire and herbivory) can be associated with high risk of collapse because they can generate positive feedbacks with vegetation. It is also important to note that threats differ in terms of their impacts on ecosystems, referred to as 'stresses' by the CMP. Salafsky *et al.* (2008) classify stresses on ecosystems as follows:

- *Ecosystem conversion* – direct and complete conversion of the ecosystem, e.g., clear-cutting or flooding a forest, eliminating a stream, removing a coral reef.
- *Ecosystem degradation* – direct damage to an ecosystem's biotic and/or abiotic condition, e.g. selective removal of species, removal of top predators, altered fire or hydrological regime.
- *Indirect ecosystem effects* – indirect damage to an ecosystem, e.g. fragmentation or isolation of an ecosystem, impacts of a threat on the food resources of a target species.

The distinction here between ecosystem degradation and conversion is perhaps useful in the context of collapse, although conversion could

Table 5.8 *Some speculations regarding the impact of different threatening processes on ecosystems. The threats are those identified by IPBES (2019), in declining order of current importance.*

(a) Impact of threat on risk of different 'stresses' occurring (*sensu* Salafsky et al., 2008). For details, see text

Threatening process	Ecosystem conversion	Ecosystem degradation	Indirect ecosystem effects
Land-use/sea-use change	High	Low	Low
Direct exploitation	Low	High	High
Climate change	Moderate	Moderate	High
Pollution	Low	Moderate	High
Invasive alien species	Low	Moderate	High

(b) Relative magnitude of impacts of threats on different ecosystem components

Threatening process	Ecosystem structure	Ecosystem composition	Ecosystem function
Land-use/sea-use change	High	Low	Low
Direct exploitation	Low	High	Moderate
Climate change	Moderate	Moderate	High
Pollution	Low	Moderate	High
Invasive alien species	Low	Moderate	High

simply be seen as degradation that is complete. Different threats will be associated with different stresses; they will also vary in their impacts on different components of an ecosystem. In an attempt to identify some broad generalisations, one might speculate what these impacts might be (Table 5.8). Obviously, the impacts will vary with location, context and ecosystem type; will likely change over time; and will be affected by interactions between different threats. So these generalisations are likely to be of limited value. Nevertheless, they are offered here in the hope that they might sufficiently antagonise someone to conduct some proper research on the topic. Even if these suggestions are of little merit, the idea that threats differ in their impacts on ecosystems is surely valid, despite the lack of attention given to this issue in the literature. To support effective conservation management, there is a clear need to identify which threats and interactions are most likely to be associated with ecosystem collapse (see Propositions 2, 3 and 4 on Table 5.1, noting that threats are there referred to as disturbances).

One of the principal lessons I derived from compiling this book is the overriding importance of climate change, which regularly featured in the case studies profiled in Chapters 3 and 4, including all the 'Big Five' mass extinction events. I confess that until I read the literature presented here, I had considered climate change to be 'just another threat', which currently accounts for less biodiversity loss than other threats such as land-use/sea-use change – as indicated by the IPBES (2019) survey. I was not alone in thinking this. In a commentary in *Nature*, for example, Maxwell *et al.* (2016) said very much the same thing, encouraging conservationists to 'refocus their efforts on the enemies of old' rather than the emerging threat of climate change. Similarly, Noss *et al.* (2012) exhorted conservationists to focus on the 'greatest threat', namely land-use change; they suggested that the effects of climate change are currently 'dwarfed' by comparison. There have also been complaints that increasing focus on climate change has led to declining concern about biodiversity loss, which has led to a reduction in funding for conservation (Veríssimo *et al.*, 2014).

When viewed from the perspective of ecosystem collapse, however, climate change appears to be a uniquely powerful driver. One only needs to consider what is happening to the Great Barrier Reef to be convinced of this. But why is this? Why does climate change appear to have such a high capacity to collapse ecosystems? The answer doesn't seem to emerge from Table 5.8, so let's consider the issue further. First, as we saw in Chapter 4, it is partly an issue of scale; while most threats typically operate at the local or landscape scales, climate change can affect all the ecosystems in entire regions. Secondly, climate change is not just a single threat, but rather it comprises a suite of different drivers (e.g. total rainfall, rainfall distribution, mean temperature, maximum temperature etc.), each of which can individually influence different ecosystem attributes or mechanisms of ecosystem impact (Peters *et al.*, 2011). Thirdly, climate change can alter some of the abiotic components of an ecosystem – such as the availability, temperature or acidity of water. Few threats demonstrate this capacity. Fourthly, as we have seen, climate change can interact with all other threats. It can therefore act as a kind of meta-threat.

Furthermore, it is well established that species respond individualistically to climate change. Much of the research on climate change impacts on biodiversity has employed modeling approaches to simulate future changes in species ranges, extinction or abundance (Bellard et al., 2012). Correlative distribution models have proved particularly popular; these relate species' location data to contemporary climate variables and then extrapolate these relationships to project distributions under future climates (Urban, 2019). For example, when the distributions of some 80,000 species were projected under climate scenarios, results suggested that extirpations would double from 25% to 50% if projected temperature rises were to increase from 2°C to 4.5°C (Warren et al., 2018). Results of such studies have shown that life-history traits play a major role in determining the impact of climate change on species; traits that are particularly important include dispersal ability, reproductive rate, habitat and dietary requirements, as well as physiological tolerances (Pacifici et al., 2015). As illustration, Schloss et al. (2012) found that 87% of Western Hemisphere terrestrial mammals are expected to experience a reduction in range size as a result of climate change, with 20% of these species displaying range reductions owing to limited dispersal ability.

Empirical observations of both altitudinal and latitudinal range shifts also demonstrate differences between species in their rate and pattern of movement in response to climate change (Walther, 2010). As species do not respond synchronously, the patterns of dominance of species within communities will change with climate (Walther, 2010). In addition, non-analogue communities may develop, in which species occur in new combinations (Keith et al., 2009; Williams and Jackson, 2007). As we saw in Chapter 3, some of the communities that developed in prehistory have no modern analogues; an example is provided by the late-glacial plant communities of North America. These were closely linked to 'novel' climates that also have no modern analogue, characterised by high seasonality of temperature (Williams and Jackson, 2007). This provides a strong indication that future climate change may result in the formation of communities that are unlike any that are currently extant. Contemporary examples where non-analogue communities appear to be forming include Marion Island in the sub-Antarctic, and European rocky shores (Walther, 2010).

Despite most research focusing on what might happen in the future, such field evidence reminds us that the impacts of climate change are already being widely observed today. For example, in a meta-analysis of terrestrial organisms, Chen et al. (2011) demonstrated that species are moving towards higher latitudes at a median rate of 17 km per decade;

furthermore, population-level extirpations attributable to climate are now widespread (Cahill *et al.*, 2012; Pacifici *et al.*, 2017). Impacts on entire ecosystems are less well understood; it is only recently that research has begun to extend its focus beyond the responses of individual species to include interactions between them (e.g. Romero *et al.*, 2018). Scheffers *et al.* (2016) examined a set of 94 ecological processes, divided between terrestrial, marine and freshwater ecosystems, and found that 82% show evidence of impact from climate change in the peer-reviewed literature. These changes have occurred with just 1°C of mean warming. In her review of climate change impacts on tropical ecosystems, Sheldon (2019) noted that carbon storage began increasing in tropical forests towards the end of the twentieth century, which may have been driven by increased CO_2 concentrations. On the other hand, there is also evidence that the rate of increase in carbon accumulation in the Amazon is declining, perhaps because of increased climate variability (Brienen *et al.*, 2015). In their review for the IPCC, Settele *et al.* (2014) suggest that increases in the frequency or intensity of ecosystem disturbances such as droughts, wind storms, fires and pest outbreaks have been observed in many parts of the world; in some cases, these are attributed to the effects of climate change.

Given the scale and magnitude of projected climate change impacts, the challenges facing conservation managers seem very daunting. Addressing the causes of climate change requires collective global action, which is proving hard to achieve (Ross *et al.*, 2016). Lindenmayer and Sato (2018) provide some pragmatic guidance to conservation managers: focus on addressing threats at the local scale; targeting these threats may contribute to alleviating climate change as well. In particular, focus on breaking the feedbacks and interactions between multiple stressors, as these are key contributors to ecosystem collapse. For example, in the case of forest ecosystems such as the eucalypt forests of southern Australia or the Amazonian rain forest, management interventions that reduce the opening of the forest canopy help to break the interactive feedback processes among logging, fire and climate change (Lindenmayer and Sato, 2018). These authors also note that land managers may often not be aware of three key points: (i) the risk of ecosystem collapse is higher when threats interact; (ii) it can take a long time for ecosystems to recover from degradation caused by interacting threats; and (iii) relatively drastic management interventions may be required to avert ecosystem collapse (compared with a system with a single threat or with multiple threats that do not interact). These three points are strongly supported by the evidence presented in this book.

Lindenmayer et al. (2016) are perhaps the only authors to have explicitly considered what conservation managers might do in practice to limit the risks of ecosystem collapse. Based on their studies of eucalypt forests in Australia, they propose a framework comprised of the following 11 elements:

- *Carefully define ecosystem collapse.* As collapse is ecosystem-specific, so ecosystem-specific management actions will be required. Therefore it is necessary to define what constitutes collapse for a given ecosystem, relative to reference conditions.
- *Identify potential pathways to collapse.* This should be based on current understanding of the ecosystem and potential mechanisms of collapse. Identification of collapse pathways might usefully be informed by models of ecosystem dynamics (e.g. state-and-transition models; see Chapter 2).
- *Conserve biodiversity.* As biodiversity plays critical roles in ecosystem function, dynamics and stability, conserving biodiversity is a key part of reducing the risk of collapse.
- *Conserve key structural attributes of the ecosystem.* These may be those attributes that play keystone roles but are sensitive to change, are hard to replace or take a long time to develop.
- *Maintain a balance of seral stages.* Management should aim to maintain important life-cycle components of dominant species.
- *Consider potentially negative and nonadditive effects of multiple stressors.* The risks of collapse may be increased by interactions between multiple threatening processes, which should be factored into management plans.
- *Be aware of the potential risks facing highly simplified ecosystems.* Structurally simplified ecosystems, such as plantation forests, may support fewer species and may be more vulnerable to collapse.
- *Adopt a conservative approach to natural resource management.* The aim should be to avoid overharvesting of species, which can increase the risk of collapse.
- *Conduct long-term ecosystem research and monitoring.* This is to ensure that incipient collapse is detected at an early stage. Lindenmayer et al. (2010) highlight the value of both research and monitoring for detecting ecological surprises. Monitoring is also required to support adaptive management.
- *Ensure that ecosystem management has well-defined trigger points for action.* Trigger points are 'management thresholds' that instigate a change in management (Hunter et al., 2009), for example, if a particular proportion of an area is burned.

- *Conduct small-scale experiments and other studies to quantify the risks of ecosystem collapse.* Such approaches would help improve our understanding of the future impacts of known or novel disturbances.

This framework surely provides a useful starting point for the development of appropriate management responses to ecosystem collapse. However, it is interesting to consider to what extent it might be relevant to other types of ecosystem, or indeed to other types of forest. Some of these suggestions represent basic good practice, which many conservation managers are likely to be implementing already. The key challenge presented by collapse, though, is that ecosystem decline may occur abruptly and with little warning. If monitoring indicates that such a rapid decline is occurring, what then? Conservationists are beginning to consider this type of crisis situation. For example, Derocher *et al.* (2013) provide an overview of proactive management options for conservation of polar bears, which are facing catastrophic declines in habitat owing to the loss of Arctic sea ice. Here, preplanning, consultation and the need to coordinate management responses are identified as key priorities, together with advance consideration of the costs, legality, logistical difficulties and likelihood of success of different management options. This implies that scenario planning (Peterson *et al.*, 2003) might be of particular value for conservation managers faced with the possibility of ecosystem collapse. Some other approaches with potential value are listed in Table 5.9. An illustration of what might happen if these approaches fail is provided in Figure 5.7.

Supporting Ecological Recovery

In contrast to the situation with ecosystem collapse, management approaches for supporting ecosystem recovery are relatively well established. The science and practice of ecological restoration, which aims to facilitate ecosystem recovery, have developed rapidly in recent decades. Billions of dollars are now being invested annually in restoration actions throughout the world (Goldstein *et al.*, 2008), supported by international policy commitments such as the Convention on Biological Diversity. As a result, ecological restoration is now making a significant contribution to sustainable development (Nellemann and Corcoran, 2010). Practical guidance to implementing ecological restoration is now widely available (e.g. Clewell and Aronson, 2013) and is supported by the development of international principles and standards (Gann *et al.*, 2019).

Table 5.9 *Some potential management responses to the risk of ecosystem collapse*

Management response	Reference
Need to adopt a variety of complementary and synergistic approaches, including protected areas and conservation payment mechanisms; be adaptive, learn by doing and embrace innovation.	Barlow et al. (2018)
Allow for variation, learning and flexibility while observing change; avoid actions that decrease variance over short timescales.	Carpenter et al. (2015)
Define clear objectives, use scenarios, emphasise pattern analysis and ensure greater scope for creative and decentralised decision making.	Game et al. (2014)
Management interventions will increasingly need to be decided on quickly without full understanding of the ecological consequences. Focused consideration and planning, and cross-disciplinary dialogues, are needed for successful mitigation strategies.	Harris et al. (2018)
Model-based analysis of mutualistic networks suggests that maintaining the abundance of some species at a constant level can prevent sudden drastic changes in the system.	Jiang et al. (2019)
Manage both external and internal threats to protected areas; establish sizeable buffer zones around reserves, maintain substantial connectivity between reserves and promote lower-impact land uses near reserves by engaging and benefiting local communities.	Laurance et al. (2012)
Create a strong evidence-based science–stakeholder dialogue concerning the importance of regime shifts at multiple scales, to identify the actions required to avoid them and to implement appropriate restoration and adaptation measures if those shifts occur.	Leadley et al. (2014)
Reduce uncertainty about species responses to management.	Nicol et al. (2019)
Increase resilience of key variables to drivers of change, for example by maintaining diversity and managing connectivity; identify thresholds of potential concern; monitor slowly changing drivers.	Pace et al. (2015)
Understanding driver–state relationships likely to influence an ecosystem are critical to identifying causes of abrupt state changes. Field studies that relate ecosystem state to a range of driver conditions over time and space can test whether current patterns are consistent with expected driver–response relationships over time. Long-term data are particularly valuable. Empirical studies of abrupt change must carefully consider alternative mechanisms for observed changes (i.e. changes in driver means, driver variability, disturbance regimes and/or interactions among these processes).	Ratajczak et al. (2018)

Table 5.9 (cont.)

Management response	Reference
Design management strategies that target bundles of drivers instead of independent variables, in order to avoid cascading ecosystem collapses	Rocha et al. (2018)
Substantial increases will be needed in established conservation practices and policies, such as protected areas, as well as proactive approaches such as national land-use planning and crop-yield increases that reduce both habitat fragmentation and the demand for land clearing.	Tilman et al. (2017)
Waiting for a signal that the threat of a regime shift is imminent may very well be a fool's errand, given what we now know about ecological dynamics. Approaches based on underlying uncertainty, such as the 'precautionary principle', should not be abandoned.	Vandermeer (2011)

Techniques and guidance are therefore available to enable ecological restoration, and these approaches are being widely implemented, but are they effective? Results from meta-analyses provide evidence that ecological restoration actions can often be at least partly effective in enabling ecosystem recovery (e.g. Crouzeilles et al., 2016; Meli et al., 2017; Rey Benayas et al., 2009). For example, following restoration actions in China, Huang et al. (2019) reported mean increases of 30%, 73% and 48% for biodiversity measures of vascular plants, soil microorganisms and soil invertebrates, respectively, based on a meta-analysis of 103 publications. Similar values have been obtained at the global scale; Rey Benayas et al. (2009) reported a mean increase of 44% in biodiversity measures following restoration actions across a range of different ecosystem types.

Evidence presented in Chapters 3 and 4 provides support for a number of propositions relating explicitly to ecosystem recovery (i.e. Propositions 4b, 6, 12 and 14; see Table 5.1). These have a number of implications for management practice. Ecological restoration is conceptually simple – if the causes of environmental degradation are removed, then the ecosystem should recover naturally. In reality, the situation is far more complex; as highlighted in the two previous chapters, there are many examples where recovery has been slow or non-existent following restoration actions. In many cases, a lack of recovery can be attributed to ongoing press disturbances (Proposition 4b); other potential factors

1. Current (a) and projected (b) cross-sectional profile of a transect across central South America

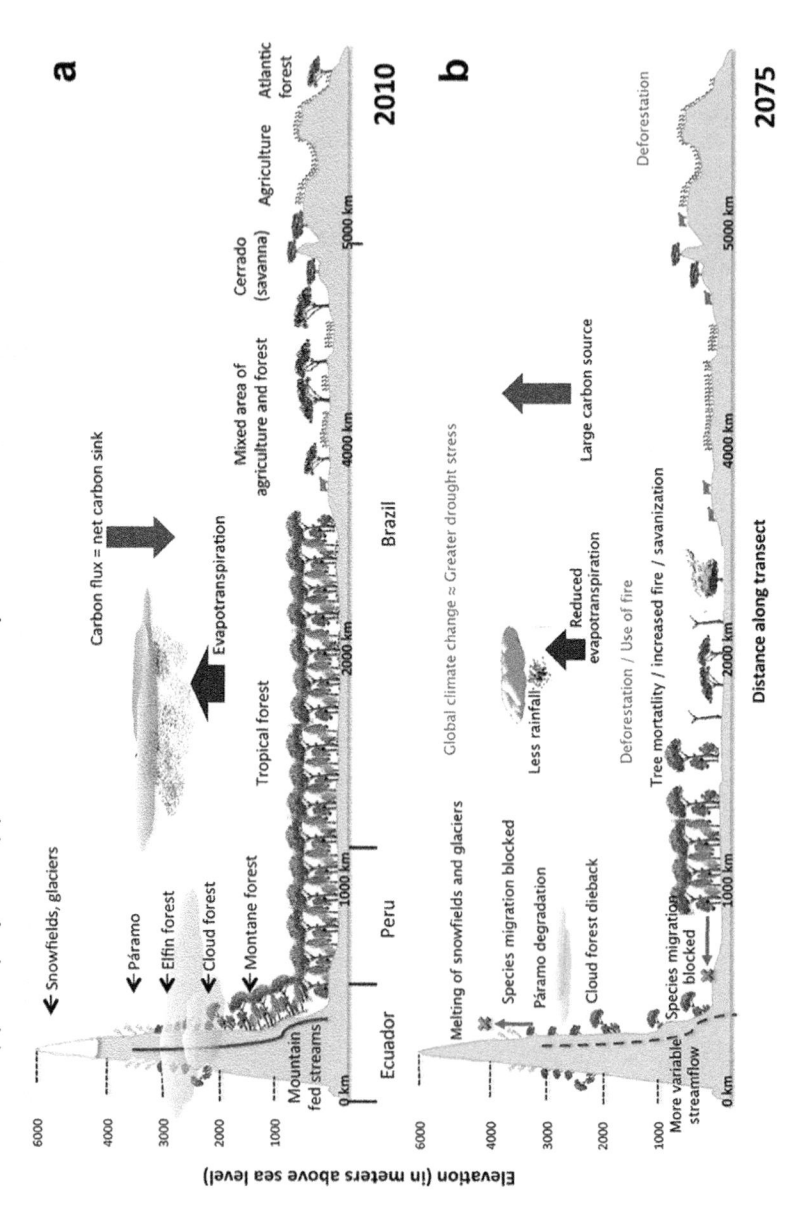

2. Current (a) and projected (b) cross-sectional profile of a transect across the Indo-pacific region

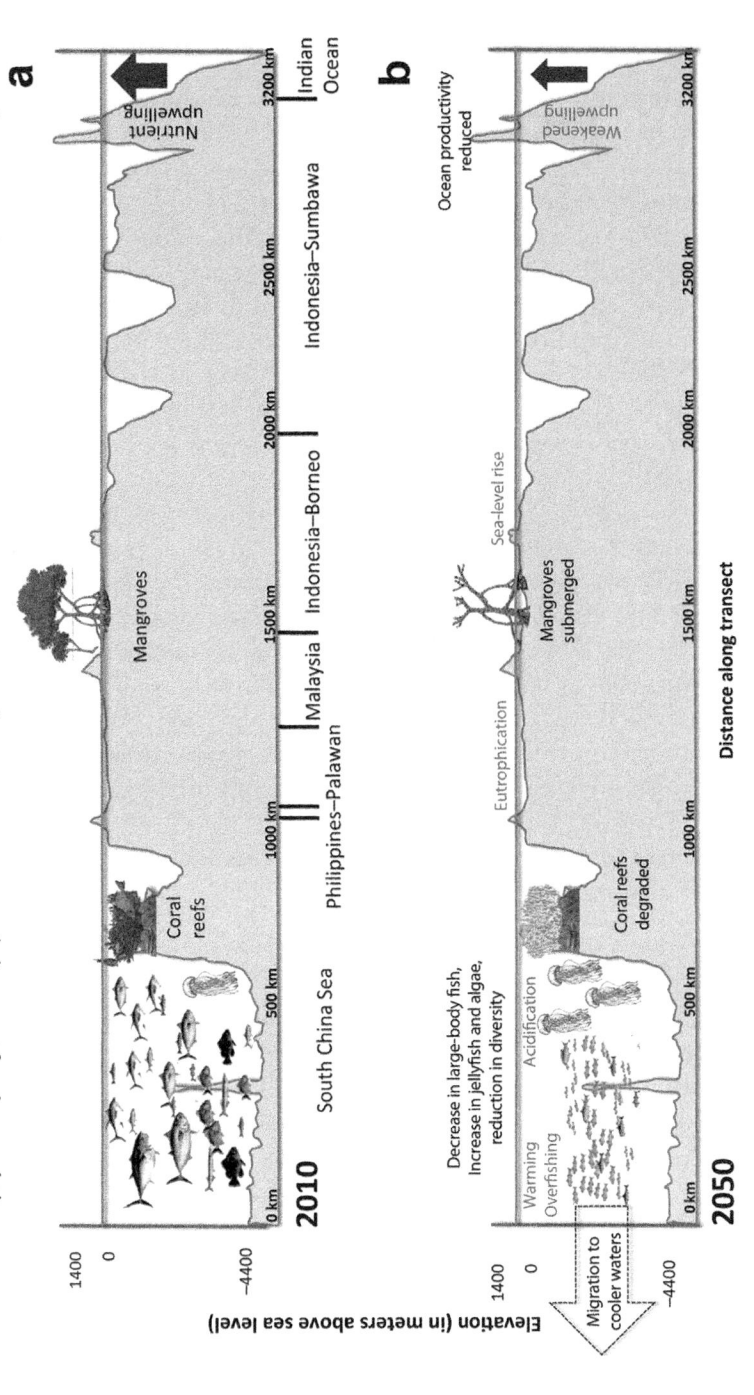

Figure 5.7 Plausible scenarios of ecosystem change as a result of high and rapidly growing rates of resource exploitation, land-use change and climate change (i.e. business-as-usual scenarios). Major drivers of regime shifts are indicated with paler text. Reproduced from Leadley *et al.* (2014), with permission from Oxford University Press.

include stabilising feedback mechanisms and other intrinsic processes (see Section 5.1; Propositions 6 and 12). In practice, the reasons for a lack of recovery are often unclear, and this presents a significant challenge to restoration management. Research may therefore be required to identify the mechanisms preventing ecosystem recovery, for effective restoration interventions to be identified.

In her review of ecological restoration activities, Suding (2011) notes that few systematic assessments of restoration success have been undertaken; this partly reflects the fact that relatively few restoration projects quantitatively monitor project outcomes. It is therefore difficult to generalise about causes of failure. Suding (2011) suggests that many cases of non-recovery may be the result of feedbacks that thwart restoration actions, citing the example of seagrasses in the Dutch Wadden Sea as an example. Despite many restoration attempts the seagrasses have never recovered, apparently owing to a negative feedback between turbidity and the presence of the plants (van der Heide *et al.*, 2007). In their survey of 240 restoration studies, Jones and Schmitz (2009) found that 30% reported no recovery of any measured variable. This lack of success was attributed to insufficient duration of monitoring in about half of the cases. A much lower proportion (5%) of studies provided evidence of a shift into a persistent alternative state, which may have prevented recovery. Other failures were attributed to a lack of suitable reference sites or pre-perturbation data with which to evaluate success.

On the basis of her review, Suding (2011) identifies some recommendations to help improve restoration success:

- *Evaluate project outcomes.* Knowledge of how to achieve successful ecosystem recovery is being hindered by the general lack of assessment and exchange of information regarding project outcomes.
- *Move beyond demonstration to decision-making science.* Conservation decision making requires prioritisation: deciding where and which processes are most important. It is important to invest in restoration measures that address the most influential controls on ecosystems. Such decision-making can be informed by experimental research or multivariate analysis.
- *Provide evidence-based assessments.* Quantitative evidence-based practice reviews are rare in restoration but are important for understanding when and why a given technique or approach is successful.

- *Form partnerships.* In particular, there is a need for greater collaboration between practitioners and scientists, to help identify the mechanisms of ecosystem recovery and to target restoration actions accordingly.

Some further suggestions are presented in Table 5.10.

Although ecological restoration might appear to be relatively straightforward conceptually, identifying the best way of approaching it has proved to be surprisingly contentious. This partly reflects the profound implications of environmental change, with climate change in particular. We noted in Chapter 1 the recent debate associated with the goals of restoration, given increasing recognition that many ecosystems are currently being transformed into new systems that have no historical analogue (Hobbs *et al.*, 2009; Keith *et al.*, 2009). As a consequence, it is often not possible to specify a single endpoint of ecosystem recovery. Traditionally, the goals of ecological restoration have often focused on restoring historical assemblages of species, but this increasingly appears to be untenable in a world affected by climate change. Instead, a focus on novel ecosystems has been proposed (Hobbs *et al.*, 2006, 2009).

This debate is still ongoing. Higgs *et al.* (2018a) note that rapid environmental change is shifting the scope of traditional restoration, just when it is becoming popular as a 'go-to approach' for addressing environmental challenges. These authors express concern about overly restrictive definitions of restoration becoming incorporated into environmental policy, citing the UNCCD as an example. Specifically, Higgs *et al.* (2018a,b) suggest that restoration definitions need to be broad enough to encompass ongoing and future change (i.e. including novel ecosystems), rather than solely focusing on the recovery of native ecosystems that are present in a given area. Conversely, Aronson *et al.* (2018) suggest that no changes are needed to existing definitions of restoration. Rather, they see an ongoing need to differentiate between ecological restoration, with its focus on recovery of species composition and community structure, from other approaches (such as rehabilitation) that focus on restoring ecosystem function to improve the provision of benefits to people.

At first glance, this argument over definitions may appear to be another example of what Governor Tarkin might have referred to as pointless academic bickering. Yet the point is an important one. As we saw in Chapter 4, there has been similar conflict over the definition of

Table 5.10 *Selected suggestions regarding how to successfully support ecosystem recovery through ecological restoration actions*

Suggested action	Ecosystem type	Reference
Integrate restoration action with conservation management and sustainable development initiatives.	All	Aronson et al. (2006)
Understand ecosystem function (e.g. physical and biological conditions required for the ecosystem to thrive); remove anthropogenic stressors (e.g. pollution, eutrophication, altered hydrology, physical damage); use clearly defined criteria for the measurement of restoration success; undertake long-term monitoring of 15–20 years; involve a broad range of stakeholders, including local communities, in the decision-making process.	Marine (coastal)	Bayraktarov et al. (2016)
View restoration as an integrated socio-ecological system.	Forest	Chazdon and Brancalion (2019)
Principles and attributes of ecological restoration, originally formulated for terrestrial and coastal ecosystems, can be applied to the deep sea. The scope for unassisted ('passive') restoration should be assessed for each type of deep-sea ecosystem; practices can be developed to facilitate this 'natural', relatively low-cost restoration approach. Governance should be in place to protect restored areas against new damage.	Marine (deep sea)	Van Dover et al. (2014)
Restoration schemes need clearly defined target states. They should take a process-oriented and stepwise adaptive management approach judging success against reference or control sites. Societal and political expectations need to be managed, and restoration schemes should not promise too much. Even minor rehabilitation of degraded ecosystems can be of value. Sometimes 'ersatz' ecosystems are better than nothing and the best that can be achieved, especially in urban settings.	Freshwater	Geist and Hawkins (2016)

Table 5.10 (cont.)

Suggested action	Ecosystem type	Reference
The ecology of the target ecosystem should be established, the need for restoration carefully assessed and the outcome properly monitored. Management will also need to cope with unpredictability, seek to maintain connectivity in time and space, assess functionality, manage conflicting interests and social restrictions and ensure adequate funding.	Forest (boreal, European)	Halme et al. (2013)
Achieve bottom-up engagement of stakeholders to set restoration goals tailored to regional ecological and socioeconomic conditions; develop, evaluate and manage restoration practices that are cost-effective and practical at large scales.	Forest (tropical)	Holl (2017)
Conduct rigorous analyses of the costs and benefits of restoration using economic tools and principles.	All	Iftekhar et al. (2016)
Major drivers of recovery include the reduction of human impacts, especially exploitation, habitat loss and pollution, combined with favourable life history and environmental conditions. Awareness, legal protection and enforcement of management plans are also crucial.	Marine	Lotze et al. (2011)
Crucial ecosystem functions can be maintained through a range of fisheries restrictions, allowing coral reef recovery that meets conservation and livelihood objectives in areas where marine reserves are not feasible solutions.	Coral reefs	MacNeil et al. (2015)
Managers should critically diagnose the stressors impacting an impaired stream and first invest resources in repairing those problems most likely to limit restoration. While many projects focus primarily on physical channel characteristics, this is not a wise investment.	Freshwater (streams and rivers)	Palmer et al. (2010)

(cont.)

Table 5.10 (*cont.*)

Suggested action	Ecosystem type	Reference
The scale of environmental changes requires the adoption of a landscape perspective in restoration, which requires coordination of, and cooperation among, multiple landholders and managers, each with potentially conflicting goals and approaches. Include increasing connectivity among restoration goals.	All	Perring *et al.* (2015)

what comprises a forest, stemming from concern about the negative environmental and socioeconomic impacts of industrial forest plantations. This is a conflict that is intensifying, as major oil companies including Eni and Shell are currently investing in major afforestation programmes as a way of offsetting their carbon emissions, an approach that has been attacked as greenwash (REDD Monitor, 2019). As noted in Chapter 4, afforestation can actually result in ecosystem collapse rather than recovery. Any loosening of restoration definitions in environmental policy could therefore provide a mandate for pursuing environmentally damaging land cover change. An illustration of the type of thing that can happen is presented by Newton (2016), who shows how in the United Kingdom, introduction of non-native tree species into ancient woodland ecosystems has recently been included among environmental policy objectives. This represents a threat to native ecosystems, and could contribute to their collapse, but has been justified in terms of increasing their resilience to climate change.

Despite such risks, Hobbs *et al.* (2009) are surely right to suggest that current rates of environmental change have profound implications for restoration practice. Specifically, these authors consider situations where biotic and/or abiotic changes have forced a transition to a novel ecosystem. In some cases, as we observed in Chapter 4, recovery from an altered to a less modified state may be difficult because of thresholds or feedbacks. When an ecosystem has been altered to such an extent that it is considered novel, traditional restoration outcomes are therefore unlikely. In this situation, Hobbs *et al.* (2009) suggest that a range of other management options nevertheless exist that could result in beneficial outcomes for biodiversity. Management goals might need to recognise the potential value of novel ecosystems and focus on maintaining or enhancing them. The focus of management might also need to shift from damage control to ecosystem engineering or development of 'designer ecosystems'.

It is not surprising that the concept of novel ecosystems has attracted so much opprobrium. Transformation into a novel ecosystem is actually equivalent to a form of ecosystem collapse; therefore, Hobbs et al. (2009) are essentially advocating acceptance of such collapse among conservation goals. According to Hobbs (2013), ecological restoration is essentially futile where ecosystems have been significantly transformed and is largely motivated by sentimentality and grief about what is happening to the environment. Unsurprisingly, the suggestion that restorationists are engaged in a futile practice owing to their 'psychological impairment' has not been well received (Murcia et al., 2014). In response, concepts of novel ecosystems have been accused of being ill-defined, based on faulty assumptions and driven by a 'hubristic, managerial mindset' that will lead to undesirable outcomes in ecosystem management, such as a 'domesticated Earth' (Aronson et al., 2014; Murcia et al., 2014). Rather, these authors argue that it is too early to give up on ecological restoration. We do not know for any ecosystem what the point of 'no return' actually is; information is lacking on ecological thresholds that might prevent recovery (Miller and Bestelmeyer, 2016). In fact, what limits restoration is usually not an ecological factor, but the financial cost and the political will to meet this cost (Murcia et al., 2014).

So what should restoration practitioners do in the face of environmental change? Murcia et al. (2014) recommend the following:

- Avoid using the term 'novel ecosystem'.
- Increase efforts to avoid further loss of biodiversity and ecosystem function, by conserving and restoring ecosystems in new and more effective ways.
- Do not abandon inherent biodiversity values in favour of 'functional values' such as provision of ecosystem services.
- Consider the economic costs of restoring together with the socio-economic costs of not restoring when making management decisions.

Alternatively, rather than abandon the novel ecosystem concept entirely, Miller and Bestelmeyer (2016) suggest that it can usefully be integrated into existing restoration frameworks (Figure 5.8). These authors suggest that the concept is useful because it does not equate 'altered' with 'degraded'. In other words, ecosystems that have been transformed by human action can still have value; the novel ecosystem concept therefore allows managers to define and justify alternative management goals when

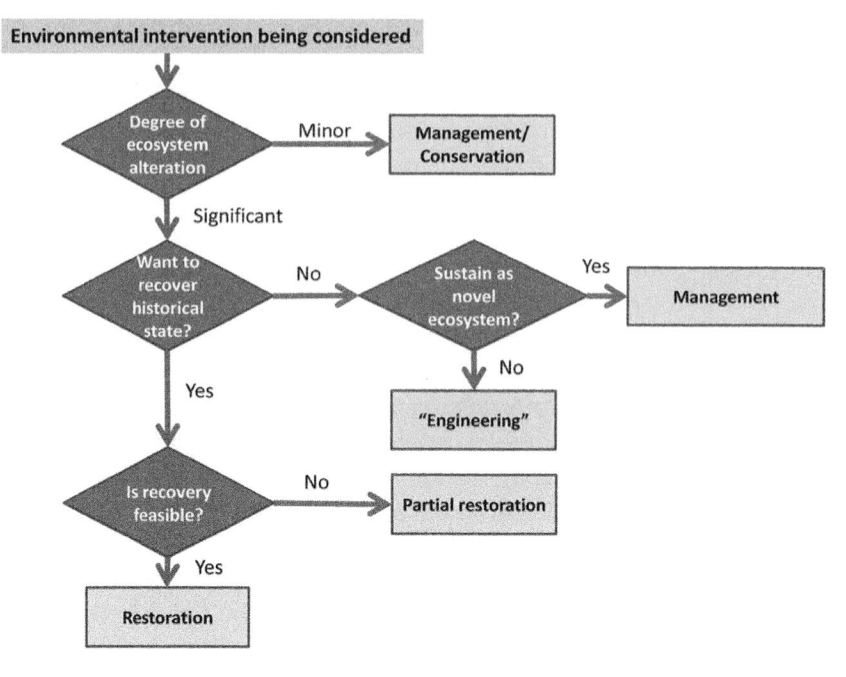

Figure 5.8 Management decision tree integrating elements of the novel ecosystem concept with a range of restoration alternatives. Reproduced from Miller and Bestelmeyer (2016) with permission from Wiley, © 2016. Society for Ecological Restoration.

restoration to a historical target is not practicable or desirable (Miller and Bestelmeyer, 2016). They find less favour with the suggestion that novel ecosystems are associated with irreversible thresholds and transitions to an alternative stable state, ideas that (needless to say) originate from dynamical systems theory (Hobbs *et al.*, 2006, 2009).

Whatever one's degree of psychological impairment, it is challenging to develop effective guidance for ecological restoration when baselines are shifting and the climate is changing. Even if all other anthropogenic pressures are removed, potentially enabling recovery to occur, ecosystems will still have climate change to contend with. The recent revision of the SER principles and standards (Gann *et al.*, 2019) represents an attempt to address this challenge. Interestingly, novel ecosystems aren't mentioned at all in this document. In the face of changes in community composition that are anticipated as a result of climate change, management interventions are encouraged that: (i) optimise genetic diversity and the potential for populations to adapt; (ii) prevent extirpations from current habitat areas; and (iii) promote migration of species to new areas. The problem of 'insurmountable environmental change' is also highlighted, specifically in relation to identification of target or reference ecosystems. Project managers are encouraged to consider using alternative native ecosystems as restoration goals in areas where the degree of change has been 'substantial and insurmountable' (e.g. when there have been hydrological changes from saline to fresh water, or the converse). Gann et al. (2019) also offer some words of hope, noting that successful restoration has sometimes been achieved in areas where it was assumed by some to be impossible (a veiled attack on the novel ecosystem concept, perhaps?). If the potential for recovery is in doubt, but recovery is highly desirable, Gann et al. (2019) make the excellent suggestion of conducting small-scale management trials to evaluate their effectiveness.

As Corlett (2016) puts it, the most important restoration debates in coming years will probably not be about baselines, targets and techniques, but about if and when to intervene. Managers may need support to make this type of decision, for example through the development of appropriate decision criteria. One particularly interesting approach in this context is rewilding, which typically focuses on limited intervention by allowing natural processes to predominate (Corlett, 2016). In the United Kingdom, the idea received significant impetus following the publication of George Monbiot's influential book *Feral* (Monbiot, 2013); in many parts of Europe, abandonment of agricultural land is providing large-scale opportunities for ecological recovery through rewilding (Navarro and Pereira, 2012). In such situations, might ecosystem recovery best be achieved simply by letting nature take its course? Or is rewilding actually a Pandora's box (Nogués-Bravo et al., 2016)? If the latter, let us remember what was left in Pandora's box after all the evils had escaped: hope.

Implications for People

After surfing this swathe of the ecological literature about ecosystem collapse, we now come to the crunch question: *does it really matter?* In other words, if an ecosystem collapses, what are the implications for people? Might it lead to the collapse of human society? Conversely, could ecosystem recovery lead to improvements in human well-being? These are big questions, which potentially have implications for all of us. Given this, it is surprising that they haven't attracted more attention from scientists. There is surely tremendous scope for further research in this area.

The ecological power couple Ann and Paul Ehrlich have never been afraid to tackle big issues head-on, and this is no exception. Their recent paper entitled *'Can a collapse of global civilization be avoided?'* (Ehrlich and Ehrlich, 2013) reaches the stark conclusion that such a collapse appears likely; the odds of avoiding it this century are deemed to be low. Although their publication doesn't refer to ecosystem collapse explicitly, environmental degradation is clearly identified as a key factor increasing the risk of societal collapse. Specifically, in addition to climate change, mention is made of accelerating extinction of animal and plant populations and species, which could lead to a loss of ecosystem functions and services essential for human survival; land degradation and land-use change; ocean acidification and eutrophication (dead zones); and the global spread of toxic compounds (Ehrlich and Ehrlich, 2013). These are all themes that have appeared repeatedly throughout this book. Perhaps the only grain of comfort that can be extracted from this paper is that the record of these authors as soothsayers has been somewhat patchy, to say the least (Sabin, 2013; Simon, 1982). The Ehrlichs are not the only ecologists to consider the imminent demise of human society, however. For example, Matthews and Boltz (2012) similarly ask 'Are we doomed yet?' and reach only slightly less depressing conclusions, based primarily on an assessment of trends in resource use. Newbold *et al.* (2016) indicate that land-use change and related pressures have reduced biodiversity values beyond a proposed planetary boundary across 58.1% of Earth's surface. Although these authors stop short of suggesting that this will contribute to societal collapse, according to Steffen *et al.* (2015) transgression of planetary boundaries could drive the Earth system into a much less hospitable state for human societies.

It's important to note that the Ehrlichs' prognostications of doom focus on the collapse of civilisation and not on the extinction of *Homo sapiens* as a species. The same is true of Jared Diamond's best-selling book, *Collapse* (Diamond, 2005) (see McAnany and Yoffee, 2010a for an extensive critique of the latter text). As Ehrlich and Ehrlich (2013) note, over-exploitation of the environment has long been thought to have contributed to the collapse of different civilisations throughout human history, such as those of Easter Island, the Classic Maya and those of Egypt and China. Indeed, as they say, virtually all civilisations have collapsed eventually. Such collapses typically refer to the abrupt decline of sociopolitical organisation (e.g. governance and power structures) rather than declines in human population size. Although these two are sometimes linked, often they are not (McAnany and Yoffee, 2010a,b).

In his review of five Old World case studies drawn from Egypt and Mesopotamia, Butzer (2012) found that environmentally triggered collapse of civilisations in Egypt tended to be relatively short-term and have limited demographic impact, whereas in Mesopotamia the demographic impacts were sometimes catastrophic. Overall, poor leadership, incompetence, administrative dysfunction and institutional failure (e.g. corruption and loss of economic networks) were found to be consistently associated with civilisation collapse, with war or climatic perturbations sometimes serving as trigger mechanisms. These factors are spookily relevant to the modern world. However, environmental degradation was not found to be a consistent cause of societal collapse in these cases (Butzer, 2012).

There is good evidence that some civilisations collapsed because of climate change (Weiss and Bradley, 2001). For example, collapse of the Classic Mayan civilisation in Central America, the ancient Pueblo (the 'Anasazi') in south-western USA, the Akkadian in Mesopotamia, the Moche in Peru and the Norse settlements in Greenland have all been attributed to climate change as a primary cause, especially extreme events like droughts (Caseldine and Turney, 2010). However, evidence for societal collapse being driven by environmental degradation is less clear-cut. Examples that have been suggested include Late Bronze Age Greek polities, Chaco in the American south-west, the Classic Maya and Mesopotamian Ur III. Tentative evidence is also available for the European Neolithic (Downey *et al.*, 2016). The best-known example of collapse is probably Rapa Nui (Easter Island), which is believed to have undergone a demographic, ecological and cultural disaster. Consistent factors in these different examples are deforestation and soil erosion and sometimes increased salinisation from irrigation and agricultural intensification (Middleton, 2012).

Middleton (2012) provides a detailed review of archaeological evidence of societal collapse and examines the widespread view that complex societies tend to degrade and outstrip their local ecosystems, in effect committing 'ecocide'. Overall, he concludes that neither climate change nor environmental degradation is usually the sole or direct cause of societal collapse. Similarly, McAnany and Yoffee (2010a) suggest that environmentally driven societal collapse is rare; Lawler (2010) reaches the same conclusion. Even the celebrated example of Rapa Nui has been challenged by new evidence (Rull *et al.*, 2013). According to Middleton (2012), many societal collapses can best be seen as the consequences of conflicts between and within different groups of people, which can be triggered or exacerbated by a wide range of factors, including those related to environmental change. For Costanza *et al.* (2007) collapse is just one of many alternative ways in which a society might respond to environmental change; for example, extreme drought has triggered social collapse in some cases, but technical innovation (e.g. irrigation) in others. One of the most striking lessons from the archaeological record is just how resilient and adaptable people are when confronted with environmental change (Middleton, 2012; McAnany and Yoffee, 2010a). Nevertheless, a number of potential mechanisms of societal collapse have been identified (Cumming and Peterson, 2018; Table 5.11).

What about the examples of prehistoric collapse considered in Chapter 3? It is a fascinating question to consider how our ancestors were affected by the extinction of megafauna in the late Quaternary. The evidence is of course limited, but Broughton and Weitzel (2018) have attempted to reconstruct both megafaunal and human population sizes in North America. Results provide some suggestion that human numbers may indeed have declined locally or regionally in the Younger Dryas, a period of rapid climatic cooling around 12,000 years ago, which was coincident with some megafaunal extinctions. However, in none of the cases they examined did humans become even locally extinct. In South America, humans initially spread rapidly throughout the continent, but remained at low population sizes for some 8,000 years. Intriguingly, this included a 4,000-year period of 'boom-and-bust' oscillations in human populations, including a marked decline after 11 ka; this may reflect the loss of a main food resource, such as megafauna (Goldberg *et al.*, 2016).

Human colonisation of the Sahul, or what O'Connell and Allen (2012) refer to as the 'restaurant at the end of the universe', may have been driven by local depletion of prey such as shellfish, which stimulated further migration. Using a radiocarbon database, Williams (2013) reconstructed the prehistoric population of Australia and noted some demographic changes in response to major climatic events, most strikingly during the Last Glacial Maximum (21–18 ka) when the population declined by about 60%. This is significantly later than the demise of most megafauna (see Chapter 3). The data provide no evidence of significant impacts of animal extinctions on human demography. In contrast, in New Zealand there is clear evidence of a human population decline having occurred following overexploitation of faunal resources (Brown and Crema, 2019).

Considered together, this research provides little evidence that ecosystem collapse can lead to collapse of human populations. Even in New Zealand, where humans underwent something of a population crash, numbers eventually recovered. Human history over the past 200,000

Table 5.11 *Summary of hypothesised mechanisms that might lead to collapse of socioecological systems, based on ideas presented in the literature in relation to different case studies and informed by complex systems theory*

Family of mechanism	Specific mechanism	Summary of mechanism
Top-heavy mechanisms	Overshoot	Ecological degradation and excessive resource consumption; collapse caused by climate change or other impact on productivity.
	Complexity threshold	Complexity creates problems that only more complexity can solve; diminishing marginal returns mean burden becomes too great for society to support, and collapse occurs.
	Elite capture	Wealthy become parasitic on the poor. Resentment, revolution or technological change can cause collapse.
	Overspecialisation and inability to adapt	Specialisation on a particular resource, sunk cost effects and/or a lack of diversity create other vulnerabilities that lead to collapse.
Mismatch mechanisms	Scale mismatch	Scales of environmental variation and governance, or production and regulation, become misaligned. This can cause system dysfunction and collapse.
	Upscaling	Obtaining resources remotely can detach people from environmental degradation, creating an overconsumption feedback and potential for collapse.
	Speculation	Success leads to a decreasing investment in regulation; returns to speculation exceed those on investments in productive capacity. If expectations about future growth are threatened, abrupt collapse of speculation and general economic activity due to borrowing can occur.
Lateral flow mechanisms	Collapse by contagion	Perturbation or negative impact is transmitted through lateral connections.
	Collapse by fragmentation	Loss of modularity and reliance on connections result in collapse if connections are broken.

(*cont.*)

Table 5.11 (*cont.*)

Family of mechanism	Specific mechanism	Summary of mechanism
Obliteration	External disruption	A force from outside the system destroys or undermines it.
	Grinding down	Gradual depletion of key resources, such as biodiversity or soil fertility, eventually leads to collapse.
Transition and boundary mechanisms	Vulnerability threshold	Systems (or individual components) grow from less vulnerable sizes through more vulnerable sizes and may collapse during a vulnerable stage.
	Leakage	Semipermeable boundaries that are important for sustainability become permeable, leading to loss of key resources and/or influx of problem-causing agents.

Source: Reproduced from Cumming and Peterson (2018), with permission from Elsevier.

years is testament to the versatility and adaptability of our species; we have colonised the entire world and made use of virtually all of its ecosystems. Despite the Ehrlich's gloomy predictions, perhaps we are not doomed yet, at least not as a result of environmental degradation. Climate change might be another matter, however (Weiss and Bradley, 2001). Somewhat comfortingly, the annual probability of natural extinction for *Homo sapiens* has recently been estimated at less than one in 870,000 (Snyder-Beattie *et al.*, 2019), although somewhat less comfortingly, the potential impact of climate change was not factored into this calculation.

How else might people be affected? If ecosystem collapse is unlikely to lead to the collapse of civilisation, and even less likely to cause human extinction, might it have less drastic impacts on human societies? There is stronger evidence available here. The question is best addressed by briefly examining the literature on ecosystem services, which have become a major focus of ecological research during the last 20 years (Costanza *et al.*, 2017). Ecosystem services can be thought of as the benefits provided by ecosystems to people and include essentials such as clean air, water and food. Provision of ecosystem services can be reduced by ecosystem degradation, providing a potential link between human well-being and the condition and functioning of ecosystems.

Ecosystem services have a somewhat tortuous history as a research theme. Originally, the idea of assessing the value of different services was presented as a way of supporting more rational decision making regarding the use of ecosystems, while also strengthening the case for their conservation (Ehrlich and Mooney, 1983; Westman, 1977). Since then, the relationship between ecosystem services and biodiversity conservation has become rather fraught. For example, in 2005, the Millennium Ecosystem Assessment provided the first global assessment of the relationship between ecosystems and human well-being. Results showed that of a group of 24 ecosystem services examined, 60% are being degraded by human activity, and this is exacerbating poverty for some groups of people (Millennium Ecosystem Assessment, 2005a). However, the assessment struggled with how best to accommodate biodiversity conservation as an issue and in the end treated it separately as a cross-cutting theme (Millennium Ecosystem Assessment, 2005b). As it is generally agreed that ecosystems are an element of biodiversity, and ecosystems were the main focus of the assessment, this approach is rather mystifying.

By 2012, the relationship between biodiversity and ecosystem services had become so confused that it was damaging efforts to create coherent policy. After having identified this problem, Mace *et al.* (2012) proposed a possible solution, by calling for the following roles of biodiversity to be recognised: (i) something that regulates the ecosystem processes underlying the provision of ecosystem services, (ii) an ecosystem service itself and (iii) a type of economic good subject to valuation. Did this clarify the situation, or just increase confusion levels still further? In fact, by this time management for provision of ecosystem services had increasingly become a goal in its own right. Approaches dubbed the 'new conservation' promote economic development and poverty alleviation over traditional approaches to biodiversity conservation, such as designation of protected areas and management of endangered species (Kareiva and Marvier, 2012; Soulé, 2013). These ideas have sparked an acrimonious debate (Soulé, 2013; Tallis *et al.*, 2014), which is still ongoing (Sandbrook *et al.*, 2019). They have also led to a shift in focus of some major conservation organisations towards meeting the needs of people rather than solely those of wildlife (Doak *et al.*, 2014).

Unsurprisingly, given all this acrimony, ecosystem services are now considered to be a contested concept (Schröter *et al.*, 2014). The best illustration of this is what has happened within IPBES (the Intergovernmental Panel for Biodiversity and Ecosystem Services). This is an independent intergovernmental body (not strictly part of the UN) that was recently established to strengthen the international science-policy interface for biodiversity and ecosystem services. Borie and Hulme (2015) document how the ecosystem services bandwagon, which has primarily been driven by Western scientists, was derailed during development of the conceptual framework of IPBES. Some developing countries (notably Bolivia) challenged ecosystem services as being a biased, Western concept that pursues a particular vision of biodiversity that focuses on the commodification of nature. As a result, the IPBES framework that was finally agreed refers to 'Nature's contributions to people' rather than to ecosystem services (Díaz et al., 2015, 2018). This is a remarkable outcome, given that 'ecosystem services' is incorporated within the very name of IPBES. The Bolivians were of course correct; ecosystem services represent a neoliberal concept that has latterly become the dominant ideology in conservation, even though its application can sometimes lead to negative outcomes for biodiversity (Büscher et al., 2012; Fletcher et al., 2019).

Needless to say, the suggestion that ecosystem services might be supplanted in the pantheon of scientific ideas (Díaz et al., 2018) has deeply annoyed many (mostly Western) researchers (Masood, 2018; Peterson et al., 2018). The best response was that of Leon Braat, who was clearly affronted by the notion that his six years of honest academic endeavour as editor-in-chief of the journal *Ecosystem Services* might suddenly be deemed irrelevant (Braat, 2018). His response? To propose that readers quickly forget that they had ever read the paper by Díaz et al. (2018). Recommending selective amnesia of publications in a competing journal surely goes beyond the boundaries of normal editorial protocol (is this a scientific first?). Lest I be accused of being biased on this topic, I should make it clear that I jumped on the ecosystem services bandwagon myself. My personal position was less ideological than purely pragmatic: I felt that if the concept could help conserve biodiversity, then it was worth exploring. I found it to be a useful approach for assessing the cost-effectiveness of ecological restoration and landscape-scale conservation management (Birch et al., 2010; Bullock et al., 2011; Cordingley et al., 2015a,b; Hodder et al., 2014; Newton et al., 2012a; Rey Benayas et al., 2009). But if neoliberalism is another of those zombie ideas that we encountered in Chapter 4, as some have suggested (Blake,

2015; Peck, 2010), then perhaps the Bolivians have provided us with a better way forward.

Despite these shenanigans, IPBES has produced some authoritative statements on the relationship between ecosystem services and human well-being. In their assessment report on land degradation, IPBES (2018) indicate that changes in ecosystem service provision resulting from land degradation have a disproportionate effect on people in vulnerable situations, including women, indigenous peoples and lower-income groups. Through this mechanism, ecosystem degradation can exacerbate income inequality, especially for those living in economically poor areas and in more marginal environments; such people are often particularly dependent on ecosystem services. For example, ecosystem degradation can reduce the availability of wild-harvested goods that provide a buffer for vulnerable households during times of hardship; the poor are also more likely to depend on ecosystem-derived fuels, such as wood, charcoal and dung, to meet their energy needs. Negative impacts of degradation on ecosystem services can also often act together with other stressors, such as socioeconomic change, climate variability, political instability and ineffective institutions, to decrease livelihood security among the most vulnerable members of society. These findings are described by IPBES (2018) as 'well established', based on their assessment of available evidence.

The IPBES assessment report on biodiversity provides further evidence, although here the approach is based on 'Nature's contributions to people' rather than ecosystem services (IPBES, 2019). The report notes that since 1970, trends in agricultural production, fish harvest, bioenergy production and harvest of materials have increased, but most (14/18) of the contributions of nature that were assessed have declined. It is suggested that these negative trends will undermine achievement of 80% of the targets associated with the Sustainable Development Goals. Again, because of their strong dependency on nature and its contributions, many of the world's indigenous peoples and poorest communities are likely to be hardest hit by these trends (Chaplin-Kramer et al., 2019; IPBES, 2019).

Although they make no explicit reference to ecosystem collapse, the results of these assessments imply that collapse could have a negative impact on people, especially the poor. However, evidence on this aspect is lacking; no studies have yet examined the dynamics of ecosystem services as an ecosystem collapses and how these might then affect human well-being. Available evidence suggests that the impacts of collapse might be difficult to predict; trends are likely to differ among services and will depend strongly on local context (Newton *et al.*, 2012a). Furthermore, the relationships between different measures of biodiversity (including both composition and structural attributes) and the provision of ecosystem services are highly complex and uncertain (Balvanera *et al.*, 2014; Cardinale *et al.*, 2012). In some cases, such as provision of fresh water, the relationships can be negative (Harrison *et al.*, 2014). As a result, there can often be trade-offs between biodiversity and different ecosystem services (Cordingley *et al.*, 2015a,b). The most widespread of these trade-offs is between biodiversity and the service of food production, as agricultural intensification typically has a negative impact on ecosystem condition and is a major contributor to biodiversity loss (Deguines *et al.*, 2014; Maes *et al.*, 2012).

Such trade-offs relate to the 'environmentalist paradox': at the global scale, human well-being has increased in recent decades despite large global declines in most ecosystem services (Raudsepp-Hearne *et al.*, 2010). This suggests that human well-being is not closely coupled to ecosystem service provision and that, at the global scale, the benefits of food production currently outweigh the costs of declines in other ecosystem services. This is despite the fact that at smaller scales, the loss of some services (e.g. mitigation of flood risk, or prevention of soil erosion) can have significant direct effects on human well-being (Raudsepp-Hearne *et al.*, 2010). While we have a reasonably good understanding of the impacts of human activity on ecosystems, the paradox also shows that we have a much poorer understanding of how changes in ecosystems affect human well-being (Suich *et al.*, 2015).

Delgado and Marín (2016) suggest that the environmentalist paradox can be resolved simply by not considering food production as an ecosystem service. However, this ignores the fact that human well-being is not solely dependent on ecosystem services, but on many other factors. In fact human well-being is itself a contested concept; some perspectives suggest that engagement in economic activity, for example through paid employment, is of key importance for many people (Daniels *et al.*, 2018). The role of ecosystem services in employment and economic growth have been strangely neglected by researchers, despite their importance to most people's livelihoods. The limited evidence available suggests that the influence of ecosystem services on the performance of most businesses, and their contribution to the overall economy, is limited (Watson and Newton, 2018). The environmentalist paradox can therefore be resolved by recognising the short-term economic gains and associated livelihood benefits that can be obtained by liquidating natural assets. Declines in ecosystems and the services they provide can also be viewed as by-products of economic growth achieved through industrialisation, which, despite causing negative environmental impacts, has benefitted millions of people (Szirmai and Verspagen, 2015).

It is clear that ecosystem collapse can sometimes have profound negative impacts on people; consider the tens of thousands of jobs that have been lost because of the collapse of fisheries, such as those of the Aral Sea and eastern Canada (see Chapter 4). Or consider the value of Australia's Great Barrier Reef, which made an economic contribution of more than AUS$5 billion in 2012 and supported around 69,000 jobs, mostly through tourism (Deloitte Access Economics, 2013). These livelihoods are potentially all at risk if the reef continues to collapse. Some other estimates of the economic value of the reef have been far higher. Of course, like all ecosystems, the Great Barrier Reef is of 'value' by and of itself – irrespective of any market price attached to it (Stoeckl *et al.*, 2011).

Yet the impacts of ecosystem collapse on livelihoods will not always be negative. It is possible that ecosystem collapse could actually increase the flow of some services, or (to jump on the new bandwagon) nature's contributions to people. For example, consider the transitions between forest and grassland that can occur in the savanna biome, as described in Chapter 4. An increase in forest cover could increase provision of services such as carbon storage and timber supply, whereas increasing grassland cover would likely provide greater support to livelihoods based on livestock husbandry. If food production is viewed as an ecosystem service, then conversion of any ecosystem to agricultural land will increase provision of this benefit. Similarly, introduction of an invasive exotic species could increase the value of an ecosystem for food production; let's not forget that Nile perch, which had such a devastating impact on the biodiversity of Lake Victoria (see Chapter 4), supports a commercial fishery. (Ironically, Nile perch itself is now being overfished – see Mkumbo and Marshall, 2014). None of this is surprising if we remember that the imperative of increasing short-term economic income is one of the main drivers of ecosystem collapse, even if this negatively affects the livelihoods of future generations (de Groot *et al.*, 2012).

How does ecosystem recovery affect people? Evidence from meta-analysis indicates that ecological restoration can increase provision of ecosystem services; for example, in a global survey across many different ecosystem types, Rey Benayas *et al.* (2009) found that the mean values of services increased by 25% following restoration. A meta-analysis of restoration actions conducted in agroecosystems in 20 countries also recorded an increase in the supply of many ecosystem services, with supporting services increasing by a mean of 42% and regulating services by a mean of 120% relative to pre-restoration values (Barral et al., 2015). Similarly, in a review of 225 restoration case studies from around the world, de Groot et al. (2013) found that the consequent increases in service provision typically exceeded costs, except for coral reefs. Highest benefit-cost ratios (35:1) were recorded in grasslands. However, during restoration, conflicts can arise among different ecosystem services, especially if individual services or benefits are targeted as a management goal (Bullock et al., 2011).

From this evidence, one might predict that the impacts of ecological restoration on livelihoods would generally be positive, but here information is lacking. Adams et al. (2016) provide a review of the livelihood impacts of large-scale forest restoration indicating a wide range of socioeconomic impacts, which were found to be dependent on variables such as job availability, land tenure, household characteristics, markets for forest products and, above all, the prevailing governance system. Overall, the authors note that livelihood impacts are not clear for many situations, indicating a need for further monitoring. This lack of evidence is echoed by a number of other authors (e.g. Erbaugh and Oldekop, 2018; Reed et al., 2017; Sayer et al., 2016). However, there are cases where restoration has had a demonstrably negative impact on livelihoods. For example, in western Inner Mongolia, nomadic pastoralists were subjected to a resettlement programme, partly with the intention of reducing grassland degradation resulting from overgrazing. Research has shown that resettlement greatly increased the usage of water resources, reduced the efficiency of water use and exacerbated regional water shortages. While household income increased as a result of the resettlement programme, subsistence costs also increased because of the water shortages. As a result, net income and quality of life were reduced (Fan et al., 2015). Restoration actions can also have negative impacts on surrounding communities; for example, wetland restoration can increase mosquito populations, grassland restoration can increase populations of agricultural weeds or pests, and river restoration can impact negatively on surrounding land uses by altering the hydrology (Buckley and Crone, 2008).

In conclusion, evidence suggests that ecosystem collapse is likely to have negative impacts on people, but sometimes these impacts might be positive, whereas the converse is true for ecosystem recovery. Further

5.2 Living with Ecosystem Collapse and Recovery · 373

research is clearly required into both collapse and recovery to better understand which of these different outcomes is more likely, and why. A key issue in this context is equity: it is often the case that some people benefit from ecosystem use or exploitation, whereas others do not. This is illustrated by experience with PES (Payment for Ecosystem Service) schemes, which have been widely introduced as a way of incentivising conservation (or monetising nature – take your pick). These schemes have had variable results (Börner et al., 2017; Muradian et al., 2013). Recent reviews suggest that PES schemes are generally successful in environmental terms, but are often unfair to some people (Calvet-Mir et al., 2015). While PES schemes may deliver benefits to the relatively rich and powerful, local people may lose access to significant livelihood resources without receiving significant benefits and may have little influence over the terms of trade (McDermott et al., 2013). On the other hand, PES schemes can sometimes be socially progressive and help mitigate preexisting economic inequality (Wang et al., 2017). This reminds us that poverty and inequality are the result of unequal power (Phillips, 2017). Understanding power structures and relationships is already an enormous academic endeavour, which should hopefully inform our understanding of the human impacts of both ecosystem collapse and recovery.

What are the implications for environmental policy? Ecosystem recovery has already been the focus of significant policy attention, which has generated a number of commitments to action. Examples include (Suding et al., 2015):

- The New York Declaration on Forests, which identified restoration of degraded ecosystems as a potential solution to climate change; in response, countries have pledged to restore more than 350 million ha by 2030.
- The Aichi Target of the Convention on Biological Diversity (CBD) to restore at least 15% of the world's degraded ecosystems by 2020.
- The Bonn Challenge to restore 150 million ha of degraded forest by 2020 and 350 million ha by 2030. To date, 45 million ha have been placed under restoration, and over 354,000 jobs created.
- The United Nations General Assembly declaration of 2021–2030 as the UN Decade on Ecosystem Restoration.

Much less policy attention has been given to ecosystem collapse. This suggests a need for future policy development to explicitly consider the issue. But what should the policy recommendations be? This is a singular time to be asking this question, as conservation is at something of a crossroads (Sandbrook *et al.*, 2019). While there is now widespread recognition of the global biodiversity crisis, and some positive steps have been taken to address it, efforts are falling short. As illustration, most of the CBD's Aichi Targets for 2020 have not been met, and international dialogue is now beginning to consider what should be done next. At the same time, the international conservation community is deeply divided about the best way forward. Debates around novel ecosystems and 'new conservation' approaches, referred to earlier in this chapter, provide examples of this.

Your views on what should be done will depend on your ethical and philosophical standpoint. The central issue is how the needs of people can best be reconciled with those of wildlife. If you are swayed by the arguments in favour of 'new conservation', in which human needs take priority, consider the points made by Miller *et al.* (2014) in their stinging attack on the concept:

- The assumption that managing nature for human benefit will preserve ecosystems is ungrounded and does not address the root causes of biological destruction, such as economic growth, consumption and increasing human population size.
- The idea rests more on delusion and faith than on evidence. The underlying ethics are utilitarian and are influenced by neoliberal economic philosophy. In fact, increasing affluence magnifies, rather than reduces, human impacts on nature.
- The cumulative, unrelenting impacts of development and economic growth usually prevent ecological recovery. In the face of human population growth, increases in technology and consumption and an economic paradigm of constant growth, it is guaranteed that these impacts will continue.
- Traditional conservation focuses on the preservation of biodiversity for ecosystem function and evolutionary potential. Above all, this requires networks of large protected areas (Laurance *et al.*, 2012). Also, there is need for a broader conservation politic that motivates people to care for nature and mobilises them to act on that belief.

This is also a time when scientists are increasingly grouping together to better communicate their views, including those relating to policy. Barnosky *et al.* (2013, 2014) provide an example: a statement of scientific consensus intended for policymakers, since endorsed by more than 3,000 people. Although the statement makes little explicit reference to ecosystem collapse, it does highlight the problem of widespread ecosystem

Table 5.12 *Some examples of steps that humanity can take to transition to sustainability (as proposed by Ripple et al., 2017)*

- Prioritise the enactment of connected, well-funded and well-managed reserves for a significant proportion of the world's terrestrial, marine, freshwater and aerial habitats;
- maintain nature's ecosystem services by halting the conversion of forests, grasslands and other native habitats;
- restore native plant communities at large scales, particularly forest landscapes;
- rewild regions with native species, especially apex predators, to restore ecological processes and dynamics;
- develop and adopt adequate policy instruments to remedy defaunation, the poaching crisis and the exploitation and trade of threatened species;
- reduce food waste through education and better infrastructure;
- promote dietary shifts towards mostly plant-based foods;
- further reduce fertility rates by ensuring that women and men have access to education and voluntary family-planning services, especially where such resources are still lacking;
- increase outdoor nature education for children, as well as the overall engagement of society in the appreciation of nature;
- divest monetary investments and purchases to encourage positive environmental change;
- devise and promote new green technologies and massively adopt renewable energy sources while phasing out subsidies to energy production through fossil fuels;
- revise our economy to reduce wealth inequality and ensure that prices, taxation and incentive systems take into account the real costs that consumption patterns impose on our environment;
- estimate a scientifically defensible, sustainable human population size for the long term while rallying nations and leaders to support that vital goal.

degradation and transformation. Proposed solutions to this issue emphasise the need to value natural capital and safeguard provision of ecosystem services (Barnosky *et al.*, 2013), for example by:

- slowing and ultimately stopping the encroachment of agriculture into currently uncultivated areas through regulatory policies and incentives for conservation;
- increasing food production in an environmentally sound way through: (a) improving yields in the world's currently less productive farmlands; (b) more efficiently using the water, energy and fertiliser necessary to increase yields; (c) eating less meat; (d) reducing food waste through better infrastructure, distribution and more efficient consumption;

- avoiding losing more land to urban expansion by emphasising development plans that provide higher-density housing and more efficient infrastructure in existing built-up areas;
- enhancing fisheries management, through sustainable aquaculture that focuses on species for which farming does not consume more protein than is produced; and reducing pollution, especially along coasts;
- investing in vital 'green infrastructure', for example through restoring wetlands, oyster reefs and forests to secure water quality, flood control and increase access to recreational benefits;
- keep climate change to a minimum.

An even more impressive number of scientists (15,364 to be precise) recently signed the second 'World Scientists' Warning to Humanity' (Ripple et al., 2017). This apparently represents a world record in terms of the numbers of cosignatories of a scientific paper. Again, no explicit mention of ecosystem collapse is given, but the statement does refer to the need to restore ecosystems, along with a number of other recommendations for action (Table 5.12). Other suggestions have been made that do explicitly refer to ecosystem collapse; for example, Alaniz et al. (2019) provide a series of recommendations regarding how the IUCN RLE could be used to inform environmental policy, based around the five assessment criteria. The RLE could also be used as a basis for assessing threats to provision of ecosystems services (Maron et al., 2017).

So we know what needs to be done (Newton and Cantarello, 2014). What is required is the political will to do it. Perhaps we need to convince the public of the importance of ecosystem conservation and restoration, so that policymakers are pressured to adopt the long-term environmental policies that are needed (Rose et al., 2018). And perhaps, as scientists, we can also help to strengthen this political will by communicating our understanding of ecosystem collapse and the risks that it presents to humanity.

6 · *Conclusions*

I started planning this book while the world's largest coral reef ecosystem, the Great Barrier Reef, was experiencing an unprecedented catastrophe. The bleaching event that occurred in 2016 and 2017, induced by high water temperatures, eventually killed half of the reef's corals (Morrison *et al.*, 2019). Much of the reef's north coast has subsequently become skeletal and barren, with little prospect of recovery. Human responses have included 'last-chance tourism' and 'ecological grief', as well as numerous calls for action (Morrison *et al.*, 2019). While I was writing the text, international news media provided almost daily reminders that many of the world's ecosystems are similarly suffering increasing damage (Table 6.1). In mid-2019, for example, extensive media coverage of fires in Brazilian Amazonia instigated a political crisis; a shift in government policy had doubled the forest area that burned compared with the previous year (Lizundia-Loiola *et al.*, 2020). In fact the increase in the burned forest area was not limited to Brazil but extended to many neighbouring countries; particularly concerning was the destruction of many forest areas that had previously experienced little or no fire occurrence (Lizundia-Loiola *et al.*, 2020). By the time I was writing the final chapters in late 2019, media attention had switched back to Australia, which was experiencing the largest mega-fires on record. More than 3.8 million ha of forest burned in New South Wales alone (Nolan *et al.*, 2020), killing an estimated 1 billion animals (Dickman, 2020) and at least 34 people. The fires were associated with ash rain, dust storms and flash floods, creating an increasing sense of an impending apocalypse.

The past two years have also witnessed increasing usage of the term 'ecosystem collapse' by news media (Table 6.1) and by environmental activists. Notable among the latter is Greta Thunberg, a Swedish teenager whose individual action led to strikes by hundreds of thousands of schoolchildren worldwide in September 2018; her wonderful, emotional speeches regularly refer to ecological collapse (Thunberg, 2019). The school strikes have been accompanied by the emergence of other new

Table 6.1 *A selection of headlines featured in the* Guardian *relating to ecosystem collapse and recovery, which appeared during September 2019–January 2020 while the final sections of this book were being written*

Headline	Date
Australia launches emergency relocation of fish, as largest river system faces collapse	9 September 2019
Fishery collapse 'confirms Silent Spring pesticide prophecy'	31 October 2019
Europe must act on intensive farming to save wildlife, scientists say	5 November 2019
Climate crisis: 11,000 scientists warn of 'untold suffering'	5 November 2019
Pacific seals at risk as Arctic ice melt lets deadly disease spread from Atlantic	17 November 2019
Climate emergency: world 'may have crossed tipping points'	27 November 2019
Countries from Siberia to Australia are burning: the age of fire is the bleakest warning yet	29 November 2019
Revealed: 'monumental' NSW bushfires have burnt 20% of Blue Mountains world heritage area	3 December 2019
Plastic pollution kills half a million hermit crabs on remote islands	5 December 2019
Oceans losing oxygen at unprecedented rate, experts warn	7 December 2019
Mekong basin's vanishing fish signal tough times ahead in Cambodia	16 December 2019
Climate of chaos: the suffocating firestorm engulfing Australia	20 December 2019
'Mother Nature recovers amazingly fast': reviving Ukraine's rich wetlands	27 December 2019
A warm welcome? The wildlife visitors warning of climate disaster	2 January 2020
Urgent new 'roadmap to recovery' could reverse insect apocalypse	6 January 2020
The Australian fires are a harbinger of things to come. Don't ignore their warning	7 January 2020
Ocean temperatures hit record high as rate of heating accelerates	13 January 2020
UN draft plan sets 2030 target to avert Earth's sixth mass extinction	13 January 2020
Beetles and fire kill dozens of 'indestructible' giant sequoia trees	18 January 2020
Bushfires, ash rain, dust storms and flash floods: two weeks in apocalyptic Australia	24 January 2020

Source: www.theguardian.com

environmentalist movements, notably Extinction Rebellion, which has similarly mobilised mass protests aiming to ensure that action is taken to address climate change, biodiversity loss and ecological collapse. According to Molyneux (2019), these actions might represent the beginning of a global mass movement aiming to avert environmental catastrophe. These efforts were stimulated by the IPCC report published in 2018, which suggested that the world had only 12 years in which to act (IPCC, 2018; Molyneux, 2019).

I found it deeply unnerving to be writing a text whose relevance seemed to increase day by day, while I was writing it. However, the scientific community has perhaps been a little slow on the uptake. Although Sato and Lindenmayer (2017) report a steady increase in the number of scientific papers referring to ecosystem collapse, the number is still low. Furthermore, most major environmental assessments, such as those of IPBES, IPCC, CBD and GEO, fail to make much reference to it. I believe they are at risk of being overtaken by events. The world is changing very rapidly. The main conclusion I draw from writing this book is that ecosystem collapse is a genuine phenomenon, which is increasingly happening throughout the world. It surely merits greater attention from researchers, conservation practitioners and policymakers, if only to better anticipate what might happen in the future. I am not alone in thinking this: in a recent survey of 221 scientists about catastrophic risks to society, the most common answers overwhelmingly focused on ecosystem collapse, biodiversity loss and climate change (Future Earth, 2020).

But what of hope? At the outset of this book, I referred to the increasing focus on hope and optimism in conservation, to help inspire engagement and action. Kidd *et al.* (2019) have even described it as currently the 'dominant paradigm' in conservation, which is perhaps something of an exaggeration. Whether or not a more positive narrative will succeed in delivering greater support for conservation is itself a moot point; Kidd *et al.* (2019) note that there is a lack of evidence either way. In the face of what is currently happening to the natural world, it is difficult to remain optimistic. Knight (2013) suggests that to remain positive, we need to replace 'unhelpful catalogues of despair' with studies that show how research can be applied to support effective conservation action. I have tried to ensure that if this book is a catalogue of despair, it is at least a helpful one. There is clearly more to be done, though, in using our understanding of ecosystem collapse to improve the practice of conservation.

For me, the most powerful source of hope lies in the remarkable ability of many ecosystems to recover, as illustrated by some of the case studies profiled in previous chapters. It is good to see ecological restoration receiving increased media attention (Table 6.1); in particular, the popularity of rewilding continues to grow. There is increasing awareness of the importance of ecological recovery for addressing climate change, as well as improving human livelihoods. Scientifically, as these pages attest, we perhaps have a better understanding of ecosystem recovery than we do of collapse. The challenge is to scale up the restoration successes achieved to date. How might this be achieved? According to Blignaut and Aronson (2020), the answer is as follows:

Over the next 30 years, ecological economics, ecological engineering, ecological restoration, and supporting disciplines and professions must work together, synergistically, to blaze the trails and build the pathways of system-wide healing, nurtured by the restoration narratives of an emerging restoration culture.

Now that's what I call hope.

The following sections provide a brief summary of some of the key findings of this book, first by considering the answers to questions posed at the beginning, then via a series of summary statements and provocations.

6.1 Answers

At the outset of this book, I posed a series of scientific questions that I hoped to answer while writing it (see Table 1.3). The hunt for these answers guided my subsequent search through the scientific literature, which inevitably revealed a wide spectrum of evidence and opinion. It is foolhardy to attempt to summarise this evidence concisely, given its richness and complexity. Yet, in the interests of achieving some semblance of narrative structure, that is what is presented in the following section. If the answers in this brief synopsis appear to be cursory or glib, as they surely will, please refer to the more detailed consideration presented in the preceding chapters.

What is ecosystem collapse; how should it be defined and assessed?
Ecosystem collapse is the result of environmental degradation, which IPBES (2018) defines as 'the state of land that results from the persistent

6.1. Answers · 381

decline or loss in biodiversity and ecosystem functions and services that cannot fully recover unaided within decadal timescales'. This describes a state that is persistent, because recovery is impeded or impaired. I believe that this provides a useful working definition of an ecosystem that has collapsed (although 'land' should be extended to include 'sea'). It is a definition that could be applied at a variety of scales, including the local or landscape scales relevant to practical conservation management.

Ecosystem collapse can also be considered as a process of decline, as well as its endpoint. I suggest that the term 'ecosystem collapse' should be limited to those ecosystems that have been degraded rapidly and that have undergone abrupt change. The choice of timescale by which 'abrupt' might be defined is essentially arbitrary, but in the spirit of the IPBES (2018) definition, perhaps 'decadal timescales' would be appropriate. This would ensure relevance to human lifespans. Given that biodiversity, ecosystem function and services do not necessarily covary (see Chapter 5), ecosystem collapse could therefore be defined as follows:

An ecosystem in which biodiversity, ecosystem functions and/or services abruptly decline or are lost over decadal timescales, and for which these losses are persistent, such that they cannot fully recover unaided within decadal timescales.

Put even more simply, ecosystem collapse refers to a situation where an ecosystem has become completely degraded very quickly in, say, less than 50 years and has been unable to recover. It might be so degraded that it has been transformed into a different kind of ecosystem.

Note that this definition differs from that employed by the IUCN Red List of Ecosystems (RLE) in a number of ways. First, it does not require replacement by a different ecosystem type; it could just refer to a loss of defining features, without necessarily involving a transformation of identity. Second, it could be applied to individual occurrences of an ecosystem, such as those within a protected area, and would not need to apply to all occurrences of a particular ecosystem type. Third, it specifies that decline is abrupt, whereas the RLE definition includes situations where ecosystem decline is gradual. These differences partly reflect contrasting objectives: the RLE is designed to enable risk assessments to be conducted throughout the geographical range of an ecosystem. The definition proposed here is one that could potentially support practical conservation management at the local scale.

Do the RLE criteria provide a robust basis for assessing collapse risk? Currently the answer has to be no, at least not at the global scale. The reservations expressed by Boitani *et al.* (2015) have not all been addressed; the RLE criteria and thresholds need firmer scientific foundations. Yet these can be improved in the light of experience, as happened with the Red List of Threatened Species. They will only be improved by applying the criteria in practice and revising them in the light of the results obtained. For assessing collapse risk at the global scale, the RLE is the best approach that is currently available; it has also been of enormous value in raising awareness of ecosystem collapse as an issue. So I wish it all success.

For conservation practitioners interested in assessing the risk of ecosystem collapse at more local scales, the key approach is monitoring, both of ecosystem attributes (such as composition, structure and function) and of the threatening processes that affect them. Monitoring data can potentially provide early warning of collapse, by enabling rapid change to be detected and providing an insight into its possible causes.

How, why and when does ecosystem collapse occur?

Any ecosystem can potentially collapse at any time. Generalisations about the process of collapse are difficult to provide, given that each ecosystem is essentially unique, and is currently being subjected to a particular combination of anthropogenic pressures. Perhaps ecosystem collapse can best be understood on a case-by-case basis, taking local context into account. Nevertheless, evidence suggests that:

- Ecosystem collapse is principally caused by extrinsic factors (i.e. anthropogenic pressures), but it can also be caused by a combination of extrinsic and intrinsic factors, such as a breakdown in stabilising feedbacks, or positive feedbacks among ecological processes.
- Collapse is most likely when ecosystems are subjected to multiple anthropogenic pressures, especially if there are positive interactions between them.
- Collapse often occurs because ecological recovery is impeded, typically by chronic anthropogenic disturbance.
- Collapse may occur when species are lost that are highly connected to many others in ecological networks. These might include generalist species and those at the top or bottom of food chains.

If one ecosystem collapses, might others follow?

Yes, there is evidence for this, although few examples have been documented to date. Also, the situations where this is likely to occur are not well understood. Evaluation of the risk of collapse cascades, and the underlying mechanisms is therefore an urgent priority for future research.

What are the different mechanisms that can cause ecosystem collapse, and are these the same in different types of ecosystems?

A very wide variety of different mechanisms have been identified, which can differ not only between different types of ecosystem but also between different examples of the same ecosystem type. In other words, there are many different ways in which an ecosystem can collapse. Multiple mechanisms may arise even within a single anthropogenic pressure. For example, climate change could affect the ecosystems occurring in a given area in a variety of different ways, depending on their relative tolerances of higher temperatures, drought, flooding and so forth. There may be some systematic differences between ecosystem types; for example, interactions between multiple types of disturbance may be less likely to drive collapse in freshwater ecosystems than in other ecosystem types. However, this has not been systematically investigated. Note that the most widespread mechanism of collapse is complete destruction of an ecosystem resulting from human activities such as land-use change. The effects of this are similar in any terrestrial ecosystem.

Are some ecosystems more at risk of collapse than others? If so, why is this?

Early results from the RLE indicate that some ecosystems are at greater risk of collapse than others. This can largely be attributed to the fact that anthropogenic pressures vary spatially in terms of their intensity and frequency. However, ecosystems may also differ in their ability to tolerate a specific form of disturbance; for example, a forest that is fire-adapted will be better able to tolerate being burned than one that isn't. There is some evidence that novel disturbances are more likely to cause collapse, but this has not been investigated systematically. Ecosystems may also differ in their capacity for recovery, which will influence their collapse risk, but again, comparative evidence is lacking.

Is it possible to provide early warning of imminent collapse, and if so, how?

Yes, this is possible, but it isn't easy. This topic has been the focus of a major research effort, which has delivered both successes and failures. Much of this research has focused on use of models with the aim of developing early-warning indicators. The limited field evidence available suggests that such indicators are rarely effective. Rather than base management decisions on early-warning indicators, it may often be preferable to use frequent monitoring of key ecosystem attributes to provide early warning of any changes that are occurring.

What are the implications of ecosystem collapse for biodiversity, ecosystem function and the provision of ecosystem benefits to people?

Impacts on biodiversity will almost always be negative in the short term, although some species (e.g. those that are 'weedy') might benefit. In the longer term, there can sometimes be positive outcomes as a result of new opportunities for evolutionary diversification being created. Impacts on ecosystem functions and services are more complex and will depend on how the ecosystem is transformed during the process of collapse. Some functions and services can increase as a result of collapse; for example, carbon storage can increase if a forest plantation is established on grassland. If food production is considered as an ecosystem benefit, then ecosystem collapse is widely associated with improved provision of some benefits to people, as conversion to agriculture is one of the principal drivers of collapse.

How does collapse relate to ecosystem recovery? Can recovery occur after collapse, and if so, how?

Collapse is typically associated with limited capacity for recovery. If some factor impedes the recovery processes, then a degraded ecosystem state may become persistent rather than transient, thereby constituting collapse. Possible reasons for a lack of recovery include stabilising feedback processes maintaining an ecosystem in a degraded state. There are field situations where such feedbacks have been identified, but it is not clear how widespread they are. More typically, recovery is prevented by ongoing, chronic disturbance. However, the reasons for a lack of recovery are often unclear. Yes, recovery can sometimes occur after collapse, but it may require human intervention, for example through application of ecological restoration techniques. Without such intervention, and the removal of anthropogenic pressures, recovery can often require long timescales. In other situations, for example if key species have been extirpated or environmental conditions have changed, recovery may

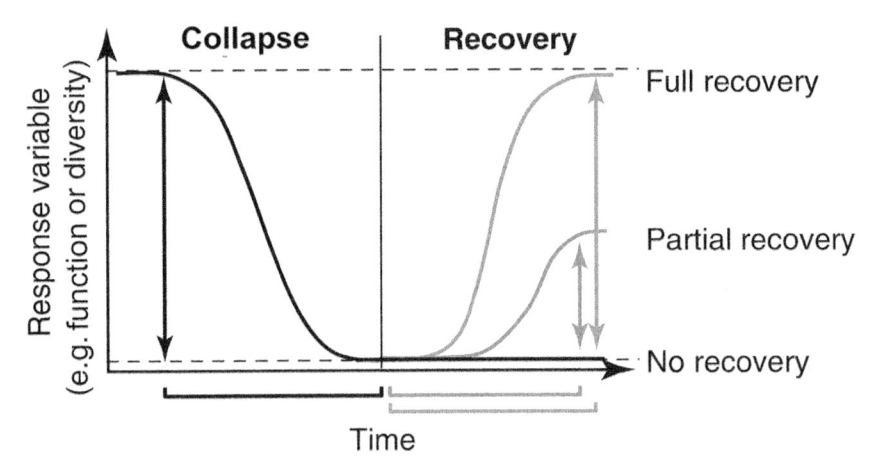

Figure 6.1 The relationship between trajectories of ecosystem collapse and recovery. Adapted from Lotze *et al.* (2011), with permission from Elsevier.

not be possible. The situations where recovery is impossible have not been defined in detail.

Are the trajectories and mechanisms of recovery related to those of collapse?

No. The trajectory of collapse will typically be non-linear (Figure 6.1), although this has not been examined systematically. A wide variety of different recovery trajectories have been reported, although asymptotic non-linear responses appear to be widespread (Figure 6.1). The shape of the recovery trajectory can vary according to the pattern and extent of degradation, the characteristics of the disturbance regime and the attributes of the ecosystem that are being considered, among other factors. Functional attributes such as biomass can sometimes recover more rapidly and to a greater extent than measures of species richness or composition. The mechanisms of recovery are generally not related to those of collapse. For example, many ecosystems have collapsed because of the loss of top predators through a trophic cascade. Recovery of populations of top predators will likely take much longer than many other elements of the ecosystem. Evidence consistently indicates that the durations needed for ecosystem recovery are longer than those of collapse, which is consistent with different mechanisms being involved.

What are the mechanisms of ecosystem recovery, and are these the same in different types of ecosystem?

As with collapse, a wide variety of different mechanisms have been identified, which can differ not only between different types of ecosystem but also between different examples of the same ecosystem type. In other words, there are many different ways in which an ecosystem can recover. Recovery is critically dependent on intrinsic factors, namely interactions of organisms between each other and with the physical environment. Key processes can include reproduction, establishment, growth, dispersal, succession, nutrient dynamics and development of critical mutualisms. Often, some elements of an ecosystem might recover, while others do not, indicating that recovery does not have a single dimension.

How can ecosystem collapse be prevented?

The key priority is to remove anthropogenic pressures, or the threatening processes that are responsible for ecosystem degradation. This is a source of hope: anthropogenic pressures are under human control, and therefore it is potentially within our power to reduce them. Unfortunately, some pressures, such as climate change, are proving very difficult to address. For the foreseeable future, all conservation actions will need to be undertaken within the context of a changing climate, and this will limit our ability to prevent collapse. Some environmental changes responsible for ecosystem collapse, such as the warming of the oceans and the spread of diseases, are now sufficiently advanced that they have effectively moved beyond human control. In such cases, management actions may need to focus on mitigating damage and supporting recovery.

How can ecological recovery be supported by appropriate environmental management?

Removal of anthropogenic pressures should enhance the ability of an ecosystem to recover naturally. However, in many cases it will not be possible to completely remove all pressures, and degradation may have compromised the ecological processes responsible for recovery. In such cases, human intervention may be required. Substantial progress has been made in developing ecological restoration techniques for different

Table 6.2 *Some new questions about ecosystem collapse and recovery, which emerged during the writing of this book*

- What are the risks of ecosystem collapse cascades, and what are their underlying mechanisms?
- Are novel disturbances more likely to cause ecosystem collapse?
- Are some ecosystem types intrinsically more vulnerable to collapse than others?
- What are the relationships between the trajectory of ecosystem decline and its endpoint(s)?
- How widespread are stabilising feedback processes that maintain an ecosystem in a degraded state?
- What is the relative importance of intrinsic and extrinsic factors in limiting ecosystem recovery?
- What are the thresholds of degradation beyond which ecosystem recovery is not possible?
- How can ecosystem recovery occur in the presence of ongoing disturbance, such as climate change?
- Can the recovery of one ecosystem support the recovery of another?
- Are there feedbacks between ecosystems and the biosphere, or the entire Earth system, which can act as a mechanism for collapse and influence recovery?
- Is rewilding the most effective way of achieving ecosystem recovery?

ecosystems, supported by practical guidance. Evidence indicates that these techniques can often be successful in enabling ecological recovery, although some ecosystem attributes will recover more rapidly, and to a greater extent, than others. Recovery may also require long timescales and significant financial investment.

Emerging Questions

One of the wonderful features of scientific research is that the process of answering questions always results in some new ones being asked. In fact, asking questions is part of the joy of science (Vale, 2013). Fortunately it is a pleasure that we can all engage in. I therefore encourage you to consider what new questions you might ask as a result of reading this book. To help get you started, I present some suggestions in Table 6.2.

6.2 Summaries

I present here a series of statements that summarise some of the points made in the preceding chapters. This set of summaries does not pretend to be either systematic or comprehensive; rather it attempts to identify some emerging issues and provide some take-home messages. I apologise if this selection appears a little idiosyncratic; inevitably it reflects my own interests and biases.

1. *We are living in an era of ecosystem collapse.* I hope that the examples in this book have convinced you that ecosystem collapse is a genuine phenomenon. It is widespread, and becoming more so. Throughout the world, coral reefs are dying, fisheries are being overharvested, forests are being cut and burned, rivers are being dammed, lakes are being poisoned, wetlands are being drained and grasslands are being overgrazed. Around three-quarters of the terrestrial environment and 66% of the marine environment have been significantly degraded by human actions; more than a third of the world's land surface and nearly 75% of freshwater resources are now devoted to crop or livestock production (IPBES, 2018, 2019).

2. *It's much, much worse than you think.* After dipping into the scientific literature, as we have here, it is easy to form an impression that we know a lot about what is happening in the world, and we know which ecosystems are collapsing. Of course, this is an illusion. While the scientific community is working strenuously to find out what it can, only a tiny fraction of the world's ecosystems have actually been investigated. We know that ecosystem collapse can often be driven by interactions between multiple stressors, but our understanding of these interactions is in its infancy. Similarly, the extinction cascades and positive feedbacks that can drive collapse are hard to observe. Ecosystem collapse can be cryptic and difficult to detect; it may be entirely hidden from us. Functionally important components of ecosystems, such as insects, fungi and bacteria, receive very little research attention. As illustration of how little we know, Janzen and Hallwachs (2019) suggest that there may have been a catastrophic decline in populations of tropical insects in recent decades. Their suggestion is based on anecdotal observations they have accumulated during many decades of fieldwork, rather than quantitative scientific evidence. This implies that a major loss of biodiversity has occurred without it being scientifically recorded. Until recently, a major biodiversity crisis in North America had similarly been overlooked: the loss of nearly 3 billion birds, representing nearly a third of those present in 1970 (Rosenberg *et al.*, 2019). The effects of this decline on food webs and ecosystem function are unknown.

3. *Climate change changes everything.* The phrase 'it's much, much worse than you think' is actually the memorable first line of *The Uninhabitable Earth*, a best-selling compendium of climate-change impacts compiled by David Wallace-Wells (2019). As I mentioned in Chapter 5, climate change is an exceptionally powerful driver of ecosystem collapse, partly because of its scale and its ability to change both biotic and abiotic components of an ecosystem. Given its ability to interact with other threats, it can also be considered as a meta-threat. The potential impacts of climate change on ecosystems are difficult to comprehend. Nolan *et al.* (2018) provide an assessment of what might happen, based on an analysis of changes in terrestrial vegetation since the last glacial period using a compilation of palaeoecological evidence. Results indicate that terrestrial ecosystems are highly sensitive to changes in temperature, which are projected to exceed anything experienced over the past 2 Myr. If greenhouse gas emissions continue to increase as projected, terrestrial ecosystems worldwide are at risk of major transformation – or, put another way, collapse. The authors conclude that many novel ecosystems will emerge, many of which will be ephemeral, given that the climate is likely to keep on changing. Communities will disintegrate and reorganise, dominant or keystone species will be replaced, ripple effects will transform species interactions and pass through different trophic levels and large changes will occur to carbon sources and sinks, as well as to atmospheric moisture recycling and other climate feedbacks.

One of the most profound insights I gained from writing this book is the fact that fossil fuels are the residue of mass extinction events (see Chapter 3). We are now burning the very same carbon that was emitted through volcanism in Earth's prehistory, which then accumulated in ocean sediments owing to the anoxic events caused by global warming. Most of the world's petroleum and natural gas reserves were formed in the Mesozoic during such anoxic events, which prevented the decomposition of dead organisms; instead they accumulated in ocean sediments. In other words, to use a Gaian analogy (Lovelock and Margulis, 1974), it is as if the Earth system counteracted the carbon dioxide emissions that drove the mass extinctions in the fossil record, by burying the carbon in rock. By burning fossil fuels, we are releasing that same carbon back into the atmosphere, where it is likely to contribute to another mass extinction event.

Climate change is becoming the main issue for conservation. Ecosystems are already changing radically in many parts of the world. It is likely that these changes are going to intensify with time, not diminish. As Wallace-Wells (2019) puts it, there is no new normal. The world our children will inherit will be very different from our own. Conserving any ecosystems under such circumstances represents an immense challenge. The academic arguments that I took delight in reporting throughout this book are going to seem increasingly irrelevant as ecosystems are transformed before our eyes.

4. *Collapse is mainly caused by extrinsic factors.* The scientific literature on ecosystem collapse is currently dominated by concepts associated with dynamical systems theory, particularly critical transitions and alternative stable states. According to this theory, state transitions are driven by intrinsic factors, such as positive feedbacks between key ecological processes. There is clear evidence that this can happen in nature, but it is less clear how widespread this phenomenon is. Many positive feedback mechanisms have been identified, but their role in driving ecosystem collapse is often unclear. These theoretical ideas are undoubtedly fascinating and they have generated a great deal of valuable research. However, there is a risk that preoccupation with these mechanisms could blind us to what is actually happening in nature. Most ecosystem collapse is not driven by intrinsic processes but solely by extrinsic factors, such as the clear-felling of a forest, the draining of a lake or the conversion of a grassland to cropland. Conservation action needs to focus on addressing these extrinsic threats.

5. *Collapse is not an endpoint.* There is a risk that incorporation of ecosystem collapse in environmental assessments, such as the IUCN Red List of Ecosystems, could lead to collapse being viewed solely as a negative outcome that should always be avoided, perhaps comparable to the extinction of a species. In reality, the situation is more nuanced. Ecosystem collapse represents a transformation of one ecosystem type into another. It is possible that a collapsed ecosystem may itself be of conservation value, for example when a grassland is replaced by natural forest. Collapse can also be serial, with one ecosystem transforming into another, which then transforms into a third type, and so on. Each of these ecosystems might have different values for wildlife and people. Recovery is not necessarily an endpoint either; it also represents a process of

6.2 Summaries · 391

transformation. Recovery of one ecosystem could be associated with collapse of another, for example if restoration of grassland is achieved through the removal of forest. Ecosystem collapse and recovery are therefore inextricably linked, like the *yin* and *yang* referred to at the start of this book. For example, collapse is partly understandable as a lack of recovery, which is often caused by ongoing chronic disturbance.

6. *Collapse and recovery are theory rich.* Until I wrote Chapter 2, I had never really appreciated the fact that ecological science is theory rich. But it is. Ecologists have generated a wealth of theoretical ideas, many of which are relevant to understanding ecological collapse and recovery (see Chapter 2). As noted earlier, the discourse around these topics is currently dominated by various flavours of dynamical systems theory. But let's broaden our palette and draw on useful theoretical ideas wherever we encounter them. A wide range of ideas could be used to develop a consolidated body of theory that relates explicitly to ecosystem collapse and recovery, something that we don't currently possess. Hopefully this book will stimulate developments in this direction.

One of the reasons for taking a more pluralist approach is that researchers have been overenthusiastic about invoking dynamical systems theory. Many examples of ecosystem collapse have been interpreted as critical transitions between alternative stable states, but empirical evidence has often failed to support these suggestions. This state of affairs was encountered repeatedly in Chapter 4, for example in coral reefs (Dudgeon *et al.*, 2010), freshwater ecosystems (Capon *et al.*, 2015) and savannas (Lloyd and Veenendaal, 2016), the very same ecosystems that are most often cited as supporting the theory. Researchers went further and accused the theory's proponents of being biased, suppressing evidence, presenting logical fallacies, engaging in wishful thinking and falsely associating the theory with results that do not support it (see Chapter 4). If the theory is being applied dogmatically and uncritically, this risks undermining those situations where it is undoubtedly proving to be of value, such as in the development of early warning indicators (Chapter 5). But the take-home message is clear: most ecosystem collapses are not critical transitions between alternative stable states.

Some authors also conflate regime shifts with transitions between alternative stable states, although these are clearly not the same thing

(Scheffer, 2009). Whereas a regime shift represents a change in the state of a system in response to a persistent change in environmental conditions, alternative stable states represent different configurations of a system under the same environment (Dudgeon et al., 2010). Unlike critical transitions, regime shifts can be driven by a change in the external environment, without invoking any intrinsic positive feedback mechanisms (Scheffer, 2009). Many examples of ecosystem collapse could therefore be considered to be regime shifts. However, regime shifts can also occur in elements of the global climate system, without involving any ecosystems (Rocha et al., 2018). Rather than employing terminology derived from dynamical systems theory, I therefore recommend that abrupt transformations of ecosystems are described as examples of what they are: ecosystem collapse or recovery.

7. *We need new metaphors.* As noted in Chapter 2, the ball- and-cup model is widely used in the literature to illustrate the concepts of alternative stable states and critical transitions. Although it may be useful in this context, it is less appropriate for communicating ecosystem collapse despite suggestions to the contrary (Keith et al., 2015). Its main limitation is that it does not convey the enormity of what ecosystem collapse entails. As the ball rolls happily around its state space, perhaps occasionally coming to rest in different basins of attraction, it essentially remains unchanged. Yet when an ecosystem collapses, it can be utterly transformed. The unchanging ball implies a level of continuity and coherence that is not actually present; in reality the system is being disassembled or completely destroyed. Perhaps if the ball changed shape during its travels, or was smashed to pieces and then ground into dust, it would better convey what actually happens to ecosystems when they collapse.

Other suggestions have been made. In their analysis of ecosystem recovery, Lake et al. (2007) compare a collapsing ecosystem to Humpty Dumpty falling off a wall. At least here the ecosystem has been seriously damaged, although the fact that this damage is irreparable isn't quite right either. Thunberg (2019) uses the imagery of a house on fire, a metaphor that has gained traction with some ecologists (Janzen and Hallwachs, 2019). Given that the word 'ecology' literally means the study of houses, this feels right. Yet our house is not only burning. It might also be flooded, baked, blown down or bulldozed flat, then have something inferior built in its place. It might be rendered uninhabitable because of poisonous fumes or toxic waste, or collapse because structural pillars have been removed. Its foundations may be being eroded or gnawed away. Or more insidiously it might be transformed piece by piece, as if its bricks were removed and replaced with other materials, such as timber or mud.

I also like Thunberg's reference to 'cathedral thinking' (Thunberg, 2019). She uses this metaphor to illustrate the size of the environmental challenge that we face, analogous to the task that our medieval ancestors addressed when building a cathedral – one that could take generations to complete. Rather than houses, perhaps we should imagine ecosystems as cathedrals, places that inspire awe and wonder, full of baroque complexity. When a cathedral burns down, it is a source of national grief. This is illustrated by what happened in France in 2019, when the Notre-Dame de Paris caught fire while (ironically) undergoing restoration work. People reported being shocked, heartbroken, devastated, as if 'something inside of them had collapsed'; crowds gathered outside the ruined building to weep. Fortunately, resurrection is a central motif in Christianity, and this offered hope to those who mourned; more than €1 billion was rapidly pledged to support its restoration. Why don't people respond to the collapse of an ecosystem in a similar way?

8. *Some threats are more important than others.* I noted previously the pervasive effects of climate change. There are also important differences between the other threatening processes affecting ecosystems. In particular, threats such as fire and herbivory can create positive feedbacks with vegetation, as some plant species are adapted to these pressures. It is no coincidence that some of the most persistent examples of ecosystem collapse, such as those of New Zealand and Madagascar (see Chapter 3), were driven by fire. Threats characterised by positive feedbacks can drive the abrupt transitions between ecosystem states envisaged by dynamical systems theory. Yet, as noted in Chapter 3, this may not be the only reason why fire is so damaging; it can also cause persistent edaphic changes, for example in soil structural, physical and chemical properties. There is a need to understand why some threats are more important than others in different systems, for example invasive species in freshwater ecosystems and hypoxia in benthic marine environments.

9. *There is still much that we don't understand.* It is something of a cliché, or even an old joke, that researchers always end their publications with a statement that 'more research is needed' (Maldonado and Poole, 1999). The converse situation, that no further research is needed, at all, anywhere, ever (Psyphago, 2013), is too awful for most researchers to contemplate. Yet in the case of both ecosystem collapse and recovery, there is strong justification for further research, not least because of their societal importance. While these phenomena provide a useful lens through which to view the ecological literature, as this book hopefully demonstrates, it is important to note that much of the available evidence is inferred or circumstantial. Very few studies have documented the process of ecosystem collapse while it happened, and fewer still have also observed the process of subsequent recovery. The relationships between collapse and recovery (e.g. Figure 6.1) are therefore largely based on supposition. There is no shortage of questions that need answering (Table 6.2). In particular, there is a need for research that explicitly examines the mechanisms underlying both collapse and recovery. Conceivably, new research approaches might need to be developed to fully understand these mechanisms, owing to the magnitude, scale and complexity of ecosystem dynamics.

10. *Ecosystem collapse and recovery need to be addressed by conservation policy and practice.* While awareness of ecosystem collapse is increasing, it has not received much explicit consideration by international policy forums such as the CBD, UNCCD and UNFCCC, and their supporting environmental assessments. Some assessments, such as the FAO FRA, are actively obscuring it (see Chapter 4). This needs to change. In particular, there is a need to better communicate the risks of ecosystem collapse to society. At the same time, both collapse and recovery present significant challenges to conservation practitioners. These challenges are compounded by the uncertainties surrounding environmental change and the best ways to respond to it. Practitioners may require support from the research community to help them operationalise ecosystem collapse as a practical concept, including its early detection and the identification of appropriate management responses. To achieve this, we might need greater linkages between research and conservation practice, and perhaps even new kinds of academia (Keeler *et al.*, 2017). Janzen and Hallwachs (2019) suggest that if our house is on fire, we need

the fire department more than we need a thermometer. This is a helpful call to conservation action. Yet because ecosystem collapse can often be cryptic, we may need the thermometer to know that it's on fire. In other words, improved implementation of monitoring approaches should be a high priority for any conservation organisation.

6.3 Provocations

In the spirit of Dada (Picabia, 2007), I here offer some provocations. These are speculative, personal observations, designed to stimulate further discussion rather than represent definitive statements of knowledge. I don't pretend that they're all true – but some of them might be. Hopefully they will provoke someone into proving them wrong by further research. Note that this does not pretend to be a comprehensive list. Indeed, readers of this book are encouraged to engage in a little Dada activity themselves and come up with provocations of their own. You too can be a beautiful monster.

- Alternative stable states are rare in nature.
- Most examples of ecosystem collapse are not critical transitions.
- Ecosystem collapse is not the same as a regime shift.
- Climate change will become the principal cause of ecosystem collapse.
- Most ecosystem collapse is cryptic.
- Ecosystems can be committed to collapse, in a similar way to a species being committed to extinction.
- Ecosystem collapse cascades are widespread.
- Ecosystem collapse does not always have negative impacts on people.
- Ecosystems may be antifragile: they may gain from shocks, challenges and disorder (Equihua Zamora *et al.*, 2019).
- Collapsed ecosystems can be of conservation value.
- Recovery of one ecosystem can cause the collapse of another.
- Ecosystem recovery is nearly always possible, at least in theory.
- Ecosystem recovery does not always provide benefits to people.
- Ecosystem collapse is a bigger issue than the extinction crisis.
- Conservation action should focus on maintaining species composition and richness rather than ecosystem functions and services.
- Ecosystem collapse should be featured in ecology textbooks, instead of being ignored entirely by those who should know better (e.g. Eichhorn, 2016).

6.4 Coda

If this book is not quite an 'unhelpful catalogue of despair' (Knight, 2013), conservationists will find plenty within its pages to despair about. Many ecosystems appear to be unravelling before our eyes, and in the future, the situation is only likely to get worse, perhaps much worse. Profound change is on the way, and for many ecosystems, this will compound the extensive changes that have already happened. According to Greta Thunberg, we shouldn't be hopeful about this situation; we should be panicking. We should be acting as if our house is on fire, because it is (Thunberg, 2019). On the other hand, there is a competing narrative that hope is a useful thing in conservation, as it could inspire people to act. Humanity has an innate attraction to optimism, and greater use of it could encourage scientists, conservation practitioners and others to work more effectively together to address environmental issues, such as ecosystem degradation and collapse (McAfee et al., 2019). Therefore, in this spirit of optimism, I close this book with a message of hope.

Remarkably, the most complex object in the known universe lies between our ears. The extraordinary problem-solving ability of our collective grey matter surely provides us with a measure of hope. Unfortunately, during the course of our evolution, that same grey matter has developed an internal reward system, which leads some individuals to acquire resources at the expense of others (Pedroni et al., 2014). Arguably, this is the most significant underlying driver of ecosystem collapse. The only solution to this problem is for the rest of society to develop social sanctions against such behaviour (Pedroni et al., 2014), which sadly hasn't happened yet. But it could. Perhaps we need a social movement to bring this about.

Another potential source of hope is that political systems and power structures, like ecosystems, can collapse (Geddes et al., 2018). According to dynamical systems theory, this can happen when increasing societal complexity is combined with a lack of economic growth (Livni, 2019). Factors such as the declining financial health and population well-being of a state can spark revolutions, which can deliver rapid societal change (Goldstone, 2001). Given the prevalence of these conditions, it is perhaps not surprising that many parts of the world are currently experiencing political upheaval. Some academics have predicted that political instability will soon spread to the USA and Western Europe (Turchin, 2010); arguably, it has already arrived. Academics have been saying this sort of thing for more than 2,000 years, of course (Tainter, 2004) – but sometimes they've been right. Hopefully, any such societal change will provide an opportunity to defeat the powerful vested interests surrounding the use of fossil fuels and other natural resources (Molyneux, 2019) and deliver more equitable and environmentally sustainable societies for everyone. Initiatives such as a Green New Deal (Klein, 2020) could help achieve this positive vision.

Two personal anecdotes illustrate where I find my own hope. The first is an ecological restoration project focusing on a South American conifer, alerce (*Fitzroya cupressoides*). This was one of the first tree species listed on CITES owing to its exceptional timber value. It is a magnificent 40 m tall tree, a southern counterpart of the giant redwoods of California (to which it is related), and one of the longest-lived organisms on the planet. It also suffered one of the most rapid deforestation events ever recorded in South America (Veblen, 2007). After European migrants colonised southern Chile in the mid-nineteenth century, where the species is endemic, they proceeded to fell almost every individual tree in the area. I had the great pleasure of working with colleagues in Chile and Argentina, who were the first to locate remnant stands of alerce forest in Chile's Central Valley and demonstrate their value for conservation (Allnutt *et al.*, 1999; Premoli *et al.*, 2003; Silla *et al.*, 2002). Even more wonderfully, we worked with local landowners, descendants of the original colonists, who were proud to establish the first attempts at ecological restoration of this forest ecosystem (Newton, 2007b) – a symbolic gesture, perhaps, but one that demonstrated an important shift in how this ecosystem is valued. When I first visited the region in the mid-1980s, *Fitzroya* was still being logged in the Andean forests further south; recently these have been incorporated in a newly designated protected area, which is one of the largest on Earth.

My second example is in southern Scotland, where a local community of which I was a part initiated Europe's first ever community-based rewilding project. The Carrifran Wildwood Project had the ambitious aim of restoring native forest to a landscape in one of the continent's most deforested regions. After successfully raising enough money to purchase an area of land, the group developed plans to re-establish woodland by planting trees raised from seed collected in local forest remnants. Once established, the woodland would be allowed to develop with minimal human interference. On the first day of the new millennium, community members planted the first trees according to the restoration plan that they had developed together. I am proud to have been a member of this inspiring group, and 20 years on, the formerly degraded hillsides are now developing into a fully functional forest ecosystem (Adair, 2016; Ashmole and Ashmole, 2009). Once the trees had established, other elements of the ecosystem, such as woodland birds and fungi, have successfully colonised the site. What is particularly impressive is that some of these species have perhaps been absent for some 6,000 years, when the site was first deforested by Neolithic farmers. Is this the longest ever interval between ecosystem collapse and recovery? In its humble way, this project shows what is possible when communities work together. It also tells us that it is never too late to begin the process of ecological recovery.

References

Aber, J. D., Nadelhoffer, K. J., Steudler, P. and Melillo, J. M. (1989). Nitrogen saturation in northern forest ecosystems. *BioScience*, 39(6), 378–386.

Abernethy, K. A., Coad, L., Taylor, G., Lee, M. E. and Maisels, F. (2013). Extent and ecological consequences of hunting in Central African rainforests in the twenty-first century. *Philosophical Transactions of the Royal Society B: Biological Sciences*, 368, 20120303.

Adair, S. (2016). Carrifran: Ecological restoration in the Southern Uplands: New native woodland and vegetation succession in the Moffat Hills. *Scottish Forestry*, 70(1), 30–40.

Adams, C., Rodrigues, S. T., Calmon, M. and Kumar, C. (2016). Impacts of large-scale forest restoration on socioeconomic status and local livelihoods: What we know and do not know. *Biotropica*, 48, 731–744.

Ahl, V. and Allen, T. F. H. (1996). *Hierarchy Theory: A Vision, Vocabulary and Epistemology*. Columbia University Press, New York.

Alaniz, A. J., Pérez-Quezada, J. F., Galleguillos, M., Vásquez, A. E. and Keith, D. A. (2019). Operationalizing the IUCN Red List of Ecosystems in public policy. *Conservation Letters*, 12(5), e12665. https://doi.org/10.1111/conl.12665.

Algeo, T. J., Chen, Z. Q., Fraiser, M. L. and Twitchett, R. J. (2011). Terrestrial–marine teleconnections in the collapse and rebuilding of Early Triassic marine ecosystems. *Palaeogeography, Palaeoclimatology, Palaeoecology*, 308(1–2), 1–11.

Allan, J. D., Abell, R., Hogan, Z., et al. (2005). Overfishing of inland waters. *BioScience*, 55(12), 1041–1051.

Allen, C. D., Macalady, A. K., Chenchouni, H., et al. (2010). A global overview of drought and heat-induced tree mortality reveals emerging climate change risks for forests. *Forest Ecology and Management*, 259, 660–684.

Allen, C. D., Breshears, D. D. and McDowell, N. G. (2015). On underestimation of global vulnerability to tree mortality and forest die-off from hotter drought in the Anthropocene. *Ecosphere*, 6(8), 129. https://dx.doi.org/10.1890/ES15-00203.1.

Allen, J. R. M., Hickler, T., Singarayer, J. S., Sykes, M. T., Valdes, P. J. and Huntley, B. (2010). Last glacial vegetation of northern Eurasia. *Quaternary Science Reviews*, 29, 2604–2618.

Allendorf, F. W. (1997). The conservation biologist as Zen student. *Conservation Biology*, 11, 1045–1046.

Allesina, S. and Tang, S. (2012). Stability criteria for complex ecosystems. *Nature*, 483, 205–208.

Allesina, S., Bodini, A. and Pascual, M. (2009). Functional links and robustness in food webs. *Philosophical Transactions of the Royal Society B: Biological Sciences*, 364 (1524), 1701–1709.

Allison, G. (2004). The influence of species diversity and stress intensity on community resistance and resilience. *Ecological Monographs*, 74, 117–134.

Allnutt, T. R., Newton, A. C., Lara, A., et al. (1999). Genetic variation in *Fitzroya cupressoides* (alerce), a threatened South American conifer. *Molecular Ecology*, 8, 975–987.

Altieri, A. H., Harrison, S. B., Seemann, J., Collin, R., Diaz, R. J. and Knowlton, N. (2017). Tropical dead zones and mass mortalities on coral reefs. *Proceedings of the National Academy of Sciences of the United States of America*, 114, 3660–3665.

Alvarez, L. W., Alvarez, W., Asaro, F. and Michel, H. V. (1980). Extraterrestrial cause for the Cretaceous–Tertiary extinction. *Science*, 208(4448), 1095–1108.

Andersen, K. H. and Pedersen, M. (2010). Damped trophic cascades driven by fishing in model marine ecosystems. *Proceedings of the Royal Society B: Biological Sciences*, 277, 795–802.

Anderson, P. (1999). Complexity theory and organization science. *Organization Science*, 10(3), 216–232.

Anderson, S. H., Kelly, D., Ladley, J. J., Molloy, S. and Terry, J. (2011). Cascading effects of bird functional extinction reduce pollination and plant density. *Science*, 331(6020), 1068–1071.

Andrén, H. (1994). Effects of habitat fragmentation on birds and mammals in landscapes with different proportions of suitable habitat: A review. *Oikos*, 71, 355–366.

Angeli, D., Ferrell, J. E. and Sontag, E. D. (2004). Detection of multistability, bifurcations, and hysteresis in a large class of biological positive-feedback systems. *Proceedings of the National Academy of Sciences of the United States of America*, 101(7), 1822–1827.

Aragão, L. E. O. C., Anderson, L. O., Fonseca, M. G., et al. (2018). 21st century drought-related fires counteract the decline of Amazon deforestation carbon emissions. *Nature Communications*, 9(1), 536.

Araujo, B. B. A., Oliveira-Santos, L. G. R., Lima-Ribeiro, M. S., Diniz-Filho, J. A. F. and Fernandez, F. A. S. (2017). Bigger kill than chill: The uneven roles of humans and climate on late Quaternary megafaunal extinctions. *Quaternary International*, 431, 216–222.

Arens, N. C. and West, I. D. (2008). Press-pulse: A general theory of mass extinction? *Paleobiology*, 34(4), 456–471.

Aronson, J., Clewell, A. F., Blignaut, J. N. and Milton, S. J. (2006). Ecological restoration: A new frontier for nature conservation and economics. *Journal for Nature Conservation*, 14(3–4), 135–139.

Aronson, J., Murcia, C., Kattan, G. H., Moreno-Mateos, D., Dixon, K. and Simberloff, D. (2014). The road to confusion is paved with novel ecosystem labels: A reply to Hobbs et al. *Trends in Ecology & Evolution*, 29(12), 646–647.

Aronson, J. C., Simberloff, D., Ricciardi, A. and Goodwin, N. (2018). Restoration science does not need redefinition. *Nature Ecology & Evolution*, 2(6), 916.

Arranz-Otaegui, A., Gonzalez Carretero, L., Ramsey, M. N., Fuller, D. Q. and Richter, T. (2018). Archaeobotanical evidence reveals the origins of bread

14,400 years ago in northeastern Jordan. *Proceedings of the National Academy of Sciences of the United States of America*, 115(31), 7925–7930.

Arroyo-Rodríguez, V., Melo, F. P. L., Martínez-Ramos, M., et al. (2017). Multiple successional pathways in human-modified tropical landscapes: New insights from forest succession, forest fragmentation and landscape ecology research. *Biological Reviews*, 92, 326–340.

Ashmole, M. and Ashmole, P. (2009). *The Carrifran Wildwood Story: Ecological Restoration from the Grass Roots*. Borders Forest Trust, Jedburgh, Scotland.

Ashwin, P., Wieczorek, S., Vitolo, R. and Cox, P. (2012). Tipping points in open systems: Bifurcation, noise-induced and rate-dependent examples in the climate system. *Philosophical Transactions of the Royal Society A: Mathematical, Physical, and Engineering Sciences*, 370(1962), 1166–1184.

Asner, G. P., Broadbent, E. N., Oliveira, P. J. C., Keller, M., Knapp, D. E. and Silva, J. N. M. (2006). Condition and fate of logged forests in the Brazilian Amazon. *Proceedings of the National Academy of Sciences of the United States of America*, 103 (34), 12947–12950.

Aubin, D. and Dahan Dalmedico, A. (2002). Writing the history of dynamical systems and chaos: Longue durée and revolution, disciplines and cultures. *Historia Mathematica*, 29, 273–339.

Baker, A. C., Glynn, P. W. and Riegl, B. (2008). Climate change and coral reef bleaching: An ecological assessment of long-term impacts, recovery trends and future outlook. *Estuarine, Coastal and Shelf Science*, 80(4), 435–471.

Bakun, A. (2017). Climate change and ocean deoxygenation within intensified surface-driven upwelling circulations. *Philosophical Transactions of the Royal Society A: Mathematical, Physical, and Engineering Sciences*, 375, 20160327.

Bakun, A. and Weeks, S. J. (2004). Greenhouse gas buildup, sardines, submarine eruptions, and the possibility of abrupt degradation of intense marine upwelling ecosystems. *Ecology Letters*, 7, 1015–1023.

(2006). Adverse feedback sequences in exploited marine systems: Are deliberate interruptive actions warranted? *Fish and Fisheries*, 7, 316–333.

Balaguer, L., Escudero, A., Martín-Duque, J. F., Mola, I. and Aronson, J. (2014). The historical reference in restoration ecology: Re-defining a cornerstone concept. *Biological Conservation*, 176, 12–20.

Balch, J. K., Nepstad, D. C., Brando, P. M., et al. (2008). Negative fire feedback in a transitional forest of southeastern Amazonia. *Global Change Biology*, 14, 2276–2287.

Balme, J. (2013). Of boats and string: The maritime colonisation of Australia. *Quaternary International*, 285, 68–75.

Balmford, A. (2012). *Wild Hope. On the Front Lines of Conservation Success*. The University of Chicago Press, Chicago, IL.

Balmford, A. and Knowlton, N. (2017). Why Earth Optimism? *Science*, 356(6335), 225.

Balmford, A., Carey, P., Kapos, V., et al. (2009). Capturing the many dimensions of threat: Comment on Salafsky et al. *Conservation Biology*, 23, 482–487.

Balvanera, P., Pfisterer, A. B., Buchmann, N., et al. (2006). Quantifying the evidence for biodiversity effects on ecosystem functioning and services. *Ecology Letters*, 9(10), 1146–1156.

Balvanera, P., Siddique, I., Dee, L., et al. (2014). Linking biodiversity and ecosystem services: Current uncertainties and the necessary next steps. *BioScience*, 64(1), 49–57.

Bambach, R. K., Knoll, A. H. and Wang, S. C. (2004). Origination, extinction, and mass depletions of marine diversity. *Paleobiology*, 30, 522–542.

Ban, S. S., Graham, N. A. J. and Connolly, S. R. (2014). Evidence for multiple stressor interactions and effects on coral reefs. *Global Change Biology*, 20(3), 681–697.

Bardgett, R. D. and Wardle, D. A. (2010). *Aboveground-Belowground Linkages. Biotic Interactions, Ecosystem Processes, and Global Change.* Oxford University Press, Oxford.

Barlow, C. (2002). *The Ghosts of Evolution: Nonsensical Fruit, Missing Partners, and Other Ecological Anachronisms.* Basic Books, New York.

Barlow, J. and Peres, C. A. (2008). Fire-mediated dieback and compositional cascade in an Amazonian forest. *Philosophical Transactions of the Royal Society B: Biological Sciences*, 363, 1787–1794.

Barlow, J., Gardner, T. A., Araujo, I. S., et al. (2007). Quantifying the biodiversity value of tropical primary, secondary, and plantation forests. *Proceedings of the National Academy of Sciences of the United States of America*, 104(47), 18555–18560.

Barlow, J., Lennox, G. D., Ferreira, J., et al. (2016). Anthropogenic disturbance in tropical forests can double biodiversity loss from deforestation. *Nature*, 535 (7610), 144–147.

Barlow, J., França, F., Gardner, T. A., et al. (2018). The future of hyperdiverse tropical ecosystems. *Nature*, 559(7715), 517–526.

Barnosky, A. D., Koch, P. L., Feranec, R. S., Wing, S. L. and Shabel, A. B. (2004). Assessing the causes of late Pleistocene extinctions on the continents. *Science*, 306, 70–75.

Barnosky, A. D., Matzke, N., Tomiya, S., et al. (2011). Has the Earth's sixth mass extinction already arrived? *Nature*, 471, 51–57.

Barnosky, A. D., Hadly, E. A., Bascompte, J., et al. (2012). Approaching a state shift in Earth's biosphere. *Nature*, 486, 52–58.

Barnosky, A. D., Brown, J. H., Daily, G. C., et al. (2013). *Scientific Consensus on Maintaining Humanity's Life Support Systems in the 21st Century: Information for Policy Makers.* Department of Integrative Biology and Museum of Paleontology, University of California-Berkeley, Berkeley.

 et al. (2014). Introducing the *Scientific consensus on maintaining humanity's life support systems in the 21st century: Information for policy makers. The Anthropocene Review*, 1 (1), 78–109.

Barnosky, A. D., Lindsey, E. L., Villavicencio, N. A., et al. (2016). Variable impact of late-Quaternary megafaunal extinction in causing ecological state shifts in North and South America. *Proceedings of the National Academy of Sciences of the United States of America*, 113(4), 856–861.

Barral, M. P., Rey Benayas, J. M., Meli, P. and Maceira, N. O. (2015). Quantifying the impacts of ecological restoration on biodiversity and ecosystem services in agroecosystems: A global meta-analysis. *Agriculture, Ecosystems & Environment*, 202, 223–231.

Bartlett, L. J., Williams, D. R., Prescott, G. W., et al. (2016). Robustness despite uncertainty: Regional climate data reveal the dominant role of humans in explaining global extinctions of Late Quaternary megafauna. *Ecography*, 39, 152–161.
Bascompte, J. and Stouffer, D. B. (2009). The assembly and disassembly of ecological networks. *Philosophical Transactions of the Royal Society B: Biological Sciences*, 364 (1524), 1781–1787.
Bastin, J.-F., Finegold, Y., Garcia, C., et al. (2019). The global tree restoration potential. *Science*, 365(6448), 76–79.
Batt, R. D., Carpenter, S. R., Cole, J. J., Pace, M. L. and Johnson, R. A. (2013). Changes in ecosystem resilience detected in automated measures of ecosystem metabolism during a whole-lake manipulation. *Proceedings of the National Academy of Sciences of the United States of America*, 110(43), 17398–17403.
Battisti, C., Poeta, G. and Fanelli, G. (2016). *An Introduction to Disturbance Ecology. A Road Map for Wildlife Management and Conservation*. Springer International, Switzerland.
Baxter-Gilbert, J. H., Riley, J. L., Neufeld, C. J. H., et al. (2015). Road mortality potentially responsible for billions of pollinating insect deaths annually. *Journal of Insect Conservation*, 19(5), 1029–1035.
Bayraktarov, E., Saunders, M. I., Abdullah, S., et al. (2016). The cost and feasibility of marine coastal restoration. *Ecological Applications*, 26(4), 1055–1074.
BCT (Bat Conservation Trust). (2017). *The State of the UKs Bats 2017*. Bat Conservation Trust / JNCC, London.
Beisner, B., Haydon, D. and Cuddington, K. (2003). Alternative stable states in ecology. *Frontiers in Ecology and the Environment*, 1, 376–382.
Bellard, C., Bertelsmeier, C., Leadley, P., Thuiller, W. and Courchamp, F. (2012). Impacts of climate change on the future of biodiversity. *Ecology Letters*, 15, 365–377.
Bello, C., Galetti, M., Pizo, M. A., et al. (2015). Defaunation affects carbon storage in tropical forests. *Science Advances*, 1(11), e1501105.
Bellwood, D. R., Hughes, T. P., Folke, C. and Nyström, M. (2004). Confronting the coral reef crisis. *Nature*, 429, 827–833.
Bellwood, D. R., Hughes, T. P. and Hoey, A. S. (2006). Sleeping functional group drives coral reef recovery. *Current Biology*, 16, 2434–2439.
Bellwood, D. R., Baird, A. H., Depczynski, M., et al. (2012). Coral recovery may not herald the return of fishes on damaged reefs. *Oecologia*, 170, 567–573.
Belovsky, G. E., Botkin, D. B., Crowl, T. A., et al. (2004). Ten suggestions to strengthen the science of ecology. *BioScience*, 54(4), 345–351.
Belyazid, S., Westling, O. and Sverdrup, H. (2006). Modelling changes in forest soil chemistry at 16 Swedish coniferous forest sites following deposition reduction. *Environmental Pollution*, 144, 596–609.
Belyea, L. R. and Lancaster, J. (1999). Assembly rules within a contingent ecology. *Oikos*, 86, 402–416.
Bender, E. A., Case, T. J. and Gilpin, M. E. (1984). Perturbation experiments in community ecology: Theory and practice. *Ecology*, 65, 1–13.
BenDor, T., Shoemaker, D. A., Thill, J.-C., Dorning, M. A. and Meentemeyer, R. K. (2014). A mixed-methods analysis of socialecological feedbacks between

urbanization and forest persistence. *Ecology and Society*, 19(3), 3. https://dx.doi.org/10.5751/ES-06508-190303.

Benítez-López, A., Santini, L., Schipper, A. M., Busana, M. and Huijbregts, M. A. J. (2019). Intact but empty forests? Patterns of hunting-induced mammal defaunation in the tropics. *PLoS Biology*, 17(5), e3000247.

Bennion, H., Simpson, G. and Goldsmith, B. (2015). Assessing degradation and recovery pathways in lakes impacted by eutrophication using the sediment record. *Frontiers in Ecology and Evolution*, 3, 94. https://doi.org/10.3389/fevo.2015.00094.

Benton, M. J., Tverdokhlebov, V. P. and Surkov, M. V. (2004). Ecosystem remodelling among vertebrates at the Permian-Triassic boundary in Russia. *Nature*, 432, 97–100.

Berg, S., Pimenov, A., Palmer, C., Emmerson, M. and Jonsson, T. (2015). Ecological communities are vulnerable to realistic extinction sequences. *Oikos*, 124(4), 486–496.

Berumen, M. L. and Pratchett, M. S. (2006). Recovery without resilience: Persistent disturbance and long-term shifts in the structure of fish and coral communities at Tiahura Reef, Moorea. *Coral Reefs*, 25, 647–653.

Beschta, R. L., Painter, L. E. and Ripple, W. J. (2018). Trophic cascades at multiple spatial scales shape recovery of young aspen in Yellowstone. *Forest Ecology and Management*, 413, 62–69.

  (2019). Trophic cascades and Yellowstone's aspen: A reply to Fleming (2019). *Forest Ecology and Management*, 454, 117344.

Bestelmeyer, B. T., Ellison, A. M., Fraser, W. R., *et al.* (2011). Analysis of abrupt transitions in ecological systems. *Ecosphere*, 2(12), art129. https://doi.org/10.1890/es11-00216.1.

Bestelmeyer, B. T., Duniway, M. C., James, D. K., Burkett, L. M. and Havstad, K. M. (2013). A test of critical thresholds and their indicators in a desertification-prone ecosystem: More resilience than we thought. *Ecology Letters*, 16, 339–345.

Bestelmeyer, B. T., Ash, A., Brown, J. R., *et al.* (2017). State and transition models: Theory, applications, and challenges. In: D. D. Briske (ed.), *Rangeland Systems. Processes, Management and Challenges*. Springer Nature, Cham, Switzerland, pp. 303–346.

Betts, R. A., Cox, P. M., Collins, M., Harris, P. P., Huntingford, C. and Jones, C. D. (2004). The role of ecosystem-atmosphere interactions in simulated Amazonian precipitation decrease and forest dieback under global climate warming. *Theoretical and Applied Climatology*, 78(1–3), 157–175.

Binney, H., Edwards, M., Macias-Fauria, M., *et al.* (2017). Vegetation of Eurasia from the last glacial maximum to present: Key biogeographic patterns. *Quaternary Science Reviews*, 157, 80–97.

Birch, J., Newton, A. C., Alvarez Aquino, C., *et al.* (2010). Cost-effectiveness of dryland forest restoration evaluated by spatial analysis of ecosystem services. *Proceedings of the National Academy of Sciences of the United States of America*, 107 (50), 21925–21930.

Bird, M. I., Hutley, L. B., Lawes, M. J., *et al.* (2013). Humans, megafauna and environmental change in tropical Australia. *Journal of Quaternary Science*, 28(5), 439–452.

References

Bird, M. I., O'Grady, D. and Ulm, S. (2016). Humans, water, and the colonization of Australia. *Proceedings of the National Academy of Sciences of the United States of America*, 113(41), 11477–11482.

Birks, H. H. and Birks, H. J. B. (2004). The rise and fall of forests. *Science*, 305(5683), 484–485.

Blake, L. (2015). 'Are we worth saving? You tell me': Neoliberalism, zombies and the failure of free trade. *Gothic Studies*, 17(2), 26–41.

Blanchon, P. and Shaw, J. (1995). Reef drowning during the last deglaciation: Evidence for catastrophic sea-level rise and ice-sheet collapse. *Geology*, 23(1), 4–8.

Bland, L. M., Keith, D. A., Miller, R. M., Murray, N. J. and Rodríguez, J. P. (eds.). (2017a). *Guidelines for the Application of IUCN Red List of Ecosystems Categories and Criteria, Version 1.1.* IUCN, Gland, Switzerland.

Bland, L. M., Regan, T. J., Dinh, M. N., et al. (2017b). Using multiple lines of evidence to assess the risk of ecosystem collapse. *Proceedings of the Royal Society B: Biological Sciences*, 284(1863), 20170660.

Bland, L. M., Rowland, J. A., Regan, T. J., et al. (2018a). Developing a standardized definition of ecosystem collapse for risk assessment. *Frontiers in Ecology and the Environment*, 16(1), 29–36.

Bland, L. M., Watermeyer, K. E., Keith, D. A., Nicholson, E., Regan, T. J. and Shannon, L. J. (2018b). Assessing risks to marine ecosystems with indicators, ecosystem models and experts. *Biological Conservation*, 227, 19–28.

Blew, R. D. (1996). On the definition of ecosystem. *Bulletin of the Ecological Society of America*, 77, 171–173.

Bliege Bird, R., Bird, D. W., Codding, B. F., Parker, C. H. and Jones, J. H. (2008). The 'fire stick farming' hypothesis: Australian Aboriginal foraging strategies, biodiversity, and anthropogenic fire mosaics. *Proceedings of the National Academy of Sciences of the United States of America*, 105(39), 14796–14801.

Blignaut, J. and Aronson, J. (2020). Developing a restoration narrative: A pathway towards system-wide healing and a restorative culture. *Ecological Economics*, 168, 106483.

Boast, A. P., Weyrich, L. S., Wood, J. R., et al. (2018). Coprolites reveal ecological interactions lost with the extinction of New Zealand birds. *Proceedings of the National Academy of Sciences of the United States of America*, 115, 1546–1551.

Bobbink, R., Hicks, K., Galloway, J., et al. (2010). Global assessment of nitrogen deposition effects on terrestrial plant diversity: A synthesis. *Ecological Applications*, 20, 30–59.

Boettiger, C. and Hastings, A. (2012). Early warning signals and the prosecutor's fallacy. *Proceedings of the Royal Society B: Biological Sciences*, 279, 4734–4739.

(2013). From patterns to predictions. *Nature*, 493, 157–158.

Boitani, L., Mace, G. M. and Rondinini, C. (2015). Challenging the scientific foundations for an IUCN Red List of ecosystems. *Conservation Letters*, 8(2), 125–131.

Bonan, G. B., Pollard, D. and Thompson, S. L. (1992). Effects of Boreal forest vegetation on global climate. *Nature*, 359, 716–718.

Bond, D. P. G. and Grasby, S. E. (2017). On the causes of mass extinctions. *Palaeogeography, Palaeoclimatology, Palaeoecology*, 478, 3–29.

Bond, W. J. (2008). What limits trees in C4 grasslands and savannas? *Annual Review of Ecology, Evolution, and Systematics*, 39(1), 641–659.

Bond, W. J. and Midgley, J. J. (2012). Fire and the angiosperm revolutions. *International Journal of Plant Science*, 173, 569–583.

Bond, W. J., Woodward, F. I. and Midgley, G. F. (2005). The global distribution of ecosystems in a world without fire. *New Phytologist*, 165, 525–538.

Borer, E. T., Seabloom, E. W., Shurin, J. B., et al. (2005). What determines the strength of a trophic cascade? *Ecology*, 86, 528–537.

Borie, M. and Hulme, M. (2015). Framing global biodiversity: IPBES between mother earth and ecosystem services. *Environmental Science & Policy*, 54, 487–496.

Börner, J., Baylis, K., Corbera, E., et al. (2017). The effectiveness of payments for environmental services. *World Development*, 96, 359–374.

Boulton, A. J. (2003). Parallels and contrasts in the effects of drought on stream macroinvertebrate assemblages. *Freshwater Biology*, 48, 1173–1185.

Bowman, D. M. J. S., Perry, G. L. W. and Marston, J. B. (2015). Feedbacks and landscape-level vegetation dynamics. *Trends in Ecology & Evolution*, 30(5), 255–260.

Braat, L. C. (2018). Five reasons why the *Science* publication 'Assessing nature's contributions to people' (Diaz et al. 2018) would not have been accepted in *Ecosystem Services*. *Ecosystem Services*, 30, A1–A2.

Bradshaw, A. D. (1984). Ecological principles and land reclamation practice. *Landscape Planning*, 11, 35–48.

Bradshaw, R. and Mitchell, F. J. G. (1999). The palaeoecological approach to reconstructing former grazing–vegetation interactions. *Forest Ecology and Management*, 120, 3–12.

Branch, T. A. (2013). Citation patterns of a controversial and high-impact paper: Worm et al. (2006). "Impacts of biodiversity loss on ocean ecosystem services". *PLoS One*, 8(2), e56723. https://doi.org/10.1371/journal.pone.0056723.

  (2015). Fishing impacts on food webs: Multiple working hypotheses. *Fisheries*, 40, 373–375.

  (2016). Books and papers cited most often by fisheries scientists. https://sites.google.com/a/uw.edu/most-cited-fisheries/home (accessed on 14 March 2019).

Branch, T. A., Watson, R., Fulton, E. A., et al. (2010). The trophic fingerprint of marine fisheries. *Nature*, 468, 431–435.

Brand, F. S. and Jax, K. (2007). Focusing the meaning(s) of resilience: Resilience as a descriptive concept and a boundary object. *Ecology and Society*, 12(1), 23. www.ecologyandsociety.org/vol12/iss1/art23/.

Brando, P. M., Balch, J. K., Nepstad, D. C., et al. (2014). Abrupt increases in Amazonian tree mortality due to drought-fire interactions. *Proceedings of the National Academy of Sciences of the United States of America*, 111(17), 6347–6352.

Brannen, P. (2017). *The Ends of the World. Volcanic Apocalypses, Lethal Oceans and Our Quest to Understand the Earth's Part Mass Extinctions*. Oneworld Publications, London.

Brault, M., Mysak, L., Matthews, H. and Simmons, C. (2013). Assessing the impact of late Pleistocene megafaunal extinctions on global vegetation and climate. *Climate of the Past*, 9, 1761–1771.

Bremer, L. L. and Farley, K. A. (2010). Does plantation forestry restore biodiversity or create green deserts? A synthesis of the effects of land-use transitions on plant species richness. *Biodiversity and Conservation*, 19, 3893–3915.

Brienen, R. J. W., Phillips, O. L., Feldpausch, T. R., et al. (2015). Long-term decline of the Amazon carbon sink. *Nature*, 519, 344–348.

Brierley, C., Manning, K. and Maslin, M. (2018). Pastoralism may have delayed the end of the green Sahara. *Nature Communications*, 9(1), 4018.

Briggs, J. C. (2014). Global biodiversity gain is concurrent with declining population sizes. *Biodiversity Journal*, 5(4), 447–452.

Brinck, K., Fischer, R., Groeneveld, J., et al. (2017). High resolution analysis of tropical forest fragmentation and its impact on the global carbon cycle. *Nature Communications*, 8, 14855.

Briske, D. D., Fuhlendorf, S. D. and Smeins, F. E. (2006). A unified framework for assessment and application of ecological thresholds. *Rangeland Ecology & Management*, 59(3), 225–236.

Briske, D. D., Washington-Allen, R. A., Johnson, C. R., et al. (2010). Catastrophic thresholds: A synthesis of concepts, perspectives and applications. *Ecology and Society*, 15(3), 37. www.ecologyandsociety.org/vol15/iss3/art37/.

Brockerhoff, E. G., Jactel, H., Parrotta, J. A. and Ferraz, S. F. B. (2013). Role of eucalypt and other planted forests in biodiversity conservation and the provision of biodiversity-related ecosystem services. *Forest Ecology and Management*, 301, 43–50.

Brodie, J. F., Aslan, C. E., Rogers, H. S., et al. (2014). Secondary extinctions of biodiversity. *Trends in Ecology & Evolution*, 29(12), 664–672.

Brook, B. W., Sodhi, N. S. and Ng, P. K. L. (2003). Catastrophic extinctions follow deforestation in Singapore. *Nature*, 424, 420–426.

Brook, B. W., Bradshaw, C. J. A., Pin Koh, L. and Sodhi, N. S. (2006). Momentum drives the crash: Mass extinction in the tropics. *Biotropica*, 38, 302–305.

Brook, B. W., Sodhi, N. S. and Bradshaw, C. J. A. (2008). Synergies among extinction drivers under global change. *Trends in Ecology & Evolution*, 23(8), 453–460.

Brook, B. W., Ellis, E. C., Perring, M. P., Mackay, A. W. and Blomqvist, L. (2013). Does the terrestrial biosphere have planetary tipping points? *Trends in Ecology & Evolution*, 28(7), 396–401.

Brooks, M. L., D'Antonio, C. M., Richardson, D. M., et al. (2004). Effects of invasive alien plants on fire regimes. *BioScience*, 54(7), 677–688.

Broughton, J. M. and Weitzel, E. M. (2018). Population reconstructions for humans and megafauna suggest mixed causes for North American Pleistocene extinctions. *Nature Communications*, 9(1), 5441.

Brovkin, V., Claussen, M., Petoukhov, V. and Ganopolski, A. (1998). On the stability of the atmosphere-vegetation system in the Sahara / Sahel region. *Journal of Geophysical Research*, 103, 31613–31624.

Brown, A. A. and Crema, E. R. (2019). Māori population growth in pre-contact New Zealand: Regional population dynamics inferred from summed probability distributions of radiocarbon dates. *The Journal of Island and Coastal Archaeology*, 1–19.

Bruelheide, H. and Luginbühl, U. (2009). Peeking at ecosystem stability: Making use of a natural disturbance experiment to analyze resistance and resilience. *Ecology*, 90, 1314–1325.

Bruno, J. F., Sweatman, H., Precht, W. F., Selig, E. R. and Schutte, V. G. (2009). Assessing evidence of phase shifts from coral to macroalgal dominance on coral reefs. *Ecology*, 90, 1478–1484.

Buckley, M. C. and Crone, E. E. (2008). Negative off-site impacts of ecological restoration: Understanding and addressing the conflict. *Conservation Biology*, 22 (5), 1118–1124.

Bull, K. R. (1995). Critical loads – Possibilities and constraints. *Water, Air, and Soil Pollution*, 85(1), 201–212.

Bullock, J. M., Aronson, J., Newton, A. C., Pywell, R. F. and Rey-Benayas, J. M. (2011). Restoration of ecosystem services and biodiversity: Conflicts and opportunities. *Trends in Ecology & Evolution*, 26, 541–549.

Buma, B. (2015). Disturbance interactions: Characterization, prediction, and the potential for cascading effects. *Ecosphere*, 6(4), 70.

Burke, L., Reytar, K., Spalding, M. and Perry, A. (2011). *Reefs at Risk Revisited*. World Resources Institute, Washington, DC.

Burkepile, D. E. and Hay, M. E. (2006). Herbivore vs. nutrient control of marine primary producers: Context-dependent effects. *Ecology*, 87, 3128–3139.

Burnett, J. L. (2019). *Regime Detection Measures for the Practical Ecologist*. Dissertations and Theses in Natural Resources, p. 299. https://digitalcommons.unl.edu/natresdiss/299.

Burney, D. A., Robinson, G. S. and Burney, L. P. (2003). *Sporormiella* and the late Holocene extinctions in Madagascar. *Proceedings of the National Academy of Sciences of the United States of America*, 100(19), 10800–10805.

Burney, D. A, Pigott Burney, L., Godfrey, L. R., et al. (2004). A chronology for late prehistoric Madagascar. *Journal of Human Evolution*, 47(1–2), 25–63.

Burns, D. A., Blett, T., Haeuber, R. and Pardo, L. H. (2008). Critical loads as a policy tool for protecting ecosystems from the effects of air pollutants. *Frontiers in Ecology and the Environment*, 6(3), 156–159.

Burthe, S. J., Henrys, P. A., Mackay, E. B., et al. (2016). Do early warning indicators consistently predict nonlinear change in long-term ecological data? *Journal of Applied Ecology*, 53(3), 666–676.

Busby, P. E. and Canham, C. D. (2011). An exotic insect and pathogen disease complex reduces aboveground tree biomass in temperate forests of eastern North America. *Canadian Journal of Forest Research*, 41, 401–411.

Büscher, B., Sullivan, S., Neves, K., Igoe, J. and Brockington, D. (2012). Towards a synthesized critique of neoliberal biodiversity conservation. *Capitalism Nature Socialism*, 23(2), 4–30.

Butchart, S. H. M., Walpole, M., Collen, B., et al. (2010). Global biodiversity: Indicators of recent declines. *Science*, 328, 1164–1168.

Butzer, K. W. (2012). Collapse, environment, and society. *Proceedings of the National Academy of Sciences of the United States of America*, 109(10), 3632–3639.

Buxton, T. H., Buffington, J. M., Tonina, D., Fremier, A. K. and Yager, E. M. (2015). Modeling the influence of salmon spawning on hyporheic exchange of marine-derived nutrients in gravel stream beds. *Canadian Journal of Fisheries and Aquatic Sciences*, 72(8), 1146–1158.

Cafaro, P. (2015). Three ways to think about the sixth mass extinction. *Biological Conservation*, 192, 387–393.

Cahill, A. E., Aiello-Lammens, M. E., Fisher-Reid, M. C., et al. (2012). How does climate change cause extinction? *Proceedings of the Royal Society B: Biological Sciences*, 280, 20121890.

Calvet-Mir, L., Corbera, E., Martin, A., Fisher, J. and Gross-Camp, N. (2015). Payments for ecosystem services in the tropics: A closer look at effectiveness and equity. *Current Opinion in Environmental Sustainability*, 14, 150–162.

Cambridge Dictionary. (2019). *Cambridge Dictionary*. Cambridge University Press. https://dictionary.cambridge.org/ (accessed on 11 December 2019).

Campos-Arceiza, A. and Blake, S. (2011). Megagardeners of the forest – The role of elephants in seed dispersal. *Acta Oecologica*, 37, 542–553.

Cantarello, E., Newton, A. C., Hill, R. A., et al. (2011). Simulating the potential for ecological restoration of dryland forests in Mexico under different disturbance regimes. *Ecological Modelling*, 222(5), 1112–1128.

Cantarello, E., Newton, A. C., Martin, P. A., Evans, P. M., Gosal, A. and Lucash, M. S. (2017). Quantifying resilience of multiple ecosystem services and biodiversity in a temperate forest landscape. *Ecology and Evolution*, 7(22), 9661–9675.

Capon, S. J., Lynch, A. J. J., Bond, N., et al. (2015). Regime shifts, thresholds and multiple stable states in freshwater ecosystems; a critical appraisal of the evidence. *Science of the Total Environment*, 534, 122–130.

Cardinale, B. J., Duffy, J. E., Gonzalez, A., et al. (2012). Biodiversity loss and its impact on humanity. *Nature*, 486, 59–67.

Cardoso, P., Branco, V. V., Chichorro, F., Fukushima, C. S. and Macías-Hernández, N. (2019). Can we really predict a catastrophic worldwide decline of entomofauna and its drivers? *Global Ecology and Conservation*, 20, e00621.

Carlson, A. K., Taylor, W. W., Liu, J. and Orlic, I. (2018). Peruvian anchoveta as a telecoupled fisheries system. *Ecology and Society*, 23, art35. https://doi.org/10.5751/ES-09923-230135.

Carlsson, N. O. L., Brönmark, C. and Hansson, L.-A. (2004). Invading herbivory: The golden apple snail alters ecosystem functioning in Asian wetlands. *Ecology*, 85, 1575–1580.

Carnicer, J., Coll, M., Ninyerola, M., Pons, X., Sánchez, G. and Peñuelas, J. (2011). Widespread crown condition decline, food web disruption, and amplified tree mortality with increased climate change-type drought. *Proceedings of the National Academy of Sciences of the United States of America*, 108(4), 1474–1478.

Carpenter, J. K., Wood, J. R., Wilmshurst, J. M. and Kelly, D. (2018). An avian seed dispersal paradox: New Zealand's extinct megafaunal birds did not disperse large seeds. *Proceedings of the Royal Society B: Biological Sciences*, 285, 20180352.

Carpenter, S. R. (2005). Eutrophication of aquatic ecosystems: Bistability and soil phosphorus. *Proceedings of the National Academy of Sciences of the United States of America*, 102(29), 10002–10005.

Carpenter, S. R., Cole, J. J., Hodgson, J. R., et al. (2001). Trophic cascades, nutrients, and lake productivity: Whole-lake experiments. *Ecological Monographs*, 71, 163–186.

Carpenter, S. R., Brock, W. A., Cole, J. J., Kitchell, J. F. and Pace, M. L. (2008). Leading indicators of trophic cascades. *Ecology Letters*, 11, 128–138.

Carpenter, S. R., Cole, J. J., Pace, M. L., et al. (2011). Early warnings of regime shifts: A whole-ecosystem experiment. *Science*, 332(6033), 1079–1082.

Carpenter, S. R., Brock, W. A., Folke, C., van Nes, E. H. and Scheffer, M. (2015). Allowing variance may enlarge the safe operating space for exploited ecosystems. *Proceedings of the National Academy of Sciences of the United States of America*, 112(46), 14384–14389.

Cascales-Miñana, B. and Cleal, C. J. (2011). Plant fossil record and survival analyses. *Lethaia*, 45, 71–82.

Caseldine, C. J. and Turney, C. (2010). The bigger picture: Towards integrating palaeoclimate and environmental data with a history of societal change. *Journal of Quaternary Science*, 25(1), 88–93.

Casey, J. M., Baird, A. H., Brandl, S. J., et al. (2017). A test of trophic cascade theory: Fish and benthic assemblages across a predator density gradient on coral reefs. *Oecologia*, 183, 161–175.

Casini, M., Lövgren, J., Hjelm, J., et al. (2008). Multi-level trophic cascades in a heavily exploited open marine ecosystem. *Proceedings of the Royal Society of London Series B: Biological Sciences*, 275, 1793–1801.

Casini, M., Hjelm, J., Molinero, J.-C., et al. (2009). Trophic cascades promote threshold-like shifts in pelagic marine ecosystems. *Proceedings of the National Academy of Sciences of the United States of America*, 106(1), 197–202.

Caughley, G. (1994). Directions in conservation biology. *Journal of Animal Ecology*, 63, 215–244.

CBD. (2012). *Quick Guide to the Aichi Biodiversity Targets. T15.* CBD Secretariat, Montreal, Canada. www.cbd.int/doc/strategic-plan/targets/T15-quick-guide-en.pdf.

Ceballos, G., Ehrlich, P. R., Barnosky, A. D., García, A., Pringle, R. M. and Palmer, T. M. (2015). Accelerated modern human–induced species losses: Entering the sixth mass extinction. *Science Advances*, 1(5), e1400253.

Ceballos, G., Ehrlich, P. R. and Dirzo, R. (2017). Biological annihilation via the ongoing sixth mass extinction signaled by vertebrate population losses and declines. *Proceedings of the National Academy of Sciences of the United States of America*, 114(30), E6089–E6096.

Cerrano, C., and Bavestrello, G. (2008). Medium-term effects of die-off of rocky benthos in the Ligurian Sea. What can we learn from gorgonians? *Chemistry and Ecology*, 24(sup1), 73–82.

Chapman, D. A., Lickel, B. and Markowitz, E. M. (2017). Reassessing emotion in climate change communication. *Nature Climate Change*, 7(12), 850–852.

Chapin, F. S. III, Matson, P. A. and Mooney, H. A. (2002). *Principles of Terrestrial Ecosystem Ecology*. Springer-Verlag, New York. 394 p.

Chapin, F. S. III, Callaghan, T. V., Bergeron, Y., et al. (2004). Global change and the boreal forest: Thresholds shifting states or gradual change? *AMBIO: A Journal of the Human Environment*, 33, 361–365.

Chaplin-Kramer, R., Sharp, R. P., Weil, C., et al. (2019). Global modeling of nature's contributions to people. *Science*, 366(6462), 255–258.

Chazdon, R. L. (2003). Tropical forest recovery: Legacies of human impact and natural disturbances. *Perspectives in Plant Ecology, Evolution and Systematics*, 6, 51–71.

(2008). Beyond deforestation: Restoring forests and ecosystem services on degraded lands. *Science*, 320, 1458–1460.

(2014). *Second Growth: The Promise of Tropical Forest Regeneration in an Age of Deforestation*. The University of Chicago Press, Chicago, IL.

Chazdon, R. and Arroyo, J. P. (2013). Tropical forests as complex adaptive systems. In: C. Messier, K. J. Puettmann and K. Coates (eds.), *Managing Forests as Complex Adaptive Systems. Building Resilience to the Challenge of Global Change*. Earthscan / Routledge, London and New York, pp. 35–59.

Chazdon, R. L. and Brancalion, P. (2019). Restoring forests as a means to many ends. *Science*, 365(6448), 24–25.

Chazdon, R. L. and Guariguata, M. R. (2016). Natural regeneration as a tool for large-scale forest restoration in the tropics: Prospects and challenges. *Biotropica*, 48, 716–730.

Chen, D., Qing, H. and Li, R. (2005). The Late Devonian Frasnian–Famennian (F/F) biotic crisis: Insights from $\delta^{13}C_{carb}$, $\delta^{13}C_{org}$ and $^{87}Sr/^{86}Sr$ isotopic systematics. *Earth Planetary Science Letters*, 235, 151–166.

Chen, I.-C., Hill, J. K., Ohlemuller, R., Roy, D. B. and Thomas, C. D. (2011). Rapid range shifts of species associated with high levels of climate warming. *Science*, 333, 1024–1026.

Chen, M.-F. (2015). Impact of fear appeals on pro-environmental behavior and crucial determinants. *International Journal of Advertising*, 35(1), 74–92.

Chen, Z. Q. and Benton, M. J. (2012). The timing and pattern of biotic recovery following the end-Permian mass extinction. *Nature Geoscience*, 5, 375–383.

Choi, Y. D. (2004). Theories for ecological restoration in changing environment: Toward 'futuristic' restoration. *Ecological Research*, 19, 75–81.

(2017). Considering the future. Anticipating the need for ecological restoration. In: S. K. Allison and S. D. Murphy (eds.), *Routledge Handbook of Ecological and Environmental Restoration*. Routledge, Abingdon and New York, pp. 7–15.

Christie, M., Holland, S. M. and Bush, A. M. (2013). Contrasting the ecological and taxonomic consequences of extinction. *Paleobiology*, 39(4), 538–559.

Chytrý, M., Horsák, M., Danihelka, J., *et al.* (2018). A modern analogue of the Pleistocene steppe-tundra ecosystem in southern Siberia. *Boreas*. https://doi.org/10.1111/bor.12338.

Çilingiroğlu, Ç. (2005). The concept of 'Neolithic package': Considering its meaning and applicability. *Documenta Praehistorica*, 32, 1–13.

Clapham, M. E. (2016). Organism activity levels predict marine invertebrate survival during ancient global change extinctions. *Global Change Biology*, 23, 1477–1485.

Clark, D. B., Hurtado, J. and Saatchi, S. S. (2015). Tropical rain forest structure, tree growth and dynamics along a 2700-m elevational transect in Costa Rica. *PLoS One*, 10, e0122905.

Clark, J. S. (1989). Ecological disturbance as a renewal process: Theory and application to fire history. *Oikos*, 56, 17–30.

(1991). Disturbance and tree life history on the shifting mosaic landscape. *Ecology*, 72, 1102–1118.

Clarkson, C., Jacobs, Z., Marwick, B., *et al.* (2017). Human occupation of northern Australia by 65,000 years ago. *Nature*, 547, 306–310.

References

Claussen, M., Kubatzki, C., Brovkin, V., Ganopolski, A., Hoelzmann, P. and Pachur, H.-J. (1999). Simulation of an abrupt change in Saharan vegetation in the mid-Holocene. *Geophysical Research Letters*, 26, 2037–2040.

Claussen, M., Dallmeyer, A. and Bader, J. (2017). Theory and modeling of the African Humid Period and the Green Sahara. In: *Oxford Research Encyclopedias of Climate Science*. https://doi.org/10.1093/acrefore/9780190228620.013.532.

Clements, C. F. and Ozgul, A. (2018). Indicators of transitions in biological systems. *Ecology Letters*, 21(6), 905–919.

Clements, C. F., McCarthy, M. A. and Blanchard, J. L. (2019). Early warning signals of recovery in complex systems. *Nature Communications*, 10(1), 1681 https://doi.org/10.1038/s41467-019-09684-y.

Clewell, A. F. and Aronson, J. (2013). *Ecological Restoration: Principles, Values and Structure of an Emerging Profession*, 2nd edition. Island Press, Washington, DC.

Clifford, C. C. and Heffernan, J. B. (2018). Artificial aquatic ecosystems. *Water*, 10, 1096.

Cline, T. J., Seekell, D. A., Carpenter, S. R., et al. (2014). Early warnings of regime shifts: Evaluation of spatial indicators from a whole-ecosystem experiment. *Ecosphere*, 5(8), art102. https://doi.org/10.1890/es13-00398.1.

Cole, L. E. S., Bhagwat, S. A. and Willis, K. J. (2014). Recovery and resilience of tropical forests after disturbance. *Nature Communications*, 5, 3906.

Colwell, R. K., Dunn, R. R. and Harris, N. C. (2012). Coextinction and persistence of dependent species in a changing world. *Annual Review of Ecology and Systematics*, 43, 183–203.

Connell, J. H. and Sousa, W. P. (1983). On the evidence needed to judge ecological stability or persistence. *The American Naturalist*, 121, 789–824.

Coomes, D. A., Allen, R. B., Forsyth, D. M. and Lee, W. G. (2003). Factors preventing the recovery of New Zealand forests following control of invasive deer. *Conservation Biology*, 17, 450–459.

Cooper, A., Turney, C., Hughen, K. A., Brook, B. W., McDonald, H. G. and Bradshaw, C. J. A. (2015). Abrupt warming events drove Late Pleistocene Holarctic megafaunal turnover. *Science*, 349(6248), 602–606.

Cordingley, J. E., Newton, A. C., Rose, R. J., Clarke, R. and Bullock, J. M. (2015a). Can landscape-scale approaches to conservation management resolve biodiversity – Ecosystem service tradeoffs? *Journal of Applied Ecology*, 53(1), 96–105.

(2015b). Habitat fragmentation intensifies trade-offs between biodiversity and ecosystem services. *PLoS One*, 10(6), e0130004. https://doi.org/10.1371/journal.pone.0130004.

Corlett, R. T. (2016). Restoration, reintroduction, and rewilding in a changing world. *Trends in Ecology & Evolution*, 31(6), 453–462.

Cortina, J., Maestre, F. T., Vallejo, R., Baeza, M. J., Valdecantos, A. and Pérez-Devesa, M. (2006). Ecosystem structure, function, and restoration success: Are they related? *Journal for Nature Conservation*, 14(3), 152–160.

Costanza, R., Graumlich, L., Steffen, W., et al. (2007). Sustainability or collapse: What can we learn from integrating the history of humans and the rest of nature? *AMBIO: A Journal of the Human Environment*, 36(7), 522–527.

Costanza, R., de Groot, R., Braat, L., *et al.* (2017). Twenty years of ecosystem services: How far have we come and how far do we still need to go? *Ecosystem Services*, 28, 1–16.

Costello, C., Ovando, D., Clavelle, T., *et al.* (2016). Global fishery prospects under contrasting management regimes. *Proceedings of the National Academy of Sciences of the United States of America*, 113(18), 5125–5129.

Côté, I. M., Darling, E. S. and Brown, C. J. (2016). Interactions among ecosystem stressors and their importance in conservation. *Proceedings of the Royal Society B: Biological Sciences*, 283, 20152592.

Côté, S., Rooney, T., Tremblay, J., Dussault, C. and Waller, D. (2004). Ecological Impacts of Deer Overabundance. *Annual Review of Ecology, Evolution, and Systematics*, 35, 113–147.

Cox, M. P., Nelson, M. G., Tumonggor, M. K., Ricault, F.-X. and Sudoyo, H. (2012). A small cohort of Island Southeast Asian women founded Madagascar. *Proceedings of the Royal Society B: Biological Sciences*, 279(1739), 2761–2768.

Cox, P., Betts, R., Collins, M., *et al.* (2004). Amazonian forest dieback under climate-carbon cycle projections for the 21st century. *Theoretical and Applied Climatology*, 78(1–3), 137–156.

Cox, P. M., Pearson, D., Booth, B. B., *et al.* (2013). Sensitivity of tropical carbon to climate change constrained by carbon dioxide variability. *Nature*, 494, 341–344.

Crain, C. M., Kroeker, K. and Halpern, B. S. (2008). Interactive and cumulative effects of multiple human stressors in marine systems. *Ecology Letters*, 11, 1304–1315.

Crawford, R. J. M. (2007). Food, fishing and seabirds in the Benguela upwelling system. *Journal of Ornithology*, 148, 253–260.

Crossman, N. D., Burkhard, B., Nedkov, S., *et al.* (2013). A blueprint for mapping and modelling ecosystem services. *Ecosystem Services*, 4(Supplement C), 4–14.

Crouzeilles, R. and Curran, M. (2016). Which landscape size best predict the influence of forest cover on restoration success? A global meta-analysis on the scale of effect. *Journal of Applied Ecology*, 53, 440–448.

Crouzeilles, R., Curran, M., Ferreira, M. S., Lindenmayer, D. B., Grelle, C. E. V. and Rey Benayas, J. M. (2016). A global meta-analysis on the ecological drivers of forest restoration success. *Nature Communications*, 7, 11666.

Crouzeilles, R., Ferreira, M. S., Chazdon, R. L., *et al.* (2017). Ecological restoration success is higher for natural regeneration than for active restoration in tropical forests. *Science Advances*, 3(11), e1701345.

Crutzen, P. J. (2006). The 'Anthropocene'. In: E. Ehlers and T. Krafft (eds.), *Earth System Science in the Anthropocene*. Springer, Berlin/Heidelberg, pp. 13–18.

Cruz, I. C. S., Waters, L. G., Kikuchi, R. K. P., Leão, Z. M. A. N. and Turra, A. (2018). Marginal coral reefs show high susceptibility to phase shift. *Marine Pollution Bulletin*, 135, 551–561.

Cui, Y., Bercovici, A., Yu, J., *et al.* (2017). Carbon cycle perturbation expressed in terrestrial Permian–Triassic boundary sections in South China. *Global and Planetary Change*, 148, 272–285.

Cumming, G. S. and Peterson, G. D. (2017). Unifying research on social and ecological resilience and collapse. *Trends in Ecology & Evolution*, 32(9), 695–713.

Cunsolo, A. and Ellis, N. R. (2018). Ecological grief as a mental health response to climate change-related loss. *Nature Climate Change*, 8(4), 275–281.

Curran, M., Hellweg, S. and Beck, J. (2014). Is there any empirical support for biodiversity offset policy? *Ecological Applications*, 24(4), 617–632.

Curtis, P. G., Slay, C. M., Harris, N. L., Tyukavina, A. and Hansen, M. C. (2018). Classifying drivers of global forest loss. *Science*, 361(6407), 1108–1111.

Curtsdotter, A., Binzer, A., Brose, U., *et al.* (2011). Robustness to secondary extinctions: Comparing trait-based sequential deletions in static and dynamic food webs. *Basic and Applied Ecology*, 12(7), 571–580.

Cury, P., Bakun, A., Crawford, R. J. M., *et al.* (2000). Small pelagics in upwelling systems: Patterns of interaction and structural changes in 'wasp-waist' ecosystems. *ICES Journal of Marine Science*, 57, 603–618.

Czembor, C. A. and Vesk, P. A. (2009). Incorporating between-expert uncertainty into state-and-transition simulation models for forest restoration. *Forest Ecology and Management*, 259, 165–175.

Daan, N., Gislason, H., Pope, J. G. and Rice, J. C. (2011). Apocalypse in world fisheries? The reports of their death are greatly exaggerated. *ICES Journal of Marine Science*, 68, 1375–1378.

Dakos, V. and Bascompte, J. (2014). Critical slowing down as early warning for the onset of collapse in mutualistic communities. *Proceedings of the National Academy of Sciences of the United States of America*, 111, 17546–17551.

Dakos, V., Carpenter, S. R., Brock, W. A., *et al.* (2012). Methods for detecting early warnings of critical transitions in time series illustrated using simulated ecological data. *PLoS One*, 7, e41010.

Dakos, V., Carpenter, S. R., van Nes, E. H. and Scheffer, M. (2015). Resilience indicators: Prospects and limitations for early warnings of regime shifts. *Philosophical Transactions of the Royal Society B: Biological Sciences*, 370, 20130263.

Danell, K., Bergstrom, R., Edenius, L. and Ericsson, G. (2003). Ungulates as drivers of tree population dynamics at module and genet levels. *Forest Ecology and Management*, 181, 67–76.

Daniels, K., Connolly, S., Ogbonnaya, C., *et al.* (2018). Democratisation of wellbeing: Stakeholder perspectives on policy priorities for improving national wellbeing through paid employment and adult learning. *British Journal of Guidance & Counselling*, 46(4), 492–511.

Danovaro, R., Gambi, C., Dell'Anno, A., *et al.* (2008). Exponential decline of deep-sea ecosystem functioning linked to benthic biodiversity loss. *Current Biology*, 18(1), 1–8.

Dantas, V. de L., Batalha, M. A. and Pausas, J. G. (2013). Fire drives functional thresholds on the savanna–forest transition. *Ecology*, 94(11), 2454–2463.

Dantas, V. de L., Hirota, M., Oliveira, R. S. and Pausas, J. G. (2016). Disturbance maintains alternative biome states. *Ecology Letters*, 19(1), 12–19.

Darling, E. S. and Côté, I. M. (2008). Quantifying the evidence for ecological synergies. *Ecology Letters*, 11, 1278–1286.

Daskalov, G. M., Grishin, A. N., Rodionov, S. and Mihneva, V. (2007). Trophic cascades triggered by overfishing reveal possible mechanisms of ecosystem

regime shifts. *Proceedings of the National Academy of Sciences of the United States of America*, 104, 10518–10523.
Davenas, E., Beauvais, F., Amara, J., et al. (1988). Human basophil degranulation triggered by very dilute antiserum against IgE. *Nature*, 333(6176), 816–818.
Davidson, E. A., de Araújo, A. C., Artaxo, P., et al. (2012). The Amazon basin in transition. *Nature*, 481, 321–328.
Dayton, P. K., Tegner, M. J., Edwards, P. B. and Riser, K. L. (1998). Sliding baselines, ghosts, and reduced expectations in kelp forest communities. *Ecological Applications*, 8(2), 309–322.
Deakin, M. A. B. (1990). Catastrophe modelling in the biological sciences. *Acta Biotheoretica*, 38, 3–22.
DeFries, R. S., Foley, J. A. and Asner, G. P. (2004). Land-use choices: Balancing human needs and ecosystem function. *Frontiers in Ecology and the Environment*, 2 (5), 249–257.
de Groot, R., Brander, L., van der Ploeg, S., et al. (2012). Global estimates of the value of ecosystems and their services in monetary units. *Ecosystem Services*, 1(1), 50–61.
de Groot, R. S., Blignaut, J., van der Ploeg, S., Aronson, J., Elmqvist, T. and Farley, J. (2013). Benefits of investing in ecosystem restoration. *Conservation Biology*, 27 (6), 1286–1293.
Deguines, N., Jono, C., Baude, M., Henry, M., Julliard, R. and Fontaine, C. (2014). Large-scale trade-off between agricultural intensification and crop pollination services. *Frontiers in Ecology and the Environment*, 12(4), 212–217.
de Laplante, K. (2005). Is ecosystem management a postmodern science? In: K. E. Cuddington and B. E. Beisner (eds.), *Ecological Paradigms Lost, Routes of Theory Change*. Elsevier Academic Press, San Diego, pp. 397–418.
Delgado, L. E. and Marín, V. H. (2016). Human well-being and historical ecosystems: The environmentalist's paradox revisited. *BioScience*, 67(1), 5–6.
Deloitte Access Economics. (2013). *Economic Contribution of the Great Barrier Reef*. Great Barrier Reef Marine Park Authority, Townsville.
de Menocal, P., Oritz, J., Guilderson, T., et al. (2000). Abrupt onset and termination of the African Humid Period: Rapid climate responses to gradual insolation forcing. *Quaternary Science Reviews*, 19, 347–361.
Dent, C. L., Cumming, G. S. and Carpenter, S. R. (2002). Multiple states in river and lake ecosystems. *Philosophical Transactions of the Royal Society of London Series B: Biological Sciences*, 357(1421), 635–645.
de Oliveira Roque, F., Menezes, J. F. S., Northfield, T., Ochoa-Quintero, J. M., Campbell, M. J. and Laurance, W. F. (2018). Warning signals of biodiversity collapse across gradients of tropical forest loss. *Scientific Reports*, 8(1), 1622.
DePalma, R. A., Smit, J., Burnham, D. A., et al. (2019). A seismically induced onshore surge deposit at the KPg boundary, North Dakota. *Proceedings of the National Academy of Sciences of the United States of America*, 116(17), 8190–8199.
Derocher, A. E., Aars, J., Amstrup, S. C., et al. (2013). Rapid ecosystem change and polar bear conservation. *Conservation Letters*, 6(5), 368–375.
de Visser, S. N., Freymann, B. P. and Olff, H. (2011). The Serengeti food web: Empirical quantification and analysis of topological changes under increasing human impact. *Journal of Animal Ecology*, 80, 484–494.

Diamond, J. M. (1975). Assembly of species communities. In: M. L. Cody and J. M. Diamond (eds.), *Ecology and Evolution of Communities*. Harvard University Press, Harvard, pp. 342–444.
  (1990). Biological effects of ghosts. *Nature*, 345, 769–770.
  (2005). *Collapse. How Societies Choose to Fail or Succeed*. Penguin Books, London.
  (1984). 'Normal' extinction of isolated populations. In: M. H. Nitecki (ed.), *Extinctions*. Chicago University Press, Chicago, pp. 191–246.
Diamond, J. M., Ashmole, N. P. and Purves, P. E. (1989). Present, past and future of human-caused extinctions [and discussion]. *Proceedings of the Royal Society of London Series B: Biological Sciences*, 325, 469–477.
Diaz, A., Keith, S. A., Bullock, J. M., Hooftman, D. A. P. and Newton, A. C. (2013). Conservation implications of long-term changes detected in a lowland heath metacommunity. *Biological Conservation*, 167, 325–333.
Díaz, M. F. and Armesto, J. J. (2007). Physical and biotic constraints on tree regeneration in secondary shrublands of Chiloe Island, Chile. *Revista Chilena de Historia Natural*, 80, 13–26.
Diaz, R. J. and Rosenberg, R. (2008). Spreading dead zones and consequences for marine ecosystems. *Science*, 321(5891), 926–929.
Díaz, S., Symstad, A. J., Stuart Chapin, F., Wardle, D. A. and Huenneke, L. F. (2003). Functional diversity revealed by removal experiments. *Trends in Ecology & Evolution*, 18(3), 140–146.
Díaz, S., Demissew, S., Carabias, J., et al. (2015). The IPBES Conceptual Framework – Connecting nature and people. *Current Opinion in Environmental Sustainability*, 14, 1–16.
Díaz, S., Pascual, U., Stenseke, M., et al. (2018). Assessing nature's contributions to people. *Science*, 359(6373), 270–272.
Diaz-Pulido, G., McCook, L. J., Dove, S., et al. (2009). Doom and boom on a resilient reef: Climate change, algal overgrowth and coral recovery. *PLoS One*, 4(4), e5239. https://doi.org/10.1371/journal.pone.0005239.
Dickey-Collas, M., Nash, R. D. M., Brunel, T., et al. (2010). Lessons learned from stock collapse and recovery of North Sea herring: A review. *ICES Journal of Marine Science*, 67, 1875–1886.
Dickman, C. (2020). More than one billion animals killed in Australian bushfires. University of Sydney. https://sydney.edu.au/news-opinion/news/2020/01/08/australian-bushfires-more-than-one-billion-animals-impacted.html (accessed on 28 January 2020).
Dirzo, R. and Raven, P. H. (2003). Global state of biodiversity and loss. *Annual Review of Environment and Resources*, 28, 137–167.
Dirzo, R., Young, H. S., Galetti, M., Ceballos, G., Isaac, N. J. B. and Collen, B. (2014). Defaunation in the Anthropocene. *Science*, 345(6195), 401–406.
Doak, D. F., Bigger, D., Harding, E. K., Marvier, M. A., O'Malley, R. E. and Thomson, D. (1998). The statistical inevitability of stability-diversity relationships in community ecology. *The American Naturalist*, 151(3), 264–276.
Doak, D. F., Estes, J. A., Halpern, B. S., et al. (2008). Understanding and predicting ecological dynamics: Are major surprises inevitable? *Ecology*, 89, 952–961.
Doak, D. F., Bakker, V. J., Goldstein, B. E. and Hale, B. (2014). What is the future of conservation? *Trends in Ecology & Evolution*, 29(2), 77–81.

Doncaster, C. P., Alonso Chávez, V., Viguier, C., et al. (2016). Early warning of critical transitions in biodiversity from compositional disorder. *Ecology*, 97(11), 3079–3090.

Done, T. J. (1992). Phase shifts in coral reef communities and their ecological significance. *Hydrobiologia*, 247, 121–132.

Donohue, I., Petchey, O. L., Montoya, J. M., et al. (2013). On the dimensionality of ecological stability. *Ecology Letters*, 16, 421–429.

Donohue, I., Hillebrand, H., Montoya, J. M., et al. (2016). Navigating the complexity of ecological stability. *Ecology Letters*, 19(9), 1172–1185.

Doubleday, Z. A. and Connell, S. D. (2018). Weedy futures: Can we benefit from the species that thrive in the marine Anthropocene? *Frontiers in Ecology and the Environment*, 16(10), 599–604.

Doughty, C. E. (2013). Preindustrial human impacts on global and regional environment. *Annual Review of Environment and Resources*, 38(1), 503–527.

Doughty, C. E., Wolf, A. and Field, C. B. (2010). Biophysical feedbacks between the Pleistocene megafauna extinction and climate: The first human-induced global warming? *Geophysical Research Letters*, 37, L15703.

Doughty, C. E., Wolf, A. and Malhi, Y. (2013). The legacy of the Pleistocene megafauna extinctions on nutrient availability in Amazonia. *Nature Geoscience*, 6, 761–764.

Doughty, C. E., Wolf, A., Morueta-Holm, N., et al. (2016). Megafauna extinction, tree species range reduction, and carbon storage in Amazonian forests. *Ecography*, 39, 194–203.

Downey, S. S., Haas, W. R. and Shennan, S. J. (2016). European Neolithic societies showed early warning signals of population collapse. *Proceedings of the National Academy of Sciences of the United States of America*, 113(35), 9751–9756.

Downing, A. S., van Nes, E. H., Janse, J. H., et al. (2012). Collapse and reorganization of a food web of Mwanza Gulf, Lake Victoria. *Ecological Applications*, 22, 229–239.

Droser, M. L., Bottjer, D. J., Sheehan, P. M. and McGhee, G. R. (2000). Decoupling of taxonomic and ecologic severity of Phanerozoic marine mass extinctions. *Geology*, 28, 675–678.

Duarte, C. M., Borja, A., Carstensen, J., Elliott, M., Krause-Jensen, D. and Marbà, N. (2015). Paradigms in the recovery of estuarine and coastal ecosystems. *Estuaries and Coasts*, 38(4), 1202–1212.

Dudgeon, S. R., Aronson, R. B., Bruno, J. F. and Precht, W. F. (2010). Phase shifts and stable states on coral reefs. *Marine Ecology Progress Series*, 413, 201–216.

Dulvy, N. K., Freckleton, R. P. and Polunin, N. V. C. (2004). Coral reef cascades and the indirect effects of predator removal by exploitation. *Ecology Letters*, 7(5), 410–416.

Dunn, R. R., Harris, N. C., Colwell, R. K., Koh, L. P. and Sodhi, N. S. (2009). The sixth mass coextinction: Are most endangered species parasites and mutualists? *Proceedings of the Royal Society B: Biological Sciences*, 276(1670), 3037–3045.

Dunne, J. and Williams, R. (2009). Cascading extinctions and community collapse in model food webs. *Philosophical Transactions of the Royal Society B: Biological Sciences*, 364, 1711–1723.

Dunne, J., Williams, R. and Martinez, N. (2002a). Network structure and biodiversity loss in food webs: Robustness increases with connectance. *Ecology Letters*, 5, 558–567.

(2002b). Food-web structure and network theory: The role of connectance and size. *Proceedings of the National Academy of Sciences of the United States of America*, 99, 12917–12922.

Dwomoh, F. K. and Wimberly, M. C. (2017). Fire regimes and forest resilience: Alternative vegetation states in the West African tropics. *Landscape Ecology*, 32 (9), 1849–1865.

Ebenman, B. and Jonsson, T. (2005). Using community viability analysis to identify fragile systems and keystone species. *Trends in Ecology & Evolution*, 20, 568–575.

Ebenman, B., Law, R. and Borvall, C. (2004). Community viability analysis: The response of ecological communities to species loss. *Ecology*, 85, 2591–2600.

Edmunds, P. J. and Carpenter, R. C. (2001). Recovery of *Diadema antillarum* reduces macroalgal cover and increases abundance of juvenile corals on a Caribbean reef. *Proceedings of the National Academy of Sciences of the United States of America*, 98(9), 5067–5071.

Edwards, D. P., Larsen, T. H., Docherty, T. D. S., et al. (2011). Degraded lands worth protecting: The biological importance of Southeast Asia's repeatedly logged forests. *Proceedings of the Royal Society B: Biological Sciences*, 278(1702), 82–90.

Edwards, D. P., Tobias, J. A., Sheil, D., Meijaard, E. and Laurance, W. F. (2014). Maintaining ecosystem function and services in logged tropical forests. *Trends in Ecology & Evolution*, 29, 511–520.

Egler, F. E. (1986). 'Physics envy' in ecology. *Bulletin of the Ecological Society of America*, 67(3), 233–235.

Ehrlich, P. R. and Ehrlich, A. H. (2013). Can a collapse of global civilization be avoided? *Proceedings of the Royal Society B: Biological Sciences*, 280, 20122845. http://dx.doi.org/10.1098/rspb.2012.2845.

Ehrlich, P. R. and Mooney, H. A. (1983). Extinction, substitution and the ecosystem services. *BioScience*, 33, 248–254.

Eichhorn, M. P. (2016). *Natural Systems: The Organisation of Life*. John Wiley & Sons, London.

Eisenhauer, N., Barnes, A. D., Cesarz, S., et al. (2016). Biodiversity-ecosystem function experiments reveal the mechanisms underlying the consequences of biodiversity change in real world ecosystems. *Journal of Vegetation Science*, 27(5), 1061–1070.

Elliott, M., Burdon, D., Hemingway, K. L. and Apitz, S. E. (2007). Estuarine, coastal and marine ecosystem restoration: Confusing management and science – A revision of concepts. *Estuarine, Coastal and Shelf Science*, 74(3), 349–366.

Ellis, E. C., Kaplan, J. O., Fuller, D. Q., Vavrus, S., Klein Goldewijk, K. and Verburg, P. H. (2013). Used planet: A global history. *Proceedings of the National Academy of Sciences of the United States of America*, 110(20), 7978–7985.

Elser, M. M., Elser, J. J. and Carpenter, S. R. (1986). Paul and Peter Lakes: A liming experiment revisited. *The American Midland Naturalist*, 116(2), 282–295.

Eltahir, E. A. and Bras, R. L. (1994). Precipitation recycling in the Amazon basin. *Quarterly Journal of the Royal Meteorological Society*, 120, 861–880.

Equihua Zamora, M., Espinosa, M., Gershenson, C., et al. (2019). Ecosystem antifragility: Beyond integrity and resilience. PeerJ Preprints, 7, e27813v1 https://doi.org/10.7287/peerj.preprints.27813v1.

Erbaugh, J. T. and Oldekop, J. A. (2018). Forest landscape restoration for livelihoods and well-being. Current Opinion in Environmental Sustainability, 32, 76–83.

Erwin, D. H. (2001). Lessons from the past: Biotic recoveries from mass extinction. Proceedings of the National Academy of Sciences of the United States of America, 8, 5399–5403.

Essington, T. E., Beaudreau, A. H. and Wiedenmann, J. (2006). Fishing through marine food webs. Proceedings of the National Academy of Sciences of the United States of America, 103(9), 3171–3175.

Estes, J. A., Terborgh, J., Brashares, J. S., et al. (2011). Trophic downgrading of planet Earth. Science, 333(6040), 301–306.

Evans, P. M., Newton, A. C., Cantarello, E., et al. (2017). Thresholds of biodiversity and ecosystem function in a forest ecosystem undergoing dieback. Scientific Reports, 7, art6775. https://doi.org/10.1038/s41598-017-06082-6.

  (2019). Testing the relative sensitivity of 102 ecological variables as indicators of woodland condition in the New Forest, UK. Ecological Indicators, 107, 105575.

Ewald, J., Wheatley, C. J., Aebsicher, N. J., et al. (2015). Influences of extreme weather, climate and pesticide use on invertebrates in cereal fields over 42 years. Global Change Biology, 21, 3931–3950.

Fa, J. E., Peres, C. A. and Meeuwig, J. (2002). Bushmeat exploitation in tropical forests: An intercontinental comparison. Conservation Biology, 16, 232–237.

Fahrig, L. (2017). Ecological responses to habitat fragmentation per se. Annual Review of Ecology, Evolution, and Systematics, 48(1), 1–23.

Fahrig, L., Arroyo-Rodríguez, V., Bennett, J. R., et al. (2019). Is habitat fragmentation bad for biodiversity? Biological Conservation, 230, 179–186.

Fan, J., Shen, S., Erwin, D. H., et al. (2020). A high-resolution summary of Cambrian to Early Triassic marine invertebrate biodiversity. Science, 367 (6475), 272–277.

Fan, M., Li, Y. and Li, W. (2015). Solving one problem by creating a bigger one: The consequences of ecological resettlement for grassland restoration and poverty alleviation in Northwestern China. Land Use Policy, 42, 124–130.

FAO. (2018). The State of World Fisheries and Aquaculture. Meeting the Sustainable Development Goals. Food and Agriculture Organisation of the United Nations, Rome.

Farquharson, L. M., Romanovsky, V. E., Cable, W. L., Walker, D. A., Kokelj, S. V. and Nicolsky, D. (2019). Climate change drives widespread and rapid thermokarst development in very cold permafrost in the Canadian High Arctic. Geophysical Research Letters, 46, 6681–6689.

Fauth, J. E. (1997). Working toward operational definitions in ecology: Putting the system back into ecosystem. Bulletin of the Ecological Society of America, 78, 295–297.

Favier, C., Aleman, J., Bremond, L., Dubois, M. A., Freycon, V. and Yangakola, J.-M. (2012). Abrupt shifts in African savanna tree cover along a climatic gradient. Global Ecology and Biogeography, 21, 787–797.

Field, D. J., Bercovici, A., Berv, J. S., *et al.* (2018). Early evolution of modern birds structured by global forest collapse at the End-Cretaceous mass extinction. *Current Biology*, 28(11), 1825–1831.

Filotas, E., Parrott, L., Burton, P. J., *et al.* (2014). Viewing forests through the lens of complex systems science. *Ecosphere*, 5, 1–23.

Fisher, R., O'Leary, R. A., Low-Choy, S., *et al.* (2015). Species richness on coral reefs and the pursuit of convergent global estimates. *Current Biology*, 25(4), 500–505.

Fitzsimmons, A. K. (1996). Stop the parade. *BioScience*, 46(2), 78–79.

Flannery, T. F. (1990). Pleistocene faunal loss: Implications of the aftershock for Australia's past and future. *Archaeology in Oceania*, 25, 45–55.

(1994). *The Future Eaters. An Ecological History of the Australasian Lands and People.* Reed New Holland, Sydney.

Fleming, P. J. S. (2019). They might be right, but Beschta et al. (2018) give no strong evidence that "trophic cascades shape recovery of young aspen in Yellowstone National Park": A fundamental critique of methods. *Forest Ecology and Management*, 454, 117283.

Fletcher, M.-S., Wood, S. W. and Haberle, S. G. (2014). A fire-driven shift from forest to non-forest: Evidence for alternative stable states? *Ecology*, 95(9), 2504–2513.

Fletcher, R. J., Didham, R. K., Banks-Leite, C., *et al.* (2018). Is habitat fragmentation good for biodiversity? *Biological Conservation*, 226, 9–15.

Fletcher, R., Dressler, W. H., Anderson, Z. R. and Büscher, B. (2019). Natural capital must be defended: Green growth as neoliberal biopolitics. *The Journal of Peasant Studies*, 46(5), 1068–1095.

Folke, C., Carpenter, S. R., Walker, B., *et al.* (2004). Regime shifts, resilience and biodiversity in ecosystem management. *Annual Review of Ecology and Systematics*, 35, 557–581.

Folke, C., Carpenter, S. R., Walker, B., Scheffer, M., Chapin, T. and Rockström, J. (2010). Resilience thinking: Integrating resilience, adaptability and transformability. *Ecology and Society*, 15(4), 20. www.ecologyandsociety.org/vol15/iss4/art20/.

Ford, E. D. (2000). *Scientific Method for Ecological Research*. Cambridge University Press, Cambridge.

Fordham, D. A., Brook, B. W., Hoskin, C. J., Pressey, R. L., VanDerWal, J. and Williams, S. E. (2016). Extinction debt from climate change for frogs in the wet tropics. *Biology Letters*, 12(10), 20160236.

Forsyth, D. M., Wilmshurst, J. M., Allen, R. B. and Coomes, D. A. (2010). Impacts of introduced deer and extinct moa on New Zealand ecosystems. *New Zealand Journal of Ecology*, 34(1), 48–65.

Fortuna, M. A., Krishna, A. and Bascompte, J. (2013). Habitat loss and the disassembly of mutalistic networks. *Oikos*, 122(6), 938–942.

Fox, J. (2011). Zombie ideas in ecology. *Oikos Blog*, June 17. https://oikosjournal.wordpress.com/2011/06/17/zombie-ideas-in-ecology/

Fox, R. (2012). The decline of moths in Great Britain: A review of possible causes. *Insect Conservation and Diversity*, 6(1), 5–19.

Fraiser, M. L. and Bottjer, D. J. (2007). When bivalves took over the world. *Paleobiology*, 33(3), 397–413.

Frank, K. T., Petrie, B., Choi, J. S. and Leggett, W. C. (2005). Trophic cascades in a formerly cod-dominated ecosystem. *Science*, 308, 1621–1623.

Frank, K. T., Petrie, B. and Shackell, N. L. (2007). The ups and downs of trophic control in continental shelf ecosystems. *Trends in Ecology & Evolution*, 22(5), 236–242.

Frank, K. T., Petrie, B., Fisher, J. A. D. and Leggett, W. C. (2011). Transient dynamics of an altered large marine ecosystem. *Nature*, 477, 86–91.

Franklin, J. F. (1993). Preserving biodiversity: Species, ecosystems, or landscapes? *Ecological Applications*, 3, 202–205.

Frederiksen, M., Wanless, S., Harris, M. P., Rothery, P. and Wilson, L. J. (2004). The role of industrial fisheries and oceanographic change in the decline of North Sea black-legged kittiwakes. *Journal of Applied Ecology*, 41, 1129–1139.

Frelich, L. E. and Reich, P. B. (1995). Spatial patterns and succession in a Minnesota southern-boreal forest. *Ecological Monographs*, 65, 325–346.

   (1999). Neighborhood effects, disturbance severity, and community stability in forests. *Ecosystems*, 2, 151–166.

Friedel, M. H. (1991). Range condition assessment and the concept of thresholds – A viewpoint. *Journal of Range Management*, 44, 422–426.

Fu, R., Yin, L., Li, W., et al. (2013). Increased dry-season length over southern Amazonia in recent decades and its implication for future climate projection. *Proceedings of the National Academy of Sciences of the United States of America*, 110 (45), 18110–18115.

Fukami, T. and Nakajima, M. (2011). Community assembly: Alternative stable states or alternative transient states? *Ecology Letters*, 14, 973–984.

Fuller, R. J., Gregory, R. D., Gibbons, D. W., et al. (1995). Population declines and range contractions among lowland farmland birds in Britain. *Conservation Biology*, 9, 1425–1441.

Future Earth. (2020). *Our Future on Earth 2020.* www.futureearth.org/publications/our-future-on-earth (accessed on 7 February 2020).

Fyfe, R. M., Twiddle, C., Sugita, S., et al. (2013). The Holocene vegetation cover of Britain and Ireland: Overcoming problems of scale and discerning patterns of openness. *Quaternary Science Reviews*, 73, 132–148.

Gale, G. (2003). *Prehistoric Dorset.* The History Press, Cheltenham.

Galetti, M. (2004). Parks of the Pleistocene: Recreating the cerrado and the Pantanal with megafauna. *Natureza e Conservação*, 2(1), 93–100.

Galetti, M. and Dirzo, R. (2013). Ecological and evolutionary consequences of living in a defaunated world. *Biological Conservation*, 163, 1–6.

Galetti, M., Guevara, R., Côrtes, M. C., et al. (2013). Functional extinction of birds drives rapid evolutionary changes in seed size. *Science*, 340, 1086–1090.

Gallardo, B. and Aldridge, D. C. (2013). Evaluating the combined threat of climate change and biological invasions on endangered species. *Biological Conservation*, 160, 225–233.

Gallardo, B., Clavero, M., Sánchez, M. I. and Vilà, M. (2015). Global ecological impacts of invasive species in aquatic ecosystems. *Global Change Biology*, 22(1), 151–163.

Game, E. T., Meijaard, E., Sheil, D. and McDonald-Madden, E. (2014). Conservation in a wicked complex world; challenges and solutions. *Conservation Letters*, 7, 271–277.

Gamfeldt, L., Lefcheck, J. S., Byrnes, J. E. K., Cardinale, B. J., Duffy, J. E. and Griffin, J. N. (2015). Marine biodiversity and ecosystem functioning: What's known and what's next? *Oikos*, 124(3), 252–265.

Gann, G. D., McDonald, T., Walder, B., et al. (2019). *International Principles and Standards for the Practice of Ecological Restoration*, 2nd edition. Society for Ecological Restoration, Washington, DC.

Gauthier, S., Bernier, P., Kuuluvainen, T., Shvidenko, A. Z. and Schepaschenko, D. G. (2015). Boreal forest health and global change. *Science*, 349(6250), 819–822.

Geddes, B., Wright, J. and Frantz, E. (2018). *How Dictatorships Work. Power, Personalization and Collapse*. Cambridge University Press, Cambridge.

Genkai-Kato, M., Vadeboncoeur, Y., Liboriussen, L. and Jeppesen, E. (2012). Benthic–planktonic coupling, regime shifts, and whole-lake primary production in shallow lakes. *Ecology*, 93, 619–631.

Geist, J. (2011). Integrative freshwater ecology and biodiversity conservation. *Ecological Indicators*, 11(6), 1507–1516.

Geist, J. and Hawkins, S. J. (2016). Habitat recovery and restoration in aquatic ecosystems: Current progress and future challenges. *Aquatic Conservation: Marine and Freshwater Ecosystems*, 26(5), 942–962.

Ghazoul, J. and Chazdon, R. (2017). Degradation and recovery in changing forest landscapes: A multiscale conceptual framework. *Annual Review of Environment and Resources*, 42, 161–188.

Ghazoul, J., Burivalova, Z., Garcia-Ulloa, J. and King, L. A. (2015). Conceptualizing forest degradation. *Trends in Ecology & Evolution*, 30(10), 622–632.

Gibson, L., Lee, T. M., Koh, L. P., et al. (2011). Primary forests are irreplaceable for sustaining tropical biodiversity. *Nature*, 478, 378–381.

Gibson, L., Lynam, A. J., Bradshaw, C. J. A., et al. (2013). Near-complete extinction of native small mammal fauna 25 years after forest fragmentation. *Science*, 341 (6153), 1508–1510.

Gill, J. L. (2014). Ecological impacts of the late Quaternary megaherbivore extinctions. *New Phytologist*, 201(4), 1163–1169.

Gilmour, J. P., Smith, L. D., Heyward, A. J., Baird, A. H. and Pratchett, M. S. (2013). Recovery of an isolated coral reef system following severe disturbance. *Science*, 340, 69–71.

Ginzburg, L. R. and Jensen, C. X. J. (2004). Rules of thumb for judging ecological theories. *Trends in Ecology & Evolution*, 19(3), 121–126.

Glasby, T. M. and Underwood, A. J. (1996). Sampling to differentiate between pulse and press perturbations. *Environmental Monitoring and Assessment*, 42, 241–252.

Glenn-Lewin, D. C. and van der Maarel, E. (1992). Patterns and processes of vegetation dynamics. In: D. C. Glenn-Lewin, R. K. Peet and T. T. Veblen (eds.), *Plant Succession. Theory and Prediction*. Chapman & Hall, London, pp. 11–59.

Goldberg, A., Mychajliw, A. M. and Hadly, E. A. (2016). Post-invasion demography of prehistoric humans in South America. *Nature*, 532, 232–235.

Goldschmidt, T., Witte, F. and Wanink, J. (1993). Cascading effects of the introduced Nile Perch on the detritivorous/ phytoplantivorous species in the sublittoral areas of Lake Victoria. *Conservation Biology*, 7(3), 686–700.

Goldstein, J. H., Pejchar, L and Daily, G. C. (2008). Using return-on-investment to guide restoration: A case study from Hawaii. *Conservation Letters*, 1, 236–243.

Goldstone, J. A. (2001). Toward a fourth generation of revolutionary theory. *Annual Review of Political Science*, 4, 139–187.

Golley, F. B. (1993). *The History of the Ecosystem Concept in Ecology*. Yale University Press, New Haven, CT.

Good, P., Jones, C., Lowe, J., Betts, R., Booth, B. and Huntingford, C. (2011). Quantifying environmental drivers of future tropical forest extent. *Journal of Climate*, 24(5), 1337–1349.

Götzenberger, L., de Bello, F., Bråthen, K. A., et al. (2012). Ecological assembly rules in plant communities – Approaches, patterns and prospects. *Biological Reviews*, 87, 111–127.

Graham, N. A. J., Nash, K. L. and Kool, J. T. (2011). Coral reef recovery dynamics in a changing world. *Coral Reefs*, 30, 283–295.

Graham, N. A. J., Bellwood, D. R., Cinner, J. E., Hughes, T. P., Norström, A. V. and Nyström, M. (2013). Managing resilience to reverse phase shifts in coral reefs. *Frontiers in Ecology and the Environment*, 11, 541–548.

Graham, N. A. J., Jennings, S., MacNeil, M. A., Mouillot, D. and Wilson, S. K. (2015). Predicting climate-driven regime shifts versus rebound potential in coral reefs. *Nature*, 518, 94–97.

Grant, M. J., Hughes, P. D. M. and Barber, K. E. (2014). Climatic influence upon early to mid-Holocene fire regimes within temperate woodlands: A multiproxy reconstruction from the New Forest, southern England. *Journal of Quaternary Science*, 29(2), 175–188.

Grayson, D. K. (2001). The archaeological record of human impacts on animal populations. *Journal of World Prehistory*, 15, 1–68.

Greaver, T. L., Sullivan, T. J., Herrick, J. D., et al. (2012). Ecological effects of nitrogen and sulfur air pollution in the US: What do we know? *Frontiers in Ecology and the Environment*, 10, 365–372.

Gregory, R., Failing, L., Harstone, M., Long, G., McDaniels, T. and Ohlson, D. (2012). *Structured Decision Making: A Practical Guide to Environmental Management Choices*. Wiley-Blackwell, Chichester.

Griffin, J., O'Gorman, E., Emmerson, M., et al. (2009). Biodiversity and the stability of ecosystem functioning. In: S. Naeem, D. E. Bunker, A. Hector, M. Loreau and C. Perrings (eds.), *Biodiversity, Ecosystem Functioning and Human Wellbeing: An Ecological and Economic Perspective*. Oxford University Press, Oxford, pp. 78–93.

Grill, G., Lehner, B., Thieme, M., et al. (2019). Mapping the world's free-flowing rivers. *Nature*, 569, 215–221.

Grimm, V. and Calabrese, J. M. (2011). What is resilience? A short introduction. In: G. Deffuant and N. Gilbert (eds.), *Viability and Resilience of Complex Systems*.

Concepts, Methods and Case Studies from Ecology and Society.. Kluwer Academic Publishers, Dordrecht, pp. 3–16.

Grimm, V. and Wissel, C. (1997). Babel, or the ecological stability discussions: An inventory and analysis of terminology and a guide for avoiding confusion. *Oecologia*, 109, 323–334.

Grimm, N. B., Pickett, S. T. A., Hale, R. L. and Cadenasso, M. L. (2017). Does the ecological concept of disturbance have utility in urban social–ecological–technological systems? *Ecosystem Health and Sustainability*, 3(1), e01255. https://doi.org/10.1002/ehs2.1255.

Griscom, B. W., Adams, J., Ellis, P. W., et al. (2017). Natural climate solutions. *Proceedings of the National Academy of Sciences of the United States of America*, 114 (44), 11645–11650.

Griscom, H. P. and Ashton, M. S. (2011). Restoration of dry tropical forests in Central America: A review of pattern and process. *Forest Ecology and Management*, 261, 1564–1579.

Groffman, P. M., Baron, J. S., Blett, T., et al. (2006). Ecological thresholds: The key to successful environmental management or an important concept with no practical application? *Ecosystems*, 9(1), 1–13.

Gsell, A. S., Scharfenberger, U., Özkundakci, D., et al. (2016). Evaluating early warning indicators of critical transitions in natural aquatic ecosystems. *Proceedings of the National Academy of Sciences of the United States of America*, 113, 8089–8095.

Gurevitch, J., Koricheva, J., Nakagawa, S. and Stewart, G. (2018). Meta-analysis and the science of research synthesis. *Nature*, 555, 175–182.

Guerra-Doce, E. J. (2015). The origins of inebriation: Archaeological evidence of the consumption of fermented beverages and drugs in Prehistoric Eurasia. *Journal of Archaeological Method and Theory*, 22, 751–782.

Guimarães, J., Paulo, R., Galetti, M. and Jordano, P. (2008). Seed dispersal anachronisms: Rethinking the fruits extinct megafauna ate. *PLoS One*, 3, 1–13.

Gunderson, L. H. (2000). Ecological resilience – In theory and application. *Annual Review of Ecology and Systematics*, 31, 425–439.

(2007). Ecology: A different route to recovery for coral reefs. *Current Biology*, 17 (1), 27–28.

Guo, Q. and Ren, H. (2014). Productivity as related to diversity and age in planted versus natural forests. *Global Ecology and Biogeography*, 23(12), 1461–1471.

Haddad, N. M., Brudvig, L. A., Clobert, J., et al. (2015). Habitat fragmentation and its lasting impact on Earth's ecosystems. *Science Advances*, 1(2), e1500052.

Haeussler, S., Canham, C. and Coates, K. (2013). Complexity in temperate forest dynamics. In: C. Messier, K. J. Puettmann and K. Coates (eds.), *Managing Forests as Complex Adaptive Systems. Building Resilience to the Challenge of Global Change*. Earthscan / Routledge, London and New York, pp. 60–78.

Hagstrom, G. I. and Levin, S. A. (2017). Marine ecosystems as Complex Adaptive Systems: Emergent patterns, critical transitions, and public goods. *Ecosystems*, 20 (3), 458–476.

Haller, B. C. (2014). Theoretical and empirical perspectives in ecology and evolution: A survey. *BioScience*, 64(10), 907–916.

Hallmann, C. A., Sorg, M., Jongejans, E., et al. (2017). More than 75 percent decline over 27 years in total flying insect biomass in protected areas. *PLoS One*, 12 (10), e0185809.

Halme, P., Allen, K. A., Auniņš, A., et al. (2013). Challenges of ecological restoration: Lessons from forests in northern Europe. *Biological Conservation*, 167, 248–256.

Hampton, S. E., Strasser, C. A., Tewksbury, J. J., et al. (2013). Big data and the future of ecology. *Frontiers in Ecology and the Environment*, 11, 156–162.

Hanan, N. P., Tredennick, A. T., Prihodko, L., Bucini, G. and Dohn, J. (2014). Analysis of stable states in global savannas. *Global Ecology and Biogeography*, 23, 259–263.

Hansen, M. C., Potapov, P. V., Moore, R., et al. (2013). High-resolution global maps of 21st-century forest cover change. *Science*, 342(6160), 850–853.

Hansson, S. O. (2006). Falsificationism falsified. *Foundations of Science*, 11, 275–286.

Harris, N., Petersen, R., Davis, C. and Payne, O. (2016). *Global Forest Watch and the Forest Resources Assessment, Explained in 5 Graphics*. World Resources Institute. https://blog.globalforestwatch.org/data-and-research/global-forest-watch-and-the-forest-resources-assessment-explained-in-5-graphics-2 (accessed on 2 July 2019).

Harris, R. M. B., Beaumont, L. J., Vance, T. R., et al. (2018). Biological responses to the press and pulse of climate trends and extreme events. *Nature Climate Change*, 8(7), 579–587.

Harrison, P. A., Berry, P. M., Simpson, G., et al. (2014). Linkages between biodiversity attributes and ecosystem services: A systematic review. *Ecosystem Services*, 9, 191–203.

Harrison, R. D., Tan, S., Plotkin, J. B., et al. (2013). Consequences of defaunation for a tropical tree community. *Ecology Letters*, 16(5), 687–694.

Hartvigsen, G., Kinzig, A. and Peterson, G. (1998). Complex adaptive systems: Use and analysis of complex adaptive systems in ecosystem science: Overview of special section. *Ecosystems*, 1(5), 427–430.

Hastings, A. and Wysham, D. B. (2010). Regime shifts in ecological systems can occur with no warning. *Ecology Letters*, 13(4), 464–472.

Hastings, A., McCann, K. S. and de Ruiter, P. C. (2016). Introduction to the special issue: Theory of food webs. *Theoretical Ecology*, 9, 1.

Haysom, K. A., Jones, G., Merrett, D. and Racey, P. A. (2010). Bats. In: N. Maclean (ed.), *Silent Summer: The State of Wildlife in Britain and Ireland*. Cambridge University Press, Cambridge, pp. 259–280.

Hebert, P. D. N., Cywinska, A., Ball, S. L. and deWaard, J. R. (2003). Biological identifications through DNA barcodes. *Proceedings of the Royal Society of London Series B: Biological Sciences*, 270(1512), 313 LP-321.

Heffernan, J. B., Soranno, P. A., Angilletta, M. J., et al. (2014). Macrosystems ecology: Understanding ecological patterns and processes at continental scales. *Frontiers in Ecology and the Environment*, 12(1), 5–14.

Heino, J., Virkkala, R. and Toivonen, H. (2009). Climate change and freshwater biodiversity: Detected patterns, future trends and adaptations in northern regions. *Biological Reviews*, 84, 39–54.

Heithaus, M. R., Frid, A., Wirsing, A. J. and Worm, B. (2008). Predicting ecological consequences of marine top predator declines. *Trends in Ecology & Evolution*, 23 (4), 202–210.

Helfield, J. M. and Naiman, R. J. (2006). Keystone interactions: Salmon and bear in riparian forests of Alaska. *Ecosystems*, 9, 167–180.

Henderson, K. A., Bauch, C. T. and Anand, M. (2016). Alternative stable states and the sustainability of forests, grasslands, and agriculture. *Proceedings of the National Academy of Sciences of the United States of America*, 113(51), 14552–14559.

Henehan, M. J., Ridgwell, A., Thomas, E., et al. (2019). Rapid ocean acidification and protracted Earth system recovery followed the end-Cretaceous Chicxulub impact. *Proceedings of the National Academy of Sciences of the United States of America*, 116(45), 22500–22504.

Henke, S. E. and Bryant, F. C. (1999). Effects of coyote removal on the faunal community in western Texas. *Journal of Wildlife Management*, 63(4), 1066–1081.

Hettelingh, J. P., Sverdrup, H. and Zhao, D. (1995). Deriving critical loads for Asia. *Water, Air, and Soil Pollution*, 85(4), 2565–2570.

Higgins, P. A. T., Mastrandrea, M. D. and Schneider, S. H. (2002). Dynamics of climate and ecosystem coupling: Abrupt changes and multiple equilibria. *Philosophical Transactions of the Royal Society B: Biological Sciences*, 357, 647–655.

Higgs, E. S., Harris, J. A., Heger, T., Hobbs, R. J., Murphy, S. D. and Suding, K. N. (2018a). Keep ecological restoration open and flexible. *Nature Ecology & Evolution*, 2(4), 580.

Higgs, E. S., Harris, J. A., Murphy, S. D., et al. (2018b). On principles and standards in ecological restoration. *Restoration Ecology*, 26, 399–403.

Hilborn, R. (2006). Faith-based fisheries. *Fisheries*, 31, 554–555.

Hilborn, R. and Ludwig, D. (1993). The limits of applied ecological research. *Ecological Applications*, 3, 550–552.

Hilderbrand, G., Hanley, T., Robbins, C., et al. (1999). Role of brown bears (*Ursus arctos*) in the flow of marine nitrogen into a terrestrial ecosystem. *Oecologia*, 121, 546–550.

Hirota, M., Holmgren, M., van Nes, E. H. and Scheffer, M. (2011). Global resilience of tropical forest and savanna to critical transitions. *Science*, 334, 232–235.

Hobbs, R. J. (2013). Grieving for the past and hoping for the future: Balancing polarizing perspectives in conservation and restoration. *Restoration Ecology*, 21 (2), 145–148.

Hobbs, R. J. and Norton, D. A. (1996). Toward a conceptual framework for restoration ecology. *Restoration Ecology*, 4, 93–110.

Hobbs, R. J. and Suding, K. N. (2009). *New Models for Ecosystem Dynamics and Restoration*. Island Press, Washington, DC.

Hobbs, R. J., Arico, S., Aronson, J., et al. (2006). Novel ecosystems: Theoretical and management aspects of the new ecological world order. *Global Ecology and Biogeography*, 15, 1–7.

Hobbs, R. J., Higgs, E. and Harris, J. A. (2009). Novel ecosystems: Implications for conservation and restoration. *Trends in Ecology & Evolution*, 24(11), 599–605.

Hodder, K. H., Newton, A. C., Cantarello, E. and Perrella, L. (2014). Does landscape-scale conservation management enhance the provision of ecosystem

services? *International Journal of Biodiversity Science, Ecosystem Services and Management*, 10(1), 71–83.
Hoffmann, W. A., Geiger, E. L., Gotsch, S. G., et al. (2012). Ecological thresholds at the savanna-forest boundary: How plant traits, resources and fire govern the distribution of tropical biomes. *Ecology Letters*, 15, 759–768.
Hofmanová, Z., Kreutzer, S., Hellenthal, G., et al. (2016). Early farmers from across Europe directly descended from Neolithic Aegeans. *Proceedings of the National Academy of Sciences of the United States of America*, 113(25), 6886–6891.
Holdaway, R. N., Holdaway, R. N., Allentoft, M. E., et al. (2014). An extremely low-density human population exterminated New Zealand moa. *Nature Communications*, 5(1), 5436.
Holl, K. D. (1998). Effects of above and below ground competition of shrubs and grass on *Calophyllum brasiliense* seedling growth in abandoned tropical pasture. *Forest Ecology and Management*, 109, 187–195.
  (2017). Restoring tropical forests from the bottom up. *Science*, 355(6324), 455–456.
Holl, K. D., Loik, M. E., Lin, E. H. V. and Samuels, I. A. (2000). Tropical montane forest restoration in Costa Rica: Overcoming barriers to dispersal and establishment. *Restoration Ecology*, 8, 339–349.
Holland, S. M. and Patzkowsky, M. E. (2015). The stratigraphy of mass extinction. *Palaeontology*, 58(5), 903–924.
Holling, C. (1978). The spruce-budworm/forest-management problem. In: C. Holling (ed.), *Adaptive Environmental Assessment and Management*. International Series on Applied Systems Analysis. John Wiley & Sons, New York, pp. 143–182.
Holtgrieve, G. W., Schindler, D. E. and Jewett, P. K. (2009). Large predators and biogeochemical hotspots: Brown bear (*Ursus arctos*) predation on salmon alters nitrogen cycling in riparian soils. *Ecological Research*, 24(5), 1125–1135.
Hooper, D. U., Chapin, F. S., Ewel, J. J., et al. (2005). Effects of biodiversity on ecosystem functioning: A consensus of current knowledge. *Ecological Monographs*, 75, 3–35.
Hooper, D. U., Adair, E. C., Cardinale, B. J., et al. (2012). A global synthesis reveals biodiversity loss as a major driver of ecosystem change. *Nature*, 486, 105–108.
Hope, G., Kershaw, A. P., Kaars, S., et al. (2004). History of vegetation and habitat change in the Austral-Asian region. *Quaternary International*, 118–119, 103–126.
Horan, R. D., Fenichel, E. P., Drury, K. L. S. and Lodge, D. M. (2011). Managing ecological thresholds in coupled environmental–human systems. *Proceedings of the National Academy of Sciences of the United States of America*, 108(18), 7333–7338.
Hsieh, C., Glaser, S. M., Lucas, A. J. and Sugihara, G. (2005). Distinguishing random environmental fluctuations from ecological catastrophes for the North Pacific Ocean. *Nature*, 435(7040), 336–340.
Huang, C., Zhou, Z., Peng, C., Teng, M. and Wang, P. (2018). How is biodiversity changing in response to ecological restoration in terrestrial ecosystems? A meta-analysis in China. *Science of the Total Environment*, 650, 1–9.

Huang, J.-P., Kraichak, E., Leavitt, S. D., Nelsen, M. P. and Lumbsch, H. T. (2019). Accelerated diversifications in three diverse families of morphologically complex lichen-forming fungi link to major historical events. *Scientific Reports*, 9(1), 8518.

Hubbell, S. P. (2001). *The Unified Neutral Theory of Biodiversity and Biogeography*. Princeton University Press, Princeton, NJ.

Huggett, A. J. (2005). The concept and utility of the "ecological thresholds" in biodiversity conservation. *Biological Conservation*, 124, 301–310.

Hughes, T. P. (1994). Catastrophes, phase shifts, and large-scale degradation of a Caribbean coral reef. *Science*, 265, 1547–1551.

Hughes, T. P., Bellwood, D. R., Folke, C., Steneck, R. S. and Wilson, J. (2005). New paradigms for supporting the resilience of marine ecosystems. *Trends in Ecology & Evolution*, 20(7), 380–386.

Hughes, T. P., Carpenter, S., Rockström, J., Scheffer, M. and Walker, B. (2013). Multiscale regime shifts and planetary boundaries. *Trends in Ecology & Evolution*, 28(7), 389–395.

Hughes, T. P., Barnes, M. L., Bellwood, D. R., et al. (2017a). Coral reefs in the Anthropocene. *Nature*, 546, 82–90.

Hughes, T. P., Kerry, J. T., Álvarez-Noriega, M., et al. (2017b). Global warming and recurrent mass bleaching of corals. *Nature*, 543, 373–378.

Hughes, T. P., Anderson, K. D., Connolly, S. R., et al. (2018a). Spatial and temporal patterns of mass bleaching of corals in the Anthropocene. *Science*, 359(6371), 80–83.

Hughes, T. P., Kerry, J. T., Baird, A. H., et al. (2018b). Global warming transforms coral reef assemblages. *Nature*, 556(7702), 492–496.

Hull, P. (2015). Life in the aftermath of mass extinctions. *Current Biology*, 25(19), R941–R952.

Hull, V., Tuanmu, M.-N. and Liu, J. (2015). Synthesis of human-nature feedbacks. *Ecology and Society*, 20(3), 17. http://dx.doi.org/10.5751/ES-07404-200317.

Humphries, P. and Baldwin, D. S. (2003). Drought and aquatic ecosystems: An introduction. *Freshwater Biology*, 48, 1141–1146.

Hunter, M. L., Bean, M. J., Lindenmayer, D. B., et al. (2009). Thresholds and the mismatch between environmental laws and ecosystems. *Conservation Biology*, 23, 1053–1055.

Huntingford, C., Zelazowski, P., Galbraith, D., et al. (2013). Simulated resilience of tropical rainforests to CO_2-induced climate change. *Nature Geoscience*, 6, 268–273.

Hutchings, J. A. and Reynolds, J. D. (2004). Marine fish population collapses: Consequences for recovery and extinction risk. *BioScience*, 54(4), 297–309.

Ibáñez, J. J., González-Urquijo, J., Teira-Mayolini, L. C. and Lazuén, T. (2018). The emergence of the Neolithic in the Near East: A protracted and multi-regional model. *Quaternary International*, 470, 226–252.

Ibelings, B. W., Portielje, R., Lammens, E. H. R. R., et al. (2007). Resilience of alternative stable states during the recovery of shallow lakes from eutrophication: Lake Veluwe as a case study. *Ecosystems*, 10(1), 4–16.

Ibisch, P. L., Hoffmann, M. T., Kreft, S., et al. (2016). A global map of roadless areas and their conservation status. *Science*, 354(6318), 1423–1427.

Iftekhar, M. S., Polyakov, M., Ansell, D., Gibson, F. and Kay, G. M. (2016). How economics can further the success of ecological restoration. *Conservation Biology*, 31(2), 261–268.

Ingrisch, J. and Bahn, M. (2018). Towards a comparable quantification of resilience. *Trends in Ecology & Evolution*, 33, 251–259.

Intergovernmental Panel on Climate Change (IPCC). (2018). *An IPCC Special Report on the Impacts of Global Warming of 1.5°C above Pre-industrial Levels and Related Global Greenhouse Gas Emission Pathways.* IPCC, Geneva, Switzerland. www.ipcc.ch/sr15/.

IPBES. (2018). *The IPBES Assessment Report on Land Degradation and Restoration*, ed. L. Montanarella, R. Scholes and A. Brainich. Secretariat of the Intergovernmental Science-Policy Platform on Biodiversity and Ecosystem Services, Bonn, Germany.

(2019). *Global Assessment Report on Biodiversity and Ecosystem Services of the Intergovernmental Science-Policy Platform on Biodiversity and Ecosystem Services*, ed. E. S. Brondizio, J. Settele, S. Díaz and H. T. Ngo. IPBES Secretariat, Bonn, Germany. https://ipbes.net/global-assessment-report-biodiversity-ecosystem-services (accessed on 18 December 2019).

Isbell, F. I., Polley, H. W. and Wilsey, B. J. (2009). Biodiversity, productivity and the temporal stability of productivity: Patterns and processes. *Ecology Letters*, 12, 443–451.

Isbell, F., Calcagno, V., Hector, A., et al. (2011). High plant diversity is needed to maintain ecosystem services. *Nature*, 477(7363), 199–202.

IUCN (World Conservation Union). (2001). *IUCN Red List Categories and Criteria: Version 3.1.* IUCN Species Survival Commission, IUCN, Gland, Switzerland and Cambridge. www.iucnredlist.org/technical-documents/categories-and-criteria (accessed on November 2017).

(2012). *IUCN Red List Categories and Criteria: Version 3.1*, 2nd edition. IUCN, Gland, Switzerland and Cambridge.

(2016). *An Introduction to the IUCN Red List of Ecosystems: The Categories and Criteria for Assessing Risks to Ecosystems.* IUCN, Gland, Switzerland.

(2019). *Red List of Ecosystems.* www.iucn.org/theme/ecosystem-management/our-work/red-list-ecosystems (accessed 11 December 2019).

Ives, A. R. (2007). Diversity and stability in ecological communities. In: R. M. May and A. R. McLean (eds.), *Theoretical Ecology. Principles and Applications.* Oxford University Press, Oxford, pp. 98–110.

Jablonski, D. (1994). Extinctions in the fossil record. *Philosophical Transactions of the Royal Society of London Series B: Biological Sciences*, 344(1307), 11–17.

(2005). Mass extinctions and macroevolution. *Paleobiology*, 31(2), 192–210.

Jackson, J. B. (2008). Colloquium paper: Ecological extinction and evolution in the brave new ocean. *Proceedings of the National Academy of Sciences of the United States of America*, 105 Suppl. 1, 11458–11465.

Jackson, J. B. C., Kirby, M. X., Berger, W. H., et al. (2001). Historical overfishing and the recent collapse of coastal ecosystems. *Science*, 293, 629–637.

Jackson, M. C., Loewen, C. J., Vinebrooke, R. D. and Chimimba, C. T. (2016). Net effects of multiple stressors in freshwater ecosystems: A meta-analysis. *Global Change Biology*, 22, 180–189.

References

Jackson, S. T. and Blois, J. L. (2015). Community ecology in a changing environment: Perspectives from the Quaternary. *Proceedings of the National Academy of Sciences of the United States of America*, 112(16), 4915–4921.

Jansson, R., Nilsson, C. and Malmqvist, B. (2007). Restoring freshwater ecosystems in riverine landscapes: The roles of connectivity and recovery processes. *Freshwater Biology*, 52(4), 589–596.

Janzen, D. H. and Hallwachs, W. (2019). Perspective: Where might be many tropical insects? *Biological Conservation*, 233, 102–108.

Janzen, D. H. and Martin, P. S. (1982). Neotropical anachronisms: The fruit the Gomphotheres ate. *Science*, 215, 19–27.

Jasinski, J. P. P. and Payette, S. (2005). The creation of alternative stable states in the southern Boreal forest, Québec, Canada. *Ecological Monographs*, 75(4), 561–583.

Jeffers, E. S., Whitehouse, N. J., Lister, A., et al. (2018). Plant controls on Late Quaternary whole ecosystem structure and function. *Ecology Letters*, 21, 814–825.

Jentsch, A. and White, P. (2019). A theory of pulse dynamics and disturbance in ecology. *Ecology*, 100(7), e02734. https://doi.org/10.1002/ecy.2734.

Jeppesen, E., Søndergaard, M., Jensen, J. P., et al. (2005). Lake responses to reduced nutrient loading – An analysis of contemporary long-term data from 35 case studies. *Freshwater Biology*, 50(10), 1747–1771.

Jia, S., Wang, X., Yuan, Z., et al. (2018). Global signal of top-down control of terrestrial plant communities by herbivores. *Proceedings of the National Academy of Sciences of the United States of America*, 115(24), 6237–6242.

Jiang, J., Huang, Z.-G., Seager, T. P., et al. (2018). Predicting tipping points in mutualistic networks through dimension reduction. *Proceedings of the National Academy of Sciences of the United States of America*, 115(4), E639–E647.

Jiang, J., Hastings, A. and Lai, Y.-C. (2019). Harnessing tipping points in complex ecological networks. *Journal of the Royal Society Interface*, 16(158), 20190345.

Johns, K. A., Osborne, K. O. and Logan, M. (2014). Contrasting rates of coral recovery and reassembly in coral communities on the Great Barrier Reef. *Coral Reefs*, 33, 553–563.

Johnson, C. J. (2013). Identifying ecological thresholds for regulating human activity: Effective conservation or wishful thinking? *Biological Conservation*, 168, 57–65.

Johnson, C. N. (2009). Ecological consequences of Late Quaternary extinctions of megafauna. *Proceedings of the Royal Society B: Biological Sciences*, 276, 2509–2519.

Johnson, C. N., Alroy, J., Beeton, N. J., et al. (2016a). What caused extinction of the Pleistocene megafauna of Sahul? *Proceedings of the Royal Society B: Biological Sciences*, 283(1824), 20152399.

Johnson, C. N., Rule, S., Haberle, S. G., Kershaw, A. P., McKenzie, G. M. and Brook, B. W. (2016b). Geographic variation in the ecological effects of extinction of Australia's Pleistocene megafauna. *Ecography*, 39, 109–116.

Jones, C. G. and Lawton, J. H. (Eds.) (1995). *Linking Species and Ecosystems*. Chapman & Hall, New York. 387 pp.

Jones, H. P. and Schmitz, O. J. (2009). Rapid recovery of damaged ecosystems. *PLoS One*, 4(5), e5653.

Jonsson, T., Berg, S., Emmerson, M. and Pimenov, A. (2015). The context dependency of species keystone status during food web disassembly. *Food Webs*, 5, 1–10.

Jorge, M. L. S. P., Galetti, M., Ribeiro, M. C. and Ferraz, K. M. P. M. B. (2013). Mammal defaunation as surrogate of trophic cascades in a biodiversity hotspot. *Biological Conservation*, 163, 49–57.

Jørgensen, S. E., Fath, B., Bastianoni, S., et al. (2007). *A New Ecology – The Systems Perspective*. Elsevier Publishers, Amsterdam.

Kaiser, J. (2000). Rift over biodiversity divides ecologists. *Science*, 289, 1282–1283.

Kandziora, M., Burkhard, B. and Müller, F. (2013). Interactions of ecosystem properties, ecosystem integrity and ecosystem service indicators – A theoretical matrix exercise. *Ecological Indicators*, 28, 54–78.

Kaneryd, L., Borrvall, C., Berg, S., et al. (2012). Species-rich ecosystems are vulnerable to cascading extinctions in an increasingly variable world. *Ecology and Evolution*, 2, 858–874.

Kaplan, J. O., Krumhardt, K. M. and Zimmermann, N. (2009). The prehistoric and preindustrial deforestation of Europe. *Quaternary Science Reviews*, 28(27–28), 3016–3034.

Karabanov, E., Williams, D., Kuzmin, M., et al. (2004). Ecological collapse of Lake Baikal and Lake Hovsgol ecosystems during the Last Glacial and consequences for aquatic species diversity. *Palaeogeography, Palaeoclimatology, Palaeoecology*, 209 (1–4), 227–243.

Kareiva, P. and Marvier, M. (2012). What is conservation science? *BioScience*, 62(11), 962–969.

Kauffman, M. J., Brodie, J. F. and Jules, E. S. (2013). Are wolves saving Yellowstone's aspen? A landscape-level test of a behaviorally mediated trophic cascade: Reply. *Ecology*, 94(6), 1425–1431.

Keddy, P. A. (1992). Assembly and response rules: Two goals for predictive community ecology. *Journal of Vegetation Science*, 3, 157–165.

Keeler, B. L., Chaplin-Kramer, R., Guerry, A. D., et al. (2017). Society is ready for a new kind of science – Is academia? *BioScience*, 67(7), 591–592.

Kéfi, S., Rietkerk, M., Alados, C. L., et al. (2007). Spatial vegetation patterns and imminent desertification in Mediterranean arid ecosystems. *Nature*, 449, 213–217.

Kéfi, S., Guttal, V., Brock, W. A., et al. (2014). Early warning signals of ecological transitions: Methods for spatial patterns. *PLoS One*, 9, e92097.

Keith, D. A., Rodríguez, J. P., Rodríguez-Clark, K. M., et al. (2013). Scientific foundations for an IUCN Red List of Ecosystems. *PLoS One*, 8(5), e62111.

Keith, D. A., Rodríguez, J. P., Brooks, T. M., et al. (2015). The IUCN Red List of Ecosystems: Motivations, challenges, and applications. *Conservation Letters*, 8, 214–226.

Keith, S. A., Newton, A. C., Herbert, R. J. H., Morecroft, M. D. and Bealey, C. E. (2009a). Non-analogous community formation in response to climate change. *Journal of Nature Conservation*, 17, 228–235.

Keith, S. A., Newton, A. C., Morecroft, M. D., Bealey, C. E. and Bullock, J. M. (2009b). Taxonomic homogenisation of woodland plant communities over seventy years. *Proceedings of the Royal Society B: Biological Sciences*, 276(1672), 3539–3544.

Keith, S. A., Newton, A. C., Morecroft, M. D., Golicher, D. J. and Bullock, J. M. (2011). Woodland metacommunity structure remains unchanged during biodiversity loss in English woodlands. *Oikos*, 120(2), 302–331.

Keith, S. A., Baird, A. H., Hobbs, J.-P. A., et al. (2018). Synchronous behavioural shifts in reef fishes linked to mass coral bleaching. *Nature Climate Change*, 8(11), 986–991.

Kidd, L. R., Bekessy, S. A. and Garrard, G. E. (2019). Neither hope nor fear: Empirical evidence should drive biodiversity conservation strategies. *Trends in Ecology & Evolution*, 34(4), 278–282.

King, D. A., Claeys, P., Gulick, S. P. S., Morgan, J. V. and Collins, G. S. (2017). Chicxulub and the exploration of large peak-ring impact craters through scientific drilling. *GSA Today*, 27(10), 4–8.

Kirby, K. J. (2004). A model of a natural wooded landscape in Britain driven by large-herbivore activity. *Forestry*, 77, 405–420.

Kirch, P. V. (1997). Microcosmic histories: Island perspectives on "global" change. *American Anthropologist*, 99(1), 30–42.

(2005). Archaeology and global change: The Holocene record. *Annual Review of Environment and Resources*, 30(1), 409–440.

Kirchhoff, T., Brand, F. S., Hoheisel, D. and Grimm, V. (2010). The one-sidedness and cultural bias of the resilience approach. *GAIA*, 19(1), 25–32.

Kirchner, J. W. and Weil, A. (2000). Delayed biological recovery from extinctions throughout the fossil record. *Nature*, 404, 177–180.

Kitzberger, T., Raffaele, E., Heinemann, K. and Mazzarino, M. J. (2005). Effects of fire severity in a north Patagonian subalpine forest. *Journal of Vegetation Science*, 16, 5–12.

Kitzberger, T., Aráoz, E., Gowda, J. H., et al. (2012). Decreases in fire spread probability with forest age promotes alternative community states, reduced resilience to climate variability and large fire regime shifts. *Ecosystems*, 15, 97–112.

Kitzberger, T., Perry, G. L. W., Paritsis, J., et al. (2016). Fire–vegetation feedbacks and alternative states: Common mechanisms of temperate forest vulnerability to fire in southern South America and New Zealand. *New Zealand Journal of Botany*, 54(2), 247–272.

Klein, N. (2020). *On Fire. The Burning Case for a Green New Deal*. Penguin Books, London.

Knight, A. T. (2013). Reframing the theory of hope in conservation science. *Conservation Letters*, 6(6), 389–390.

Knoll, A. H. (1984). Patterns of extinction in the fossil record of vascular plants. In: M. H. Nitecki (ed.), *Extinctions*. Chicago University Press, Chicago, pp. 21–68.

Knowlton, N. (2004). Multiple "stable" states and the conservation of marine ecosystems. *Progress in Oceanography*, 60(2), 387–396.

(2008). Coral reefs. *Current Biology*, 18, R18–R21.

(2017). Doom and gloom won't save the world. *Nature*, 544, 271.

Knox, K. J. E. and Clarke, P. J. (2012). Fire severity, feedback effects and resilience to alternative community states in forest assemblages. *Forest Ecology and Management*, 265, 47–54.

Koch, P. L. and Barnosky, A. D. (2006). Late Quaternary extinctions: State of the debate. *Annual Review of Ecology, Evolution, and Systematics*, 37(1), 215–250.

Kolasa, J. (2011). Theory makes ecology evolve. In: S. M. Scheiner and M. R. Willig (eds.), *The Theory of Ecology*. The University of Chicago Press, Chicago, IL, pp. 21–49.

Kolding, J., van Zwieten, P., Mkumbo, O., Silsbe, G. and Hecky, R. (2008). Are the Lake Victoria fisheries threatened by exploitation or eutrophication? Towards an ecosystem-based approach to management. In: G. Bianchi and H. R. Skjoldal (eds.), *The Ecosystem Approach to Fisheries*. CABI Publishing, Wallingford, pp. 309–350.

Komonen, A., Halme, P. and Kotiaho, J. S. (2019). Alarmist by bad design: Strongly popularized unsubstantiated claims undermine credibility of conservation science. *Rethinking Ecology*, 4, 17–19.

Kosten, S., Vernooij, M., van Nes, E. H., Sagrario, M. Á. G., Clevers, J. G. P. W. and Scheffer, M. (2012). Bimodal transparency as an indicator for alternative states in South American lakes. *Freshwater Biology*, 57, 1191–1201.

Kuffner, I. B., Walters, L. J., Becerro, M. A., et al. (2006). Inhibition of coral recruitment by macroalgae and cynobacteria. *Marine Ecology Progress Series*, 323, 107–117.

Kumar, S. S., Hanan, N. P., Prihodko, L., et al. (2019). Alternative vegetation states in tropical forests and savannas: The search for consistent signals in diverse remote sensing data. *Remote Sensing*, 11, 815. https://doi.org/10.3390/rs11070815.

Kump, L. R., Pavlov, A. and Arthur, M. A. (2005). Massive release of hydrogen sulfide to the surface ocean and atmosphere during intervals of oceanic anoxia. *Geology*, 33(5), 397–400.

Kurten, E. L. (2013). Cascading effects of contemporaneous defaunation on tropical forest communities. *Biological Conservation*, 163, 22–32.

Kurz, W. A., Dymond, C. C., Stinson, G., et al. (2008). Mountain pine beetle and forest carbon feedback to climate change. *Nature*, 452(7190), 987–990.

Ladle, R. J., Jepson, P., Araújo, M. B. and Whittaker, R. J. (2004). Dangers of crying wolf over risk of extinctions. *Nature*, 428, 799.

Lafferty, K. D. and Hopkins, S. R. (2018). Unique parasite aDNA in moa coprolites from New Zealand suggests mass parasite extinctions followed human-induced megafauna extinctions. *Proceedings of the National Academy of Sciences of the United States of America*, 115(7), 1411–1413.

Lake, P. S. (2003). Ecological effects of perturbation by drought in flowing waters. *Freshwater Biology*, 48, 1161–1172.

Lake, P. S., Bond, N. and Reich, P. (2007). Linking ecological theory with stream restoration. *Freshwater Biology*, 52(4), 597–615.

Lamb, D., Erskine, P. D. and Parrotta, J. A. (2005). Restoration of degraded tropical forest landscapes. *Science*, 310, 1628–1632.

Larson, B. M. H. (2005). The war of the roses: Demilitarizing invasion biology. *Frontiers in Ecology and the Environment*, 3, 495–500.

(2011). *Metaphors for Environmental Sustainability. Redefining Our Relationship with Nature*. Yale University Press, New Haven, CT, and London.

(2014). The metaphorical links between ecology, ethics, and society In: R. Rozzi, S. T. A. Pickett, B. Callicott, C. Palmer and J. Armesto (eds.), *Linking Ecology and Ethics for a Changing World: Values, Philosophy, and Action*. New York: Springer, pp. 137–145.

Laundré, J. W., Hernandez, L. and Altendorf, K. B. (2001). Wolves, elk, and bison: Reestablishing the "landscape of fear" in Yellowstone National Park, USA. *Canadian Journal of Zoology*, 79, 1401–1409.

Laurance, W. F., Delamônica, P., Laurance, S. G., Vasconcelos, H. L. and Lovejoy, T. E. (2000). Rainforest fragmentation kills big trees. *Nature*, 404(6780), 836.

Laurance, W. F., Lovejoy, T. E., Vasconcelos, H. L., et al. (2002). Ecosystem decay of Amazonian forest fragments: A 22-Year investigation. *Conservation Biology*, 16, 605–618.

Laurance, W. F., Camargo, J. L. C., Luizão, R. C. C., et al. (2011). The fate of Amazonian forest fragments: A 32-year investigation. *Biological Conservation*, 144(1), 56–67.

Laurance, W. F., Carolina Useche, D., Rendeiro, J., et al. (2012). Averting biodiversity collapse in tropical forest protected areas. *Nature*, 489(7415), 290–294.

Lawler, A. (2010). Collapse? What collapse? Societal change revisited. *Science*, 330, 907–909.

Leadley, P., Proença, V., Fernández-Manjarrés, J., et al. (2014). Interacting regional-scale regime shifts for biodiversity and ecosystem services. *BioScience*, 64(8), 665–679.

Leather, S. R. (2016). Insects in flight: Whatever happened to the splatometer? https://simonleather.wordpress.com/2016/12/05/insects-in-flight-whatever-happened-to-the-splatometer/.

(2018). 'Ecological Armageddon' – More evidence for the drastic decline in insect numbers. *Annals of Applied Biology*, 172, 1–3.

Ledlie, M. H., Graham, N. A. J., Bythell, J. C., et al. (2007). Phase shifts and the role of herbivory in the resilience of coral reefs. *Coral Reefs*, 26, 641–653.

Lee, W. G., Wood, J. R. and Rogers, G. M. (2010). Legacy of avian-dominated plant/herbivore systems in New Zealand. *New Zealand Journal of Ecology*, 34, 28–47.

Legg, C. J. and Nagy, L. (2006). Why most conservation monitoring is, but need not be, a waste of time. *Journal of Environmental Management*, 78, 194–199.

Lehmann, C. E. R., Anderson, T. M., Sankaran, M., et al. (2014). Savanna vegetation-fire-climate relationships differ among continents. *Science*, 343 (6170), 548–552.

Lenton, T. M., Held, H., Kriegler, E., et al. (2008). Tipping elements in Earth's climate system. *Proceedings of the National Academy of Sciences of the United States of America*, 105, 1786–1793.

Levin, S. A. (1998). Ecosystems and the biosphere as complex adaptive systems. *Ecosystems*, 1(5), 431–436.

(1999). *Fragile Dominion: Complexity and the Commons*. Perseus Books, Reading, MA.

Levine, N. M., Zhang, K., Longo, M., et al. (2016). Ecosystem heterogeneity determines the ecological resilience of the Amazon to climate change.

Proceedings of the National Academy of Sciences of the United States of America, 113 (3), 793–797.
Lewis, S. L. (2012). We must set planetary boundaries wisely. *Nature*, 485, 417.
Lewis, S. L., Wheeler, C. E., Mitchard, E. T. and Koch, A. (2019). Restoring natural forests is the best way to remove atmospheric carbon. *Nature*, 568(7750), 25–28.
Lewontin, R. C. (1969). The meaning of stability. In: G. M. Woodwell and H. H. Smith (eds.), *Diversity and Stability in Ecological Systems*. Brookhaven Symposium of Biology, vol. 22, pp. 13–23.
Liao, C., Luo, Y., Fang, C. and Li, B. (2010). Ecosystem carbon stock influenced by plantation practice: Implications for planting forests as a measure of climate change mitigation. *PLoS One*, 5(5), e10867. https://doi.org/10.1371/journal.pone.0010867.
Lindegren, M., Dakos, V., Gröger, J. P., *et al.* (2012). Early detection of ecosystem regime shifts: A multiple method evaluation for management application. *PLoS One*, 7(7), e38410.
Lindenmayer, D. B. and Laurance, W. F. (2016a). The unique challenges of conserving large old trees. *Trends in Ecology & Evolution*, 31(6), 416–418.
  (2016b). The ecology, distribution, conservation and management of large old trees. *Biological Reviews*, 92(3), 1434–1458.
Lindenmayer, D. B. and Sato, C. (2018). Hidden collapse is driven by fire and logging in a socioecological forest ecosystem. *Proceedings of the National Academy of Sciences of the United States of America*, 115(20), 5181–5186.
Lindenmayer, D. B., Likens, G. E., Krebs, C. J. and Hobbs, R. J. (2010). Improved probability of detection of ecological 'surprises'. *Proceedings of the National Academy of Sciences of the United States of America*, 107(51), 21957–21962.
Lindenmayer, D. B., Hobbs, R. J., Likens, G. E., Krebs, C. J. and Banks, S. C. (2011). Newly discovered landscape traps produce regime shifts in wet forests. *Proceedings of the National Academy of Sciences of the United States of America*, 108, 15887–15891.
Lindenmayer, D. B., Messier, C. and Sato, C. (2016). Avoiding ecosystem collapse in managed forest ecosystems. *Frontiers in Ecology and the Environment*, 14(10), 561–568.
Lipson, M., Szécsényi-Nagy, A., Mallick, S., *et al.* (2017). Parallel palaeogenomic transects reveal complex genetic history of early European farmers. *Nature*, 551, 368–372.
Lister, B. C. and Garcia, A. (2018). Climate-driven declines in arthropod abundance restructure a rainforest food web. *Proceedings of the National Academy of Sciences of the United States of America*, 115(44), E10397–E10406.
Liu, J., Dietz, T., Carpenter, S. R., *et al.* (2007). Complexity of coupled human and natural systems. *Science*, 317, 1513–1516.
Livni, J. (2019). Investigation of collapse of complex socio-political systems using classical stability theory. *Physica A: Statistical Mechanics and Its Applications*, 524, 553–562.
Lizundia-Loiola, J., Pettinari, M. L. and Chuvieco, E. (2020). Temporal anomalies in burned area trends: Satellite estimations of the Amazonian 2019 Fire Crisis. *Remote Sensing*, 12(1), 151.

Lloyd, J. and Veenendaal, E. M. (2016). Are fire mediated feedbacks burning out of control? *Biogeosciences Discussions*. https://doi.org/10.5194/bg-2015-660.

Loehle, C. (1989). Catastrophe theory in ecology: A critical review and an example of the butterfly catastrophe. *Ecological Modelling*, 49(1), 125–152.

Loeser, M. R. R., Sisk,T. D. and Crews, T. E. (2007). Impact of grazing intensity during drought in an Arizona grassland. *Conservation Biology*, 21, 87–97.

Looy, C. V., Brugman, W. A., Dilcher, D. L. and Visscher, H. (1999). The delayed resurgence of equatorial forests after the Permian-Triassic ecologic crisis. *Proceedings of the National Academy of Sciences of the United States of America*, 96, 13857–13862.

Loreau, M. (2010a). Linking biodiversity and ecosystems: Towards a unifying ecological theory. *Philosophical Transactions of the Royal Society B: Biological Sciences*, 365(1537), 49–60.

(2010b). *From Populations to Ecosystems: Theoretical Foundations for a New Ecological Synthesis*. Princeton University Press, Princeton, NJ.

Lortie, C. J. and Bonte, D. (2016). Zen and the art of ecological synthesis. *Oikos*, 125, 285–287.

Lotka, A. J. (1956). *Elements of Mathematical Biology*. Dover Publications, New York.

Lotze, H. K. and Worm, B. (2009). Historical baselines for large marine animals. *Trends in Ecology & Evolution*, 24, 254–262.

Lotze, H. K., Lenihan, H. S., Bourque, R. H., et al. (2006). Depletion, degradation, and recovery potential of estuaries and coastal seas. *Science*, 312, 1806–1809.

Lotze, H. K., Coll, M., Magera, A. M., Ward-Paige, C. and Airoldi, L. (2011). Recovery of marine animal populations and ecosystems. *Trends in Ecology & Evolution*, 26(11), 595–605.

Louppe, D., Oattara, N. K. and Coulibaly, A. (1995). The effects of brush fires on vegetation: The Aubreville fire plots after 60 years. *Commonwealth Forestry Review*, 74, 288–292.

Lovejoy, T. E. and Nobre, C. (2018). Amazon tipping point. *Science Advances*, 4, eaat2340.

Lovelock, J. E. and Margulis, L. (1974). Atmospheric homeostasis by and for the biosphere: The Gaia hypothesis. *Tellus Series A, Stockholm International Meteorological Institute*, 26(1–2), 2–10.

Lovett, G. M. (2013). Critical issues for critical loads. *Proceedings of the National Academy of Sciences of the United States of America*, 110(3), 808–809.

Lucas, S. G. and Tanner, L. H. (2015). End-Triassic nonmarine biotic events. *Journal of Palaeogeography*, 4(4), 331–348.

Lund, H. G. (2018). *Definitions of Forest, Deforestation, Afforestation, and Reforestation*. Gainesville, VA: Forest Information Services. https://doi.org/10.13140/RG.2.1.2364.9760.

Lurgi, M., Montoya, D. and Montoya, J. M. (2016). The effects of space and diversity of interaction types on the stability of complex ecological networks. *Theoretical Ecology*, 9(1), 3–13.

Lyson, T. R., Miller, I. M., Bercovici, A. D., et al. (2019). Exceptional continental record of biotic recovery after the Cretaceous–Paleogene mass extinction. *Science*, 366(6468), 977–983.

MacDougall, A. S., McCann, K. S., Gellner, G., et al. (2013). Diversity loss with persistent human disturbance increases vulnerability to ecosystem collapse. *Nature*, 494, 86–89.
Mace, G. M., Collar, N. J., Gaston, K. J., et al. (2008). Quantification of extinction risk: IUCN's system for classifying threatened species. *Conservation Biology*, 22, 1424–1442.
Mace, G. M., Norris, K. and Fitter, A. H. (2012). Biodiversity and ecosystem services: A multilayered relationship. *Trends in Ecology & Evolution*, 27(1), 19–26.
Mace, G. M., Reyers, B., Alkemade, R., et al. (2014). Approaches to defining a planetary boundary for biodiversity. *Global Environmental Change*, 28, 289–297.
Macgregor, C. J., Williams, J. H., Bell, J. R. and Thomas, C. D. (2019). Moth biomass increases and decreases over 50 years in Britain. *Nature Ecology & Evolution*, 3, 1645–1649. https://doi.org/10.1038/s41559-019-1028-6.
Macias-Fauria, M., Forbes, B. C., Zetterberg, P. and Kumpula, T. (2012). Eurasian Arctic greening reveals teleconnections and the potential for structurally novel ecosystems. *Nature Climate Change*, 2(8), 613–618.
MacNeil, M. A., Graham, N. A. J., Cinner, J. E., et al. (2015). Recovery potential of the world's coral reef fishes. *Nature*, 520(7547), 341–344.
Maes, J., Paracchini, M. L., Zulian, G., Dunbar, M. B. and Alkemade, R. (2012). Synergies and trade-offs between ecosystem service supply, biodiversity, and habitat conservation status in Europe. *Biological Conservation*, 155, 1–12.
Maestre, F. T., Quero, J. L., Gotelli, N. J., et al. (2012). Plant species richness and ecosystem multifunctionality in global drylands. *Science*, 335(6065), 214–218.
Magrach, A., Laurance, W. F., Larrinaga, A. R. and Santamaria, L. (2014). Meta-analysis of the effects of forest fragmentation on interspecific interactions. *Conservation Biology*, 28(5), 1342–1348.
Mahootian, F. and Eastman, T. E. (2009). Complementary frameworks of scientific inquiry: Hypothetico-deductive, hypothetico-inductive, and observational-inductive. *World Futures*, 65(1), 61–75.
Maldonado, G. and Poole, C. (1999). Editorial. *Annals of Epidemiology*, 9(1), 17–18.
Malhi, Y., Roberts, J. T., Betts, R. A., Killeen, T. J., Li, W. and Nobre, C. A. (2008). Climate change, deforestation, and the fate of the Amazon. *Science*, 319 (5860), 169–172.
Malhi, Y., Aragão, L. E. O. C., Galbraith, D., et al. (2009). Exploring the likelihood and mechanism of a climate-change-induced dieback of the Amazon rainforest. *Proceedings of the National Academy of Sciences of the United States of America*, 106 (49), 20610–20615.
Malhi, Y., Gardner, T. A., Goldsmith, G. R., Silman, M. R. and Zelazowski, P. (2014). Tropical forests in the Anthropocene. *Annual Review of Environment and Resources*, 39, 125–159.
Malhi, Y., Doughty, C. E., Galetti, M., Smith, F. A., Svenning, J. C. and Terborgh, J. W. (2016). Megafauna and ecosystem function from the Pleistocene to the Anthropocene. *Proceedings of the National Academy of Sciences of the United States of America*, 113, 838–846.
Maller, C., Townsend, M., Pryor, A., Brown, P. and St Leger, L. (2006). Healthy nature healthy people: 'Contact with nature' as an upstream health promotion intervention for populations. *Health Promotion International*, 21(1), 45–54.

Mann, D. H., Rupp, T. S., Olson, M. A. and Duffy, P. A. (2012). Is Alaska's Boreal forest now crossing a major ecological threshold? *Arctic, Antarctic, and Alpine Research*, 44(3), 319–331.

Margules, C. R. and Pressey, R. L. (2000). Systematic conservation planning. *Nature*, 405, 243–253.

Mariana Morais, V., Cristina, B.-L., Leandro Reverberi, T., et al. (2019). Predicting the non-linear collapse of plant–frugivore networks due to habitat loss. *Ecography*, 42(10), 1765–1776.

Marín, V. H. (1997). General system theory and the ecosystem concept. *Bulletin of the Ecological Society of America*, 78, 102–104.

Maron, M., Mitchell, M. G. E., Runting, R. K., et al. (2017). Towards a threat assessment framework for ecosystem services. *Trends in Ecology & Evolution*, 32(4), 240–248.

Marquet, P. A., Allen, A. P., Brown, J. H., et al. (2014). On theory in ecology. *BioScience*, 64(8), 701–710.

Marsden, S. J., Whiffin, M. and Galetti, M. (2001). Bird diversity and abundance in forest fragments and *Eucalyptus* plantations around an Atlantic forest reserve, Brazil. *Biodiversity and Conservation*, 10, 737–751.

Marshall, B. E. (2018). Guilty as charged: Nile perch was the cause of the haplochromine decline in Lake Victoria. *Canadian Journal of Fisheries and Aquatic Sciences*, 75(9), 1542–1559.

Martin, P. A., Newton, A. C. and Bullock, J. M. (2013). Carbon pools recover more quickly than plant biodiversity in tropical secondary forests. *Proceedings of the Royal Society B: Biological Sciences*, 280(1773), 20132236.

Martin, P., Newton, A. C., Evans, P. and Cantarello, E. (2015a). Stand collapse in a temperate forest and its impact on forest structure and biodiversity. *Forest Ecology and Management*, 358, 130–138.

Martin, P. A., Newton, A. C., Pfeifer, M., Khoo, M. and Bullock, J. M. (2015b). Species richness and carbon storage responses to reduced impact logging in tropical forests: A meta-analysis. *Forest Ecology and Management*, 356, 224–233.

Martin, P., Newton, A. C., Cantarello, E. and Evans, P. M. (2017). Analysis of ecological thresholds in a temperate forest undergoing dieback. *PLoS One*, 12(12), e0189578. https://doi.org/10.1371/journal.pone.0189578.

Martin, P. S. (1984). Prehistoric overkill: The global model. In: P. S. Martin and R. G. Klein (eds.), *Quaternary Extinctions: A Prehistoric Revolution*. University of Arizona Press, Tucson, pp. 354–403.

Masood, E. (2018). The battle for the soul of biodiversity. *Nature*, 560(7719), 423–425.

Matthews, B., Narwani, A., Hausch, S., et al. (2011). Toward an integration of evolutionary biology and ecosystem science. *Ecology Letters*, 14, 690–701.

Matthews, J. H. and Boltz, F. (2012). The shifting boundaries of sustainability science: Are we doomed yet? *PLoS Biology*, 10, e1001344.

Matthews, J. W., Spyreas, G. and Endress, A. G. (2009). Trajectories of vegetation-based indicators used to assess wetland restoration progress. *Ecological Applications*, 19, 2093–2107.

Matusick, G., Ruthrof, K. X., Brouwers, N. C., et al. (2013). Sudden forest canopy collapse corresponding with extreme drought and heat in a Mediterranean-type

eucalypt forest in southwestern Australia. *European Journal of Forest Research*, 132, 497–510.
Maureaud, A., Gascuel, D., Colléter, M., et al. (2017). Global change in the trophic functioning of marine food webs. *PLoS One*, 12(8), e0182826.
Maurer, B. A. (2000). Ecology needs theory as well as practice. *Nature*, 408, 768.
Maxwell, P. S., Eklöf, J. S., van Katwijk, M. M., et al. (2017). The fundamental role of ecological feedback mechanisms for the adaptive management of seagrass ecosystems – A review. *Biological Reviews*, 92(3), 1521–1538.
Maxwell, S., Fuller, R., Brooks, T., et al. (2016). Biodiversity: The ravages of guns, nets and bulldozers. *Nature*, 536, 143–145.
May, R. M. (1976). Simple mathematical models with very complicated dynamics. *Nature*, 261, 459–67.
  (1977). Thresholds and breakpoints in ecosystems with a multiplicity of stable states. *Nature*, 269, 471–477.
McAfee, D., Doubleday, Z. A., Geiger, N. and Connell, S. D. (2019). Everyone loves a success story: Optimism inspires conservation engagement. *BioScience*, 69(4), 274–281.
McAnany, P. A. and Yoffee, N. (eds.) (2010a). *Questioning Collapse: Human Resilience, Ecological Vulnerability, and the Aftermath of Empire*. Cambridge University Press, Cambridge.
  (2010b). Questioning how different societies respond to crises. *Nature*, 464, 977.
McCann, K. S. (2000). The diversity–stability debate. *Nature*, 405(6783), 228–233.
McCook, L. J. (1999). Macroalgae, nutrients and phase shifts on coral reefs: Scientific issues and management consequences for the Great Barrier Reef. *Coral Reefs*, 18(4), 357–367.
McCrackin, M. L., Jones, H. P., Jones, P. C. and Moreno-Mateos, D. (2017). Recovery of lakes and coastal marine ecosystems from eutrophication: A global meta-analysis. *Limnology and Oceanography*, 62, 507–518.
McDonald-Madden, E., Sabbadin, R., Game, E. T., Baxter, P. W. J., Chadès, I. and Possingham, H. P. (2016). Using food-web theory to conserve ecosystems. *Nature Communications*, 7, 10245, 1–8.
McDowell, N. G., Michaletz, S. T., Bennett, K. E., et al. (2018). Predicting chronic climate-driven disturbances and their mitigation. *Trends in Ecology & Evolution*, 33(1), 15–27.
McElwain, J. C. and Punyasena, S. W. (2007). Mass extinction events and the plant fossil record. *Trends in Ecology & Evolution*, 22(10), 548–557.
McGhee, G. R. (1988). The Late Devonian extinction event: Evidence for abrupt ecosystem collapse. *Paleobiology*, 14(3), 250–257.
McGhee, G. R., Sheehan, P. M., Bottjer, D. J. and Droser, M. L. (2004). Ecological ranking of Phanerozoic biodiversity crises: Ecological and taxonomic severities are decoupled. *Palaeogeography, Palaeoclimatology, Palaeoecology*, 211(3), 289–297.
McGhee, G. R., Clapham, M. E., Sheehan, P. M., Bottjer, D. J. and Droser, M. L. (2013). A new ecological-severity ranking of major Phanerozoic biodiversity crises. *Palaeogeography, Palaeoclimatology, Palaeoecology*, 370, 260–270.
McGlone, M. S. and Clarkson, B. D. (1993). Ghost stories: Moa, plant defences and evolution in New Zealand. *Tuatara*, 32, 1–21.

McGovern, P., Jalabadze, M., Batiuk, S., et al. (2017). Early Neolithic wine of Georgia in the South Caucasus. *Proceedings of the National Academy of Sciences of the United States of America*, 114(48), E10309–E10318.

McClanahan, T. R. and Muthiga, N. A. (1998). An ecological shift in a remote coral reef atoll of Belize over 25 years. *Environmental Conservation*, 25, 122–130.

McCulloch, M., Fallon, S., Wyndham, T., Hendy, E., Lough, J. and Barnes, D. (2003). Coral record of increased sediment flux to the inner Great Barrier Reef since European settlement. *Nature*, 421(6924), 727–730.

McDermott, M., Mahanty, S. and Schreckenberg, K. (2013). Examining equity: A multidimensional framework for assessing equity in payments for ecosystem services. *Environmental Science & Policy*, 33, 416–427.

McDowell, N. G. and Allen, C. D. (2015). Darcy's law predicts widespread forest mortality under climate warming. *Nature Climate Change*, 5, 669.

McIntyre, P. B., Jones, L. E., Flecker, A. S. and Vanni, M. J. (2007). Fish extinctions alter nutrient recycling in tropical freshwaters. *Proceedings of the National Academy of Sciences of the United States of America*, 104, 4461–4466.

McLauchlan, K. K., Williams, J. J., Craine, J. M. and Jeffers, E. S. (2013). Changes in global nitrogen cycling during the Holocene epoch. *Nature*, 495, 352–355.

McManus, J. W. and Polsenberg, J. F. (2004). Coral–algal phase shifts on coral reefs: Ecological and environmental aspects. *Progress in Oceanography*, 60, 263–279.

McNaughton, S. J. (1984). Grazing lawns: Animals in herds, plant form, and coevolution. *The American Naturalist*, 124(6), 863–886.

McWethy, D. B., Whitlock, C., Wilmshurst, J. M., McGlone, M. S. and Li, X. (2009). Rapid deforestation of South Island, New Zealand, by early Polynesian fires. *Holocene*, 19(6), 883–897.

McWethy, D. B., Whitlock, C., Wilmshurst, J. M., et al. (2010). Rapid landscape transformation in South Island, New Zealand, following initial Polynesian settlement. *Proceedings of the National Academy of Sciences of the United States of America*, 107, 21343–21348.

McWethy, D. B., Wilmshurst, J. M., Whitlock, C., Wood, J. R. and McGlone, M. S. (2014). A high-resolution chronology of rapid forest transitions following Polynesian arrival in New Zealand. *PLoS One*, 9(11), e111328.

Mee, L. D., Friedrich, J. and Gomoiu, M. T. (2005). Restoring the Black Sea in times of uncertainty. *Oceanography*, 18(2), 100–111.

Mehner, T., Diekmann, M., Gonsiorczyk, T., et al. (2008). Rapid recovery from eutrophication of a stratified lake by disruption of internal nutrient load. *Ecosystems*, 11(7), 1142–1156.

Meijer, M.-L., de Boos, I., Scheffer, M., Portielje, R. and Hosper, H. (1999). Biomanipulation in shallow lakes in the Netherlands: An evaluation of 18 case studies. *Hydrobiologia*, 408/409, 13–30.

Meli, P., Holl, K. D., Rey Benayas, J. M., et al. (2017). A global review of past land use, climate, and active vs. passive restoration effects on forest recovery. *PLoS One*, 12(2), e0171368.

Memmott, J., Waser, N. and Price, M. (2004). Tolerance of pollination networks to species extinctions. *Proceedings of the Royal Society B: Biological Sciences*, 271, 2605–2611.

Mentis, M. T. (1988). Hypothetico-deductive and inductive approaches in ecology. *Functional Ecology*, 2, 5–14.

Messier, C., Puettmann, K. J. and Coates, K. D. (eds.). (2013). *Managing Forests as Complex Adaptive Systems: Building Resilience to the Challenge of Global Change*. Routledge, Abingdon.

Metcalf, J. L., Turney, C., Barnett, R., et al. (2016). Synergistic roles of climate warming and human occupation in Patagonian megafaunal extinctions during the Last Deglaciation. *Science Advances*, 2(6), e1501682.

Middleton, G. D. (2012). Nothing lasts forever: Environmental discourses on the collapse of past societies. *Journal of Archaeological Research*, 20(3), 257–307.

Millennium Ecosystem Assessment. (2005a). *Ecosystems and Human Well-being: Synthesis*. Island Press, Washington, DC. www.millenniumassessment.org/en/index.html.

(2005b). *Ecosystems and Human Well-being: Biodiversity Synthesis*. World Resources Institute, Washington, DC.

Miller, B., Soulé, M. E. and Terborgh, J. (2014). 'New conservation' or surrender to development? *Animal Conservation*, 17, 509–515.

Miller, G. H., Magee, J. W., Johnson, B. J., et al. (1999). Pleistocene extinction of *Genyornis newtoni*: Human impact on Australian megafauna. *Science*, 283, 205–208.

Miller, G. H., Fogel, M. L., Magee, J. W., et al. (2005). Ecosystem collapse in Pleistocene Australia and a human role in megafaunal extinction. *Science*, 309, 287–290.

Miller, G. H., Fogel, M. L., Magee, J. W. and Gagan, M. K. (2016a). Disentangling the impacts of climate and human colonization on the flora and fauna of the Australian arid zone over the past 100 ka using stable isotopes in avian eggshell. *Quaternary Science Reviews*, 151, 27–57.

Miller, G., Magee, J., Smith, M., et al. (2016b). Human predation contributed to the extinction of the Australian megafaunal bird *Genyornis newtoni* ~47 ka. *Nature Communications*, 7, 10496.

Miller, G. S. and Magee, J. W. (1992). *Drought in the Australian Outback: Anthropogenic Impacts on Regional Climate*. American Geophysical Union, fall meeting, p. 104.

Miller, J. R. and Bestelmeyer, B. T. (2016). What's wrong with novel ecosystems, really? *Restoration Ecology*, 24(5), 577–582.

Miller-Rushing, A. J., Primack, R. B., Devictor, V., et al. (2019). How does habitat fragmentation affect biodiversity? A controversial question at the core of conservation biology. *Biological Conservation*, 232, 271–273.

Mills, L. S., Soulé, M. E. and Doak, D. F. (1993). The keystone-species concept in ecology and conservation. *BioScience*, 43(4), 219–224.

Mitsch, W. J. and Day, J. W. (2004). Thinking big with whole-ecosystem studies and ecosystem restoration – A legacy of H.T. Odum. *Ecological Modelling*, 178(1), 133–155.

Mittelbach, G. G., Garcia, E. A. and Taniguchi, Y. (2006). Fish reintroductions reveal smooth transitions between lake community states. *Ecology*, 87, 312–318.

Mizukami, T., Kaiho, K. and Oba, M. (2013). Significant changes in land vegetation and oceanic redox across the Cretaceous/Paleogene boundary. *Palaeogeography, Palaeoclimatology, Palaeoecology*, 369, 41–47.

Mkumbo, O. and Marshall, B. (2014). The Nile perch fishery of Lake Victoria: Current status and management challenges. *Fisheries Management and Ecology*, 22, 56–63.

Moberg, F. and Folke, C. (1999). Ecological goods and services of coral reef ecosystems. *Ecological Economics*, 29(2), 215–233.

Möllmann, C. and Diekmann, R. (2012). Marine ecosystem regime shifts induced by climate and overfishing. *Advances in Ecological Research*, 47, 303–347.

Molyneux, J. (2019). The environmental crisis and the new environmental revolt. *Irish Marxist Review*, 8(24), 38–42.

Monbiot, G. (2013). *Feral: Searching for Enchantment on the Frontiers of Rewilding*. Penguin Books, London.

Montoya, J. M., Donohue, I. and Pimm, S. L. (2018). Planetary boundaries for biodiversity: Implausible science, pernicious policies. *Trends in Ecology & Evolution*, 33(2), 71–73.

Mooney, S. D., Harrison, S. P., Bartlein, P. J., et al. (2011). Late quaternary fire regimes of Australasia. *Quaternary Science Reviews*, 30, 28–46.

Moore, J. K., Fu, W., Primeau, F., et al. (2018). Sustained climate warming drives declining marine biological productivity. *Science*, 359, 1139–1143.

Mora, C., Aburto-Oropeza, O., Ayala Bocos, A., et al. (2011). Global human footprint on the linkage between biodiversity and ecosystem functioning in reef fishes. *PLoS Biology*, 9(4), e1000606.

Moreno-Mateos, D., Barbier, E. B., Jones, P. C., et al. (2017). Anthropogenic ecosystem disturbance and the recovery debt. *Nature Communications*, 8(1), 14163.

Mori, A. S. (2011). Ecosystem management based on natural disturbances: Hierarchical context and non-equilibrium paradigm. *Journal of Applied Ecology*, 48, 280–292.

Moritz, M. A., Morais, M. E., Summerell, L. A., Carlson, J. M. and Doyle, J. (2005). Wildfires, complexity, and highly optimized tolerance. *Proceedings of the National Academy of Sciences of the United States of America*, 102, 17912–17917.

Morrison, T. H., Hughes, T. P., Adger, W. N., Brown, K., Barnett, J. and Lemos, M. C. (2019). Save reefs to rescue all ecosystems. *Nature*, 573(7774), 333–336.

Morueta-Holme, N., Engemann, K., Sandoval-Acuña, P., Jonas, J. D., Segnitz, R. M. and Svenning, J.-C. (2015). Strong upslope shifts in Chimborazo's vegetation over two centuries since Humboldt. *Proceedings of the National Academy of Sciences of the United States of America*, 112, 12741–12745.

Mouchet, M. A., Villéger, S., Mason, N. W. and Mouillot, D. (2010). Functional diversity measures: An overview of their redundancy and their ability to discriminate community assembly rules. *Functional Ecology*, 24, 867–876.

Mougi, A. and Kondoh, M. (2012). Diversity of interaction types and ecological community stability. *Science*, 337, 349–351.

Mouillot, D., Mason, N. W. H. and Wilson, J. B. (2007). Is the abundance of species determined by their functional traits? A new method with a test using plant communities. *Oecologia*, 152, 729–737.

Mumby, P. J. and Steneck, R. S. (2008). Coral reef management and conservation in light of rapidly evolving ecological paradigms. *Trends in Ecology & Evolution*, 23(10), 555–563.

Muradian, R., Arsel, M., Pellegrini, L., *et al.* (2013), Payments for ecosystem services and the fatal attraction of win-win solutions. *Conservation Letters*, 6, 274–279.

Murcia, C., Aronson, J., Kattan, G. H., Moreno-Mateos, D., Dixon, K. and Simberloff, D. (2014). A critique of the novel ecosystem concept. *Trends in Ecology & Evolution*, 29(10), 548–553.

Murphy, B. P. and Bowman, D. M. J. S. (2012). What controls the distribution of tropical forest and savanna? *Ecology Letters*, 15, 748–758.

Muscente, A. D., Prabhu, A., Zhong, H., *et al.* (2018). Quantifying ecological impacts of mass extinctions with network analysis of fossil communities. *Proceedings of the National Academy of Sciences of the United States of America*, 115 (20), 5217–5222.

Myers, R. A., Baum, J. K., Shepherd, T. D., Powers, S. P. and Peterson, C. H. (2007). Cascading effects of the loss of apex predatory sharks from a coastal ocean. *Science*, 315, 1846–1850.

Myers-Smith, I. H., Forbes, B. C., Wilmking, M., *et al.* (2011). Shrub expansion in tundra ecosystems: Dynamics, impacts and research priorities. *Environmental Research Letters*, 6(4), 45509.

Myers-Smith, I. H., Trefry, S. A. and Swarbrick, V. J. (2012). Resilience: Easy to use but hard to define. *Ideas in Ecology and Evolution*, 5, 44–53.

Naeem, S. (2008). Advancing realism in biodiversity research. *Trends in Ecology & Evolution*, 23, 414–416.

Naeem, S., Duffy, J. E. and Zavaleta, E. (2012). The functions of biological diversity in an age of extinction. *Science*, 336(6087), 1401–1406.

Nagel, E. (1979). *Teleology Revisited and Other Essays in the Philosophy and History of Science*. Columbia University Press, New York.

Nagelkerken, I. and Munday, P. L. (2016). Animal behaviour shapes the ecological effects of ocean acidification and warming: Moving from individual to community-level responses. *Global Change Biology*, 22, 974–989.

Nauta, A. L., Heijmans, M. M. P. D., Blok, D., *et al.* (2014). Permafrost collapse after shrub removal shifts tundra ecosystem to a methane source. *Nature Climate Change*, 5, 67.

Navarro, L. M. and Pereira, H. M. (2012). Rewilding abandoned landscapes in Europe. *Ecosystems*, 15, 900–912.

Nellemann, C. and Corcoran, E. (eds.). (2010). *Dead Planet, Living Planet – Biodiversity and Ecosystem Restoration for Sustainable Development. A Rapid Response Assessment*. United Nations Environment Programme, GRID-Arendal, Norway.

Nepstad, D., Carvalho, G., Cristina Barros, A., *et al.* (2001). Road paving, fire regime feedbacks, and the future of Amazon forests. *Forest Ecology and Management*, 154(3), 395–407.

Nepstad, D. C., Stickler, C. M., Soares-Filho, B. and Merry, F. (2008). Interactions among Amazon land use, forests and climate: Prospects for a near-term forest tipping point. *Philosophical Transactions of the Royal Society B: Biological Sciences*, 363(1498), 1737–1746.

Neubauer, P., Jensen, O. P., Hutchings, J. A. and Baum, J. K. (2013). Resilience and recovery of overexploited marine populations. *Science*, 340(6130), 347–349.

Newbold, T., Hudson, L. N., Arnell, A. P., et al. (2016). Has land use pushed terrestrial biodiversity beyond the planetary boundary? A global assessment. *Science*, 353(6296), 288–291.

Newton, A. C. (2007a). *Forest Ecology and Conservation. A Handbook of Techniques*. Oxford University Press, Oxford.

(ed.). (2007b). *Biodiversity Loss and Conservation in Fragmented Forest Landscapes. The Forests of Montane Mexico and Temperate South America*. CABI Publishing, Wallingford.

(2016). Biodiversity risks of adopting resilience as a policy goal. *Conservation Letters*, 9(5), 369–376.

Newton, A. C. and Cantarello, E. (2014). *An Introduction to the Green Economy*. Earthscan, Abingdon.

(2015). Restoration of forest resilience: An achievable goal? *New Forests*, 46, 645–668.

Newton, A. C. and Echeverría, C. (2014). Analysis of anthropogenic impacts on forest biodiversity as a contribution to empirical theory. BES Symposium volume. In: D. A. Coomes, D. F. R. P. Burslem and W. D. Simonson (eds.), *Forests and Global Change*. Cambridge University Press, Cambridge, pp. 417–446.

Newton, A. C. and Oldfield, S. (2008). Red Listing the world's tree species: A review of recent progress. *Endangered Species Research*, 6, 137–147.

Newton, A. C., Cantarello, E., Tejedor, N. and Myers, G. (2013a). Dynamics and conservation management of a wooded landscape under high herbivore pressure. *International Journal of Biodiversity*, 2013, 15. https://doi.org/10.1155/2013/273948.

Newton, A. C., Cantarello, E., Lovegrove, A., Appiah, D. and Perrella, L. (2013b). The influence of grazing animals on tree regeneration and woodland dynamics in the New Forest, England. In: I. Rotherham (ed.), *Trees, Forested Landscapes and Grazing Animals – A European Perspective on Woodlands and Grazed Treescapes*. Routledge, Oxford, pp. 163–179.

Newton, A. C., Hodder, K., Cantarello, E., et al. (2012a). Cost-benefit analysis of ecological networks assessed through spatial analysis of ecosystem services. *Journal of Applied Ecology*, 49(3), 571–580.

Newton, A. C., Walls, R. M., Golicher, D., Keith, S. A., Diaz, A. and Bullock, J. M. (2012b). Structure, composition and dynamics of a calcareous grassland metacommunity over a seventy year interval. *Journal of Ecology*, 100(1), 196–209.

Newton, A. C., Boscolo, D., Ferreira, P. A., Lopes, L. E. and Evans, P. (2018). Impacts of deforestation on plant-pollinator networks assessed using an agent based model. *PLoS One*, 13(12), e0209406.

Newton, A. C., Watson, S., Evans, P., et al. (2019). *Trends in Natural Capital, Ecosystem Services and Economic Development in Dorset*. Bournemouth University, Poole.

Nicol, S., Brazill-Boast, J., Gorrod, E., McSorley, A., Peyrard, N. and Chadès, I. (2019). Quantifying the impact of uncertainty on threat management for

biodiversity. *Nature Communications*, 10(1), 3570. https://doi.org/10.1038/s41467-019-11404-5.
Nijp, J. J., Temme, A. J. A. M., van Voorn, G. A. K., et al. (2019). Spatial early warning signals for impending regime shifts: A practical framework for application in real-world landscapes. *Global Change Biology*, 25, 1905–1921.
Nilsson, J. and Grennfelt, P. (1988). *Critical Levels for Sulphur and Nitrogen*. Copenhagen, Denmark: Nordic Council of Ministers.
Nimmo, D. G., MacNally, R., Cunningham, S. C., Haslem, A. and Bennett, A. F. (2015). Vive la resistance: Reviving resistance for 21st century conservation. *Trends in Ecology & Evolution*, 30, 516–523.
Nobre, C. A. and Borma, L. D. S. (2009). 'Tipping points' for the Amazon forest. *Current Opinion in Environmental Sustainability*, 1(1), 28–36.
Nogués-Bravo, D., Simberloff, D., Rahbek, C. and Sanders, N. J. (2016). Rewilding is the new Pandora's box in conservation. *Current Biology*, 26(3), R87–R91.
Nolan, C., Overpeck, J. T., Allen, J. R. M., et al. (2018). Past and future global transformation of terrestrial ecosystems under climate change. *Science*, 361 (6405), 920–923.
Nolan, R. H., Boer, M. M., Collins, L., et al. (2020). Causes and consequences of eastern Australia's 2019–20 season of mega-fires. *Global Change Biology*, 26(3), 1039–1041. https://doi.org/10.1111/gcb.14987.
Norden, N., Chazdon, R. L., Chao, A., Jiang, Y.-H. and Vilchez-Alvarado, B. (2009). Resilience of tropical rain forests: Tree community reassembly in secondary forests. *Ecology Letters*, 12, 385–394.
Norman, K., Inglis, J., Clarkson, C., Faith, J. T., Shulmeister, J. and Harris, D. (2018). An early colonisation pathway into northwest Australia 70-60,000 years ago. *Quaternary Science Reviews*, 180, 229–239.
Norgaard, R. B. (2010). Ecosystem services: From eye-opening metaphor to complexity blinder. *Ecological Economics*, 69, 1219–1227.
Noss, R. F. (1996). Ecosystems as conservation targets. *Trends in Ecology & Evolution*, 11(8), 351.
Noss, R. F., Dobson, A. P., Baldwin, R., et al. (2012). Bolder thinking for conservation. *Conservation Biology*, 26, 1–4.
Nowacki, G. J. and Abrams, M. D. (2008). The demise of fire and 'Mesophication' of forests in the eastern United States. *BioScience*, 58, 123–138.
Nyström, M., Norström, A. V., Blenckner, T., et al. (2012). Confronting feedbacks of degraded marine ecosystems. *Ecosystems*, 15, 695–710.
O'Connell, J. F. and Allen, J. (2012). The restaurant at the end of the universe: Modelling the colonisation of Sahul. *Austral Archaeology*, 74, 5–31.
O'Connor, N. E. and Crowe, T. P. (2005). Biodiversity loss and ecosystem functioning: Distinguishing between number and identity of species. *Ecology*, 86, 1783–1796.
Odum, H. T. (1971). *Environment, Power, and Society*. Wiley, New York.
O'Gorman, E. J. and Emmerson, M. C. (2009). Perturbations to trophic interactions and the stability of complex food webs. *Proceedings of the National Academy of Sciences of the United States of America*, 106, 13393–13398.

Oguz, T. and Gilbert, D. (2007). Abrupt transitions of the top-down controlled Black Sea pelagic ecosystem during 1960–2000: Evidence for regime-shifts under strong fishery exploitation and nutrient enrichment modulated by climate-induced variations. *Deep Sea Research Part I: Oceanographic Research Papers*, 54, 220–242.

Oguz, T. and Velikova, V. (2010). Abrupt transition of the northwestern Black Sea shelf ecosystem from a eutrophic to an alternative pristine state. *Marine Ecology Progress Series*, 405, 231–242.

Olesen, J., Bascompte, J., Dupont, Y. and Jordano, P. (2007). The modularity of pollination networks. *Proceedings of the National Academy of Sciences of the United States of America*, 104, 19891–19896.

Oliveras, I. and Malhi, Y. (2016). Many shades of green: The dynamic tropical forest–savannah transition zones. *Philosophical Transactions of the Royal Society B: Biological Sciences*, 371(1703), 20150308.

Oliver, T. H., Heard, M. S., Isaac, N. J. B., et al. (2015). Biodiversity and resilience of ecosystem functions. *Trends in Ecology & Evolution*, 30(11), 673–684.

Ollerton, J., Erenler, H., Edwards, M. and Crockett, R. (2014). Extinctions of aculeate pollinators in Britain and the role of large-scale agricultural changes. *Science*, 346(6215), 1360–1362.

Olsson, L., Jerneck, A., Thoren, H., Persson, J. and O'Byrne, D. (2015). Why resilience is unappealing to social science: Theoretical and empirical investigations of the scientific use of resilience. *Science Advances*, 1(4). http://dx.doi.org/10.1126/sciadv.1400217.

Owen-Smith, R. N. (1988). *Megaherbivores: The Influence of Very Large Body Size on Ecology*. Cambridge University Press, Cambridge.

O'Neill, R. V. (1999). Recovery in complex ecosystems. *Journal of Aquatic Ecosystem Stress and Recovery*, 6, 181–187.

  (2001). Is it time to bury the ecosystem concept? (With full military honors, of course!). *Ecology*, 82(12), 3275–3284.

O'Neill, R. V., DeAngelis, D. L., Waide, J. B. and Allen, T. F. H. (1986). *A Hierarchical Concept of Ecosystems*. Monographs in population biology, vol. 23. Princeton University Press, Princeton, NJ. 253 pp.

O'Neill, S. and Nicholson-Cole, S. (2009). 'Fear won't do it': Promoting positive engagement with climate change through visual and iconic representations. *Science Communication*, 30(3), 355–379.

Ormerod, S. J., Dobson, M., Hildrew, A. G. and Townsend, C. R. (2010). Multiple stressors in freshwater ecosystems. *Freshwater Biology*, 55, 1–4.

Österblom, H., Hansson, S., Larsson, U., Hjerne, O., Wulff, F., Elmgren, R. and Folke, C. (2007). Human-induced trophic cascades and ecological regime shifts in the Baltic sea. *Ecosystems*, 10, 877–889.

Osuri, A. M., Ratnam, J., Varma, V., et al. (2016). Contrasting effects of defaunation on aboveground carbon storage across the global tropics. *Nature Communications*, 7, 11351.

Pace, M. L. (2001). Prediction and the aquatic sciences. *Canadian Journal of Fisheries and Aquatic Sciences*, 58, 63–72.

Pace, M. L., Carpenter, S. R., Johnson, R. and Kurtzweil, J. (2013). Zooplankton provide early warnings of a regime shift in a whole lake manipulation. *Limnology and Oceanography*, 58, 525–532.

Pace, M. L., Carpenter, S. R. and Cole, J. J. (2015). With and without warning: Managing ecosystems in a changing world. *Frontiers in Ecology and the Environment*, 13(9), 460–467.

Pacifici, M., Foden, W. B., Visconti, P., et al. (2015). Assessing species vulnerability to climate change. *Nature Climate Change*, 5(3), 215–224.

Pacifici, M., Visconti, P., Butchart, S. H. M., Watson, J. E. M., Cassola, F. M. and Rondinini, C. (2017). Species' traits influenced their response to recent climate change. *Nature Climate Change*, 7(3), 205–208.

Paine, R. T., Tegner, M. J. and Johnson, E. A. (1998). Compounded perturbations yield ecological surprises. *Ecosystems*, 1, 535–545.

Palmer, M. A., Menninger, H. L. and Bernhardt, E. (2010). River restoration, habitat heterogeneity and biodiversity: A failure of theory or practice? *Freshwater Biology*, 55, 205–222.

Paquette, A. and Messier, C. (2010). The role of plantations in managing the world's forests in the Anthropocene. *Frontiers in Ecology and the Environment*, 8(1), 27–34.

Pardi, M. I. and Smith, F. A. (2016). Biotic responses of canids to the terminal Pleistocene megafauna extinction. *Ecography*, 39, 141–151.

Pardini, R., de Arruda Bueno, A., Gardner, T. A., Prado, P. I. and Metzger, J. P. (2010). Beyond the fragmentation threshold hypothesis: Regime shifts in biodiversity across fragmented landscapes. *PLoS One*, 5, e13666.

Pardo, L. H., Fenn, M. E., Goodale, C. L., et al. (2011). Effects of nitrogen deposition and empirical nitrogen critical loads for ecoregions of the United States. *Ecological Applications*, 21, 3049–3082.

Park Williams, A., Allen, C. D., Macalady, A. K., et al. (2012). Temperature as a potent driver of regional forest drought stress and tree mortality. *Nature Climate Change*, 3, 292.

Parr, C. L., Gray, E. F. and Bond, W. J. (2012). Cascading biodiversity and functional consequences of a global change-induced biome switch. *Diversity and Distributions*, 18, 493–503.

Parsons, T. R. and Lalli, C. M. (2002). Jellyfish population explosions: Revisiting a hypothesis of possible causes. *La Mer (Paris)*, 40, 111–121.

Pascual, M. and Guichard, F. (2005). Criticality and disturbance in spatial ecological systems. *Trends in Ecology & Evolution*, 20(2), 88–95.

Patricola, C. M. and Cook, K. H. (2007). Dynamics of the West African Monsoon under mid-Holocene processional forcing: Regional climate model simulations. *Journal of Climate*, 20(4), 694–716.

Patten, B. C., Straškraba, M. and Jørgensen, S. E. (1997). Ecosystem emerging. 1. Conservation. *Ecological Modelling*, 96, 221–284.

   (2011). Ecosystems emerging. 5: Constraints. *Ecological Modelling*, 222(16), 2945–2972.

Pauly, D. (2007). The *Sea around Us* Project: Documenting and communicating global fisheries impacts on marine ecosystems. *AMBIO: A Journal of the Human Environment*, 36(4), 290–295.

   (2008). Global fisheries: A brief review. *Journal of Biological Research – Thessaloniki*, 9, 3–9.

   (2009). Aquacalypse now: The end of fish. *The New Republic*, 28 September 2009. www.tnr.com/article/environment-energy/aquacalypse-now.

(2016). On the importance of fisheries catches, with a rationale for their reconstruction. In: D. Pauly and D. Zeller (eds.), *Global Atlas of Marine Fisheries: A Critical Appraisal of Catches and Ecosystem Impacts*. Island Press, Washington, DC, pp. 11–18.

Pauly, D. and Zeller, D. (2016). Catch reconstructions reveal that global marine fisheries catches are higher than reported and declining. *Nature Communications*, 7(10244), 1–9.

Pauly, D., Christensen, V., Dalsgaard, R., Froese, R. and Torres, F. C. (1998). Fishing down marine food webs. *Science*, 279(5352), 860–863.

Pauly, D., Hilborn, R. and Branch, T. A. (2013). Fisheries: Does catch reflect abundance? *Nature*, 494, 303.

Pausas, J. G. and Dantas, V. de L. (2016). Scale matters: Fire-vegetation feedbacks are needed to explain tropical tree cover at the local scale. *Global Ecology and Biogeography*, 26(4), 395–399.

Pawson, S. M., McCarthy, J. K., Ledgard, N. J. and Didham, R. K. (2010). Density-dependent impacts of exotic conifer invasion on grassland invertebrate assemblages. *Journal of Applied Ecology*, 47, 1053–1062.

Payne, J. L. and Finnegan, S. (2007). The effect of geographic range on extinction risk during background and mass extinction. *Proceedings of the National Academy of Sciences of the United States of America*, 104, 10506–10511.

Payne, R. J., Dise, N. B., Stevens, C. J. and Gowing, D. J. (2013). Impact of nitrogen deposition at the species level. *Proceedings of the National Academy of Sciences of the United States of America*, 110(3), 984–987.

Peck, J. (2010). Zombie neoliberalism and the ambidextrous state. *Theoretical Criminology*, 14(1), 104–110.

Pedroni, A., Eisenegger, C., Hartmann, M. N., Fischbacher, U. and Knoch, D. (2014). Dopaminergic stimulation increases selfish behavior in the absence of punishment threat. *Psychopharmacology*, 231(1), 135–141.

Peng, C., Ma, Z., Lei, X., et al. (2011). A drought-induced pervasive increase in tree mortality across Canada's boreal forests. *Nature Climate Change*, 1, 467.

Penn, J. L., Deutsch, C., Payne, J. L. and Sperling, E. A. (2018). Temperature-dependent hypoxia explains biogeography and severity of end-Permian marine mass extinction. *Science*, 362(6419), eaat1327. https://doi.org/10.1126/science.aat1327.

Pereira, H. M. and Navarro, L. M. (2015). *Rewilding European Landscapes*. Springer, London.

Peres, C. A., Emilio, T., Schietti, J., Desmoulière, S. J. M. and Levi, T. (2016). Dispersal limitation induces long-term biomass collapse in overhunted Amazonian forests. *Proceedings of the National Academy of Sciences of the United States of America*, 113, 892–897.

Perring, M. P., Standish, R. J. and Hobbs, R. J. (2013). Incorporating novelty and novel ecosystems into restoration planning and practice in the 21st century. *Ecological Processes*, 2(1), 18.

Perring, M. P., Standish, R. J., Price, J. N., et al. (2015). Advances in restoration ecology: Rising to the challenges of the coming decades. *Ecosphere*, 6(8), 131.

Perry, G. and Pianka, E. R. (1997). Animal foraging: Past, present and future. *Trends in Ecology & Evolution*, 12(9), 360–364.

Perry, G. L. W., Wilmshurst, J. M., McGlone, M. S., McWethy, D. B. and Whitlock, C. (2012). Explaining firedriven landscape transformation during the Initial Burning Period of New Zealand's prehistory. *Global Change Biology*, 18, 1609–1621.

Pershing, A. J., Mills, K. E., Record, N. R., et al. (2015). Evaluating trophic cascades as drivers of regime shifts in different ocean ecosystems. *Philosophical Transactions of the Royal Society B: Biological Sciences*, 370(1659), 20130265. http://doi.org/10.1098/rstb.2013.0265.

Persson, L., de Roos, A. M., Claessen, D., et al. (2003). Gigantic cannibals driving a whole-lake trophic cascade. *Proceedings of the National Academy of Sciences of the United States of America*, 100, 4035–4039.

Petchey, O. L., Eklöf, A., Borrvall, C. and Ebenman, B. (2008). Trophically unique species are vulnerable to cascading extinction. *The American Naturalist*, 171(5), 568–579.

Peters, D. P. C., Lugo, A. E., Chapin III, F. S., et al. (2011). Cross-system comparisons elucidate disturbance complexities and generalities. *Ecosphere*, 2, 1–26.

Peters, R. H. (1991). *A Critique for Ecology*. Cambridge University Press, Cambridge.

Peterson, C. H. (1984). Does a rigorous criterion for environmental identity preclude the existence of multiple stable points? *The American Naturalist*, 124, 127–133.

Peterson, G. (2002). Forest dynamics in the Southeastern United States: Managing multiple stable states. In: L. Gunderson and L. Pritchard (eds.), *Resilience and the Behavior of Large-Scale Ecosystems*. Island Press, Washington, DC, pp. 227–246.

Peterson, G., Pope, S., de Leo, G. A., et al. (1997). Ecology, ethics, and advocacy. *Conservation Ecology*, 1(1), 17. www.consecol.org/vol1/iss1/art17/.

Peterson, G., Cumming, G. and Carpenter, S. (2003). Scenario planning: A tool for conservation in an uncertain world. *Conservation Biology*, 17, 358–366.

Peterson, G. D., Harmackova, Z. V., Meacham, M., et al. (2018). Welcoming different perspectives in IPBES: 'Nature's contributions to people' and "Ecosystem services". *Ecology and Society*, 23(1), 39. https://doi.org/10.5751/ES-10134-230139.

Petraitis, P. S. (2013). *Multiple Stable States in Natural Ecosystems*. Oxford University Press, Oxford.

Petraitis, P. S. and Dudgeon, S. R. (2004). Detection of alternative stable states in marine communities. *Journal of Experimental Marine Biology and Ecology*, 300(1), 343–371.

   (2015). Cusps and butterflies: Multiple stable states in marine systems as catastrophes. *Marine and Freshwater Research*, 67(1), 37–46.

Petraitis, P. S. and Hoffman, C. (2010). Multiple stable states and relationship between thresholds in processes and states. *Marine Ecology Progress Series*, 413, 189–200.

Pettitt, P. and Bahn, P. (2015). An alternative chronology for the art of Chauvet cave. *Antiquity*, 89(345), 542–553.

Pfeifer, M., Lefebvre, V., Peres, C. A., et al. (2017). Creation of forest edges has a global impact on forest vertebrates. *Nature*, 551, 187–191.

Phillips, J. D. (2011). Predicting modes of spatial change from state-and-transition models. *Ecological Modelling*, 222, 475–484.

Phillips, N. (2017). Power and inequality in the global political economy. *International Affairs*, 93(2), 429–444.
Phillips, O. L., Aragão, L. E. O. C., Lewis, S. L., et al. (2009). Drought sensitivity of the Amazon rainforest. *Science*, 323(5919), 1344–1347.
Picabia, F. (2007). *I Am a Beautiful Monster. Poetry, Prose and Provocation.* Massachusetts Institute of Technology, Cambridge.
Pickett, S. T. A. and Cadenasso, M. L. (2002). The ecosystem as a multidimensional concept: Meaning, model, and metaphor. *Ecosystems*, 5, 1–10.
Pickett, S. T. A. and White, P. S. (eds.) (1985). *The Ecology of Natural Disturbance and Patch Dynamics*. Academic Press, New York.
Pickett, S. T. A., Parker, V. T. and Fiedler, P. L. (1992). The new paradigm in ecology: Implications for conservation above the species level. In: P. L. Fiedler and S. K. Jain (eds.), *Conservation Biology: The Theory and Practice of Nature Conservation, Preservation and Management*. Chapman & Hall, New York, pp. 65–88.
Pickett, S. T. A., Kolasa, J. and Jones, C. G. (2007). *Ecological Understanding: The Nature of Theory and the Theory of Nature*, 2nd edition. Elsevier, Amsterdam/Boston/Heidelberg/London.
Pickett, S. T. A., Meiners, S. J. and Cadenasso, M. L. (2013). Domain and propositions of succession theory. In: S. M. Scheiner and M. R. Willig (eds.), *The Theory of Ecology*. The University of Chicago Press, Chicago, IL, pp. 185–216.
Pimm, S. L. (1984). The complexity and stability of ecosystems. *Nature*, 307, 321–326.
Pinsky, M. L. and Byler, D. (2015). Fishing, fast growth and climate variability increase the risk of collapse. *Proceedings of the Royal Society B: Biological Sciences*, 282, 20151053.
Pinsky, M. L., Jensen, O. P., Ricard, D. and Palumbi, S. R. (2011). Unexpected patterns of fisheries collapse in the world's oceans. *Proceedings of the National Academy of Sciences of the United States of America*, 108(20), 8317–8322.
Pires, M. M., Galetti, M., Donatti, C. I., et al. (2014). Reconstructing past ecological networks: The reconfiguration of seed-dispersal interactions after megafaunal extinction. *Oecologia*, 175(4), 1247–1256.
Pires, M. M., Koch, P. L., Fariña, R. A., de Aguiar, M. A. M., dos Reis, S. F. and Guimarães, P. R. (2015). Pleistocene megafaunal interaction networks became more vulnerable after human arrival. *Proceedings of the Royal Society B: Biological Sciences*, 282(1814), 20151367.
Pires, M. M., Guimarães, P. R., Galetti, M. and Jordano, P. (2018). Pleistocene megafaunal extinctions and the functional loss of long-distance seed-dispersal services. *Ecography*, 41, 153–163.
Popper, K. R. (1959). *The Logic of Scientific Discovery*. Hutchinson, London.
Possingham, H. P., Andelman, S. J., Burgman, M. A., Medellín, R. A., Master, L. L. and Keith, D. A. (2002). Limits to the use of threatened species lists. *Trends in Ecology & Evolution*, 17(11), 503–507.
Post, D. M., Doyle, M. W., Sabo, J. L. and Finlay, J. C. (2007). The problem of boundaries in defining ecosystems: A potential landmine for uniting geomorphology and ecology. *Geomorphology*, 89(1–2 SPEC. ISS.), 111–126.

Potts, S. G., Biesmeijer, J. C., Kremen, C., Neumann, P., Schweiger, O. and Kunin, W. E. (2010). Global pollinator declines: Trends, impacts and drivers. *Trends in Ecology & Evolution*, 25(6), 345–353.

Poulsen, J. R., Clark, C. J., and Palmer, T. M. (2013). Ecological erosion of an Afrotropical forest and potential consequences for tree recruitment and forest biomass. *Biological Conservation*, 163, 122–130.

Powney, G. D., Carvell, C., Edwards, M., et al. (2019). Widespread losses of pollinating insects in Britain. *Nature Communications*, 10(1). https://doi.org/10.1038/s41467-019-08974-9.

Prach, K. and Pyšek, P. (1999). How do species dominating in succession differ from others? *Journal of Vegetation Science*, 10, 383–392.

Prada, F., Caroselli, E., Mengoli, S., et al. (2017). Ocean warming and acidification synergistically increase coral mortality. *Scientific Reports*, 7, 40842.

Premoli, A. C., Vergara, R. A, Souto, C. P., Lara, A. and Newton, A. C. (2003). Lowland valleys shelter the ancient conifer *Fitzroya cupressoides* in the Central Depression of southern Chile. *Journal of the Royal Society of New Zealand*, 33(3), 623–631.

Price, T. D. and Bar-Yosef, O. (2011). The origins of agriculture: New data, new ideas. *Current Anthropology*, 52(4), S163–S174.

Proctor, J. D. and Larson, B. M. H. (2005). Ecology, complexity and metaphor. *BioScience*, 55, 1065–1068.

Psyphago, Dr. (2013). Scientists conclude: 'No further research is needed'. https://collectivelyunconscious.wordpress.com/2013/01/16/scientists-conclude-no-further-research-is-needed/ (accessed on 6 February 2020).

Pullin, A. S. and Stewart, G. B. (2006). Guidelines for systematic review in conservation and environmental management. *Conservation Biology*, 20, 1647–1656.

Pulsford, S. A., Lindenmayer, D. B. and Driscoll, D. A. (2016). A succession of theories: Purging redundancy from disturbance theory. *Biological Reviews*, 91, 148–167.

Queirós, A. M., Fernandes, J. A., Faulwetter, S., et al. (2015). Scaling up experimental ocean acidification and warming research: From individuals to the ecosystem. *Global Change Biology*, 21, 130–143.

Quince, C., Higgs, P. G. and McKane, A. J. (2005). Deleting species from model food webs. *Oikos*, 110, 283–296.

Rabanus-Wallace, M. T., Wooller, M. J., Zazula, G. D., et al. (2017). Megafaunal isotopes reveal role of increased moisture on rangeland during late Pleistocene extinctions. *Nature Ecology & Evolution*, 1, 125.

Racki, G. (2012). The Alvarez impact theory of mass extinction; limits to its applicability and the 'great expectations syndrome'. *Acta Palaeontologica Polonica*, 57(4), 681–702.

Radford, J. Q., Bennett, A. F. and Cheers, G. J. (2005). Landscape-level thresholds of habitat cover for woodland-dependent birds. *Biological Conservation*, 124, 317–337.

Raffa, K. F., Aukema, B. H., Bentz, B. J., et al. (2008). Cross-scale drivers of natural disturbances prone to anthropogenic amplification: The dynamics of bark beetle eruptions. *BioScience*, 58, 501–517.

Rahel, F. J. and Olden, J. D. (2008). Assessing the effects of climate change on aquatic invasive species. *Conservation Biology*, 22, 521–533.

Ramirez-Llodra, E., Tyler, P. A., Baker, M. C., et al. (2011). Man and the last great wilderness: Human impact on the deep sea. *PLoS One*, 6(8), e22588–e22588.

Rasher, D. B. and Hay, M. E. (2010). Chemically rich seaweeds poison corals when not controlled by herbivores. *Proceedings of the National Academy of Sciences of the United States of America*, 107, 9683–9688.

Rasher, D. B., Hoey, A. S. and Hay, M. E. (2017). Cascading predator effects in a Fijian coral reef ecosystem. *Scientific Reports*, 7(1), 15684.

Ratajczak, Z., Nippert, J. B. and Ocheltree, T. W. (2014). Abrupt transition of mesic grassland to shrubland: Evidence for thresholds, alternative attractors, and regime shifts. *Ecology*, 95(9), 2633–2645.

Ratajczak, Z., Carpenter, S. R., Ives, A. R., et al. (2018). Abrupt change in ecological systems: Inference and diagnosis. *Trends in Ecology & Evolution*, 33 (7), 513–526.

Raudsepp-Hearne, C., Peterson, G. D., Tengö, M., et al. (2010). Untangling the environmentalist's paradox: Why is human well-being increasing as ecosystem services degrade? *BioScience*, 60(8), 576–589.

Raymond, C. M., Singh, G. G., Benessaiah, K., et al. (2013). Ecosystem services and beyond: Using multiple metaphors to understand human–environment relationships. *BioScience*, 63(7), 536–546.

REDD Monitor. (2019). NGOs oppose the oil industry's Natural Climate Solutions and demand that Eni and Shell keep fossil fuels in the ground. https://redd-monitor.org/2019/05/14/ (accessed on 9 January 2020).

Redford, K. H. (1992). The empty forest. *BioScience*, 42, 412–422

Redford, K. H., Amato, G., Baillie, J., et al. (2011). What does it mean to successfully conserve a (vertebrate) species? *BioScience*, 61, 39–48.

Redford, K. H., Hulvey, K. B., Williamson, M. A. and Schwartz, M. W. (2018). Assessment of the conservation measures partnership's effort to improve conservation outcomes through adaptive management. *Conservation Biology*, 32, 926–937.

Reed, J., van Vianen, J., Barlow, J. and Sunderland, T. (2017). Have integrated landscape approaches reconciled societal and environmental issues in the tropics? *Land Use Policy*, 63, 481–492.

Reeves, J. M., Barrows, T. T., Cohen, T. J., et al. (2013). Climate variability over the last 35,000 years recorded in marine and terrestrial archives in the Australian region: An OZ-INTIMATE compilation. *Quaternary Science Review*, 74, 21–34.

Reich, P. B., Tilman, D., Isbell, F., et al. (2012). Impacts of biodiversity loss escalate through time as redundancy fades. *Science*, 336, 589–593.

Resilience Alliance. (2010). *Assessing Resilience in Social-Ecological Systems: Workbook for Practitioners*. Version 2.0. www.resalliance.org/3871.php.

Resilience Alliance and Santa Fe Institute. (2004). Thresholds and alternate states in ecological and social-ecological systems. www.resalliance.org/index.php/data base (accessed on 6 May 2015).

Revenga, C., Brunner, J., Henninger, N., Kassem, K. and Payne, R. (2000). *Pilot Analysis of Global Ecosystems. Freshwater Systems*. World Resources Institute, Washington, DC.

Rey Benayas, J. M., Newton, A. C., Diaz, A. and Bullock, J. M. (2009). Enhancement of biodiversity and ecosystem services by ecological restoration: A meta-analysis. *Science*, 325(5944), 1121–1124.
Richards, M. A., Alvarez, W., Self, S., et al. (2015). Triggering of the largest Deccan eruptions by the Chicxulub impact. *GSA Bulletin*, 127(11–12), 1507–1520.
Richardson, A. J., Bakun, A., Hays, G. C. and Gibbons, M. J. (2009). The jellyfish joyride: Causes, consequences and management responses to a more gelatinous future. *Trends in Ecology & Evolution*, 24(6), 312–322.
Richardson, S. J., Peltzer, D. A., Allen, R. B., McGlone, M. S. and Parfitt, R. L. (2004). Rapid development of phosphorus limitation in temperate rainforest along the Franz Josef soil chronosequence. *Oecologia*, 139, 267–276.
Rickles, D., Hawe, P. and Shiell, A. (2007). A simple guide to chaos and complexity. *Journal of Epidemiology and Community Health*, 61(11), 933–937.
Rigler, F. H. (1982). Recognition of the possible: An advantage of empiricism in ecology. *Canadian Journal of Fisheries and Aquatic Sciences*, 39, 1323–1331.
Ripple, W. J., Rooney, T. P. and Beschta, R. L. (2010). Large predators, deer, and trophic cascades in boreal and temperate ecosystems. In: J. Terborgh and J. A. Estes (eds.), *Trophic Cascades: Predators, Prey, and the Changing Dynamics of Nature*. Island Press, Washington, DC, pp. 141–161.
Ripple, W. J., Estes, J. A., Beschta, R. L., et al. (2014). Status and ecological effects of the world's largest carnivores. *Science*, 343, 1241484.
Ripple, W. J., Estes, J. A., Schmitz, O. J., et al. (2016a). What is a trophic cascade? *Trends in Ecology & Evolution*, 31(11), 842–849.
Ripple, W. J., Abernethy, K., Betts, M. G., et al. (2016b). Bushmeat hunting and extinction risk to the world's mammals. *Royal Society Open Science*, 3(10), 160498.
Ripple, W. J., Wolf, C., Newsome, T. M., et al. (2017). World Scientists' Warning to Humanity: A Second Notice. *BioScience*, 67(12), 1026–1028.
Roberts, R. G., Flannery, T. F., Ayliffe, L. K., et al. (2001). New ages for the last Australian megafauna: Continent-wide extinction about 46,000 years ago. *Science*, 292(5523), 1888–1892.
Robinson, G. S., Burney, L. P. and Burney, D. A. (2005). Landscape paleoecology and megafaunal extinction in southeastern New York state. *Ecological Monographs*, 75, 295–315.
Robinson, R. A. and Sutherland, W. J. (2002). Post-war changes in arable farming and biodiversity in Great Britain. *Journal of Applied Ecology*, 39, 157–176.
Rocha, J. C., Peterson, G., Bodin, Ö. and Levin, S. (2018). Cascading regime shifts within and across scales. *Science*, 362(6421), 1379–1383.
Rockström, J., Steffen, W., Noone, K., et al. (2009a). A safe operating space for humanity. *Nature*, 461, 472–475.
  (2009b). Planetary boundaries: Exploring the safe operating space for humanity. *Ecology and Society*, 14(2), 32. www.ecologyandsociety.org/vol14/iss2/art32/.
Rockström, J., Richardson, K., Steffen, W. and Mace, G. (2018). Planetary boundaries: Separating fact from fiction. A response to Montoya et al. *Trends in Ecology & Evolution*, 33(4), 233–234.

Rodrigues, A. S. L., Pilgrim, J. D., Lamoreux, J. F., Hoffmann, M. and Brooks, T. M. (2006). The value of the IUCN Red List for conservation. *Trends in Ecology & Evolution*, 21, 71–76.

Rodríguez, J. P., Blach, J. K. and Rodríguez-Clark, K. M. (2007). Assessing extinction risk in the absence of species-level data: Quantitative criteria for terrestrial ecosystems. *Biodiversity and Conservation*, 16, 183–209.

Rodríguez, J. P., Rodríguez-Clark, K. M., Baillie, J. E., et al. (2011). Establishing IUCN Red List Criteria for threatened ecosystems. *Conservation Biology*, 25, 21–29.

Rodríguez, J. P., Rodríguez-Clark, K. M., Keith, D. A., et al. (2012). IUCN Red List of Ecosystems. *S.A.P.I.E.N.S.*, 5(2). sapiens.revues.org/1286.

Rodriguez-Cabal, M. A., Barrios-Garcia, M. N., Amico, G. C., Aizen, M. A. and Sanders, N. J. (2013). Node-by-node disassembly of a mutualistic interaction web driven by species introductions. *Proceedings of the National Academy of Sciences of the United States of America*, 110(41), 16503–16507.

Rodriguez Iglesias, R. M. and Kothmann, M. M. (1997). Structure and causes of vegetation change in state and transition model applications. *Journal of Range Management*, 50, 399–408.

Roff, G. and Mumby, P. J. (2012). Global disparity in the resilience of coral reefs. *Trends in Ecology & Evolution*, 27(7), 404–413.

Rogers-Bennett, L. and Catton, C. A. (2019). Marine heat wave and multiple stressors tip bull kelp forest to sea urchin barrens. *Scientific Reports*, 9(1), 15050.

Rolett, B. and Diamond, J. (2004). Environmental predictors of pre-European deforestation on Pacific islands. *Nature*, 431(7007), 443–446.

Romero, G. Q., Gonçalves-Souza, T., Kratina, P., et al. (2018). Global predation pressure redistribution under future climate change. *Nature Climate Change*, 8 (12), 1087–1091.

Roopnarine, P. D. (2006). Extinction cascades and catastrophe in ancient food webs. *Paleobiology*, 32(1), 1–19.

Roopnarine, P. D., Angielczyk, K. D., Wang, S. C. and Hertog, R. (2007). Trophic network models explain instability of Early Triassic terrestrial communities. *Proceedings of the Royal Society B: Biological Sciences*, 274, 2077–2086.

Röpke, C. P., Amadio, S., Zuanon, J., et al. (2017). Simultaneous abrupt shifts in hydrology and fish assemblage structure in a floodplain lake in the central Amazon. *Scientific Reports*, 7, 40170.

Rose, D. C., Sutherland, W. J., Amano, T., et al. (2018). The major barriers to evidence-informed conservation policy and possible solutions. *Conservation Letters*, 11(5), e12564.

Rosenberg, K. V., Dokter, A. M., Blancher, P. J., et al. (2019). Decline of the North American avifauna. *Science*, 366(6461), 120–124.

Ross, L., Arrow, K., Cialdini, R., et al. (2016). The climate change challenge and barriers to the exercise of foresight intelligence. *BioScience*, 66(5), 363–370.

Rowe, J. S. (1997). Defining the ecosystem. *Bulletin of the Ecological Society of America*, 78, 95–97.

Rowe, J. S. and Barnes, B. V. (1994). Geo-ecosystems and bio-ecosystems. *Bulletin of the Ecological Society of America*, 75, 40–41.

Rubenstein, D. R., Rubenstein, D. I., Sherman, P. W. and Gavin, T. A. (2006). Pleistocene park: Does re-wilding North America represent sound conservation in the 21st century? *Biological Conservation*, 132, 232–238.

Rule, S., Brook, B. W., Haberle, S. G., Turney, C. S. M., Kershaw, A. P. and Johnson, C. N. (2012). The aftermath of megafaunal extinction: Ecosystem transformation in Pleistocene Australia. *Science*, 335, 1483–1486.

Rull, V., Cañellas-Boltà, N., Saez, A., et al. (2013). Challenging Easter Island's collapse: The need for interdisciplinary synergies. *Frontiers in Ecology and Evolution*, 1(3), 1–5.

Runyan, C. W., D'Odorico, P. and Lawrence, D. (2012). Physical and biological feedbacks of deforestation. *Reviews of Geophysics*, 50, RG4006.

Ruppert, J. L. W., Travers, M. J., Smith, L. L., Fortin, M.-J. and Meekan, M. G. (2013). Caught in the middle: Combined impacts of shark removal and coral loss on the fish communities of coral Reefs. *PLoS One*, 8(9), e74648. https://doi.org/10.1371/journal.pone.0074648.

Sabin, P. (2013). *The Bet: Paul Ehrlich, Julian Simon and Our Gamble over the Earth's Future*. Yale University Press, New Haven, CT.

Sahasrabudhe, S. and Motter, A. E. (2011). Rescuing ecosystems from extinction cascades through compensatory perturbations. *Nature Communications*, 2, 170.

Salafsky, N., Margoluis, R., Redford, K. H. and Robinson, J. G. (2002). Improving the practice of conservation: A conceptual framework and research agenda for conservation science. *Conservation Biology*, 16, 1469–1479.

Salafsky, N., Salzer, D., Stattersfield, A. J., et al. (2008). A standard lexicon for biodiversity conservation: Unified classifications of threats and actions. *Conservation Biology*, 22, 897–911.

Saltré, F., Rodríguez-Rey, M., Brook, B. W., et al. (2016). Climate change not to blame for late Quaternary megafauna extinctions in Australia. *Nature Communications*, 7, 10511.

Saltré, F., Chadoeuf, J., Peters, K. J., et al. (2019). Climate-human interaction associated with southeast Australian megafauna-extinction patterns. *Nature Communications*, 10, 5311. https://doi.org/10.1038/s41467-019-13277-0.

Salomon, A. K., Gaichas, S. K., Shears, N. T., Smith, J. E., Madin, E. M. P. and Gaines, S. D. (2010). Key features and context-dependence of fishery-induced trophic cascades. *Conservation Biology*, 24(2), 382–394.

Sánchez-Bayo, F. and Wyckhuys, K. A. G. (2019a). Worldwide decline of the entomofauna: A review of its drivers. *Biological Conservation*, 232, 8–27.

  (2019b). Response to 'Global insect decline: Comments on Sánchez-Bayo and Wyckhuys (2019)'. *Biological Conservation*, 233, 334–335.

Sanchirico, J. N., Springborn, M. R., Schwartz, M. W. and Doerr, A. N. (2014). Investment and the policy process in conservation monitoring. *Conservation Biology*, 28, 361–371.

Sandbrook, C., Fisher, J. A., Holmes, G., Luque-Lora, R. and Keane, A. (2019). The global conservation movement is diverse but not divided. *Nature Sustainability*, 2(4), 316–323.

Sand-Jensen, K. (2007). How to write consistently boring scientific literature. *Oikos*, 116, 723–727.

Sandom, C. J., Ejrnaes, R., Hansen, M. D. D. and Svenning, J.-C. (2014a). High herbivore density associated with vegetation diversity in interglacial ecosystems. *Proceedings of the National Academy of Sciences of the United States of America*, 111, 4162–4167.

Sandom, C., Faurby, S., Sandel, B. and Svenning, J.-C. (2014b). Global late Quaternary megafauna extinctions linked to humans, not climate change. *Proceedings of the Royal Society B: Biological Sciences*, 281, 20133254.

Säterberg, T., Sellman, S. and Ebenman, B. (2013). High frequency of functional extinctions in ecological networks. *Nature*, 499, 468–470.

Sato, C. F. and Lindenmayer, D. B. (2017). Meeting the global ecosystem collapse challenge. *Conservation Letters*, 11(1), e12348. https://doi.org/10.1111/conl.12348.

Sayer, J. A., Margules, C., Boedhihartono, A. K., et al. (2016). Measuring the effectiveness of landscape approaches to conservation and development. *Sustainability Science*, 12(3), 465–476.

Scharfenberger, U., Mahdy, A. and Adrian, R. (2013). Threshold-driven shifts in two copepod species: Testing ecological theory with observational data. *Limnology and Oceanography*, 58(2), 741–752.

Scheffer, M. (2009). *Critical Transitions in Nature and Society*. Princeton University Press, Princeton, NJ.

Scheffer, M. and Carpenter, S. R. (2003). Catastrophic regime shifts in ecosystems: Linking theory to observation. *Trends in Ecology & Evolution*, 18(12), 648–656.

Scheffer, M., Hosper, S. H., Meijer, M.-L., Moss, B. and Jeppesen, E. (1993). Alternative equilibria in shallow lakes. *Trends in Ecology & Evolution*, 8, 275–279.

Scheffer, M., Carpenter, S., Foley, J. A., Folke, C. and Walker, B. (2001). Catastrophic shifts in ecosystems. *Nature*, 413, 591–596.

Scheffer, M., Carpenter, S. R. and de Young, B. (2005). Cascading effects of overfishing marine systems. *Trends in Ecology & Evolution*, 20(11), 579–581.

Scheffer, M., Bascompte, J., Brock, W. A., et al. (2009). Early-warning signals for critical transitions. *Nature*, 461, 53–59.

Scheffer, M., Carpenter, S. R., Lenton, T. M., et al. (2012). Anticipating critical transitions. *Science*, 338(6105), 344–348.

Scheffer, M., Carpenter, S. R., Dakos, V. and van Nes, E. H. (2015). Generic indicators of ecological resilience: Inferring the chance of a critical transition. *Annual Review of Ecology, Evolution, and Systematics*, 46, 145–167.

Scheffers, B. R., de Meester, L., Bridge, T. C. L., et al. (2016). The broad footprint of climate change from genes to biomes to people. *Science*, 354(6313), aaf7671.

Scheiner, S. M. (2013). The ecological literature, an idea-free distribution. *Ecology Letters*, 16, 1421–1423.

Scheiner, S. M. and Willig, M. R. (2011). A general theory of ecology. In: S. M. Scheiner and M. R. Willig (eds.), *The Theory of Ecology*. The University of Chicago Press, Chicago, IL, pp. 3–18.

Schmidt, N. M., Reneerkens, J., Christensen, J. H., Olesen, M. and Roslin, T. (2019). An ecosystem-wide reproductive failure with more snow in the Arctic. *PLoS Biology*, 17(10), e3000392. https://doi.org/10.1371/journal.pbio.3000392.

References · 457

Schmitz, O. J., Hawlena, D. and Trussell, G. C. (2010). Predator control of ecosystem nutrient dynamics. *Ecology Letters*, 13, 1199–1209.

Schloss, C. A., Nuñez, T. A. and Lawler, J. J. (2012). Dispersal will limit ability of mammals to track climate change in the Western Hemisphere. *Proceedings of the National Academy of Sciences of the United States of America*, 109, 8606–8611.

Schoene, B., Eddy, M. P., Samperton, K. M., et al. (2019). U-Pb constraints on pulsed eruption of the Deccan Traps across the end-Cretaceous mass extinction. *Science*, 363(6429), 862–866.

Schröder, A. (2009). Inference about complex ecosystem dynamics in ecological research and restoration practice. In: K. N. Suding and R. J. Hobbs (eds.), *New Models for Ecosystem Dynamics and Restoration*. Society for Ecological Restoration International. Island Press, Washington, DC, pp. 50–62.

Schröder, A., Persson, L. and de Roos, A. M. (2005). Direct experimental evidence for alternative stable states: A review. *Oikos*, 110, 3–19.

(2012). Complex shifts between food web states in response to whole-ecosystem manipulations. *Oikos*, 121, 417–427.

Schröter, M., van der Zanden, E. H., van Oudenhoven, A. P., et al. (2014). Ecosystem services as a contested concept: A synthesis of critique and counter-arguments. *Conservation Letters*, 7, 514–523.

Schubert, J. K. and Bottjer, D. J. (1992). Early Triassic stromatolites as post-mass extinction disaster forms. *Geology*, 20, 883–886.

Schulte, P., Alegret, L, Arenillas, I., et al. (2010). The Chicxulub asteroid impact and mass extinction at the Cretaceous-Paleogene boundary. *Science*, 327(5970), 1214–1218.

Schwartz, M. W., Deiner, K., Forrester, T., et al. (2012). Perspectives on the Open Standards for the practice of conservation. *Biological Conservation*, 155, 169–177.

Schwartz, M. W., Cook, C. N., Pressey, R. L., et al. (2017). Decision support frameworks and tools for conservation. *Conservation Letters*, 11(2), e12385. https://doi.org/10.1111/conl.12385.

Secretariat of the Convention on Biological Diversity. (2010). Global Biodiversity Outlook 3. Secretariat of the Convention on Biological Diversity, Montréal.

(2014). Global Biodiversity Outlook 4. Secretariat of the Convention on Biological Diversity, Montréal.

Seekell, D. A., Carpenter, S. R., Cline, T. J. and Pace, M. L. (2012). Conditional heteroskedasticity forecasts regime shift in a whole-ecosystem experiment. *Ecosystems*, 15, 741–747.

Seekell, D. A., Cline, T. J., Carpenter, S. R. and Pace, M. L. (2013). Evidence of alternate attractors from a whole-ecosystem regime shift experiment. *Theoretical Ecology*, 6, 385–394.

Seersholm, F. V, Cole, T. L., Grealy, A., et al. (2018). Subsistence practices, past biodiversity, and anthropogenic impacts revealed by New Zealand-wide ancient DNA survey. *Proceedings of the National Academy of Sciences of the United States of America*, 115(30), 7771–7776.

Seidl, R., Thom, D., Kautz, M., et al. (2017). Forest disturbances under climate change. *Nature Climate Change*, 7, 395.

Sellman, S., Säterberg, T. and Ebenman, B. (2016). Pattern of function extinctions in ecological networks with a variety of interaction types. *Theoretical Ecology*, 9, 83–94.

Sephton, M. A., Visscher, H., Looy, C. V, Verchovsky, A. B. and Watson, J. S. (2009). Chemical constitution of a Permian-Triassic disaster species. *Geology*, 37(10), 875–878.

Settele, J., Scholes, R., Betts, R., et al. (2014). Terrestrial and water systems. In C. B. Field, V. R. Barros, D. J. Dokken, et al. (eds.), *Climate Change 2014: Impacts, Adaptation, Vulnerability. Part A: Global and Sectoral Aspects. Contribution of Working Group II to the Fifth Assessment Report of the IPCC*. Cambridge University Press, Cambridge, pp. 271–359.

Sewell, S. L. (2014). The spatial diffusion of beer from its Sumerian origins to today. In: M. Patterson and N. Hoalst-Pullen (eds.), *The Geography of Beer*. Springer Science and Business Media, Dordrecht, pp. 23–29.

Shackelford, N., Starzomski, B. M., Banning, N. C., et al. (2017). Isolation predicts compositional change after discrete disturbances in a global meta-study. *Ecography*, 40(11), 1256–1266.

Shanahan, T. M., McKay, N. P., Hughen, K. A., et al. (2015). The time-transgressive termination of the African Humid Period. *Nature Geoscience*, 8, 140–144.

Sheldon, K. S. (2019). Climate change in the tropics: Ecological and evolutionary responses at low latitudes. *Annual Review of Ecology, Evolution, and Systematics*, 50(1). https://doi.org/10.1146/annurev-ecolsys-110218-025005.

Shortall, C. R., Moore, A., Smith, E., Hall, M. J., Woiwod, I. P., and Harrington, R. (2009). Long-term changes in the abundance of flying insects. *Insect Conservation and Diversity*, 2, 251–260.

Shurin, J. B., Borer, E. T., Seabloom, E. W., et al. (2002). A cross-ecosystem comparison of the strength of trophic cascades. *Ecology Letters*, 5, 785–791.

Shvidenko, A., Barber, C. V., Persson, R., et al. (2005). Forest and woodland systems. In: R. Hassan, R. Scholes and N. Ash (eds.), *Ecosystems and Human Well-being: Current State and Trends*. Millennium Ecosystem Assessment, vol. 1. Island Press, Washington, DC, pp. 585–621.

Silla, F., Fraver, S., Lara, A., Allnutt, T. R. and Newton, A. C. (2002). Regeneration and stand dynamics of *Fitzroya cupressoides* (Cupressaceae) in the Central Depression of Chile. *Forest Ecology and Management*, 165, 213–224.

Silvério, D. V., Brando, P. M., Balch, J. K., et al. (2013). Testing the Amazon savannization hypothesis: Fire effects on invasion of a neotropical forest by native cerrado and exotic pasture grasses. *Philosophical Transactions of the Royal Society B: Biological Sciences*, 368, 20120427.

Simberloff, D. (2006). Invasional meltdown 6 years later: Important phenomenon, unfortunate metaphor, or both? *Ecology Letters*, 9, 912–919.

Simberloff, D. and Von Holle, B. (1999). Positive interactions of nonindigenous species: Invasional meltdown? *Biological Invasions*, 1, 21–32.

Simenstad, C., Reed, D. and Ford, M. (2006). When is restoration not?: Incorporating landscape-scale processes to restore self-sustaining ecosystems in coastal wetland restoration. *Ecological Engineering*, 26(1), 27–39.

Simon, J. L. (1982). Paul Ehrlich saying it is so doesn't make it so. *Social Science Quarterly*, 63(2), 381–385.

Skeffington, R. A. (1999). The use of critical loads in environmental policy making: A critical appraisal. *Environmental Policy Analysis*, 33, 245–252.

Smith, A. B. (2001). Large-scale heterogeneity of the fossil record: Implications for Phanerozoic biodiversity studies. *Philosophical Transactions of the Royal Society of London Series B: Biological Sciences*, 356(1407), 351–367.

Smith, F. A., Hammond, J. I., Balk, M. A., et al. (2016a). Exploring the influence of ancient and historic megaherbivore extirpations on the global methane budget. *Proceedings of the National Academy of Sciences of the United States of America*, 113, 874–879.

Smith, F. A., Tomé, C. P., Elliott Smith, E. A., et al. (2016b). Unraveling the consequences of the terminal Pleistocene megafauna extinction on mammal community assembly. *Ecography*, 39, 223–239.

Smith, J. E., Hunter, C. L. and Smith, C. M. (2010). The effects of top-down versus bottom-up control on benthic coral reef community structure. *Oecologia*, 163, 497–507.

Smith, M. D., Knapp, A. K. and Collins, S. L. (2009). A framework for assessing ecosystem dynamics in response to chronic resource alterations induced by global change. *Ecology*, 90, 3279–3289.

Smith, P., Albanito, F., Bell, M., et al. (2012). Systems approaches in global change and biogeochemistry research. *Philosophical Transactions of the Royal Society B: Biological Sciences*, 367(1586), 311 LP-321.

Snyder-Beattie, A. E., Ord, T. and Bonsall, M. B. (2019). An upper bound for the background rate of human extinction. *Scientific Reports*, 9(1), 11054.

Sober, E. (2015). Is the scientific method a myth? Perspectives from the history and philosophy of science. *MÈTODE Science Studies Journal*, 5, 195–199.

Society for Ecological Restoration International Science & Policy Working Group. (2004). *The SER International Primer on Ecological Restoration*. Society for Ecological Restoration, Tucson, AZ.

Soja, A. J., Tchebakova, N. M., French, N. H. F., et al. (2006). Climate-induced boreal forest change: Predictions versus current observations. *Global and Planetary Change*, 56, 274–296.

Solé, R. V. and Montoya, J. M. (2001). Complexity and fragility in ecological networks. *Proceedings of the Royal Society B: Biological Sciences*, 268, 2039–2045.

Solé, R. V., Montoya, J. M. and Erwin, D. H. (2002). Recovery after mass extinction: Evolutionary assembly in large-scale biosphere dynamics. *Philosophical Transactions of the Royal Society of London Series B: Biological Sciences*, 357 (1421), 697–707.

Soulé, M. (2013). The 'New Conservation'. *Conservation Biology*, 27(5), 895–897.

Soulé, M. and Lease, G. (1995). *Reinventing Nature?* Island Press, Washington, DC.

Sousa, W. P. (1984). The role of disturbance in natural communities. *Annual Review of Ecology and Systematics*, 15, 353–391.

Sousa, W. P. and Connell, J. H. (1985). Further comments on the evidence for multiple stable points in natural communities. *The American Naturalist*, 125(4), 612–615.

Souter, D. W. and Lindén, O. (2000). The health and future of coral reef systems. *Ocean and Coastal Management*, 43(8), 657–688.

Spake, R., Ezard, T., Martin, P., Newton, A. C. and Doncaster, C. P. (2015). A meta-analysis of functional group responses to forest recovery outside of the tropics. *Conservation Biology*, 29(6), 1695–1703.

References

Spears, B. M., Futter, M. N., Jeppesen, E., et al. (2017). Ecological resilience in lakes and the conjunction fallacy. *Nature Ecology & Evolution*, 1(11), 1616–1624.

Spiesman, B. J. and Inouye, B. D. (2013). Habitat loss alters the architecture of plant–pollinator interaction networks. *Ecology*, 94(12), 2688–2696.

Spracklen, D. V., Baker, J. C. A., Garcia-Carreras, L. and Marsham, J. H. (2018). The effects of tropical vegetation on rainfall. *Annual Review of Environment and Resources*, 43, 193–218.

Springer, A. M., van Vliet, G. B., Bool, N., et al. (2018). Transhemispheric ecosystem disservices of pink salmon in a Pacific Ocean macrosystem. *Proceedings of the National Academy of Sciences of the United States of America*, 115, E5038–E5045.

Staal, A., van Nes, E. H., Hantson, S., et al. (2018). Resilience of tropical tree cover: The roles of climate, fire, and herbivory. *Global Change Biology*, 24(11), 5096–5109. https://doi.org/10.1111/gcb.14408.

Standage, T. (2005). *A History of the World in 6 Glasses*. Walker and Company, New York

Standish, R. J., Hobbs, R. J., Mayfield, M. M., et al. (2014). Resilience in ecology: Abstraction, distraction, or where the action is? *Biological Conservation*, 177, 43–51.

Stanley, E. H., Powers, S. M. and Lottig, N. R. (2010). The evolving legacy of disturbance in stream ecology: Concepts, contributions and coming challenges. *Journal of the North American Benthological Society*, 29(1), 67–83.

Staver, A. C. and Bond, W. J. (2014). Is there a "browse trap"? Dynamics of herbivore impacts on trees and grasses in an African savanna. *Journal of Ecology*, 102(3), 595–602.

Staver, A. C. and Hansen, M. C. (2015). Analysis of stable states in global savannas: Is the CART pulling the horse? – a comment. *Global Ecology and Biogeography*, 24, 985–987.

Staver, A. C., Archibald, S. and Levin, S. A. (2011). The global extent and determinants of savanna and forest as alternative biome states. *Science*, 334(6053), 230–232.

Steadman, D. W. (1995). Prehistoric extinctions of Pacific Island birds: Biodiversity meets zooarchaeology. *Science*, 267, 1123–1130.

Steffen, W., Richardson, K., Rockström, J., et al. (2015). Planetary boundaries: Guiding human development on a changing planet. *Science*, 347(6223), 736.

Steffen, W., Rockström, J., Richardson, K., et al. (2018). Trajectories of the Earth System in the Anthropocene. *Proceedings of the National Academy of Sciences of the United States of America*, 115(33), 8252–8259.

Stephens, S. S. and Wagner, M. R. (2007). Forest plantations and biodiversity: A fresh perspective. *Journal of Forestry*, 105(6), 307–313.

Stevenson, P. R. and Guzman-Caro, D. C. (2010). Nutrient transport within and between habitats through seed dispersal processes by woolly monkeys in northwestern Amazonia. *American Journal of Primatology*, 72, 992–1003.

Stigall, A. L. (2012). Speciation collapse and invasive species dynamics during the Late Devonian "Mass Extinction". *GSA Today*, 22(1), 4–9.

Stoeckl, N., Hicks, C. C., Mills, M., et al. (2011). The economic value of ecosystem services in the Great Barrier Reef: Our state of knowledge. *Annals of the New York Academy of Sciences*, 1219(1), 113–133.

Stokstad, E. (2009). Detente in the fisheries war. *Science*, 324, 170–171.
Strayer, D. L. (2010). Alien species in fresh waters: Ecological effects, interactions with other stressors, and prospects for the future. *Freshwater Biology*, 55, 152–174.
Strona, G. and Lafferty, K. D. (2016). Environmental change makes robust ecological networks fragile. *Nature Communications*, 7, 12462.
Stuart, A. J. and Lister, A. M. (2012). Extinction chronology of the woolly rhinoceros *Coelodonta antiquitatis* in the context of late Quaternary megafaunal extinctions in northern Eurasia. *Quaternary Science Reviews*, 51, 1–17.
Stuart-Smith, R. D., Brown, C. J., Ceccarelli, D. M. and Edgar, G. J. (2018). Ecosystem restructuring along the Great Barrier Reef following mass coral bleaching. *Nature*, 560(7716), 92–96.
Suding, K. N. (2011). Toward an era of restoration in ecology: Successes, failures, and opportunities ahead. *Annual Review of Ecology, Evolution, and Systematics*, 42, 465–487.
Suding, K. and Gross, K. (2006). The dynamic nature of ecological systems: Multiple states and restoration trajectories. In: D. Falk, M. Palmer and J. Zedler (eds.), *Foundations of Restoration Ecology*. Island Press, Washington, DC, pp. 190–209.
Suding, K. N. and Hobbs, R. J. (2009). Models of ecosystem dynamics as frameworks for restoration ecology. In: K. N. Suding and R. J. Hobbs (eds.), *New Models for Ecosystem Dynamics and Restoration*. Society for Ecological Restoration International. Island Press, Washington, DC, pp. 3–21.
Suding, K. N., Gross, K. L. and Houseman, G. R. (2004). Alternative states and positive feedbacks in restoration ecology. *Trends in Ecology & Evolution*, 19(1), 46–53.
Suding, K., Higgs, E., Palmer, M., *et al.* (2015). Committing to ecological restoration. *Science*, 348, 638–640.
Suich, H., Howe, C. and Mace, G. (2015). Ecosystem services and poverty alleviation: A review of the empirical links. *Ecosystem Services*, 12, 137–147.
Sutherland, J. P. (1990). Perturbations, resistance, and alternative views of the existence of multiple stable points in nature. *The American Naturalist*, 136, 270–275.
Sutherland, W. J. (2000). *The Conservation Handbook. Research, Management and Policy*. Blackwell Science, Oxford.
Sutherland, W. J., Pullin, A. S., Dolman, P. M. and Knight, T. M. (2004). The need for evidence-based conservation. *Trends in Ecology & Evolution*, 19, 305–308.
Sutherland, W. J., Freckleton, R. P., Godfray, H. C. J., *et al.* (2013). Identification of 100 fundamental ecological questions. *Journal of Ecology*, 101, 58–67.
Suweis, S., Simini, F., Banavar, J. R. and Maritan, A. (2013). Emergence of structural and dynamical properties of ecological mutualistic networks. *Nature*, 500, 449–452.
Swann, A. L. S., Fung, I. Y., Liu, Y. and Chiang, J. C. H. (2014). Remote vegetation feedbacks and the mid-Holocene Green Sahara. *Journal of Climate*, 27(13), 4857–4870.
Swift, T. L. and Hannon, S. J. (2010). Critical thresholds associated with habitat loss: A review of the concepts, evidence, and applications. *Biological Reviews*, 85, 35–53.

Szirmai, A. and Verspagen, B. (2015). Manufacturing and economic growth in developing countries, 1950–2005. *Structural Change and Economic Dynamics*, 34, 46–59.

Szmant, A. M. (2001). Why are coral reefs world-wide becoming overgrown by algae? Algae, algae everywhere, and nowhere a bite to eat! *Coral Reefs*, 19, 299–302.

Tainter, J. A. (2004). Plotting the downfall of society. *Nature*, 427(6974), 488–489.

Tallis, H. and Lubchenko, J.; 238 cosignatories. (2014). A call for inclusive conservation. *Nature*, 515, 27–28.

Tang, S., Pawar, S. and Allesina, S. (2014). Correlation between interaction strengths drives stability in large ecological networks. *Ecology Letters*, 17, 1094–1100.

Tanner, T., Lewis, D., Wrathall, D., *et al.* (2015). Livelihood resilience in the face of climate change. *Nature Climate Change*, 5, 23–26.

Tansley, A. G. (1935). The use and abuse of vegetational concepts and terms. *Ecology*, 16, 284–307.

Taubert, F., Fischer, R., Groeneveld, J., *et al.* (2018). Global patterns of tropical forest fragmentation. *Nature*, 554(7693), 519–522.

Taylor, J., Vanni, M. J. and Flecker, A. S. (2015). Top-down and bottom-up interactions in freshwater ecosystems: Emerging complexities. In: T. Hanley and K. La Pierre (eds.), *Trophic Ecology: Bottom-Up and Top-Down Interactions across Aquatic and Terrestrial Systems*. Ecological Reviews. Cambridge: Cambridge University Press, pp. 55–85.

Taylor, P. (2009). Re-wilding the grazers: Obstacles to the 'wild' in wildlife management. *British Wildlife*, June 2009, 50–55.

Temperton, V. M., Hobbs, R. J., Nuttle, T. and Halle, S. (eds.). (2004). *Assembly Rules and Restoration Ecology. Bridging the Gap between Theory and Practice*. Society for Ecological Restoration International. Island Press, Washington, DC, and London.

Tennant, J. P., Mannion, P. D., Upchurch, P., Sutton, M. D. and Price, G. D. (2017). Biotic and environmental dynamics through the Late Jurassic–Early Cretaceous transition: Evidence for protracted faunal and ecological turnover. *Biological Reviews*, 92, 776–814.

Tepley, A. J., Veblen, T. T., Perry, G. L. W., *et al.* (2016). Positive feedbacks to fire-driven deforestation following human colonization of the South Island of New Zealand. *Ecosystems*, 19(8), 1325–1344.

Terborgh, J. W. (2015). Toward a trophic theory of species diversity. *Proceedings of the National Academy of Sciences of the United States of America*, 112(37), 11415–11422.

Terborgh, J. and Estes, J. A. (eds.). (2010). *Trophic Cascades: Predators, Prey, and the Changing Dynamics of Nature*. Island Press, Washington, DC.

Terborgh, J., Lopez, L., Nuñez, P., *et al.* (2001). Ecological meltdown in predator-free forest fragments. *Science*, 294(5548), 1923–1926.

Thébault, E. and Fontaine, C. (2010). Stability of ecological communities and the architecture of mutualistic and trophic networks. *Science*, 329, 853–856.

Thébault, E., Huber, V. and Loreau, M. (2007). Cascading extinctions and ecosystem functioning: Contrasting effects of diversity depending on food web structure. *Oikos*, 116, 163–173.

Thomas, C. D., Moller, H., Plunkett, G. M. and Harris, R. J. (1990). The prevalence of introduced *Vespula vulgaris* wasps in a New Zealand beech forest community. *New Zealand Journal of Ecology*, 13(1), 63–72.

Thomas, C. D., Jones, T. H. and Hartley, S. E. (2019). 'Insectageddon': A call for more robust data and rigorous analyses. *Global Change Biology*, 25(6), 1891–1892. https://doi.org/10.1111/gcb.14608.

Thompson, S. L., Govindasamy, B., Mirin, A., et al. (2004). Quantifying the effects of CO_2-fertilized vegetation on future global climate and carbon dynamics. *Geophysical Research Letters*, 31, L23211.

Thunberg, G. (2019). *No One Is Too Small to Make a Difference*. Penguin Books, London.

Tickler, D., Meeuwig, J. J., Palomares, M.-L., Pauly, D. and Zeller, D. (2018). Far from home: Distance patterns of global fishing fleets. *Science Advances*, 4, r3279.

Tilman, D., Reich, P. B. and Knops, J. M. H. (2006). Biodiversity and ecosystem stability in a decade-long grassland experiment. *Nature*, 441(7093), 629–632.

Tilman, D., Isbell, F. and Cowles, J. M. (2014). Biodiversity and ecosystem functioning. *Annual Review of Ecology, Evolution, and Systematics*, 45(1), 471–493.

Tilman, D., Clark, M., Williams, D. R., Kimmel, K., Polasky, S. and Packer, C. (2017). Future threats to biodiversity and pathways to their prevention. *Nature*, 546(7656), 73–81.

Touboul, J. D., Staver, A. C. and Levin, S. A. (2018). On the complex dynamics of savanna landscapes. *Proceedings of the National Academy of Sciences of the United States of America*, 115(7), E1336–E1345.

Trueman, C. N., Field, J. H, Dortch, J., Charles, B. and Wroe, S. (2005). Prolonged co-existence of humansand megafauna in Pleistocene Australia. *Proceedings of the National Academy of Sciences of the United States of America*, 182, 8381–8385.

Turchin, P. (2010). Political instability may be a contributor in the coming decade. *Nature*, 463(7281), 608.

Turner, I. M., Tan, H. T. W., Wee, Y. C., et al. (1994). A study of plant species extinction in Singapore: Lessons for the conservation of tropical biodiversity. *Conservation Biology*, 8, 705–712.

Turvey, S. T. and Risley, C. L. (2006). Modelling the extinction of Steller's sea cow. *Biology Letters*, 2, 94–97.

Twitchett, R. J. (2007). The Lilliput effect in the aftermath of the end-Permian extinction event. *Palaeogeography, Palaeoclimatology, Palaeoecology*, 252, 132–144.

Twitchett, R. J., Looy, C. V, Morante, R., Visscher, H. and Wignall, P. B. (2001). Rapid and synchronous collapse of marine and terrestrial ecosystems during the end-Permian biotic crisis. *Geology*, 29(4), 351–354.

Twitchett, R. J., Krystyn, L., Baud, A., Wheeley, J. R. and Richoz, S. (2004). Rapid marine recovery after the end-Permian mass-extinction event in the absence of marine anoxia. *Geology*, 32, 805–808.

Tylianakis, J. M., Didham, R. K., Bascompte, J. and Wardle, D. A. (2008). Global change and species interactions in terrestrial ecosystems. *Ecology Letters*, 11(12), 1351–1363.

Tyukavina, A., Hansen, M. C., Potapov, P. V., Krylov, A. M. and Goetz, S. J. (2015). Pan-tropical hinterland forests: Mapping minimally disturbed forests. *Global Ecology and Biogeography*, 25(2), 151–163.

References

UNEP. (2012). *GEO 5 Global Environment Outlook.* United Nations Environment Programme, Nairobi, Kenya.

Urban, M. C. (2019). Projecting biological impacts from climate change like a climate scientist. *WIREs Climate Change,* 10(4), e585. https://doi.org/10.1002/wcc.585.

Vajda, V., Raine, J. I. and Hollis, C. J. (2001). Indication of global deforestation at the Cretaceous-Tertiary boundary by New Zealand fern spike. *Science,* 294, 1700–1702.

Valiente-Banuet, A. and Verdú, M. (2013). Human impacts on multiple ecological networks act synergistically to drive ecosystem collapse. *Frontiers in Ecology and the Environment,* 11, 408–413.

Valiente-Banuet, A., Aizen, M. A., Alcántara, J. M., et al. (2015). Beyond species loss: The extinction of ecological interactions in a changing world. *Functional Ecology,* 29, 299–307.

Valladares, G., Cagnolo, L. and Salvo, A. (2012). Forest fragmentation leads to food web contraction. *Oikos,* 121(2), 299–305.

Vale, R. D. (2013). The value of asking questions. *Molecular Biology of the Cell,* 24(6), 680–682.

Valone, T. J. and Barber, N. A. (2008). An empirical evaluation of the insurance hypothesis in diversity–stability models. *Ecology,* 89(2), 522–531.

van Altena, C., Hemerik, L. and de Ruiter, P. C. (2016). Food web stability and weighted connectance: The complexity-stability debate revisited. *Theoretical Ecology,* 9(1), 49–58.

van Auken, O. W. (2000). Shrub invasions of North American semiarid grasslands. *Annual Review of Ecology and Systematics,* 31, 197–215.

van de Leemput, I. A., Hughes, T. P., van Nes, E. H. and Scheffer, M. (2016). Multiple feedbacks and the prevalence of alternate stable states. *Coral Reefs* 35, 857–865.

van der Ent, R. J., Savenije, H. H. G., Schaefli, B. and Steele-Dunne, S. C. (2010). Origin and fate of atmospheric moisture over continents. *Water Resources Research,* 46(9), W09525.

van der Heide, T., van Nes, E. H., Geerling, G. W., Smolders, A. J. P., Bouma, T. J. and van Katwijk, M. M. (2007). Positive feedbacks in seagrass ecosystems: Implications for success in conservation and restoration. *Ecosystems,* 10, 1311–1322.

van der Kaars, S., Miller, G. H., Turney, C. S. M., et al. (2017). Humans rather than climate the primary cause of Pleistocene megafaunal extinction in Australia. *Nature Communications,* 8, 14142.

Vandermeer, J. (2011). The inevitability of surprise in agroecosystems. *Ecological Complexity,* 8(4), 377–382.

Vandermeer, J., de la Cerda, I. G., Perfecto, I., Boucher, D., Ruiz, J. and Kaufmann, A. (2004). Multiple basins of attraction in a tropical forest: Evidence for nonequilibrium community structure. *Ecology,* 85, 575–579.

van der Putten, W. H., Bardgett, R. D., Bever, J. D., et al. (2013). Plant–soil feedbacks: The past, the present and future challenges. *Journal of Ecology,* 101, 265–276.

van de Schootbrugge, B., Payne, J. L., Tomasovych, A., et al. (2008). Carbon cycle perturbation and stabilization in the wake of the Triassic-Jurassic boundary mass-extinction event. *Geochemistry, Geophysics, Geosystems*, 9(4). https://doi.org/10.1029/2007GC001914.

van Dover, C. L., Aronson, J., Pendleton, L., et al. (2014). Ecological restoration in the deep sea: Desiderata. *Marine Policy*, 44, 98–106.

van Gerven, L. P. A., Kuiper, J. J., Janse, J. H., et al. (2016). How regime shifts in connected aquatic ecosystems are affected by the typical downstream increase of water flow. *Ecosystems*, 20, 733–744.

van Mantgem, P. J., Stephenson, N. L., Byrne, J. C., et al. (2009). Widespread increase of tree mortality rates in the western United States. *Science*, 323, 521–524.

van Nes, E. H., Hirota, M., Holmgren, M. and Scheffer, M. (2014). Tipping points in tropical tree cover: Linking theory to data. *Global Change Biology*, 20(3), 1016–1021.

van Nes, E. H., Arani, B. M., Staal, A., et al. (2016). What do you mean, "tipping point"?. *Trends in Ecology & Evolution*, 31(12), 902–904.

van Nes, E. H., Staal, A., Hantson, S., et al. (2018). Fire forbids fifty-fifty forest. *PLoS One*, 18, e0191027.

van Strien, A. J., van Swaay, C. A. M., van Strien-van Liempt, W. T. F. H., Poot, M. J. M. and WallisDeVries, M. F. (2019). Over a century of data reveal more than 80% decline in butterflies in the Netherlands. *Biological Conservation*, 234, 116–122.

van Valkenburgh, B., Hayward, M. W., Ripple, W. J., Meloro, C. and Roth, V. L. (2015). The impact of large terrestrial carnivores on Pleistocene ecosystems. *Proceedings of the National Academy of Sciences of the United States of America*, 113(4), 862–867.

Vanni, M. J. (2002). Nutrient cycling by animals in freshwater ecosystems. *Annual Review of Ecology and Systematics*, 33, 341–370.

Vanwalleghem, T., Gómez, J. A., Infante Amate, J., et al. (2017). Impact of historical land use and soil management change on soil erosion and agricultural sustainability during the Anthropocene. *Anthropocene*, 17, 13–29.

Veblen, T. T. (2007). Temperate forests of the Southern Andean Region. In: T. T. Veblen, K. R. Young and A. R. Orme (eds.), *The Physical Geography of South America*. Oxford University Press, Oxford, pp. 217–231.

Veenendaal, E. M., Torello-Raventos, M., Miranda, H. S., et al. (2018). On the relationship between fire regime and vegetation structure in the tropics. *New Phytologist*, 218(1), 153–166.

Veldman, J. W. (2016). Clarifying the confusion: Old-growth savannahs and tropical ecosystem degradation. *Philosophical Transactions of the Royal Society B: Biological Sciences*, 371, 20150306.

Veldman, J. W., Buisson, E., Durigan, G., et al. (2015). Toward an old-growth concept for grasslands, savannas, and woodlands. *Frontiers in Ecology and the Environment*, 13(3), 154–162.

Veldman, J. W., Aleman, J. C., Alvarado, S. T., et al. (2019). Comment on 'The global tree restoration potential'. *Science*, 366(6463), eaay7976. https://doi.org/10.1126/science.aay7976.

Vellend, M., Baeten, L., Myers-Smith, I. H., et al. (2013). Global meta-analysis reveals no net change in local-scale plant biodiversity over time. *Proceedings of the National Academy of Sciences of the United States of America*, 110(48), 19456–19459.

Vera, F. W. M. (2000). *Grazing Ecology and Forest History*. CABI Publishing, Wallingford.

(2009). Large-scale nature development – The Oostvaardersplassen. *British Wildlife*, June 2009, 28–36.

Verdonschot, P. F. M., Spears, B. M., Feld, C. K., et al. (2013). A comparative review of recovery processes in rivers, lakes, estuarine and coastal waters. *Hydrobiologia*, 704(1), 453–474.

Verdy, A and Amarasekare, P. (2010). Alternative stable states in communities with intraguild predation. *Journal of Theoretical Biology*, 262(1), 116–128.

Veríssimo, D., MacMillan, D. C., Smith, R. J., Crees, J. and Davies, Z. G. (2014). Has climate change taken prominence over biodiversity conservation? *BioScience*, 64(7), 625–629.

Vermeij, G. J. (2004). Ecological avalanches and the two kinds of extinction. *Evolutionary Ecology Research*, 6, 315–337.

Veraverbeke, S., Rogers, B. M., Goulden, M. L., et al. (2017). Lightning as a major driver of recent large fire years in North American boreal forests. *Nature Climate Change*, 7(7), 529–534.

Vetter, D., Rúcker, G. and Storch, I. (2013). Meta-analysis: A need for well-defined usage in ecology and conservation biology. *Ecosphere*, 4(6), 74.

Vieira, M. C. and Almeida-Neto, M. (2014). A simple stochastic model for complex coextinctions in mutualistic networks: Robustness decreases with connectance. *Ecology Letters*, 18(2), 144–152.

Villnäs, A., Norkko, J., Hietanen, S., Josefson, A. B., Lukkari, K. and Norkko, A. (2013). The role of recurrent disturbances for ecosystem multifunctionality. *Ecology*, 94, 2275–2287.

Visscher, H., Sephton, M. A. and Looy, C. V. (2011). Fungal virulence at the time of the end-Permian biosphere crisis? *Geology*, 39(9), 883–886.

Vörösmarty, C. J., McIntyre, P. B., Gessner, M. O., et al. (2010). Global threats to human water security and river biodiversity. *Nature*, 467(7315), 555 561.

Wagner, D. L. (2019). Global insect decline: Comments on Sánchez-Bayo and Wyckhuys (2019). *Biological Conservation*, 233, 332–333.

Wake, D. B. and Vredenburg, V. T. (2008). Are we in the midst of the sixth mass extinction? A view from the world of amphibians. *Proceedings of the National Academy of Sciences of the United States of America*, 105, 11466–11473.

Walker, L. R. and del Moral, R. (2003). *Primary Succession and Ecosystem Rehabilitation*. Cambridge University Press, Cambridge.

Walker, M., Head, M. J., Berkelhammer, M., et al. (2018). Formal ratification of the subdivision of the Holocene Series/ Epoch (Quaternary System/Period): Two new Global Boundary Stratotype Sections and Points (GSSPs) and three new stages/ subseries. *Episodes. Journal of International Geoscience*, 41(4), 213–223. https://doi.org/10.18814/epiiugs/2018/018016.

Wallace-Wells, D. (2019). *The Uninhabitable Earth*. Penguin Books, London.

Walther, G.-R. (2010). Community and ecosystem responses to recent climate change. *Philosophical Transactions of the Royal Society B: Biological Sciences*, 365 (1549), 2019–2024.

Wang, P., Poe, G. L. and Wolf, S. A. (2017). Payments for ecosystem services and wealth distribution. *Ecological Economics*, 132, 63–68.

Wang, R., Dearing, J. A., Langdon, P. G., et al. (2012). Flickering gives early warning signals of a critical transition to a eutrophic lake state. *Nature*, 492 (7429), 419–422.

Wang, S. and Loreau, M. (2016). Biodiversity and ecosystem stability across scales in metacommunities. *Ecology Letters*, 19(5), 510–518.

Wardle, D. A. (2016). Do experiments exploring plant diversity–ecosystem functioning relationships inform how biodiversity loss impacts natural ecosystems? *Journal of Vegetation Science*, 27, 646–653.

Wardle, D. A. and Zackrisson, O. (2005). Effects of species and functional group loss on island ecosystem properties. *Nature*, 435, 806–810.

Wardle, D. A., Zackrisson, O., Hörnberg, G. and Gallet, C. (1997). Influence of island area on ecosystem properties. *Science*, 277, 1296–1299.

Wardle, D. A., Bonner, K. I., Barker, G. M., et al. (1999). Plant removals in perennial grassland: Vegetation dynamics, decomposers, soil biodiversity, and ecosystem properties. *Ecological Monographs*, 69, 535–568.

Wardle, D. A., Huston, M. A., Grime, J. P., et al. (2000). Biodiversity and ecosystem function: An issue in ecology. *Bulletin of the Ecological Society of America*, July 2000, 235–239.

Wardle, D. A., Bardgett, R. D., Klironomos, J. N., Setälä, H., van der Putten, W. H. and Wall, D. H. (2004a). Ecological linkages between aboveground and belowground biota. *Science*, 304, 1629–1633.

Wardle, D. A., Walker, L. R., and Bardgett, R. D. (2004b). Ecosystem properties and forest decline in contrasting long-term chronosequences. *Science*, 305 (5683), 509–513.

Wardle, D. A., Karl, B. J., Beggs, J. R., et al. (2010). Determining the impact of scale insect honeydew, and invasive wasps and rodents, on the decomposer subsystem in a New Zealand beech forest. *Biological Invasions*, 12(8), 2619–2638.

Wardle, D. A., Bardgett, R. D., Callaway, R. M. and van der Putten, W. H. (2011). Terrestrial ecosystem responses to species gains and losses. *Science*, 332(6035), 1273–1277.

Wardle, D. A., Jonsson, M., Bansal, S., Bardgett, R. D., Gundale, M. J. and Metcalfe, D. B. (2012). Linking vegetation change, carbon sequestration and biodiversity: Insights from island ecosystems in a long-term natural experiment. *Journal of Ecology*, 100, 16–30.

Warman, L. and Moles, A. T. (2009). Alternative stable states in Australia's Wet Tropics: A theoretical framework for the field data and a field-case for the theory. *Landscape Ecology*, 24, 1–13.

Warren, R., Price, J., VanDerWal, J., Cornelius, S. and Sohl, H. (2018). The implications of the United Nations Paris agreement on climate change for globally significant biodiversity areas. *Climatic Change*, 147(3–4), 395–409.

Watson, J. E. M., Evans, T., Venter, O., *et al.* (2018). The exceptional value of intact forest ecosystems. *Nature Ecology & Evolution*, 2(4), 599–610.

Watson, R. (2017). A database of global marine commercial, small-scale, illegal and unreported fisheries catch 1950–2014. *Scientific Data*, 4, 170039 https://doi.org/10.1038/sdata.2017.39.

Watson, R. and Pauly, D. (2001). Systematic distortions in world fisheries catch trends. *Nature*, 414, 534–536.

Watson, S. C. L. and Newton, A. C. (2018). Dependency of businesses on flows of ecosystem services: A case study from the county of Dorset, UK. *Sustainability*, 10, 1368.

Watt, A. (1947). Pattern and process in the plant community. *Journal of Ecology*, 35, 1–22.

(1955). Bracken versus heather, a study in plant sociology. *Journal of Ecology*, 43, 490–506.

Watt, K. E. F. (1971). Dynamics of populations: A synthesis. In: P. J. den Boer and G. R. Gradwell (eds.), *Dynamics of Populations: A Synthesis*. Centre for Agricultural Publishing and Documentation, Wageningen, Netherlands, pp. 568–580.

Wei, H., Shen, J., Schoepfer, S. D., Krystyn, L., Richoz, S. and Algeo, T. J. (2015). Environmental controls on marine ecosystem recovery following mass extinctions, with an example from the Early Triassic. *Earth-Science Reviews*, 149, 108–135.

Webster, J. M., Braga, J. C., Humblet, M., *et al.* (2018). Response of the Great Barrier Reef to sea-level and environmental changes over the past 30,000 years. *Nature Geoscience*, 11(6), 426–432.

Weed, A. S., Ayres, M. P. and Hicke, J. A. (2013). Consequences of climate change for biotic disturbances in North American forests. *Ecological Monographs*, 83, 441–470.

Weiss, H. and Bradley, R. S. (2001). What drives societal collapse? *Science*, 291, 609–610.

Weisse, M. and Goldman, E. D. (2019). The world lost a Belgium-sized area of primary rainforests last year. World Resources Institute. www.wri.org/blog/2019/04/world-lost-belgium-sized-area-primary-rainforests-last-year (accessed on 2 July 2019).

Weng, W., Luedeke, M. K. B., Zemp, D. C., Lakes, T. and Kropp, J. P. (2018). Aerial and surface rivers: Downwind impacts on water availability from land use changes in Amazonia. *Hydrology and Earth System Sciences*, 22, 911–927.

Westaway, M. C., Olley, J. and Grün, R. (2017). At least 17,000 years of coexistence: Modern humans and megafauna at the Willandra Lakes, South-Eastern Australia. *Quaternary Science Reviews*, 157, 206–211.

Westman, W. E. (1977). How much are nature's services worth? *Science*, 197, 960–964.

Westoby, M. and Burgman, M. (2006). Climate change as a threatening process. *Austral Ecology*, 31, 549–550.

Westoby, M., Walker, B. and Noy-Meir, I. (1989). Opportunistic management for rangelands not at equilibrium. *Journal of Range Management*, 42, 266–274.

Westwood, A., Reuchlin-Hugenholtz, E. and Keith, D. M. (2014). Re-defining recovery: A generalized framework for assessing species recovery. *Biological Conservation*, 172, 155–162.

Whitehouse, N. J. and Smith, D. (2010). How fragmented was the British Holocene wildwood? Perspectives on the 'Vera' grazing debate from the fossil beetle record. *Quaternary Science Reviews*, 29(3), 539–553.

Whiles, M. R., Lips, K. R., Pringle, C. M., *et al.* (2006). The effects of amphibian population declines on the structure and function of Neo-tropical stream ecosystems. *Frontiers in Ecology and the Environment*, 4, 27–34.

Whiles, M. R., Hall Jr., R. O., Dodds, W. K., *et al.* (2013). Disease-driven amphibian declines alter ecosystem processes in a tropical stream. *Ecosystems*, 16, 146–157.

White, P. S. and Pickett, S. T. A. (1985). Natural disturbance and patch dynamics: An introduction. In: S. T. A. Pickett and P. S. White (eds.), *The Ecology of Natural Disturbance and Patch Dynamics*. Academic Press, New York, pp. 3–13.

Whitt, C. (2017). Dying and drying: The case of Bolivia's Lake Poopó. NACLA. https://nacla.org/news/2017/06/30/dying-and-drying-case-bolivia%E2%80%99s-lake-poop%C3%B3.

Wicklum, D. and Davies, R. W. (1995). Ecosystem health and integrity? *Canadian Journal of Botany*, 73(7), 997–1000.

Wickramasinghe, L. P., Harris, S., Jones, G. and Vaughan, N. (2003). Bat activity and species richness on organic and conventional farms: Impact of agricultural intensification. *Journal of Applied Ecology*, 40, 984–993.

Wilkinson, C. (2008). *Status of Coral Reefs of the World: 2008*. Global Coral Reef Monitoring Network and Reef and Rainforest Research Center, Townsville.

Wilkinson, G. M., Carpenter, S. R., Cole, J. J., *et al.* (2018). Early warning signals precede cyanobacterial blooms in multiple whole-lake experiments. *Ecological Monographs*, 88(2), 188–203.

Willerslev, E., Davison, J., Moora, M., *et al.* (2014). Fifty thousand years of Arctic vegetation and megafaunal diet. *Nature*, 506, 47–51.

Williams, A. N. (2013). A new population curve for prehistoric Australia. *Proceedings of the Royal Society B: Biological Sciences*, 280(1761), 20130486.

Williams, J. W. and Jackson, S. T. (2007). Novel climates, no-analog communities, and ecological surprises. *Frontiers in Ecology and the Environment*, 5, 475–482.

Williams, J. W., Shuman, B. N. and Webb, T. (2001). Dissimilarity analyses of late-Quaternary vegetation and climate in eastern North America. *Ecology*, 82, 3346–3362.

Williams, J. W., Shuman, B. N., Webb, T. I., Bartlein, P. J. and Leduc, P. L. (2004). Late-quaternary vegetation dynamics in North America: Scaling from taxa to biomes. *Ecological Monographs*, 74, 309–334.

Willis, A. J. (1997). The ecosystem: An evolving concept viewed historically. *Functional Ecology*, 11, 268–271.

Wilmshurst, J. M., Anderson, A. J., Higham, T. F. G. and Worthy, T. H. (2008). Dating the late prehistoric dispersal of Polynesians to New Zealand using the commensal Pacific rat. *Proceedings of the National Academy of Sciences of the United States of America*, 105(22), 7676–7680.

Wilson, K., Pressey, B., Newton, A., Burgman, M., Possingham, H. and Weston, C. (2005). Measuring and incorporating vulnerability into conservation planning. *Environmental Management*, 35(5), 527–543.

Wing, S. L. (2004). Mass extinctions in plant evolution. In: P. D. Taylor (ed.), *Extinctions in the History of Life*. Cambridge University Press, Cambridge, pp. 61–97.

Winnie, J. A. (2012). Predation risk, elk, and aspen: Tests of a behaviorally mediated trophic cascade in the Greater Yellowstone Ecosystem. *Ecology*, 93, 2600–2614.

Wintle, B. A., Kujala, H., Whitehead, A., et al. (2019). Global synthesis of conservation studies reveals the importance of small habitat patches for biodiversity. *Proceedings of the National Academy of Sciences of the United States of America*, 116 (3), 909–914.

With, K. A. (1997). The theory of conservation biology. *Conservation Biology*, 11(6), 1436–1440.

Woinarski, J. C. Z., Risler, J. and Kean, L. (2004). Response of vegetation and vertebrate fauna to 23 years of fire exclusion in a tropical *Eucalyptus* open forest, Northern Territory, Australia. *Austral Ecology*, 29, 156–176.

Wookey, P. A., Aerts, R., Bardgett, R. D., et al. (2009). Ecosystem feedbacks and cascade processes: Understanding their role in the responses of Arctic and alpine ecosystems to environmental change. *Global Change Biology*, 15(5), 1153–1172.

Wootton, K. L. and Stouffer, D. B. (2016). Species' traits and food-web complexity interactively affect a food web's response to press disturbance. *Ecosphere*, 7(11), e01518. https://doi.org/10.1002/ecs2.1518.

Worm, B. (2016). Averting a global fisheries disaster. *Proceedings of the National Academy of Sciences of the United States of America*, 113(18), 4895–4897.

Worm, B. and Myers, R. A. (2003). Meta-analysis of cod-shrimp interactions reveals top-down control in oceanic food webs. *Ecology*, 84(1), 162–173.

Worm, B. and Paine, R. T. (2016). Humans as a hyperkeystone species. *Trends in Ecology & Evolution*, 31(8), 600–607.

Worm, B., Barbier, E. B., Beaumont, N., et al. (2006). Impacts of biodiversity loss on ocean ecosystem services. *Science*, 314(5800), 787–790.

Worm, B., Hilborn, R., Baum, J. K. and Branch, T. A., et al. (2009). Rebuilding global fisheries. *Science*, 325, 578–585.

Wright, D. K. (2017). Humans as agents in the termination of the African Humid Period. *Frontiers in Earth Science*, 5, 4. https://doi.org/10.3389/feart.2017.00004.

Wroe, S. (2005). On little lizards and big extinctions. *Quaternary Australasia*, 23, 8–12.

Wroe, S., Field, J. H., Archer, M., et al. (2013). Climate change frames debate over the extinction of megafauna in Sahul (Pleistocene Australia-New Guinea). *Proceedings of the National Academy of Sciences of the United States of America*, 110 (22), 8777–8781.

WRM. (2016). *How Does the FAO Forest Definition Harm People and Forests? An Open Letter to the FAO.* World Rainforest Movement. https://wrm.org.uy/actions-and-campaigns/how-does-the-fao-forest-definition-harm-people-and-forests-an-open-letter-to-the-fao/.

WWF. (2006). *Living Planet Report 2006.* WWF International, Gland, Switzerland.
  (2016). *Living Planet Report 2016.* Risk and resilience in a new era. WWF International, Gland, Switzerland.

Yang, L. H., Bastow, J. L., Spence, K. O. and Wright, A. N. (2008). What can we learn from resource pulses? Ecology, 89, 621–634.

References

Yeakel, J. D., Pires, M. M., Rudolf, L., et al. (2014). Collapse of an ecological network in Ancient Egypt. *Proceedings of the National Academy of Sciences of the United States of America*, 111(40), 14472–14477.

Yoccoz, N. G., Nichols, J. D. and Boulinier, T. (2001). Monitoring of biological diversity in space and time. *Trends in Ecology & Evolution*, 16(8), 446–453.

Young, M. N. and Leemans, R. (2006). Group report: Future scenarios of human-environment systems. In: R. Costanza, L. J. Graumlich and W. Steffen (eds.), *Sustainability or Collapse? An Integrated History and Future of People on Earth*. The MIT Press, Cambridge, MA and London, pp. 447–470.

Zalasiewicz, J., Waters, C. N., Summerhayes, C. P., et al. (2017a). The Working Group on the Anthropocene: Summary of evidence and interim recommendations. *Anthropocene*, 19, 55–60.

Zalasiewicz, J., Waters, C. N., Wolfe, A., et al. (2017b). Making the case for a formal Anthropocene Epoch: An analysis of ongoing critiques. *Newsletters on Stratigraphy*, 50(2), 205–226.

Zeder, M. A. (2011). The origins of agriculture in the Near East. *Current Anthropology*, 52(S4), S221–S235.

Zeller, D., Rossing, P., Harper, S., Persson, L., Booth, S. and Pauly, D. (2011). The Baltic Sea: Estimates of total fisheries removals 1950–2007. *Fisheries Research*, 108, 356–363.

Zemanova, M. A., Perotto-Baldivieso, H. L., Dickins, E. L., Gill, A. B., Leonard, J. P. and Wester, D. B. (2017). Impact of deforestation on habitat connectivity thresholds for large carnivores in tropical forests. *Ecological Processes*, 6(1), 21.

Zemp, D. C., Schleussner, C.-F., Barbosa, H. M. J. et al. (2014). On the importance of cascading moisture recycling in South America. *Atmospheric Chemistry and Physics*, 14(23), 13337–13359.

  (2017). Self-amplified Amazon forest loss due to vegetation-atmosphere feedbacks. *Nature Communications*, 8, 14681. https://doi.org/10.1038/ncomms14681.

Zerboni, A. and Nicoll, K. (2019). Enhanced zoogeomorphological processes in North Africa in the human-impacted landscapes of the Anthropocene. *Geomorphology*, 331(15), 22–35.

Zhang, K., Almeida Castanho, A. D., Galbraith, D. R., et al. (2015). The fate of Amazonian ecosystems over the coming century arising from changes in climate, atmospheric CO_2, and land use. *Global Change Biology*, 21(7), 2569–2587.

Zhang, Q., Yang, R., Tang, J., Yang, H., Hu, S. and Chen, X. (2010). Positive feedback between mycorrhizal fungi and plants influences plant invasion success and resistance to invasion. *PLoS One*, 5(8), e12380.

Zimov, S. A., Chuprynin, V. I., Oreshko, A. P., Chapin, F. S., Reynolds, J. F. and Chapin, M. C. (1995). Steppe–tundra transition: A herbivore-driven biome shift at the end of the Pleistocene. *The American Naturalist*, 146, 765–794.

Zimov, S. A., Zimov, N. S., Tikhonov, A. N. and Chapin, F. S. (2012). Mammoth steppe: A high-productivity phenomenon. *Quaternary Science Reviews*, 57, 26–45.

Index

acid rain, 75
Africa, 130, 149, 184, 210, 230, 241
agriculture
   current environmental impacts, 271
   environmental impacts in prehistory, 140
   intensification of, 274
   origins of, 139
Alaska, USA, 135, 194
alerce (*Fitzroya cupressoides*), 397
Algeria, 150
Allee effect, 183, 192
Amazonia, 135, 235, 241, 243. *See also* Brazil
America
   Central, 134
   North, 52, 76, 82, 109, 128, 131–132, 134, 151, 236–237, 243, 248, 271
   South, 52, 128, 134, 231, 236, 250
amphibians, 119, 209
Antarctica, 108, 110
Anthropocene, 119, 138, 142
ants, 82
Appalachian Mountains, 97
Arabian Gulf, 171
Aral Sea, 200, 329
Arctic, 138, 264
Argentina, 232, 306
Asia, 76, 137
assembly rules, 89
Atlantic Ocean, 171, 185
Atlas cedar (*Cedrus atlantica*), 236
Australia, 53, 108, 110, 120, 194, 197, 237, 253, 297, 329, 340, 348, 364

balance of nature, 12, 24
ball and cup diagrams, 66, 392
Balmford, Andrew, 29
Baltic Sea, 187, 189, 337
baselines
   shifting, 195

basin of attraction, 57, 66
bats, 274
bears, 135, 209, 322
beaver (*Castor* spp.), 208
beer
   civilising role of, 140
Belgium
   area the size of, 224
Belize, 165
Beringia, 133, 323
Betula nana, 318
bioturbation, 113
birch (*Betula* spp.), 135
birds, 89, 96, 143–144, 274, 388
'black box' approach, 12
Black Sea, 186, 190, 193
Blob, The, 192
bog, 50
Bolivia, 201
Bonn Challenge, 258, 373
boreal forest, 242
boring
   how not to be, 30
Borneo, 231, 241, 247
bottleneck, 263
boundaries, 11, 13–14
   functional, 13
   structural, 13
Braat, Leon, 368
bracken fern, 144
Brazil, 225, 231, 241–242, 247, 266
Britain. *See* United Kingdom
browsing, 51
bryophytes, 119
bushmeat, 242

California, USA, 194
Canada, 179, 185, 197, 213, 237, 243, 252
cannibalism, 144
carbon cycle, 105

Index · 473

carbon sequestration, 227
Caribbean Sea, 164–165, 169, 174
Carrifran Wildwood Project, 397
CAS. *See* complex adaptive systems
catastrophe theory
  as shipwreck, 56
cats, 81
cave paintings, 127
cereal
  breakfast
    universe in, 40
chaos theory, 56
Chauvet Cave, 137
Chazdon, Robin, 254
Chicxulub impact, 95, 99–100, 118
Chile, 225, 250, 397
China, 137, 178, 226, 236, 351
civilisation
  collapse of, 362
climate change, 71, 95, 189, 236, 299, 345
  ecosystem effects of, 347
coccolithophores, 114–115
collapse
  early-warning indicators of, 334
Colorado, 118
community ecology, 303
complex adaptive systems (CAS), 58–59
conifers, 108
connectivity, 219
conservation management, 325, 347
Conservation Measures Partnership (CMP), 341
conservation policy, 323, 328
Convention on Biological Diversity, 1, 349, 373, 457
Cook Islands, 141
coral bleaching, 162
  ecological impacts of, 163
coral reefs, 99, 105, 112, 161, 292
Costa Rica, 134, 260
critical load
  definition, 75
  estimation, 76
criticality, 57, 62
  'robust', 62
  self-organised (SOC), 62

Dada, 395
Darwin, Charles, 38–39, 81
dead zones, 193

debate
  heated, 73
Deccan Traps, 100
defaunation, 240–241
deforestation, 292
degradation
  anthropogenic, 229
  definition of, 5
  how disturbance differs from, 261
desert, 45
determinism
  nominative, 220
Diamond, Jared, 18, 146, 362
dinosaurs, 99, 111
disaster taxa, 104
disturbance
  characteristics of, 293
  definition of, 5
  key features of, 41
  press, 41, 63, 100, 169, 294
  pulse, 41, 63, 100, 170, 248, 294, 295
DNA barcoding, 24
Dorset, UK, 274
dose–response relationships, 78
driver
  definition of, 5
drought
  ecological impacts of, 215
Dumbo, 327

Earth system succession, 114
Easter Island. *See* Rapa Nui
ecological theory
  components of, 37
ecosystem
  definition of, 5, 10, 18
ecosystem collapse
  definition of, 6, 18, 21, 229, 324, 380
  early-warning indicators of, 338–339
  evidence of, 333
  mechanisms of, 383
ecosystem function, 311
  biodiversity and, 313
ecosystem recovery
  definition of, 7, 21
  forest, 256
  indicators of, 340
  mechanisms of, 89, 386
  models describing, 87
ecosystem services, 25, 366
Egypt, 153

Ehrlich, Ann and Paul, 362
elephant bird (*Aepyornis maximus*), 147
elephants, 134
Emperor
  Danish, naked, 17
England. *See* United Kingdom (UK)
Ennerdale, UK, 52
environmentalist paradox, 370
equity, 373
Erwin, Douglas, 120
Etches, Steve, 95
Eurasia, 131
Europe, 52, 75, 109, 136–137, 142, 217, 237, 271, 361
eutrophication, 202, 216
extinction
  functional, 83
  secondary, 246–247, 304
extinction apocalypse
  horsemen of, 311
extinction cascade, 83, 103, 246, 306
extinction dynamics, 102

Fahrig, Lenora, 233
fallacy
  conjunction, 337
  logical, 268
  prosecutor's, 337
FAO. *See* UN Food and Agriculture Organization
farming. *See* agriculture
feedback, 70, 74, 149, 152, 167, 173, 190, 229, 237, 250, 253, 266, 300, 318
  detection of, 302
  stabilising, 202, 384
ferns, 118
Fertile Crescent, 139
Fiji, 141, 176
firestick farming, 126
fish, 11, 95, 100, 144, 164, 202
fishing, 177
Florida Keys, USA, 166
Florida, USA, 170
food web
  coastal, 196
  Jurassic, 95
  recovery of, 221
food webs, 303, 305
foraminifera, 114, 118

forest, 45, 50, 52, 74, 77, 82, 84, 89–90, 118–119, 125, 325
  definition of, 224
forest elephant (*Loxodonta cyclotis*), 230
forest fragmentation, 231
forests, 108, 222
fossil fuels, 389
fossil record
  biases in, 107
fossils, 95, 100, 118
FRA. *See* Global Forest Resources Assessment
France, 127
freshwater ecosystems, 45, 199, 293
fruit dispersal by megaherbivores, 134
fungi, 79, 105, 107, 109, 118, 125, 132

Galápagos Islands, 163–164
Germany, 207, 218, 271, 329
GFW. *See* Global Forest Watch
Ghana, 250
ghosts
  biological effects of, 146
glass
  half full or half empty, 334
Global Biodiversity Outlook, 1, 457
Global Environmental Outlook, 1
Global Forest Resources Assessment (FRA), 224, 394
Global Forest Watch (GFW), 224, 227
Gondwana, 108
Good, Ronald, 275
grassland, 45, 51, 53, 77, 131, 136, 151, 297, 325
grazing, 52–53
Great Barrier Reef, 163, 166, 170–171, 173, 176, 327, 371
Greece, 141
Green New Deal, 397
Greenland, 74, 108–109
greenwash, 358

habitat fragmentation, 233
Hawaii, USA, 303, *See* Hawaiian Islands
Hawaiian Islands, 141, 143, 166, 175
heathland, 50
herbivores, 50, 52
herbivory, 51
Herzog, Werner, 127
hierarchical-response framework (HRF), 43

Hilborn, R., 181
Holocene, 52
Holocene epoch, 138
Homer's *Odyssey*, 222
Homo sapiens
  probability of extinction of, 366
howler monkeys, 82
HRF. *See* hierarchical-response framework
hummingbirds, 307
Humpty Dumpty model of recovery, 220
Hutchinson, G. Evelyn, 202
hysteresis, 60, 64, 173, 186, 190, 197, 203

ichthyosaurs, 95–96
ideas
  zombie, 233
ideology
  neoliberal, 368
Indian Ocean, 171
insectageddon, 272
insects, 270, 388
Intergovernmental Panel for Biodiversity and Ecosystem Services (IPBES), 368
invasional meltdown, 24
invasive species, 210
IPBES. *See* Intergovernmental Panel for Biodiversity and Ecosystem Services
Ireland, 131, 136
IUCN (International Union for Conservation of Nature), 2
IUCN Red List of Ecosystems (RLE), 2, 16, 25, 27, 52, 323, 326–328, 381
  criticisms of, 331
IUCN Red List of Threatened Species (RLTS), 2, 17, 328–329, 382

jaguars, 245
Jamaica, 165, 170
jellyfish, 192
Jurassic period, 95

Kazakhstan, 200
kelp forests, 82, 191, 194
keystone species, 24, 80, 131
Knepp Estate, 52

Lake Baikal, 200
Lake Chad, 151
Lake Poopó, 201

Lake Victoria, 211
lakes, 200, 337
landscape of fear, 238
largemouth bass (*Micropterus salmoides*), 82, 203
Leather, Simon, 273
lemurs, 147
Levant, the, 141
lichens, 77, 118
'Lilliput effect', 112
Lindeman, Raymond, 202
Lindenmayer, David, 252
livestock
  impacts of, on semi-arid ecosystems, 151
Lord of the Rings, The
  ecological perspective on, 143
lycopsids, 108, 119

Madagascar, 147
  process of ecosystem collapse in, 148
Maldives, 171
mammals, 310, 346
mammoth steppe, 131, 135, 138, 300
marine ecosystems, 45, 95, 104, 110, 177, 340
marsupials, 120
mass extinctions
  'Big Five', 96–97, 106, 113, 326, 345
  definition, 97
  mechanisms of, 101
  scale of, 107
  sixth?, 119
  winners and losers, 111
May, Robert, 59
MEA. *See* Millennium Ecosystem Assessment
Mediterranean Sea, 298
megafauna
  Australian, 120
    extinction of, 122
  Late Quaternary, 147, 364
Megaloceros giganteus, 131
Mesoamerica, 329
Mesopotamia, 141
metaphor, 23, 392–393
Mexico, 141, 310
Michigan, USA, 203, 206
Millennium Ecosystem Assessment (MEA), 9, 222, 367
moa, 144, 146

model
  difference from theory, 37
  simulation, 304
  state-and-transition, 53, 333
Monbiot, George, 361
Mongolia, 372
monsters
  beautiful, 395
  topological, 56
moose, 248
Morocco, 150
Mountain Ash (*Eucalyptus regnans*), 253
multiple stable states, 20, 61, 409, 449
mussels, 62
mutualisms, 250

Namibia, 190
Neolithic package, 139
Netherlands, The, 52, 203, 273, 354
New Caledonia, 141
New Guinea, 89
New Mexico, USA, 142
New York State, USA, 132
New Zealand, 50, 143, 202, 210, 239, 246, 250, 364
  'spot the native species' game, 143
niche filtering, 90
Nile River, 150
non-analogue communities, 346
North Atlantic, 337
Norway, 193
nose
  Cleopatra's, length of, 59
novel ecosystems, 23, 112, 324, 355, 358
nutrients
  translocation by animals, 135

Odum, Howard, 58
Oostvaardersplassen, 52
orcs. *See* wasps
Oregon, USA, 151

Pacific Ocean, 82, 171, 173, 192
palaeoecology, 50, 52
Palaeolithic, 127
Pampas, 133
Panama, 164, 209
Pangaea, 119

passenger pigeon, 120
pastoralism
  date of introduction of, 150
Patagonia, 133
Pauly, Daniel, 178, 187
Payment for Ecosystem Service. *See* PES
penguins, 144, 184
perch (*Perca fluviatilis*), 222
Permian period, 105
PES (Payment for Ecosystem Service) schemes, 373
Peters
  R. H., 34–35
Peters, R. H., 34
physics envy, 34, 38
plankton, 82
plantation, 224–225
plants
  resilience of, 108
polar bears, 349
policy commitments
  environmental, 373
Polynesia, 142
Popper, Karl, 33
positivity, 29
prairie, 151
pressure. *See* driver
propositions, 288
pteridosperms, 108
Puerto Rico, 271

rain forest, 223
rangeland, 53
Rapa Nui, 141–142, 363
Rasputin, Gregori, 126
recovery debt, 261
recovery trajectories, 22
recovery trajectory, 88
Red List of Ecosystems. *See* IUCN Red List of Ecosystems
Red List of Threatened Species. *See* IUCN Red List of Threatened Species
regime shift, 60, 83, 169, 191, 203, 266, 392
  cascading, 319
reptiles, 121
resilience
  difference from resistance, 84
  ecological, 85
  engineering, 85

restoration
    ecological, 21, 23, 87, 349, 356, 359, 371, 380
    forest, 258, 397
    river, 218
    stream, 220
rewilding, 52, 361, 380, 397
rivers, 11, 200
RLE. *See* IUCN Red List of Ecosystems
RLTS. *See* IUCN Red List of Threatened Species
rotifers, 222

Sagan, Carl, 206
Sahara
    'Green', 149–150, 152
Sahel-Sahara region, 149
salmon, 11, 135, 209
sandeels (*Ammodytes marinus*), 184
savannah, 45, 264, 310, 325
science
    difference from astrology, 92
    junk, 120
    philosophy of, 33
    rocket, 34
Scotland, 397
sea otters, 194
Seychelles, the, 166, 174
sharks, 96, 175, 185
shrubland, 50, 53
shrubs, 51
Siberia, 131, 135, 137, 200, 243
Singapore, 247
SOC. *See* criticality, self-organised
Sofia Protocol, 76
South Africa, 110, 329
Spain, 137
stability, 315
stable states
    alternative, 52, 60, 66, 74, 138, 150, 165, 190, 192, 204, 248, 252, 298
    multiple, 62–63, 65, 69, 71, 103, 105, 110
starfish, 173
Steller's sea cow
    extinction of, 194
stromatolites, 104
succession
    circular, 50
    mechanisms of, 260
    progressive, 49
    retrogressive, 50
succession theory, 48
successional transitions, 299
Sweden, 202
systems science, 17
systems theory
    dynamical, 56, 69, 92, 166, 173, 202, 248, 264, 391

Tarkin, Governor, 355
Tasmania, 250
Texas, USA, 136, 151, 240
Thailand, 240
theory
    ecological
        definitions of, 36
threat, 44, 103, 344
    definition of, 7
    taxonomy of, 45
threshold, 64, 151, 215, 245, 305
Thunberg, Greta, 392, 396
tipping point, 73, 81, 243–244
trees, 50–51
Triassic period, 105
trophic cascade, 75, 81, 185, 188, 189, 206, 238, 310
trophic triangles, 204
tundra, 136, 138, 318, 322
Twain, Mark, 180

UN Food and Agriculture Organization (FAO), 224
United Kingdom (UK), 52, 136–137, 232, 324, 340, 358
    Kimmeridge, Dorset, 95
urbanisation, 213
USA, 231

Venezuela, 82, 239
Vera, Franz, 50, 152
volcanoes, 96

Wallace, Alfred Russel, 38
Wardle, David, 317
wasps
    invasive, scarier than orcs, 144
wolf, 82

wolves, 238, 240
woodland, 53, 136
woolly monkeys, 135
woolly rhinoceros, 137
World Resources Institute, 224
Worm, B., 181

Yellowstone, USA, 238
Younger Dryas, 133, 364

zebra mussel (*Dreissena polymorpha*), 213
Zen, moment of, 39